D1187401

J. S um

SCIENCE AND TECHNOLOGY OF RUBBER

Contributors

J. R. BEATTY

R. J. CERESA

STUART L. COOPER

A. Y. CORAN

P. J. CORISH

JOHN D. FERRY

A. N. GENT

F. J. KOVAC

OLE KRAMER

GERARD KRAUS

MAURICE MORTON

MITCHEL SHEN

M. L. STUDEBAKER

G. VER STRATE

JAMES C. WEST

JAMES LINDSAY WHITE

SCIENCE AND TECHNOLOGY OF RUBBER

Edited by

FREDERICK R. EIRICH

Polytechnic Institute of New York
Brooklyn, New York

Under the auspices of the

RUBBER DIVISION OF THE AMERICAN CHEMICAL SOCIETY

ACADEMIC PRESS   New York San Francisco London   1978

A Subsidiary of Harcourt Brace Jovanovich, Publishers

COPYRIGHT © 1978, BY ACADEMIC PRESS, INC.
ALL RIGHTS RESERVED.
NO PART OF THIS PUBLICATION MAY BE REPRODUCED OR
TRANSMITTED IN ANY FORM OR BY ANY MEANS, ELECTRONIC
OR MECHANICAL, INCLUDING PHOTOCOPY, RECORDING, OR ANY
INFORMATION STORAGE AND RETRIEVAL SYSTEM, WITHOUT
PERMISSION IN WRITING FROM THE PUBLISHER.

ACADEMIC PRESS, INC.
111 Fifth Avenue, New York, New York 10003

United Kingdom Edition published by
ACADEMIC PRESS, INC. (LONDON) LTD.
24/28 Oval Road, London NW1 7DX

Library of Congress Cataloging in Publication Data

Main entry under title:

Science and technology of rubber.

    Includes bibliographies.
    1. Rubber. 2. Elastomers. I. Eirich, Frederick
Roland, Date
TS1890.S315   678'.2   77–80783
ISBN 0–12–234360–3

PRINTED IN THE UNITED STATES OF AMERICA

Contents

Chapter 3 **Structure Characterization in the Science and Technology of Elastomers**

G. Ver Strate

Chapter 4 **The Molecular and Phenomenological Basis of Rubberlike Elasticity**

Mitchel Shen

Chapter 5 **Dynamic Mechanical Properties**

Ole Kramer and John D. Ferry

Chapter 6 **Rheological Behavior of Unvulcanized Rubber**

James Lindsay White

Chapter 7   **Vulcanization**

A. Y. Coran

Chapter 8   **Reinforcement of Elastomers by Particulate Fillers**

Gerard Kraus

Chapter 9   **The Rubber Compound and Its Composition**

M. L. Studebaker and J. R. Beatty

Chapter 14   **Tire Manufacture and Engineering**

F. J. Kovac

List of Contributors

Numbers in parentheses indicate the pages on which the authors' contributions begin.

J. R. BEATTY (367), The B. F. Goodrich Research & Development Center, Brecksville, Ohio 44141

R. J. CERESA (455), Chemistry & Polymer Technology Department, Polytechnic of South Bank, London, England

STUART L. COOPER (531), Department of Chemical Engineering, University of Wisconsin, Madison, Wisconsin 53706

A. Y. CORAN* (291), Monsanto Industrial Chemicals Company, Rubber Chemicals Research Laboratories, Akron, Ohio 44313

P. J. CORISH (489), Dunlop Research Centre, Birmingham, England

JOHN D. FERRY (179), University of Wisconsin, Madison, Wisconsin 53706

A. N. GENT (1, 419), Institute of Polymer Science, The University of Akron, Akron, Ohio 44325

F. J. KOVAC† (569), Tire Reinforcing Systems, The Goodyear Tire and Rubber Company, Dayton, Ohio

OLE KRAMER (179), University of Copenhagen, Copenhagen, Denmark

GERARD KRAUS (339), Phillips Petroleum Company, Bartlesville, Oklahoma 74004

MAURICE MORTON (23), Institute of Polymer Science, The University of Akron, Akron, Ohio 44325

MITCHEL SHEN (155), Department of Chemical Engineering, University of California, Berkeley, California 94720

M. L. STUDEBAKER (367), Phillips Chemical Company, Stow, Ohio 44224

G. VER STRATE (75), Elastomers Technology Division, Exxon Chemical Company, Linden, New Jersey 07036

JAMES C. WEST* (531), Department of Chemical Engineering, University of Wisconsin, Madison, Wisconsin 53706

JAMES LINDSAY WHITE (223) The University of Tennessee, Knoxville, Tennessee 37916

*Present address: 260 Springside Drive, Akron, Ohio 44313.

†Present address: Goodyear International Tire Technical Center, Colmar-Berg, Grand-Duchy of Luxembourg.

*Present address: Allied Chemical, Morristown, New Jersey.

Preface

The continuing success of the American Chemical Society Rubber Division's correspondence course, based on Professor Morton's "Rubber Technology" persuaded the Division's Educational Committee to introduce a second, more-advanced course. This editor was commissioned to assemble a number of chapters, on the graduate to postgraduate level, stressing the continuous relation between ongoing research in synthesis, structure, physics, and mechanics and rubber technology and industry. This collection of chapters covering, to various depths, the most important aspects of rubber science and technology, and the list of authors, all leading authorities in their fields, should be of vital interest not only to those who want to expand their formal education or update and supplement their experience in the field, but to anyone interested in the unusual chemistry and physics and the outstanding properties and farflung usefulness of elastomers. The intermediate level of presentation, a mixture of theory, experiment, and practical procedures, should offer something of value to students, practitioners, and research and development managers.

It has been the bias of this editor, based on many years of teaching at Polytechnic's Institute of Polymer Chemistry, that the most successful way of teaching and learning polymer subjects is to refer continually to the special features of macromolecules. For elastomers, in particular, it is most instructive to derive the unique features of high elasticity from those of long flexible chain molecules in their matted and netted state and the changes imposed by large deformations, including the key role played by the internal viscosity as a function of temperature and rate. Swaying the authors to lean to this approach inevitably caused some overlap but, at the same time, allowed synthesis and structure, elasticity and flow, blending, filling, and cross-linking to be treated in different contexts; a more integral composition without too frequent a need for cross references to other chapters became possible. For the same reason, some variation in nomenclature was allowed, especially if it reflected differing uses in the literature.

Particular concerns in preparing this composite book have been the combination of information and instruction, and the sequence and correlation of the chapters' contents. The first ten chapters take the reader from an introduction through synthesis characterization, mechanical behavior, and flow to the major processing steps of filling, compounding, and vulcanization and to the theories and measurement of elastomeric performance, leaning strongly on the "materials" approach. The next three chapters deal with the ever broadening fields of blended, modified, and thermoplastic elastomers, while the last chapter, for reasons of space, is the only representative of the chapters originally planned on manufacturing, possibly the forerunner of another volume. All chapters, while presenting theory, mechanism, and the author's overview of the internal consistency of the material's pattern of behavior, serve also as substantial sources of data and as guides to the relevant literature and to further selfstudy. As such, this book should be suitable not only as a basis for the new course, but also as an instrument of instruction for students, teachers, and workers in all fields of polymer and, indeed, of material science.

This, in any case, was the intent of all the authors whose extensive, conscientious, and patient cooperation made this book possible. Special thanks are due to Dr. A. Gessler of the Exxon Corporation, Linden, New Jersey, and Dr. E. Kontos, Uniroyal Chemical Division, who conceived the idea of a second course and of the nature of this book and to Dr. H. Remsberg, Carlisle Tire and Rubber Company, then Chairman of the Division's Educational Committee, without whose firm backing and continuous understanding this effort could not have been concluded. Drs. Gessler, Kontos, and Remsberg were further instrumental in gathering many of the authors and offering a number of early revisions of the manuscripts.

Chapter 1

Rubber Elasticity: Basic Concepts and Behavior

A. N. GENT

INSTITUTE OF POLYMER SCIENCE
THE UNIVERSITY OF AKRON
AKRON, OHIO

I. Introduction

The single most important property of elastomers—that from which their name derives—is their ability to undergo large elastic deformations, that is, to stretch and return to their original shape in a reversible way. Theories to account for this characteristic high elasticity have passed through three distinct phases: the early development of a molecular model relating experimental observations to the known molecular features of rubbery polymers; then generalization of this approach by means of symmetry considerations taken from continuum mechanics which are independent of the molecular structure; and now a critical reassessment of the basic premises upon which these two quantitative theories are founded. In this chapter, the theoretical treatment is briefly outlined and shown to account quite successfully for the observed elastic behavior of rubbery materials. The special case of small elastic deformations is then discussed in some detail, because of its technical importance. Finally, attention is drawn to some aspects of rubber elasticity which are still little understood.

1

Copyright © 1978 by Academic Press, Inc.
All rights of reproduction in any form reserved.
ISBN 0-12-234360-3

II. Elasticity of a Single Molecule

The essential requirement for a substance to be rubbery is that it consist of long flexible chainlike molecules. The molecules themselves must therefore have a "backbone" of many noncollinear single valence bonds, about which rapid rotation is possible as a result of thermal agitation. Some representative molecular subunits of rubbery polymers are shown in Fig. 1; thousands of these units linked together into a chain constitute a typical molecule of the elastomers listed in Fig. 1. Such molecules will change their shape readily and continuously at normal temperatures by Brownian motion. They take up random conformations in a stress-free state, but assume somewhat oriented conformations if tensile forces are applied at their ends (Fig. 2). One of the first questions to consider, then, is the relationship between the applied tension f and the mean chain end separation r, averaged over time or over a large number of chains at one instant in time.

Chains in isolation will take up a wide variety of conformations,* governed by three factors: the statistics of random processes; a preference for certain sequences of bond arrangements because of steric and energetic restraints within the molecule; and the exclusion of some hypothetical conformations which would require parts of the chain to occupy the same volume in space. In addition, cooperative conformations will be preferred for space-filling reasons in concentrated solutions or in the bulk state.

Flory [1] has argued that the occupied-volume exclusion (repulsion) for an isolated chain is exactly balanced in the bulk state by the external (repulsive) environment of similar chains, and that the exclusion factor can therefore be ignored in the solid state. He has also pointed to many experimental observations which indicate that cooperative effects do not significantly affect the distribution of chain end-to-end distances in bulk, or the relation between tension and distance. It is noteworthy, for example, that modest swelling by simple liquids (say $<50\%$) does not make rubbers much softer [2], although it would certainly reduce any packing constraints. Also, direct observation of single chain dimensions in the bulk state by inelastic neutron scattering gives values fully consistent with unperturbed chain dimensions obtained for dilute solutions in theta solvents† [3, 4], although intramolecular effects may distort the local randomness of chain conformation.

Flory has again given compelling reasons for concluding that the chain end-to-end distance r in the bulk state will be distributed in accordance with

---

* Although the terms "configuration" and "conformation" are sometimes used interchangeably, the former has acquired a special meaning in organic stereochemistry and designates specific steric structures. Conformation is used here to denote a configuration of the molecule which is arrived at by rotation of single-valence bonds in the polymer backbone.

† These are (poor) solvents in which repulsion between different segments of the polymer molecule is balanced by repulsion between polymer segments and solvent molecules.

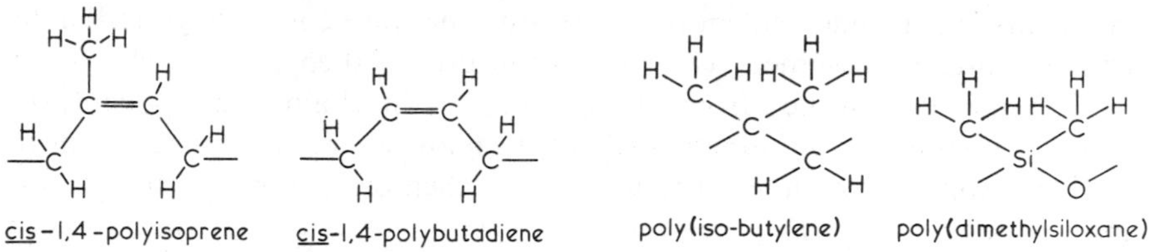

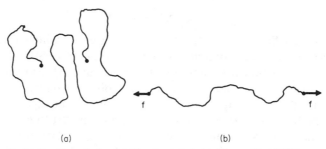

cis-1,4-polyisoprene   cis-1,4-polybutadiene   poly(iso-butylene)   poly(dimethylsiloxane)

Fig. 1. Repeat units for some common elastomer molecules.

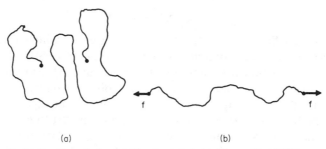

(a)   (b)

Fig. 2. (a) Random chain and (b) oriented chain. (From Gent [46].)

Gaussian statistics for sufficiently long chains, even if the chains are relatively stiff and inflexible over short lengths [1]. With this restriction to long chains it follows that the tension–displacement relation becomes a simple linear one:

$$f = Ar \tag{1}$$

where f is the tensile force, r is the average distance between the ends of the chain, and A is inversely related to the mean square end-to-end distance r_0^2 for unstressed chains;

$$A = 3kT/r_0^2 \tag{2}$$

where k is Boltzmann's constant and T is the absolute temperature.

If the real molecule is replaced by a hypothetical chain consisting of a large number n of rigid, freely jointed links, each of length l (Fig. 3), then

$$r_0^2 = nl^2 \tag{3}$$

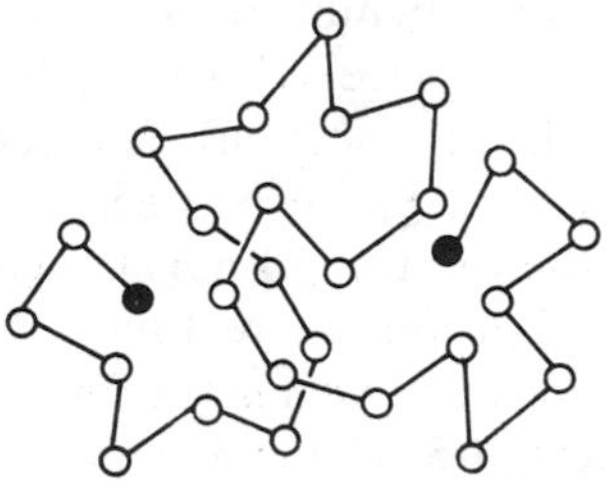

Fig. 3. Model chain of freely jointed links.

In this case $r_0{}^2$ is independent of temperature because completely random link arrangements are assumed. The tension f in Eq. (1) then arises solely from an entropic mechanism, i.e., from the tendency of the chain to adopt conformations of maximum randomness, and not from any energetic preference for one conformation over another. The tension f is then directly proportional to the absolute temperature T.

For real chains, consisting of a large number n of primary valence bonds along the chain backbone, each of length l,

$$r_0{}^2 = C_\infty n l^2 \qquad (4)$$

where the coefficient C_∞ represents the degree to which this real molecule departs from the freely jointed model. C_∞ is found to vary from 4 to 10, depending upon the chemical structure of the molecule and also upon temperature, because the energetic barriers to random bond arrangements are more easily overcome at higher temperatures [1]. $C_\infty^{1/2} l$ may thus be regarded as the effective bond length of the real chain [5], a measure of the "stiffness" of the molecule.

Equation (1) is reasonably accurate only for relatively short distances r, less than about one third of the fully stretched chain length [2]. Unfortunately, no good treatment exists for the tension in real chains at larger end separations. We must therefore revert to the model chain of freely jointed links, for which

$$f = (kT/l)L^{-1}(r/nl) \qquad (5)$$

where L^{-1} denotes the inverse Langevin function. An expansion of this relation in terms of r/nl [2],

$$f = (3kTr/nl^2)[1 + (3/5)(r/nl)^2$$
$$+ (99/175)(r/nl)^4 + (513/875)(r/nl)^6 + \cdots] \qquad (6)$$

gives a useful indication of where significant departures from Eq. (1) may be expected.

Equation (5) gives a steeply rising relation between tension and chain end separation when the chain becomes nearly taut (Fig. 4), in contrast to the Gaussian solution, Eq. (1), which becomes inappropriate for $r > \frac{1}{3}nl$. Rubber shows a similar steeply rising relation between tensile stress and elongation at high elongations. Indeed, experimental stress–strain relations closely resemble those calculated using Eq. (5) in place of Eq. (1) in the network theory of rubber elasticity (outlined in the following section). The deformation at which a small but significant departure is first found between the observed stress and that predicted by small-strain theory, using Eq. (1), yields a value for the effective length l of a freely jointed link for the real molecular chain. This provides a direct experimental measure of molecular stiffness. The values obtained are relatively large, of the order of 5–15 main-chain bonds, for the only polymer which has been examined by this method so far, cis-1,4-polyisoprene [6, 7].

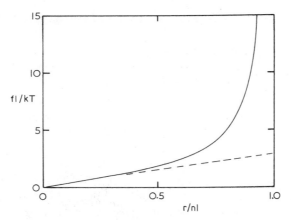

Fig. 4. Tension–displacement relation for a freely jointed chain (Eq. (5)): (–––), Gaussian solution (Eq. (1)). (From Gent [46].)

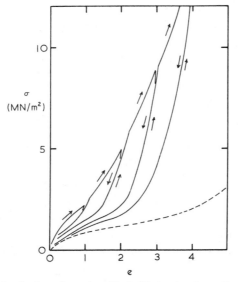

Fig. 5. Stress-induced softening of a carbon-black-filled vulcanizate of a copolymer of styrene and butadiene (25/75); (–––), stress–strain curve of a corresponding unfilled vulcanizate. (From Tobolsky and Mark [7a].)

Equation (5) has also been used to estimate the force at which a rubber molecule will become detached from a particle of a reinforcing filler, for example carbon black, when a filled rubber is deformed [8]. In this way, a general semi-quantitative treatment has been achieved for stress-induced softening (Mullins effect) of filled rubbers (shown in Fig. 5).

III. Elasticity of a Three-Dimensional Network of Polymer Molecules

Some type of permanent structure is necessary to form a coherent solid and prevent liquidlike flow of elastomer molecules. This requirement is usually met by incorporating a small number of intermolecular chemical bonds (cross-links) to make a loose three-dimensional molecular network. Such cross-links are generally assumed to form in the most probable positions, i.e., so that the long sections of molecules between them have the same spectrum of end-to-end lengths as a similar set of uncross-linked molecules would have. Under Brownian motion each molecular section takes up a wide variety of conformations, as before, but now subject to the condition that its ends lie at the cross-link sites. The elastic properties of such a molecular network are treated later. We consider first another type of interaction between molecules.

High molecular weight polymers form entanglements by molecular intertwining, with a spacing (in the bulk state) characteristic of the particular molecular structure. Some representative values of the molecular weight M_e between entanglement sites are given in Table I. Thus, a high molecular weight polymeric melt will show transient rubberlike behavior even in the absence of any permanent intermolecular bonds.

In a cross-linked rubber, many of these entanglements are permanently locked in (Fig. 6), the more so the higher the degree of cross-linking [9, 10]. If they are regarded as fully equivalent to cross-links, the effective number N of network chains per unit volume may be taken to be the sum of two terms N_e and N_c, arising from entanglements and chemical cross-links, respectively, where

$$N_e = \rho N_A / M_e, \qquad N_c = \rho N_A / M_c$$

and ρ is the density of the polymer, N_A is Avogadro's number, and M_e and M_c denote the average molecular weights between entanglements and between cross-links, respectively. However, the efficiency of entanglements in constraining the participating chains is somewhat uncertain, particularly when the number of chemical cross-links is relatively small [11–13]. Moreover, the force–

TABLE I

REPRESENTATIVE VALUES OF THE AVERAGE MOLECULAR WEIGHT M_e BETWEEN
ENTANGLEMENTS FOR POLYMERIC MELTS[a]

Polymer	M_e	Polymer	M_e
Polyethylene	4,000	Poly(iso-butylene)	17,000
cis-1,4-Polybutadiene	7,000	Poly(dimethylsiloxane)	29,000
cis-1,4-Polyisoprene	14,000	Polystyrene	35,000

[a] Obtained from flow viscosity measurements.

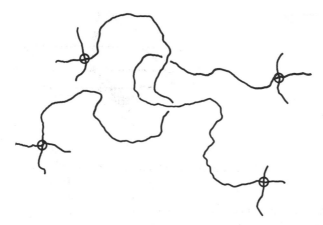

Fig. 6. Sketch of a permanent entanglement. (From Gent [46].)

extension relation for an entangled chain will differ from that for a cross-linked chain [14], being stiffer initially and nonlinear in form. The effective number N of molecular chains which lie between fixed points (i.e., cross-links or equivalent sites of molecular entanglement) is therefore a somewhat ill-defined quantity, even when the chemical structure of the network is completely specified.

It is convenient to express the elastic behavior of the network in terms of the strain energy density W per unit of unstrained volume. The strain energy w for a single chain is obtained from Eq. (1) as

$$w = Ar^2/2 \tag{7}$$

For a random network of N such chains under a general deformation characterized by extension ratios λ_1, λ_2, λ_3 (deformed dimension/undeformed dimension) in the three principal directions (Fig. 7), W is given by [2]

$$W = NAr_f^2(\lambda_1^2 + \lambda_2^2 + \lambda_3^2 - 3)/6 \tag{8}$$

where r_f^2 denotes the mean square end-to-end distance between chain ends (cross-link points or equivalent junctions) in the undeformed state. The close similarity of Eqs. (7) and (8) is evident, especially since $r^2 = (r_f^2/3)(\lambda_1^2 + \lambda_2^2 + \lambda_3^2)$.

For a random cross-linking process r_f^2 may be assumed to be equal to r_0^2, the corresponding mean square end-to-end distance for unconnected chains of the same molecular length. Because A is inversely proportional to r_0^2 (Eq. (2)), the only molecular parameter which then remains in Eq. (8) is the number N of elastically effective chains per unit volume. Thus, the elastic behavior of a molecular network under moderate deformations is predicted to depend only upon the number of molecular chains and not upon their flexibility, provided that they are long enough to obey Gaussian statistics.

Although r_f^2 and r_0^2 are generally assumed to be equal at the temperature of network formation, they may well differ at other temperatures because of

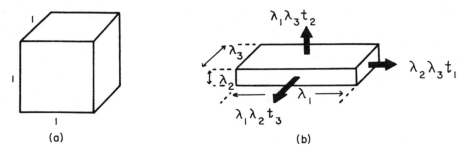

Fig. 7. (a) Undeformed and (b) deformed state.

the temperature dependence of $r_0{}^2$ for real chains (Eq. (4)). Indeed, the temperature dependence of elastic stresses in rubbery networks has been widely employed to study the temperature dependence of $r_0{}^2$, as discussed elsewhere [1, 15–17].

Another way in which $r_f{}^2$ and $r_0{}^2$ may differ is when the network is altered after formation. For example, when the network imbibes a swelling liquid, r^2 for the swollen network will be increased by a factor $\lambda_s{}^2$ in comparison to its original value, where λ_s is the linear swelling ratio. At the same time the number of chains per unit volume will be decreased by a factor $\lambda_s{}^{-3}$. Thus, the strain energy density under a given deformation will be smaller for a swollen network by a factor $\lambda_s{}^{-1}$ [2].

From the general relation for strain energy, Eq. (8), the elastic stresses required to maintain any given deformation can be obtained by means of virtual work considerations (Fig. 7),

$$\lambda_2 \lambda_3 t_1 = \partial W / \partial \lambda_1$$

with similar relations for t_2 and t_3. Because of the practical incompressibility of rubbery materials in comparison to their easy deformation in other ways, the original volume is approximately conserved under deformation. The extension ratios then obey the simple relationship

$$\lambda_1 \lambda_2 \lambda_3 = 1 \tag{9}$$

As a result, the stress–strain relations become

$$t_1 = \lambda_1(\partial W / \partial \lambda_1) - p, \qquad \text{etc.}$$

where p denotes a possible hydrostatic pressure (which has no effect on an incompressible solid). Thus, only stress differences can be written explicitly [2],

$$t_1 - t_2 = (NAr^2/3)(\lambda_1{}^2 - \lambda_2{}^2) \tag{10}$$

For a simple extension, say in the 1-direction, we set $\lambda_1 = \lambda$, and $\lambda_2 = \lambda_3 =$

$\lambda^{-1/2}$ (from Eq. (9)), and $t_2 = t_3 = 0$. Hence,

$$t(= t_1) = (NAr^2/3)(\lambda^2 - \lambda^{-1}) \qquad (11)$$

It is customary to express this result in terms of the tensile force f acting on a test piece of cross-sectional area A_0 in the unstrained state, where

$$f/A_0 = t/\lambda$$

The corresponding relation is shown in Fig. 8. It illustrates a general feature of the elastic behavior of rubbery solids: although the constituent chains obey a linear force–extension relationship (Eq. (1)), the network does not. This feature arises from the geometry of deformation of randomly oriented chains. Indeed, the degree of nonlinearity depends upon the type of deformation imposed. In simple shear, the relationship is predicted to be a linear one with a slope (shear modulus G) given by

$$t_{12} = G\gamma, \qquad G = NAr^2/3 \qquad (12)$$

where γ is the amount of shear, e.g., dx/dy.

Because rubbery materials are virtually incompressible in bulk, the value of Poisson's ratio is close to 0.5. Young's modulus E is therefore given by $3G$ to good approximation. However, the predicted relation between stress and tensile strain (extension) $e (= \lambda - 1)$ is only linear for quite small extensions (Fig. 8), so that Young's modulus is only applicable for extensions or compressions of a few percent.

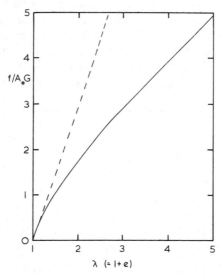

Fig. 8. Force–extension relation for simple extension: (---), linear relation obtaining at infinitesimal strains. (From Gent [46].)

For other deformations, notably the inflation of a thin spherical shell and of a cylindrical tube, the load–deformation relations become so nonlinear that in these cases the inflation pressure P passes through a maximum value when the radial expansion is only 38% and 60%, respectively, above r_0, the unstrained radius. Thereafter, the pressure decreases continuously on further inflation. This feature indicates a potentially unstable condition. Indeed, it is not possible to inflate a spherical balloon or cylindrical tube uniformly to expansions greater than these amounts because they undergo a transition from a uniform to a nonuniform deformation state at these points.

Experimental measurements for a spherical rubber balloon (initial radius $r_0 = 29$ mm, initial thickness $d_0 = 0.40$ mm, shear modulus $G = 0.35$ MN/ m^2) are shown in Fig. 9 and compared with the relation obtaine from Eqs. (9) and (10):

$$Pr_0/Gd_0 = 2(\lambda^{-1} - \lambda^{-2}) \tag{13}$$

where λ is now the radial expansion ratio r/r_0. The expansion at which P reaches a maximum value is indicated by a vertical broken line. Up to this point the deformation appears to be uniform and the experimental measurements of inflation pressure are in good agreement with Eq. (13). Beyond this point they diverge from the calculated relation and the balloon undergoes a strikingly nonuniform expansion, as revealed by a grid drawn on it in the unstrained state.

All of the stress relations given above are derived from Eq. (8). They are therefore only valid for moderate deformations of the network, i.e., for deformations sufficiently small for the chain tensions to be linearly related to their end-to-end distances r (Eq. (1)). Unfortunately, no correspondingly simple expression can be formulated for W using Eq. (5), the relationship for large strains of the constituent chains, in which the molecular stiffness parameter reappears. Instead, a variety of series approximations must be used, as in Eq. (6), to give close approximations to the behavior of rubber networks under large strains [2].

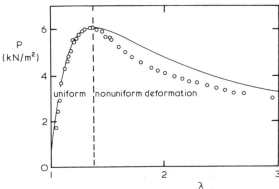

Fig. 9. Inflation of a thin-walled spherical rubber balloon. Solid curve: Eq. (13). (From Gent [46].)

IV. Comparison with Experiment

Although the treatment of rubber elasticity given in the preceding section is generally rather successful, certain discrepancies are found to occur. The first consists of observed stresses higher than predicted, e.g., by Eq. (11), and is often expressed by an additional contribution referred to as the C_2 term. This contribution is relatively large at small strains (although it is always the smaller part of the observed stress) and decreases in importance as the strain increases. It also decreases as the network is dilated by swelling with an inert liquid. Thus, the "C_2 stress" appears to reflect a non-Gaussian characteristic of network chains which is only important at small values of the chain end-to-end distance r. Indeed, Thomas [18] has shown that the magnitude of the C_2 stress and its complex dependence upon type and degree of strain, and upon degree of swelling, can all be accurately described by a simple additional term in the relation for the strain energy w for a single network chain, Eq. (7), which becomes

$$2w = Ar^2 + Br^{-2} \qquad (14)$$

The second term clearly becomes insignificant at large values of r.

Further evidence bearing on the physical nature of the discrepancy is provided by two other observations: C_2 does not appear to be strongly dependent upon temperature [19] and therefore does not appear to be associated with the energetics of chain conformations; and it is closely correlated with the tendency of the polymer chains to form molecular entanglements [20, 21]. For example, those polymers that have a high density of entanglements in the bulk state (Table I) yield rubbery networks with a relatively high C_2 stress component.

Finally, there is no evidence that isolated chains in theta solvents fail to conform to Gaussian statistics, so that the C_2 discrepancy appears to arise only when the molecular chains are tied into a network.

These varied aspects of the C_2 stress suggest that it is associated with entangled chains in networks (Fig. 6) and specifically that it arises from restrictions on the conformations available to entangled chains, different from those operating at cross-link sites. Prager and Frisch [14] have pointed out that chains involved in model entanglements are governed by different statistics; their conclusions are quite consistent with what is known of the C_2 stress, but a quantitative comparison is not yet possible.

A second discrepancy between theory and experiment is found when the Gaussian part of the measured stresses is compared with the theoretical result for an ideal network. Numerical differences of up to 50% are obtained between the density of effective chains calculated from the observed stresses and that from the chemistry of cross-linking (e.g., see Bueche [9]). This discrepancy may be due to an error in the theoretical treatment as given here [22]. James and

Guth [23] arrived at stresses only half as large as those given in Eq. (10), from a somewhat different theoretical standpoint. However, up to the present the quantitative aspects of cross-linking and the topology of networks formed by cross-linking have been difficult to measure and to control, so that an accurate test of the theory is not yet feasible.

A third and major discrepancy, already referred to, is found at large deformations when the network chains fail to obey Gaussian statistics, even approximately. Considerable success is achieved in this case by using Eq. (5) in place of Eq. (1) for chain tensions in the network [2].

Notwithstanding these discrepancies, the simple treatment of rubber elasticity outlined in this chapter has proved to be remarkably successful in accounting for the elastic properties of rubbers under moderate strains, up to about 300% of the unstrained length (depending upon the length and flexibility, and hence the extensibility, of the constituent chains). It predicts the general form of the stress–strain relationships correctly under a variety of strains, the approximate numerical magnitudes of the stresses for various chemical structures, and the effects of temperature and of swelling the rubber with an inert mobile liquid upon the elastic behavior. It also predicts novel second-order stresses, discussed later, which have no counterpart in classical elasticity theory. In summary, it constitutes a major advance in our understanding of the properties of polymeric materials.

V. Continuum Theory of Rubber Elasticity

A general treatment of the stress–strain relations of rubberlike solids was developed by Rivlin [24, 25], assuming only that the material is isotropic in elastic behavior in the unstrained state, and incompressible in bulk. It is quite surprising to note what far-reaching conclusions follow from these elementary propositions, which make no reference to molecular structure.

Symmetry considerations suggest that appropriate measures of strain are given by three strain invariants, defined as

$$I_1 = \lambda_1^2 + \lambda_2^2 + \lambda_3^2$$
$$I_2 = \lambda_1^2\lambda_2^2 + \lambda_2^2\lambda_3^2 + \lambda_3^2\lambda_1^2$$
$$I_3 = \lambda_1^2\lambda_2^2\lambda_3^2$$

Moreover, for an incompressible material I_3 is identically unity (Eq. (9)) and hence only two independent measures of strain, namely, I_1 and I_2, remain. It follows that the strain energy density W is a function of these two variables only:

$$W = f(I_1, I_2) \tag{15}$$

Furthermore, because the differences between the deformed and undeformed states $I_1 - 3$ and $I_2 - 3$ are of second order in the strains, e_1, e_2, e_3, the strain energy function at sufficiently small strains must take the form

$$W = C_1(I_1 - 3) + C_2(I_2 - 3) \qquad (16)$$

where C_1 and C_2 are constants. This particular form of strain energy function was originally proposed by Mooney [26] and is therefore often called the Mooney–Rivlin equation. It is noteworthy that the first term corresponds to the relation obtained from the molecular theory of rubber elasticity, Eq. (8), if the coefficient C_1 were identified with $NAr^2/6 = \frac{1}{2}NkT(r^2/r_0^2)$.

Eq. (16) leads to the following stress–strain relation in simple extension or compression,

$$t(= t_1) = 2C_1(\lambda^2 - \lambda^{-1}) + 2C_2(\lambda - \lambda^{-2})$$

in place of Eq. (11). In terms of the tensile or compressive force F acting on a test piece of cross-sectional area A_0 in the unstrained state, this relation becomes

$$F/A_0(\lambda^2 - \lambda^{-1}) = 2C_1 + 2C_2/\lambda \qquad (17)$$

Although Eq. (17) is necessarily valid at small strains, considerable confusion has arisen from its application at larger strains, when it no longer holds. It is almost unfortunate that the stress–strain relation obtained in simple extension *appears* to be in accord with the Mooney–Rivlin equation (16), up to moderately large strains. This fortuitous fit arises because the particular strain energy function obeyed by rubber, discussed later, depends upon strain in a certain way [27].

The strain energy function W is determined experimentally from measured stresses in terms of its derivatives $\partial W/\partial I_1$ and $\partial W/\partial I_2$, denoted hereafter W_1 and W_2. The former is found to be approximately constant, but the latter varies with the strain, primarily as a function of the strain measure I_2 [2, 27–29]. This variation may be described to a good approximation by the simple empirical relation

$$W_2 = K_2/I_2$$

where K_2 is a constant [30]. However, it should be noted that both the small dependence of W_1 and the large dependence of W_2 upon strain are described accurately by Eq. (14), the Thomas modification of Gaussian molecular theory [18].

Considerable success has also been achieved in fitting the observed elastic behavior of rubbers by strain energy functions which are formulated directly in terms of the extension ratios $\lambda_1, \lambda_2, \lambda_3$ instead of in terms of the strain invariants I_1, I_2 [28, 29, 31–35]. Although experimental results can be described economically and accurately in this way, the functions employed are empirical and the numerical parameters used as fitting constants do not appear to have any direct physical significance in terms of the molecular structure of the material. On the other hand, the molecular elasticity theory, supplemented by a simple non-Gaussian term whose molecular origin is in principle within reach, seems able to account for the observed behavior at small and moderate strains with comparable success.

VI. Second-Order Stresses

Because the strain energy function for rubber is valid at large strains, and yields stress–strain relations which are nonlinear in character, the stresses depend upon the square and higher powers of strain, rather than the simple proportionality expected at small strains. A striking example of this feature of large elastic deformations is afforded by the normal stresses t_{11}, t_{22}, t_{33} that are necessary to maintain a simple shear deformation of amount γ (in addition, of course, to simple shear stresses) [24, 25, 36]. These stresses are predicted to increase in proportion to γ^2.

They are represented schematically in Figs. 10 and 11 for two different choices of the arbitrary hydrostatic pressure p, chosen so as to give the appropriate reference (zero) stress. In Fig. 10, for example, the normal stress t_{11} in the shear direction is put equal to zero; this condition would arise near the front and rear surfaces of a sheared block. In Fig. 11, the normal stress t_{33} is put equal to zero; this condition would arise near the side surfaces of a sheared block. In each case a *compressive* stress t_{22} is found to be necessary to maintain the simple shear deformation. In its absence the block would tend to increase in thickness on shearing.

When the imposed deformation consists of an inhomogeneous shear, as in torsion, the normal forces generated (corresponding to the stresses t_{22} in Figs. 10 and 11) vary from point to point over the cross section (Fig. 12). The exact way in which they are distributed depends upon the particular form of strain energy function obeyed by the rubber, i.e., upon the values of W_1 and W_2 which obtain under the imposed deformation state [36].

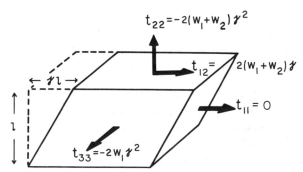

Fig. 10. Stresses required to maintain a simple shear deformation of amount γ. The normal stress t_{11} is set equal to zero. (From Gent [46].)

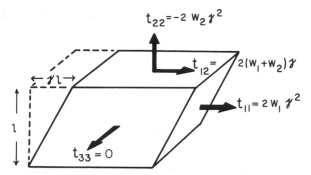

Fig. 11. Stresses required to maintain a simple shear deformation of amount γ. The normal stress t_{33} is set equal to zero. (From Gent [46].)

If a rubber block is subjected to a shearing stress while it is simultaneously compressed against a rigid frictional surface, it may undergo, when the coefficient of friction μ is high, a large shear deformation before sliding begins. If the applied compressive deformation (assumed small) is held constant, the normal stress t_{22} will then increase sharply as the amount of shear increases. Indeed, when the amount of shear γ exceeds a critical value of $(2\mu)^{-1}$ and $W_2(W_1 + W_2)^{-1}(2\mu)^{-1}$ for the stress conditions shown in Figs. 10 and 11, respectively, no sliding is possible, because the greater the applied shear stress t_{12}, the more the compressive stress t_{22} generated by the shear deformation exceeds that necessary to prevent sliding.

We therefore infer that sliding becomes completely inhibited for soft elastic materials held under a small compressive strain of approximately $0.08\mu^{-2}$ [37]. The condition for such frictional "locking" depends upon whether the material under consideration lies near front and rear surfaces of the block, or near the side surfaces, because different stress conditions prevail in these dif-

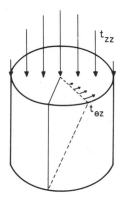

Fig. 12. Sketch of a cylindrical rod under torsion, showing the distribution of normal stress t_{zz} (corresponding to $-t_{22}$ in Figs. 10 and 11) over the cross section of the rod [36].

ferent regions (Figs. 10 and 11). When it does occur, however, the block must undergo frictional tearing, and hence abrasive wear, in order to move over the frictional surface.

These considerations illustrate again the close connection between an understanding of rubberlike elasticity and of other physical properties of rubbery materials.

VII. Elastic Behavior under Small Deformations

Under small deformations rubbers are linearly elastic solids. Because of high moduli of bulk compression, about 2000 MN/m², compared to the shear moduli G, about 0.2–5 MN/m², they may be regarded as relatively incompressible. The elastic behavior under small strains can thus be described by a single elastic constant G. Poisson's ratio is effectively 1/2, and Young's modulus E is given by $3G$, to good approximation.

A wide range of values for G can be obtained by varying the composition of the elastomer, i.e., by changing the chemistry of cross-linking, oil dilution, and filler content. However, soft materials with shear moduli of less than about 0.2 MN/m² prove to be extremely weak and are seldom used. Also, particularly hard materials made by cross-linking to high degrees prove to be brittle and inextensible. The practical range of shear modulus, from changes in degree of cross-linking and oil dilution, is thus about 0.2–1 MN/m². Stiffening by fillers increases the upper limit to about 5 MN/m², but those fillers which have a particularly pronounced stiffening action also give rise to stress-softening effects like those shown in Fig. 5, so that the modulus becomes a somewhat uncertain quantity.

It is customary to characterize the modulus, stiffness, or hardness of rubbers by measuring their elastic indentation by a rigid die of prescribed size and shape under specified loading conditions. Various nonlinear scales are employed to derive a value of hardness from such measurements [38]. Corresponding values of shear modulus G for two common hardness scales are given in Fig. 13.

Many rubber products are normally subjected to fairly small deformations, rarely exceeding 25% in extension or compression or 75% in simple shear. A good approximation for the corresponding stresses can then be obtained by conventional elastic analysis assuming linear relationships. One particularly important deformation is treated here: the compression or extension of a thin rubber block, bonded on its major surfaces to rigid plates (Fig. 14). A general treatment of such deformations has recently been reviewed [39].

It is convenient to assume that the deformation takes place in two stages: a pure homogeneous compression or extension of amount e, requiring a uniform compressive or tensile stress $\sigma_1 = Ee$, and a shear deformation restoring points in the planes of the bonded surfaces to their original positions in these

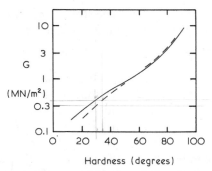

Hardness (degrees)

Fig. 13. Relations between shear modulus G and indentation hardness: (———), Shore A Scale; (———), International Rubber Hardness Scale. (From Tobolsky and Mark [7a].)

planes [40]. For a cylindrical block of radius a and thickness h, the corresponding shear stress t acting at the bonded surfaces at a radial distance r from the cylinder axis is given by

$$t = Eer/h$$

This shear stress is associated with a corresponding normal stress or pressure σ_2, given by

$$\sigma_2 = Ee(a^2/h^2)[1 - (r^2/a^2)] \tag{18}$$

These stress distributions are shown schematically in Fig. 14. Although they must be incorrect right at the edges of the block, because the assumption of a simple shear deformation cannot be valid at these points of singularity, they appear to provide satisfactory approximations over the major part of the bonded surfaces [41].

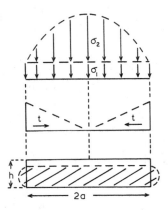

Fig. 14. Sketch of a bonded rubber block under a small compression. The distributions of normal stress σ and shear stress t acting at the bonded surfaces are represented by the upper portions of the diagram. (From Tobolsky and Mark [7a].)

By integrating the sum of the normal stresses $\sigma_1 + \sigma_2$ over the bonded surface, the total compressive force F is obtained in the form [40]

$$F/\pi a^2 e = E[1 + (a^2/2h^2)] \equiv E' \tag{19}$$

Clearly, for thin blocks of large radius the effective value E' of Young's modulus (given by the right-hand side of Eq. (19)) is much larger than the real value E, due to the restraints imposed by the bonded surfaces. Indeed, for values of the ratio a/h greater than about 10, a significant contribution to the observed displacement comes from volume compression or dilation because E' is now so large that it becomes comparable to the modulus of bulk compression [40] (Fig. 15).

When a thin bonded block is subjected to tensile loading, a state of approximately equal triaxial tension is set up in the central region of the block. The magnitude of the stress in each direction is given by the tensile stress, or negative pressure, σ_2 at $r = 0$, i.e., Eea^2/h^2, from Eq. (18). Under this outwardly directed tension a small cavity in the central region of the block will expand uniformly in size. However, the degree of expansion is predicted by the theory of rubberlike elasticity to become indefinitely large at a critical value of the tension of about $5E/6$ [42]. (This is a type of elastic instability, resembling that observed in the inflation of a balloon or tube.) Thus, if cavities are present in the interior of a bonded block, they are predicted to expand indefinitely, i.e., rupture, at a critical tensile strain e_c, given approximately by

$$e_c = 5h^2/6a^2$$

and at a corresponding critical value of the applied tensile load, obtained by substituting this value of e in Eq. (19).

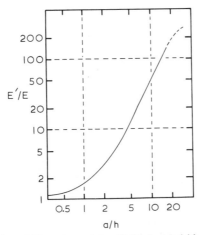

Fig. 15. Effective value of Young's modulus E' for bonded blocks versus ratio a/h of radius to thickness. (From Tobolsky and Mark [7a].)

As discussed in Chapter 10, internal cracks and voids are found to develop suddenly in bonded rubber blocks at well-defined tensile loads, which agree with those predicted by the treatment outlined above [42]. In particular, the loads are found to increase in proportion to Young's modulus E for rubbers of different hardness, in support of the proposed mechanism of fracture due to elastic instability. To avoid internal fractures of this kind it is thus necessary to restrict the mean tensile stress applied to thin bonded blocks to less than about $E/3$.

In compression, on the other hand, quite large stresses can be supported. A stress limit can be calculated by assuming that the maximum shear stress, developed at the bonded edges, should not exceed G, i.e., that the maximum shear deformation should not exceed about 100%. This yields a value for the allowable overall compressive strain of $h/3a$, corresponding to a mean compressive stress of the order of E for discs with ratios a/h between about 3 and 10. However, this calculation assumes that the approximate stress analysis outlined earlier is valid right at the edges of the block, and this is certainly incorrect. Indeed, the local stresses in these regions will depend strongly upon the detailed shape of the free surface in the neighborhood of the edge.

VIII. Some Unsolved Problems in Rubber Elasticity

We turn now to some features of the elastic response of rubbery materials which are still not fully understood:

(a)   As normally prepared, molecular networks comprise chains of a wide distribution of molecular lengths. Numerically, small chain lengths will tend to predominate. The effect of this diversity upon the elastic behavior of networks, particularly under large deformations, is not known.

A related problem concerns the elasticity of short chains. They are inevitably non-Gaussian in character and the analysis of their conformational statistics is likely to be difficult. Nevertheless, it seems necessary to carry out this analysis in order to be able to treat real networks in an appropriate way.

(b)   Insufficient attention seems to have been paid to problems of network topology, i.e., to the functionality of cross-links, their distribution in space, intramolecular loop formation, and the type and degree of molecular entanglement set up in networks. (See however Alfrey and Lloyd [43], Dusek and Prins [44] and interpenetrating networks [45].)

(c)   The effect of mutual interaction between molecular chains in the condensed state, and their probable adoption of cooperative conformations in hydrocarbon rubbers appears to be small, as discussed earlier. However, this is unlikely to be a valid generalization for all networks; some will probably interact strongly and the effect in these cases will presumably be quite significant. No analysis of this effect is known to the author.

(d)   Finally, the formulation of a satisfactory general treatment of networks under large deformations, when the chains approach their fully stretched state, would be valuable in the quantitative treatment of work hardening, fatigue, and fracture processes.

ACKNOWLEDGMENTS

The reader is referred to two excellent and authoritative surveys of rubberlike elasticity by Professor L. R. G. Treloar, one of the principal contributors to the subject [2, 16]. The present chapter is an expanded version of an earlier review, prepared on the occasion of Professor Treloar's retirement [46]. Acknowledgment is also due to the Engineering Division of the National Science Foundation for support of a program of research in the course of which this chapter was written. Reference has been made to some recent work carried out in this program [37]. The author's thanks are due to Mr. R. A. Paden for drawing the figures.

REFERENCES

1. P. J. Flory, "Statistical Mechanics of Chain Molecules." Wiley (Interscience), New York, 1969.
2. L. R. G. Treloar, "Physics of Rubber Elasticity," 2nd Ed. Oxford Univ. Press (Clarendon), London and New York, 1958.
3. J. P. Cotton, B. Farnoux, and G. Jannink, J. Chem. Phys. **57,** 290–294 (1972).
4. R. G. Kirste, W. A. Kruse, and J. Schelten, Makromol. Chem. **162,** 299–303 (1972).
5. W. H. Stockmayer and C. W. Pyun, in "Polymers in the Engineering Curriculum" (H. Markovitz, ed.), pp. 17–61. Carnegie-Mellon Univ. Press, Pittsburgh, Pennsylvania, 1971.
6. L. Mullins, J. Appl. Polym. Sci. **2,** 257–263 (1959).
7. M. C. Morris, J. Appl. Polym. Sci. **8,** 545–553 (1964).
7a. A. V. Tobolsky and H. F. Mark, eds., "Polymer Science and Materials," Chap. 13. Wiley, New York, 1971.
8. F. Bueche, J. Appl. Polym. Sci. **4,** 107–114 (1960); **5,** 271–281 (1961).
9. A. M. Bueche, J. Polym. Sci. **19,** 297–306 (1956).
10. C. G. Moore and W. F. Watson, J. Polym. Sci. **19,** 237–254 (1956).
11. L. Mullins and A. G. Thomas, J. Polym. Sci. **43,** 13–21 (1960).
12. J. Scanlan, J. Polym. Sci. **43,** 501–508 (1960).
13. N. R. Langley, Macromolecules **1,** 348–352 (1968).
14. S. Prager and H. L. Frisch, J. Chem. Phys. **46,** 1475–1483 (1967).
15. W. R. Krigbaum and R. J. Roe, Rubber Chem. Technol. **38,** 1039–1069 (1965).
16. L. R. G. Treloar, Rep. Prog. Phys. **36,** 755–826 (1973).
17. J. E. Mark, Rubber Chem. Technol. **46,** 593–618 (1973).
18. A. G. Thomas, Trans. Faraday Soc. **51,** 569–582 (1955).
19. T. L. Smith, J. Polym. Sci., Part C **16,** 841–858 (1967).
20. G. Kraus and G. A. Moczvgemba, J. Polym. Sci., Part A **2,** 277–288 (1964).
21. C. Price, G. Allen, and F. de Candia, Polymer **11,** 486–491 (1970).
22. K. J. Smith, Jr. and R. J. Gaylord, J. Polym. Sci., Polym. Phys. Ed. **13,** 2069–2077 (1975).
23. H. M. James and E. Guth, J. Chem. Phys. **11,** 455–481 (1943); J. Polym. Sci. **4,** 153–182 (1949).
24. R. S. Rivlin, Philos. Trans. R. Soc. L on, Ser. A **241,** 379–397 (1948).
25. R. S. Rivlin, in "Rheology, Theory and Applications" (F. R. Eirich, ed.), Vol. 1, Ch. 10. Academic Press, New York, 1956.
26. M. Mooncy, J. Appl. Phys. **11,** 582–592 (1940).

27. R. S. Rivlin and D. W. Saunders, *Philos. Trans. R. Soc. London, Ser. A* **243,** 251–288 (1951).
28. Y. Obata, S. Kawabata, and H. Kawai, *J. Polym. Sci., Part A-2* **8,** 903–919 (1970).
29. D. F. Jones and L. R. G. Treloar, *J. Phys. D* **8,** 1285–1304 (1975).
30. A. N. Gent and A. G. Thomas, *J. Polym. Sci.* **28,** 625–628 (1958).
31. O. H. Varga, "Stress-Strain Behavior of Elastic Materials." Wiley (Interscience), New York, 1966.
32. K. C. Valanis and R. F. Landel, *J. Appl. Phys.* **38,** 2997–3002 (1967).
33. R. A. Dickie and T. L. Smith, *Trans. Soc. Rheol.* **15,** 91–110 (1971).
34. R. W. Ogden, *Proc. R. Soc., Ser. A* **326,** 565–584 (1972).
35. P. J. Blatz, S. C. Sharda, and N. W. Tschoegl, *Proc. Natl. Acad. Sci. U.S.A.* **70,** 3041–3042 (1973).
36. R. S. Rivlin, *J. Appl. Phys.* **18,** 444–449 (1947).
37. A. N. Gent, *Wear* **29,** 111–116 (1974); **31,** 189 (1975).
38. A. L. Soden, "A Practical Manual of Rubber Hardness Testing." Maclaren, London, 1952.
39. A. N. Gent and E. A. Meinecke, *Polym. Eng. Sci.* **10,** 48–53 (1970).
40. A. N. Gent and P. B. Lindley, *Proc. Inst. Mech. Eng. (London)* **173,** 111–122 (1959).
41. A. N. Gent, R. L. Henry, and M. L. Roxbury, *J. Appl. Mech.* **41,** 855–859 (1974).
42. A. N. Gent and P. B. Lindley, *Proc. R. Soc., Ser. A* **249,** 195–205 (1958).
43. T. Alfrey, Jr. and W. G. Lloyd, *J. Polym. Sci.* **62,** 159–165 (1962).
44. K. Dusek and W. Prins, *Fortschr. Hochpolym.-Forsch.* **6,** 1–102 (1969).
45. D. Klempner, K. C. Frisch, and H. L. Frisch, *J. Eleastoplast.* **1973**(5), 196–200 (1973).
46. A. N. Gent, *J. Polym. Sci., Polym. Symp.* **48,** 1–17 (1974).

Chapter 2

Polymerization

MAURICE MORTON

INSTITUTE OF POLYMER SCIENCE
THE UNIVERSITY OF AKRON
AKRON, OHIO

Copyright © 1978 by Academic Press, Inc.
All rights of reproduction in any form reserved.
ISBN 0-12-234360-3

I. Introduction

The development of synthetic rubber played a special role in the history of polymerization chemistry. This was primarily due to the fact that attempts to synthesize rubber were made long before there was even the faintest idea of the nature of polymerization reactions. Such attempts began very soon after the elegant analytical work of Greville Williams [1] in 1860, which clearly demonstrated that Hevea rubber was "composed" of isoprene. Thus, Bouchardat [2] in 1879 was actually able to prepare a rubberlike substance from isoprene (which he obtained from rubber pyrolysis), using heat and hydrogen chloride. Tilden [3] repeated this process in 1884, but used isoprene obtained from pyrolysis of turpentine to demonstrate that it was not necessary to use the "mother substance" of rubber itself. These explorations were soon followed by the work of Kondakow (1900) [4] with 2,3-dimethylbutadiene, that of Thiele (1901) [5] with piperylene, and finally that of Lebedev (1910) [6] on butadiene itself. Mention should also be made of the almost simultaneous, and apparently independent, discoveries in 1910 by Harries [7] in Germany and Matthews and Strange [8] in England of the efficient polymerization of isoprene by sodium.

Although all of these attempts had a noble purpose indeed, the means used could hardly be considered a contribution to science, since the transformation of the simple molecules of a diene into the "colloidal" substance known as rubber was then far beyond the comprehension of chemical science. As a matter of fact, the commercial production of synthetic rubber was already well established, at least in Germany and Russia, *before* Staudinger laid the basis for his macromolecular hypothesis during the 1920s. Even such relatively modern synthetic elastomers as polychloroprene and the poly(alkylene sulfides) were already in commercial production by 1930–1931. This was, of course, also before Carothers' pioneering studies on the polymerization of chloroprene!

Hence it is apparent that it was not the development of an understanding of polymerization that led to the invention of synthetic rubber, but perhaps the reverse. In contrast, it was the new science of organic macromolecules, whose foundations were established by Staudinger, which expanded rapidly during the 1930s and 1940s, and pointed the way to the synthesis of a vast array of new polymeric materials, including synthetic fibers and plastics and even new elastomers. This new science included the classical studies of polycondensation by Carothers and Flory, and the establishment of the principles governing free radical chain addition reactions by G. V. Schulz, Flory, Mayo, and others [9].

Thus it was that the paths of synthetic rubber and macromolecular science finally crossed and became one broad avenue [10]. Hence today the design of a new elastomer, or the modification of an old one, requires the same kind of molecular architecture which applies to any other polymer, and is based on an understanding of the principles of polymerization reactions.

II. Classification of Polymerization Reactions and Kinetic Considerations

Historically polymers have been divided into two broad classes: *condensation* polymers and *addition* polymers. For example, Flory [9, p. 37]* has defined these as follows:

> ... condensation polymers, in which the molecular formula of the structural unit (or units) lacks certain atoms present in the monomer from which it is formed, or to which it may be degraded by chemical means, and addition polymers, in which the molecular formula of the structural unit (or units) is identical with that of the monomer from which the polymer is derived.

Thus an example of a condensation polymer would be a polyester, formed by the condensation reaction between a glycol and a dicarboxylic acid (with the evolution of water), whereas an addition polymer is exemplified by polystyrene, formed by the self-addition of styrene monomers.

Although these earlier definitions were based on the chain structure of the polymers, they were closely related, as just described, to the *mode of formation* as well. It soon became apparent that such a classification has serious shortcomings, since so-called polycondensates could result from "addition" polymerization reactions. Thus, for example, Nylon 6 is now synthesized by the ring-opening addition polymerization of caprolactam, and this process has a profound effect on the properties of the resulting polymer. This is, of course, due basically to the magnitude of the molecular weight of the final polymer. Since it is the extraordinarily large size of the macromolecules which leads to their unusual properties, it would be most sensible to classify polymerization reactions in accordance with *the way in which they affect the molecular size and size distribution of the final product*. On this basis, there appears to be only two basic processes whereby macromolecules are synthesized:

(1) functional group polymerization, and
(2) chain addition polymerization.

A. FUNCTIONAL GROUP POLYMERIZATION

These reactions fall into two classes:
(a) *Polycondensation*

$$A—R—A + B—R'—B \rightarrow A—R—R'—B + AB \tag{1}$$

(b) *Polyaddition*

$$A—R—A + B—R'—B \rightarrow A—R—AB—R'—B \tag{2}$$

---

* Reprinted from P. J. Flory, "Principles of Polymer Chemistry." Copyright 1953 by Cornell University. Used by permission of Cornell University Press.

Here A and B are the functional end groups which react with each other. Examples of polycondensation can be seen in the formation of polyesters and polyamides, where the A and B groups would be hydroxyl and hydrogen, respectively, which would combine and split off as water. On the other hand, a polyaddition reactions, such as (2), would be exemplified by the reaction of diisocyanates with glycols to form polyurethanes. In that case, of course, no side products are formed.

Polymerizations (1) and (2) actually represent well-known reactions of small molecules, the only distinction being the minimum requirements of *difunctionality* of each molecule, which makes it possible for *the product of each reaction to participate in further reactions.* As a rule, the functional groups retain their reactivity regardless of the chain length [9, p. 75], so that these reactions follow the same kinetic rules as for simple molecules. However, in contrast to polyaddition reactions, polycondensations suffer from the serious problem of a reverse reaction (e.g., hydrolysis, or "depolymerization") due to the possible accumulation of the by-product (e.g., water), and this must be taken into account.

In both of the foregoing types of reaction, the two factors which govern the molecular weight of the polymer are (a) the stoichiometry and (b) the extent of reaction. Thus it is obvious that an excess of one type of end group will control the maximum chain length attainable, and this can be predicted if the initial ratio of functional groups is known. On the other hand, with equivalent amounts of the two types of end groups, the final chain length is theoretically limitless, i.e., infinite in size. In both cases, however, the characteristic feature of functional group polymerizations is that *the chain length increases monotonically with extent of reaction,* i.e., with time of reaction, since any molecule having the necessary functional group can react with any other molecule having the opposite functional group.

B. CHAIN ADDITION POLYMERIZATION

This type of polymerization involves the successive addition of monomers to a growing chain, which is initiated by some reactive species. Such addition reactions may involve either multiple bonds or rings. The reactive species which initiate such chain reactions must be capable of opening one of the bonds in the monomer, and may be either a radical, an electrophile, or a nucleophile. Hence these polymerizations may proceed by three possible mechanisms, viz., free radical, cationic, and anionic, as illustrated by the following equations for reactions of double bonds:

$$\text{Free radical} \quad R^{\cdot} + {>}C{=}C{<} \rightarrow R{-}\overset{|}{\underset{|}{C}}{-}\overset{|}{\underset{|}{C}}{\cdot} \rightarrow \cdots \rightarrow$$

$$\text{Cationic} \quad H^{+}A^{-} + {>}C{=}C{<} \rightarrow H{-}\overset{|}{\underset{|}{C}}{-}\overset{|}{\underset{|}{C}}{^{+}}A^{-} \rightarrow \cdots \rightarrow$$

Anionic $A^-M^+ + \;{>}C{=}C{<} \longrightarrow A{-}\overset{\displaystyle |}{\underset{\displaystyle |}{C}}{-}\overset{\displaystyle |}{\underset{\displaystyle |}{C}}{-}M^+ \longrightarrow \cdots \longrightarrow$

In these equations, HA might represent a strong acid, while AM could represent an organometallic compound. The ionic species are represented as undissociated ion pairs, rather than free ions, since the existence of the latter will depend on circumstances. The particular ionic mechanism is, of course, defined by the nature of the charge on the growing chain end.*

The significant distinction between functional group polymerization and chain addition polymerization is that in the latter, each macromolecule is formed by a "chain reaction" which is initiated by some activation step. Thus, at any given time during the polymerization, the reacting species present consist only of *growing chains and monomer molecules*, in addition to the "dead" polymer chains formed earlier. These growing chains may be very short lived (e.g., free radicals or free ions) but may attain very high chain lengths during their brief lifetimes. On the other hand, they may have very long lifetimes (e.g., "living" polymers), in which case the chain lengths may increase as a direct function of time of reactions. Hence, unlike the case of functional group polymerizations, the molecular weights in chain addition polymerization systems may, or may not, be directly related to time or extent of reaction.

III. Functional Group Polymerization

Although, as indicated earlier, functional group of polymerization did not figure prominently in the early explorations of rubber synthesis, it was one of the earliest general methods used for polymerization, due to its relative simplicity. Thus it is not surprising that the earliest truly synthetic resins and plastics were of the polycondensate type, such as phenol–formaldehyde or polyester. The concept of linking together reactive end groups to build large molecules is fairly simple to comprehend, and also lends itself to a relatively simple mathematical analysis.

As stated previously, the kinetics of these functional group polymerizations follow the same rules as the simple monofunctional reactions, since the reactivity of the functional groups is maintained [9, p. 75] regardless of chain length. The only new feature is, of course, the growth in molecular size, and this has been amenable to a mathematical analysis [9, p. 91]. Considering the type of reactions defined in Eqs. (1) and (2), in the normal case, where the number of

---

* Many of the complex organometallic initiators used in stereospecific polymerization (i.e., Ziegler–Natta catalysts) have led to such expressions as "coordinate anionic" mechanisms, because the initiators are derived from coordination compounds, sometimes involving the monomer. However, since the growing chain end generally involves the carbon–metal bond of an organometallic species, these polymerizations can still be considered as proceeding by an anionic mechanism.

A and B groups are equal, the chain lengths are easily predictable as a function of the extent of reaction. Thus, if p represents the *fraction of end groups consumed* at any given time, then the number-average number of units per chain is given by $1/(1 - p)$. Thus,

$$M_n = M_0/(1 - p) \tag{3}$$

where M_n is the number-average molecular weight and M_0 is the molecular weight of a chain unit.

Since this type of polymerization is a completely random process, with all molecules having equal probability of reacting, the *distribution of molecular weights* corresponds to the most *probable*, or *binomial*, *distribution*, which is related to the extent of polymerization as follows [9, p. 318].

$$W_x = xp^{x-1}(1 - p)^2 \tag{4}$$

$$N_x = p^{x-1}(1 - p) \tag{5}$$

where W_x is the weight fraction of x-mers (chains having x units) and N_x is the mole fraction of x-mers. This distribution function can be used to calculate M_w, the weight-average molecular weight, since $M_w = M_0 \sum x W_x$. It can be shown that the foregoing summation leads to the relation

$$M_w = \frac{1 + p}{1 - p}(M_0) \tag{6}$$

which then means that

$$M_w/M_n = 1 + p \tag{7}$$

Hence the weight/number ratio of chain lengths in these systems undergoes *a steady increase with extent of reaction*, approaching an ultimate value of 2.

Thus, we see that functional group polymerizations are characterized by the following features:

(a)   All molecules have equal probability of reacting.

(b)   The polymerization rates are essentially described by the reactivity of the functional groups.

(c)   The chain lengths are monotonic functions of the extent of reaction, and hence of time of reaction.

(d)   The attainment of high molecular weights requires a high degree of reaction, e.g., the consumption of 99% of the functional groups only leads to a number-average chain length of 100 units, i.e., a molecular weight of about 10,000.

In those cases where at least one of the monomers has *more than two* functional groups, the added feature of branching chains is introduced, eventually leading to the formation of molecular networks [9, p. 347] (i.e., gelation). This, of course, complicates the molecular size distribution but does not affect the kinetics of the polymerization.

The foregoing relationships of chain length to extent of reaction would then be expected to apply to such functional group polymerizations as are involved in the synthesis of poly(alkylene sulfides) from a dihalide and sodium polysulfide, or in the formation of the urethane polymers from glycols and diisocyanates. The polysulfide reaction is actually carried out in a suspension of the dihalide in an aqueous solution of the polysulfide, using a surfactant to stabilize the resulting polymer suspension.

The urethane polymers offer an interesting illustration of the characteristic molecular weights to be expected in this type of polymerization, which can be written as

$$HO—P—OH + OCN \quad R'—NCO \rightarrow HO[—P—O—\overset{\overset{O}{\|}}{C}—\overset{\overset{H}{|}}{N}—R'—\overset{\overset{H}{|}}{N}—\overset{\overset{O}{\|}}{C}—O]_x—P—OH$$

$$(8)$$

It should be noted that the symbol P in Eq. (8) represents a low polymer of a polyester or polyether type (MW $\sim$ 2000), so that this is really a "chain extension" reaction. It turns out that the reaction between an isocyanate group and a hydroxyl goes to a high conversion, i.e., to approximately 98% ($p = 0.98$). Hence the value of x in Eq. (8) is about 50, and the final molecular weight of the urethane polymer is about 100,000. Such high molecular weights are, of course, due solely to the fact that this reaction goes so far toward completion, i.e., where the reactive functional groups can be reduced to concentrations of the order of 10^{-2} M.

IV. Chain Addition Polymerization by Free Radical Mechanism

A. GENERAL KINETICS

This mechanism involves the usual three primary steps of any chain reaction, initiated by the formation of free radicals through the homolytic dissociation of weak bonds (e.g., in peroxides, azo compounds) or by irradiation, etc.

Initiation	$I \rightarrow 2R^{\cdot}$ or $I + M \rightarrow 2M^{\cdot}$	
Propagation	$M_j^{\cdot} + M \rightarrow M_{j+1}^{\cdot}$	
*Termination**	$M_j^{\cdot} + M_k^{\cdot} \rightarrow M_{j+k}$ or $M_j + M_k$	

* *Note:* Free radicals have been shown to terminate each other either by *combination* (coupling) or by *disproportionation*, as shown next in the polymerization of vinyl monomers.

where

$$I = \text{initiator} \qquad R^{\cdot} = \text{initial free radical}$$

$$M = \text{monomer} \qquad M_j^{\cdot} = \text{propagating free radical}$$

Combination

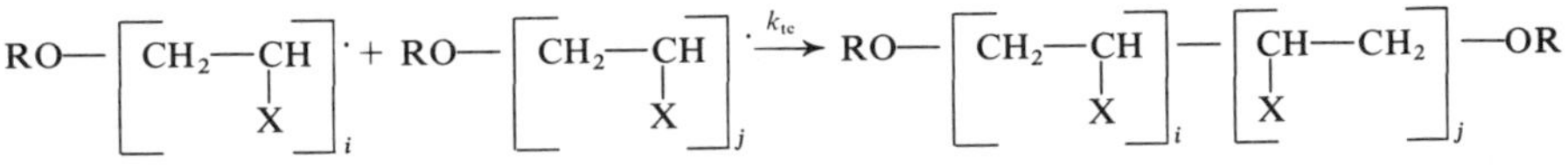

Disproportionation

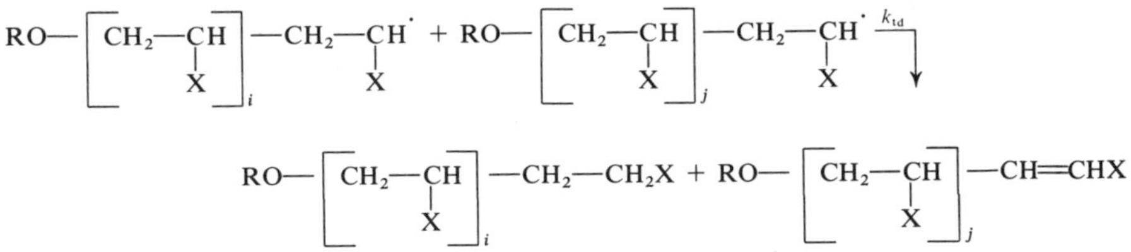

This sequence of steps then leads to the following kinetic treatment:

$$\text{Rate of initiation} \qquad R_i = 2k_i[I] \qquad \text{or} \qquad R_i = 2k_i[I][M] \qquad (9)$$

$$\text{Rate of propagation} \qquad R_p = k_p[M_j^{\cdot}][M] \qquad (10)$$

$$\text{Rate of termination} \qquad R_t = 2k_t[M_j^{\cdot}]^2 \qquad (11)$$

Assuming a *steady-state condition* where the rate of formation of radicals is equal to their rate of disappearance, i.e., $R_i = R_t$, then

$$[M_j^{\cdot}] = k_i^{1/2}k_t^{-1/2}[I]^{1/2} \qquad \text{or} \qquad [M_j^{\cdot}] = k_i^{1/2}k_t^{-1/2}[M]^{1/2}[I]^{1/2}$$

and

$$R_p = k_p k_i^{1/2}k_t^{-1/2}[M][I]^{1/2} \qquad \text{or} \qquad R_p = k_p k_i^{1/2}k_t^{-1/2}[M]^{3/2}[I]^{1/2} \quad (12)$$

Thus Eq. (12) illustrates the dependency of the overall rate of polymerization on the initiator and monomer concentrations. The half-power dependency of the rate on the initiator concentration appears to be a universal feature of the free radical mechanism, and has been used as a diagnostic test for the presence of this mechanism.

The general nature of free radical addition polymerization deserves some special attention. Because of the high reactivity of the propagating chain radical, it can only attain a very short lifetime, several seconds at best. This results in a very low stationary concentration of propagating chain radicals (about 10^{-8} M in a homogeneous medium). However, during this short lifetime, each growing radical may still have the opportunity of adding thousands of monomer units. Hence the chain length of the macromolecules formed in these systems has no direct relation to the extent of reaction, i.e., to the conversion of monomer to polymer. At all times during the polymerization, the reaction mixture contains only monomer and polymer, the latter usually of high molecular weight.

To illustrate more clearly the nature of free radical polymerization, it is instructive to examine the values of the individual rate constants for the propagation and termination steps. A number of these have been deduced, using special techniques of photopolymerization [11] and emulsion polymerization [12], and are listed in Table I. Thus, although the chain growth step can be seen to be a very fast reaction (several orders of magnitude faster than those of the end groups discussed in Section III), it is still several orders of magnitude slower than the termination step, i.e., the reaction of two radicals. It is this high ratio of k_t/k_p which leads to the very low stationary concentration of growing radicals ($\sim 10^{-8} M$) in these systems.

Although the three individual steps which combine to make up the chain reaction act as the primary control of the chain lengths, "chain transfer" reactions can occur whereby one chain is terminated, while a new one is initiated, without affecting the polymerization rate; such reactions will also, of course, affect the chain length. Chain transfer usually involves the homolytic cleavage of the most susceptible bond in molecules of solvent, monomer, impurity, etc., by the propagating radical, e.g.,

$$\text{\textasciitilde CH}_2\text{—CH}^{\cdot} + CCl_4 \longrightarrow \text{\textasciitilde CH}_2\text{—CHCl} + CCl_3^{\cdot}$$

TABLE I

PROPAGATION AND TERMINATION RATE CONSTANTS
IN RADICAL POLYMERIZATION

Monomer	k_p at 60°C (liter mole^{-1} sec^{-1})	k_t at 60°C ($\times 10^{-7}$) (liter mole^{-1} sec^{-1})
Styrene	176	3.6
Methyl methacrylate	367	1.0
Methyl acrylate	2100	0.5
Vinyl acetate	3700	7.4
Butadiene[a,b]	100	~100
Isoprene[a]	50	—

[a] Morton et al. [12].
[b] Morton and Gadkary [13].

and can be designated as follows:

$$\text{Monomer transfer} \qquad M_j^{\cdot} + M \xrightarrow{k_{trM}} M_j + M^{\cdot} \qquad (13)$$

$$\text{Solvent transfer} \qquad M_j^{\cdot} + S \xrightarrow{k_{trS}} M_j + S^{\cdot} \qquad (14)$$

$$S^{\cdot} + M \xrightarrow{k_p'} SM^{\cdot} \qquad (15)$$

Hence the chain length of the polymer being formed at *any given instant* can be expressed as the ratio of the propagation rate to the sum of all the reactions leading to termination of the chain, as follows

$$x_n = \frac{k_p[M_j^{\cdot}][M]}{(k_{tc} + 2k_{td})[M_j^{\cdot}]^2 + k_{trM}[M_j^{\cdot}][M] + k_{trS}[M_j^{\cdot}][S]}$$

or

$$\frac{1}{x_n} = \frac{(k_{tc} + 2k_{td})R_p}{k_p^2[M]^2} + \frac{k_{trM}}{k_p} + \frac{k_{trS}}{k_p} \frac{[S]}{[M]} \qquad (16)$$

where x_n is the number-average number of units per chain, k_{tc} the rate constant for termination by combination, and k_{td} the rate constant for termination by disproportionation.

B. MOLECULAR WEIGHT DISTRIBUTION

The chain length distribution in free radical addition polymerization can also be derived from simple statistics. Thus, for polymer formed at any given instant, the distribution will be the "most probable" and will be governed by the ratio of the rates of chain growth to chain termination, as follows:

$$W_x = x p^{x-1}(1 - p)^2 \qquad (17)$$

where p is the probability of propagation, and $1 - p$ the probability of termination (by disproportionation or transfer). This expression is of course identical with Eq. (4) except for the different significance of the term p. However, unlike Eq. (4), it only expresses the *instantaneous* chain length for an increment of polymer, and not the cumulative value for the total polymer obtained.

From Eq. (17) it follows that the number- and weight-average chain lengths x_n and x_w are expressed by

$$x_n = \frac{1}{1 - p} \qquad \text{and} \qquad x_w = \frac{1 + p}{1 - p} \sim \frac{2}{1 - p} \qquad (18)$$

since p must always be close to unity for high polymers. Hence it follows again that

$$x_w/x_n = 2 \qquad (19)$$

The value of x_w/x_n for the *cumulative polymer* may, of course, be much higher, depending on the changes in the value of p with increasing conversion. It should be noted, however, that this is only valid where the growing chains terminate by disproportionation or transfer, and *not* by combination. It can be shown in the latter case [9, p. 335; 14] that the increment distribution is much narrower, i.e.,

$$x_w/x_n = 1.5 \tag{20}$$

Thus, in summary, the kinetics of free radical polymerization are characterized by the following features:

(a) rate directly proportional to the half-power of the initiator concentration;

(b) molecular weight inversely proportional to the half-power of initiator concentration;

(c) short lifetime of the growing chain (several seconds) to high molecular weight, leading to formation of high polymer at the outset of reaction;

(d) no direct relation between extent of conversion and chain length;

(e) instantaneous chain length is statistical but the cumulative value can be considerably broader due to changes in relative rates of propagation and termination.

C. Special Case of Diene Polymerization

Since polydienes still constitute the backbone of the synthetic rubber industry, it is of special interest to consider the special features which dienes exhibit in free radical polymerization. Despite the fact that this type of polymerization has played and is still playing the major role in industrial production of various polymers, it has never been successful in bulk or solution polymerization of dienes. This is an outcome of the kinetic features of the free radical polymerization of dienes, as indicated in Table I. Thus the relatively high k_t/k_p ratio (as compared to the other monomers shown) leads to very low molecular weights and very slow rates for polydienes prepared in homogeneous systems, as illustrated in Table II. It can be seen from these data that even in the case of these thermal uncatalyzed polymerizations, where the molecular weight would be at a maximum compared to catalyzed systems, it is still too low by at least an order of magnitude. These systems are also complicated by a competitive Diels–Alder reaction, leading to low molecular weight compounds, i.e., to "oils."

It is therefore not surprising that the early investigators saw no promise in this mechanism of polymerization of butadiene, isoprene, etc., either by pure thermal initiation or by the use of free radical initiators, such as the peroxides. Instead they turned to sodium polymerization, which, although also rather slow and difficult to reproduce, at least yielded high molecular weight rubbery

TABLE II

THERMAL POLYMERIZATION OF DIENES[a]

Temp. (°C)	Time (hr)	Isoprene			2,3-Dimethylbutadiene		
		Yield (%)		Mol. wt. rubber	Yield (%)		Mol. wt.
		Oil	Rubber		Oil	Rubber	
85	100	7.9	16.3	4600	—	—	—
85	250	—	—	—	2.7	19.6	3500
85	900	—	35.3	5700	—	49.7	3500
145	12.5	54.7	15.6	4000	11.1	15.6	2100

[a] From Whitby and Crozier [15].

polymers from the dienes. Later on, in the 1930s, when emulsion polymerization was introduced, it was found that this system, even though it involves the free radical mechanism, leads both to fast rates and high molecular weights, conducive to the production of synthetic rubber. The special features of emulsion polymerization which lead to such surprising results are discussed in the next section.

V. Emulsion Polymerization

A. MECHANISM AND KINETICS

Polymerization in aqueous emulsions, which has been widely developed technologically, represents a special case of free radical chain addition polymerization in a heterogeneous system. Although it might be thought that the polymerization of water-insoluble monomers in an emulsified state simply involves the direct transformation of a dispersion of monomer into a dispersion of polymer, this is not really the case, as evidenced by the following features of a true emulsion polymerization.

(a)   The polymer emulsion (or latex) has a much smaller particle size than the emulsified monomer, by several orders of magnitude.

(b)   The polymerization rate is much faster than that of the undiluted monomer, by one or two orders of magnitude.

(c)   The molecular weight of the emulsion polymer is much greater than that obtained from bulk polymerization, by one or two orders of magnitude.

It is obvious from the foregoing facts that the mechanism of emulsion polymerization involves far more than the mere bulk polymerization of monomer in a finely divided state. In fact, the very small particle size of the latex, relative to that of the original monomer emulsion, indicates the presence of a special mechanism for the formation of such polymer particles.

The mechanism of emulsion polymerization, as originally proposed by Harkins [16], can best be understood by examining the components of this system, as depicted in Fig. 1, for a typical "water-insoluble" monomer such as styrene. The figure shows the various loci in which monomer is found, and which compete with each other for the available free radicals. Thus, in the initial stages, the monomer is found in three loci: (a) in aqueous solution, (b) as emulsified droplets, and (c) within the soap micelles. Both the dissolved monomer and the relatively large monomer droplets represent minor loci for reaction with the initiator radicals (except, of course, in the case of highly water-soluble monomers). However, the large number of soap micelles containing imbibed monomer represent a statistically important locus for initiation of polymerization. Thus it is not surprising that most of the polymer chains are generated within the monomer-swollen soap micelles. The very large number ($\sim 10^{15}$/ml) of very small polymer particles thus formed which are stabilized by adsorbing mono layers of soap deplete the available molecularly dissolved soap, thus destroying the soap micelles at an early stage of the polymerization ($\sim 10\%$ conversion in the usual recipe). Since all the available soap is distributed, and redistributed, over the surface of the growing particles, the amount of soap is the main factor controlling latex particle size.

During the second stage of the emulsion polymerization, therefore, the loci for available monomer consist of the dissolved monomer, the free monomer droplets, and the monomer imbibed by the numerous polymer particles. As before, the first two of these loci make a minor contribution, while the polymer–monomer particles provide a major locus for reaction with the initiator radicals

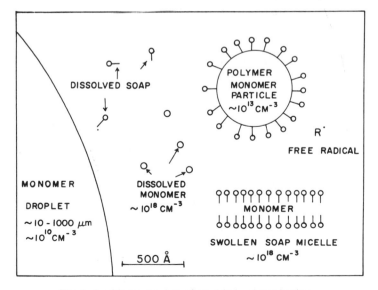

Fig. 1. Loci in mechanism of emulsion polymerization.

diffusing from the aqueous phase. The major portion of the polymerization reaction apparently occurs within this large number of latex particles isolated from each other and kept saturated [17] with monomer diffusing from the monomer droplets, and it is this aspect which leads to the unique characteristics of this system [18]. Thus, once an initiator radical enters a polymer–monomer particle and initiates a chain, the latter must continue to propagate with the available monomer until *another* radical enters the same particle. In this way, the rate of chain termination is actually controlled by the *rate of entry* of radicals into the particles, and this generally increases the lifetime of the growing chains, and hence the chain length. Furthermore, since the growing chains are all located in *different* particles, they are unable to terminate each other, leading to a higher *concentration* of growing chains and hence a *faster rate.*

In this way, emulsion polymerization systems can simultaneously achieve a much faster rate and a much higher molecular weight than homogeneous systems. A comparison of the kinetic features of bulk and emulsion polymerization of styrene is given in Table III. It is obvious at once that the main difference lies in the fact that the emulsion system is capable of raising the steady-state concentration of growing chain radicals by 2 to 3 orders of magnitude but *not* at the expense of increasing the termination rate!

The situation described earlier, i.e., where radicals entering individual latex particles successively initiate and terminate growing chains, is referred to as "ideal" emulsion polymerization, as defined by the Smith–Ewart theory. Under these conditions, the concentration of growing chains per unit volume of latex is easily predictable, since at any given time, *half of the particles will contain a growing chain.* In other words, the number of growing chains will be one half the number of particles. Since the latter is of the order of 10^{18} per liter, the concentration of growing chains is of the order of 10^{-6} M compared to 10^{-8} M for homogeneous polymerization systems. Since such growing chains are in an environment rich in monomer (within the monomer–polymer particles), it is not surprising that emulsion polymerization rates are 1 or 2 orders of magnitude higher than those of bulk polymerization, as shown in Table III.

TABLE III

FREE RADICAL POLYMERIZATION OF STYRENE

	Homogeneous	Emulsion
Monomer concentration (M)	5	5^a
Radical concentration (M)	10^{-8}	10^{-6}
Rate of polymerization at 60°C (%/hr)	~ 2	100
Molecular weight (no. av)	10^5	10^7

^{a} Within latex particles.

Furthermore, this high radical concentration does not affect the radical life-time, i.e., the chain size, which is governed solely by the availability of another free radical for termination, and thus by the period between successive entries of radicals into particles.

The foregoing situation, of course, holds only for the ideal case, as defined earlier. If the growing chain within the latex particle undergoes some side reaction which transfers the radical activity out of the particle before the next radical enters, or if termination is not rapid when two radicals occupy the same particle, then the number of growing chains, at any given time, will be, respectively, either smaller or larger than one half the number of particles. The latter case (more than one radical per particle) can occur, for example, if the particle size is sufficiently large and the termination rate too slow. The rate and molecular weight will then also be governed by other considerations than the interval between entry of successive radicals. Diagnostically, these situations can be distinguished from the ideal case by the effect of added initiator on the rate of polymerization after formation of particles is complete. Thus, in either case, an increase in initiator concentration will lead to a faster rate of entry of radicals into particles and hence an increase in the number of radicals per particle, leading to an increase in polymerization rate. In contrast, in the ideal case, an increase in frequency of radical entries into particles should not affect the rate, since the particles will still *contain a radical only half the time*, even though the periods of chain growth will be shorter, leading to a lower molecular weight.

The ideal case of the Smith–Ewart treatment actually proposes a rather elegant method for obtaining the absolute value of the propagation rate constant k_p from emulsion polymerization systems, since

$$R_p = k_p[M]N/2 \tag{21}$$

where N is the number of particles per unit volume.

Equation (21) leads to a solution for k_p from available knowledge of the rate R_p, the concentration of monomer in the monomer–polymer particles [M], and the number of particles. This method has been applied to several monomers, and has been especially useful in the case of the dienes, where the classical method of photoinitiation poses difficulties. Some of these results are shown in Table IV in the form of the usual kinetic parameters. The results obtained for styrene by photoinitiation techniques are included for comparison. It can be seen that the agreement is remarkably good, considering the widely different experimental methods used.

The data in Table IV provide evidence that the slow rates and low molecular weights obtained in homogeneous free radical polymerization of these dienes is *not* due to a low rate constant for propagation, but must be caused by a high rate constant for termination (as indicated in Table I). Hence, under the special conditions of emulsion polymerization, where the termination rate is controlled by the rate of entry of radicals into particles, it becomes possible to attain both

TABLE IV

PROPAGATION RATE CONSTANTS FROM EMULSION POLYMERIZATION

Monomer	k_p (liter mole^{-1} sec^{-1}) at 60°C	E_p (kcal mole^{-1})	A_p ($\times 10^{-7}$) (liter mole^{-1} sec^{-1})
Butadiene[a]	100	9.3	12
Isoprene[a]	50	9.8	12
2,3-Dimethyl-butadiene[b]	120	9.0	9
Styrene[c]	280	7.9	4
Styrene[d]	176	7.8	2.2

[a] Morton et al. [12].
[b] Morton and Gibbs [19].
[c] Morton et al. [12, p. 279].
[d] From photoinitiation (Matheson et al. [11]).

faster rates and higher molecular weights. It is this phenomenon which led to the rise of the emulsion polymerization system for the production of diene-based synthetic rubbers.

B. STYRENE–BUTADIENE RUBBER (SBR)

a. Kinetics and Molecular Weights

The most successful method for the production of a general-purpose synthetic rubber was the emulsion copolymerization of butadiene and styrene, which still represents the main process in use today. The general principles of copolymerization will be discussed in a later section, but it is instructive at this point to examine the other main features of this system. The types of recipes used are seen in Table V. The recipes shown are only to be considered as typical, since they are subject to many variations. It should be noted that the initiator in the 50°C recipe is the persulfate, while in the 5°C recipe it consists of a redox system comprising the hydroperoxide–iron–sulfoxylate–EDTA. In the latter case, the initiating radicals are formed by the reaction of the hydroperoxide with the ferrous iron, whose concentration is controlled by the EDTA complexing agent, while the sulfoxylate is needed to convert the oxidized ferric back to ferrous iron. The phosphate salt serves as a stabilizing electrolyte for the latex.

In both recipes, the thiol acts as a chain transfer agent to prevent the molecular weight from attaining the excessively high values possible in emulsion polymerization systems. It acts in an analogous fashion to the solvent in Eqs. (14) and (15), except that the sulfur–hydrogen bond is extremely susceptible to attack by the growing chain radical, which is thus terminated by a hydrogen atom, releasing the RS˙ radical to initiate a new chain. These thiols, which are

TABLE V

TYPICAL SBR EMULSION POLYMERIZATION RECIPES[a]

	SBR-1000[b]	SBR-1500[b]
Polymerization temp. (°C)	50	5
Time (hr)	12	12
Conversion (%)	72	60
Ingredients		
Butadiene	75	72
Styrene	25	28
Water	180	180
Soap (fatty or rosin acid)	4.5	4.0
Potassium persulfate	0.3	—
t-Dodecanethiol	0.3	0.2
p-Methane hydroperoxide	—	0.08
Trisodium phosphate ($Na_3PO_4 \cdot 10\ H_2O$)	—	0.5
Ferrous sulfate ($FeSO_4 \cdot 7\ H_2O$)	—	0.03
Sodium formaldehyde sulfoxylate	—	0.08
Tetrasodium salt of ethylene-		
diaminetetracetic acid (EDTA)	—	0.05

[a] Parts by weight.
[b] Commercial grade numbers assigned by the International Institute of Synthetic Rubber Producers to "hot" and "cold" SBR, respectively.

known as regulators, have transfer constants *greater* than 1, e.g., k_{tr}/k_p may be 3–4, so that only a small proportion is needed to reduce the molecular weight from several million to several hundred thousand. Diene-based polymers can undergo cross-linking reactions during the polymerization, which leads to the formation of insoluble "gel" rubber when the molecular weight becomes too high. Hence, thiol is used as "modifier" to prevent gel formation and keep the rubber processible.

Shortly after World War II, the American synthetic rubber industry began production of "cold" SBR, from which, it was found, superior tire rubber, especially as regards tread wear, could be prepared. Subsequent studies showed that the reduction in temperature from 50 to 5°C had little or no effect on the microstructure of the polydiene units (cis-1,4 versus trans-1,4 versus 1,2), or on the comonomer composition, but did exert a marked influence on the molecular weight distribution. It was also shown [20] that the cross-linking reaction, i.e., addition of growing chains to polymer double bonds (mainly of 1,2 chain units), was substantially reduced at these lower temperatures, thus reducing the tendency for gel formation at any given molecular weight.

Table VI shows the maximum molecular weights of polybutadiene attainable at different polymerization temperatures, prior to gelation, expressed as the critical weight average chain length, x_w, of the primary chains at the gel

TABLE VI

CROSS-LINKING PARAMETERS FOR POLYBUTADIENE[a]

Temp. (°C)	Relative cross-linking rate $r_x{}^b$ ($\times 10^4$)	x_w of Primary chains at gel point ($\times 10^{-3}$)
60	1.98	2.15
50	1.36	3.13
40	1.02	4.18
0 (calc.)	0.16	26.3

[a] From Morton and Salatiello [20].
[b] $r_x = k_x/k_p$ where k_x is the cross-linking rate constant and k_p the propagation rate constant.

point. Thus it can be seen that it is possible to increase the chain length by a factor of 9, without forming gel, by decreasing the polymerization temperature from 50 to 0°C. Hence the amount of thiol chain transfer agent can also be reduced, and this improves the overall chain length distribution by avoiding the formation of the very low fraction which results from the rapid reaction of the thiol in the early stages of the polymerization. Furthermore, this possibility of producing gel-free high molecular weight SBR at reduced polymerization temperature enabled the preparation of a high Mooney viscosity (~ 100) polymer which could be plasticized by low-cost petroleum oils ("oil-extended" rubber), and still retain its advantageous mechanical properties.

b. Chain Microstructure

As might be expected, the emulsion polymerization system does not alter the basic mechanism of free radical polymerization as regards the chain unit structure. The latter is, of course, independent of the type of free radical initiator used, in view of the "free" nature of the growing chain end radical. The temperature of polymerization does exert some influence, as shown by the data in Table VII, but not to a very great extent. It can be seen that the 1,2 side vinyl content is rather insensitive to the temperature, whereas the trans-1,4 content increases with decreasing temperature, at the expense of the cis-1,4 content. The latter almost vanishes, in fact, at $-20°C$ and the polymer then attains its highest trans-1,4 content of about 80%. Hence this type of polybutadiene is sufficiently stereoregular to undergo a substantial amount of crystallization on cooling [22, 23]. However, the introduction of the styrene comonomer is sufficient to destroy the necessary chain regularity for crystallization. Furthermore, it is the high-cis-1,4 polybutadiene which is desirable and not the trans-1,4 form, since the latter has a crystal melting point of about 150°C and is not an elastomer at ambient temperature. As can be seen from Table VII, the possibility of attaining a high cis-1,4 content at a reasonably high polymerization temperature is quite remote.

TABLE VII

CHAIN STRUCTURE OF EMULSION POLYBUTADIENE[a]

Polymerization temp. (°C)	Isomer wt %		
	cis-1,4[b]	trans-1,4	1,2
100	27.6	51.4	21.0
50	14.8	62.0	23.2
5	7.7	71.5	20.8
−20	0.8	79.6	19.6

[a] From Richardson and Sacher [21].
[b] By difference.

Hence it appears that these minor effects of temperature on the structure of the butadiene units cannot be expected to have any real influence on the properties of SBR.

C. EMULSION POLYMERIZATION OF CHLOROPRENE

a. Kinetics

The only other diene that has been used extensively for commercial emulsion polymerization is chloroprene (2-chlorobutadiene-1,3). The chlorine substituent apparently imparts a marked reactivity to this monomer, since it polymerizes very much more rapidly than butadiene, isoprene, or any other dienes. In fact, chloroprene is even more susceptible to free radical polymerization than styrene, and requires a powerful inhibitor for stabilization. It polymerizes extremely rapidly in emulsion systems, so that its rate must be carefully controlled.

Various recipes [24, 25] can be used for emulsion polymerization of chloroprene, with potassium persulfate as a popular initiator. A basic recipe [26] which illustrates several interesting features about this monomer is shown in Table VIII. Two aspects of this recipe are especially noteworthy: the use of a rosin soap, and the presence of elemental sulfur. Rosin soaps are notorious as retarders in emulsion polymerization, as are most *polyunsaturated* fatty acids. Yet complete conversion can be attained within a few hours. With *saturated* fatty acid soaps, the reaction is almost completed [24] within *one* hour at 40°C!

As for the sulfur, it apparently copolymerizes [26, 27] with the chloroprene, forming di- and polysulfide linkages in the chain. The latex is then treated with the well-known vulcanization accelerator, tetraethyl thiuram disulfide, which, by sulfur–sulfur bond interchange, degrades the cross-linked polychloroprene "gel" and renders it soluble and processible. In this way it serves the same purpose as the thiol chain transfer agents in SBR polymerization. As a matter of fact, the newer grades of polychloroprene are prepared [28] with the use of

TABLE VIII

BASIC RECIPE FOR NEOPRENE GN[a]

Ingredient	Parts by weight
Chloroprene	100
Water	150
N wood rosin	4
Sulfur	0.6
Sodium hydroxide	0.8
Potassium persulfate	0.2–1.0
Latex stabilizer[b]	0.7

[a] From Neal and Mayo [26]. Temperature, 40°C; time, several hours; conversion, 100%.
[b] A sodium salt of naphthalenesulfonic acid–formaldehyde condensation product.

thiols and other chain transfer agents. The thiols have been found [25] to yield narrower molecular weight distributions for chloroprene than for butadiene or isoprene, due to their much slower rate of disappearance in the presence of chloroprene.

b. *Chain Structure*

Another feature of the emulsion polymerization of chloroprene that distinguishes it from that of the other dienes is the fact that it leads to a predominantly trans-1,4 chain structure. Thus, even at ambient polymerization temperature, the polychloroprene contains over 80% trans-1,4 units, as shown in Table IX, which illustrates the effect of polymerization temperature on stereoregularity of the chain. As expected, the lower polymerization temperatures lead to a more stereoregular *trans*-1,4-polychloroprene. Because of the higher

TABLE IX

EFFECT OF POLYMERIZATION TEMPERATURE ON POLYCHLOROPRENE CHAIN STRUCTURE[a]

Temp. (°C)	Isomeric chain structure (%)					Crystal mp (°C)
	cis-1,4	trans-1,4	1,2	3,4	Total	
−40	5	94	0.9	0.3	100	67
10	9	84	1.1	1.0	95	55
50	11	—	—	—	—	40–45
80	—— 87 ——		2.0	2.1	91	—
100	13	71	2.4	2.4	89	—

[a] From Maynard and Mochel [29].

crystal melting point of the *trans*-1,4-polychloroprene, as compared with that of the *cis*-1,4-polyisoprene in Hevea rubber ($\sim 20°C$), even the polymer containing as little as 80% trans units crystallizes readily on cooling, or on stretching. Hence emulsion polychloroprene is the only latex polymer which resembles natural rubber, in that it is sufficiently stereoregular to exhibit strain-induced crystallization. This, then, results in high tensile strength in gum vulcanizates, without the need of reinforcing fillers, just as in the case of Hevea rubber. This makes possible the use of polychloroprene in a variety of gum rubber products, endowing them with superior oil and solvent resistance (because of its polarity), as well as high strength.

VI. Copolymerization

A. KINETICS

Copolymerization involves the simultaneous chain addition polymerization of a mixture of two or more monomers. Aside from the general kinetic considerations which govern these chain reactions, as described earlier, there is imposed an additional feature, i.e., the relative participation of the different monomers during the growth of the chain. This new parameter is most important, since it controls the composition of the copolymer. Systems involving more than two monomers are difficult to resolve in this respect, but it has been found possible to treat the case of a pair of monomers with relative ease [30, 31].

In the chain addition polymerization of *two* monomers, *regardless of the mechanism involved*, the growing chain always must make a choice of reacting with one of the two monomers. Furthermore, there are *two* kinds of growing chains, depending on which type of monomer unit occupies the growing end. Thus, using the free radical mechanism as the simplest example, the *four* types of propagation steps can be written as follows:

$$M_1{}^{\cdot} + M_1 \xrightarrow{k_{11}} M_1{}^{\cdot} \tag{22}$$

$$M_1{}^{\cdot} + M_2 \xrightarrow{k_{12}} M_2{}^{\cdot} \tag{23}$$

$$M_2{}^{\cdot} + M_2 \xrightarrow{k_{22}} M_2{}^{\cdot} \tag{24}$$

$$M_2{}^{\cdot} + M_1 \xrightarrow{k_{21}} M_1{}^{\cdot} \tag{25}$$

where M· and M refer to the growing chain and the monomer, respectively, as before, while the *subscripts* refer to the two kinds of monomer in the mixture. It can be seen that these four propagation reactions lead to *four* propagation rate constants, as shown. Hence the rate of consumption of each monomer may be expressed by the following equations:

$$d[M_1]/dt = k_{11}[M_1{}^{\cdot}][M_1] + k_{21}[M_2{}^{\cdot}][M_1] \tag{26}$$

$$d[M_2]/dt = k_{12}[M_1{}^{\cdot}][M_2] + k_{22}[M_2{}^{\cdot}][M_2] \tag{27}$$

Since it is the *relative rate* of consumption of the two monomers which will decide the composition of the chain, it can be expressed by dividing Eq. (26) by Eq. (27), leading to the expression

$$\frac{d[M_1]}{d[M_2]} = \frac{k_{11}[M_1^\cdot][M_1] + k_{21}[M_2^\cdot][M_1]}{k_{12}[M_1^\cdot][M_2] + k_{22}[M_2^\cdot][M_2]} \tag{28}$$

It is obvious at once that Eq. (28) is quite intractable for direct use. However, it is possible to simplify it considerably by utilizing the "steady-state" treatment, analogous to the one previously described. This is done by assuming the rate of *Eq. (23) to be equal to that of Eq. (25)*, and leads to the equivalence

$$[M_2^\cdot] = \left(\frac{k_{12}}{k_{21}}\right)[M_1^\cdot]\frac{[M_2]}{[M_1]}$$

which, when inserted into Eq. (28), yields Eq. (29), after appropriate rearrangements are made.

$$\frac{d[M_1]}{d[M_2]} = \frac{[M_1]}{[M_2]} \cdot \frac{r_1[M_1] + [M_2]}{r_2[M_2] + [M_1]} \tag{29}$$

where $r_1 = k_{11}/k_{12}$ and $r_2 = k_{22}/k_{21}$. The parameters r_1 and r_2 are known as the "reactivity ratios," since they express the *relative* reactivity of each of the two kinds of growing chain ends toward their "own" monomer as compared with the "other" monomer. They may in fact be considered as expressing the "homopolymerization" tendency of each type of monomer.

Equation (29), which relates the instantaneous composition of the copolymer ($d[M_1]/d[M_2]$) to the prevailing monomer concentrations, can be used to determine the values of r_1 and r_2. Many such values have been recorded [31]. Typical values of these parameters are shown in Table X, which illustrates the wide variations that prevail. The *relative reactivity* actually expresses the relative reactivity of each of the monomers shown toward the styrene radical.

TABLE X

FREE RADICAL COPOLYMERIZATION WITH STYRENE[a]

Monomer M_2	r_1	r_2	Relative reactivity $(1/r_1)$
Maleic anhydride	0.04	0	25
2,5-Dichlorostyrene	0.2	0.8	5
Methyl methacrylate	0.52	0.48	2
Vinylidene chloride	2.0	0.14	0.5
Diethyl maleate	5.0	0	0.2
Vinyl acetate	55	0.01	0.02

[a] At 60–70°C.

Thus, the r_1 and r_2 values permit some conclusions about the expected composition of the copolymer obtained at any given monomer ratio.

For example, it can be deduced from Table X that a copolymer of styrene and maleic anhydride would be strongly "alternating," since it would be *improbable* to have a sequence of two styrene units, and *impossible* to have a sequence of two maleic anhydride units. Also, it would obviously be extremely difficult to prepare a copolymer of styrene and vinyl acetate, since the latter monomer would be virtually excluded from the styrene polymerization.

It is also obvious, from Eq. (29), that the copolymer composition would not necessarily correspond to the comonomer charge, depending on the values of r_1 and r_2. An ideal system would, of course, be one in which this were the case, i.e., where the comonomers enter into the copolymer in the ratio of their concentrations; i.e., where

$$\frac{d[M_1]}{d[M_2]} = \frac{[M_1]}{[M_2]} \tag{30}$$

This is defined as an "azeotropic" copolymerization, by analogy to the distillation of two miscible liquids. Equation (30) would apply under the conditions where

$$\frac{r_1[M_1] + [M_2]}{r_2[M_2] + [M_1]} = 1 \tag{31}$$

and this would be valid, for example, where $r_1 = r_2 = 1$. In that case, Eq. (30) could apply for *all charge ratios*, i.e., the two types of growing chains show no particular preference for either of the two monomers. Also, Eq. (31) would be valid when $r_1 = r_2$ and $[M_1] = [M_2]$, i.e., an azeotropic copolymerization would result *only at equimolar* charge ratios. *In general*, Eq. (31) is valid when

$$\frac{[M_1]}{[M_2]} = \frac{(r_2 - 1)}{(r_1 - 1)} \tag{32}$$

This means that any copolymerization will be of an azeotropic type at the particular comonomer charge ratio indicated by Eq. (32). However, it also means that an azeotropic copolymerization is only possible when *both* r_1 and r_2 are either greater than 1 or less than 1.

Although the foregoing kinetic treatment has been applied here in the specific case of the free radical mechanism, it is valid for any of the other mechanisms. However, in the case of the ionic mechanisms, the type of initiator used and the nature of the medium may influence the r_1 and r_2 values. This is due to the fact that the growing chain end in ionic systems is generally associated with a counterion, so that the structure and reactivity of such chain ends can be expected to be affected by initiator and the solvent. This will be discussed in Section VIIIC.

B. Emulsion Copolymerization of Dienes

The three cases which involve copolymerizations leading to commercial synthetic rubbers are styrene–butadiene (SBR), butadiene–acrylonitrile (NBR), and chloroprene with various comonomers.

a. Butadiene–Styrene (SBR)

A large number of studies have been made of the reactivity ratios in this copolymerization, both in homogeneous and emulsion systems, and the average r values have been computed [32, 33] for butadiene and styrene, respectively, as

$$r_b = 1.6, \qquad r_s = 0.5$$

These values apply to solution and emulsion polymerization, presumably because neither monomer is particularly soluble in water, and both are quite insensitive to temperature. It appears, therefore, that the butadiene must enter the chain substantially faster than its charging ratio, and that each increment of polymer formed contains progressively more styrene. This is confirmed by the change in composition of the copolymer with conversion, as shown in Table XI. It can be seen that, at high conversion, the increment, or differential, composition becomes quite high in styrene content with concomitant loss of rubbery properties, even though the cumulative, or integral, composition still shows a low styrene content. This indicates the advisability of stopping the reaction at conversions not much higher than 60%.

It should be noted, too, that the r values for this system do not permit an azeotropic polymerization, as predicted by Eq. (32).

TABLE XI

COMONOMER COMPOSITION OF SBR[a]

Conversion (%)	Styrene in copolymer (wt %)	
	Differential	Integral
0	17.2	—
20	18.8	17.9
40	20.6	18.7
60	23.3	19.7
80	29.5	21.2
90	36.4	22.5
95	45.0	—
100	(100)	(25.0)

[a] From Fryling [34]. (Charge weight ratio, 75/25 butadiene/styrene; 50°C.)

b. Butadiene–Acrylonitrile (Nitrile Rubber)

According to Hofmann [35], the reactivity ratios of this pair of monomers at 50°C in *emulsion* polymerization are

$$r_B = 0.4, \qquad r_{AN} = 0.04$$

and they decrease somewhat at lower temperatures, but not to a great extent. These ratios are no doubt influenced by the marked water solubility of the acrylonitrile compared to that of butadiene.

The foregoing *r* values lead to the following situation. In accordance with Eq. (32), an azeotropic copolymer is formed when the acrylonitrile charge is 35–40% by weight (or by mole), so that a constant composition is maintained throughout the polymerization. If the acrylonitrile charge is *below* this value, the initial copolymer is relatively rich in acrylonitrile, which progressively *decreases* with increasing conversion. However, if the acrylonitrile charge is higher than the "azeotrope," the initial copolymer contains less acrylonitrile than charged but the acrylonitrile content *increases* with conversion. Since the commercial nitrile rubbers have nitrile contents of from 10 to 40%, these considerations have a very practical significance.

c. Chloroprene

Polychloroprene is generally prepared commercially as a homopolymer, although small amounts of a comonomer are included in several grades of this elastomer. There are two good reasons for the paucity of chloroprene-based copolymers. In the first place, the homopolymer, as stated previously, has a high-trans-1,4 chain structure and is therefore susceptible to strain-induced crystallization, much like natural rubber, leading to excellent tensile strength. It also has other favorable mechanical properties. Furthermore, chloroprene is *not* very susceptible to copolymerization by the free radical mechanism, as indicated by the *r* values in Table XII. Hence it is not surprising that the few

TABLE XII

MONOMER REACTIVITY RATIOS IN
COPOLYMERIZATION OF CHLOROPRENE[a]

Monomer M_2	r_1	r_2
Styrene	5–8	0–0.05
Isoprene	4	0.15
Acrylonitrile	6	0.01
Methyl methacrylate	4	0.2
Vinyl acetate	50	0.01

[a] From Hargreaves [28, p. 238]. Chloroprene is M_1.

copolymers of chloroprene available commercially contain only minor amounts of comonomers, which are included for their moderate effects in modifying the properties of the elastomer.

VII. Addition Polymerization by Cationic Mechanism

A. MECHANISM AND KINETICS

In these chain addition reactions, the active species is cationic in nature, initiated by strong acids, either of the protic or Lewis variety. Since most of these ionic polymerizations are carried out in nonaqueous solvents, it must be remembered that there is no presumption that the active species is necessarily a "free" ion, analogous to a free radical, but may indeed be some form of ion pair. Unfortunately, very little information is available about the exact nature of the propagating species in cationic systems. This is mainly due to the inherent experimental difficulties, caused by high reactivity and sensitivity to impurities, especially to traces of water.

The most common initiators of cationic polymerization are Lewis acids, such as $AlCl_3$, BF_3, $SnCl_4$, and $TiCl_4$, although strong protic acids such as H_2SO_4 may also be used. They generally work best with vinyl monomers with electropositive substituents, e.g., isobutylene, vinyl alkyl ethers, styrene, and other conjugated hydrocarbons. These polymerizations are characterized by rapid rates at very low temperatures, e.g., isobutylene is polymerized almost instantaneously at $-100°C$ by $AlCl_3$. The presence of a hydrogen donor, such as water or a protic acid, as a cocatalyst, is usually a prerequisite, as has been shown [36] in the case of isobutylene. On this basis, the Evans–Polanyi mechanism [37] proposed the following reaction sequence for the polymerization of isobutylene by BF_3 monohydrate:

Initiation

$$BF_3 \cdot H_2O + CH_2{=}C(CH_3)_2 \longrightarrow CH_3{-}\underset{\underset{CH_3}{|}}{\overset{\overset{CH_3}{|}}{C^+}} + BF_3 \cdot OH^- \tag{33}$$

Propagation

$$CH_3{-}\underset{\underset{CH_3}{|}}{\overset{\overset{CH_3}{|}}{C^+}} + BF_3OH^- + CH_2{=}\underset{\underset{CH_3}{|}}{\overset{\overset{CH_3}{|}}{C}} \rightarrow \cdots \rightarrow CH_3{-}\underset{\underset{CH_3}{|}}{\overset{\overset{CH_3}{|}}{C}}{-}\left[CH_2{-}\underset{\underset{CH_3}{|}}{\overset{\overset{CH_3}{|}}{C}}{-}\right]_x^+ + BF_3 \cdot OH^- \tag{34}$$

Termination

$$CH_3-\underset{\underset{CH_3}{|}}{\overset{\overset{CH_3}{|}}{C}}-\left[CH_2-\underset{\underset{CH_3}{|}}{\overset{\overset{CH_3}{|}}{C}}-\right]_x CH_2-\underset{\underset{CH_3}{|}}{\overset{\overset{CH_3}{|}}{C^+}} + BF_3\cdot OH^- \longrightarrow$$

$$CH_3-\underset{\underset{CH_3}{|}}{\overset{\overset{CH_3}{|}}{C}}-\left[CH_2-\underset{\underset{CH_3}{|}}{\overset{\overset{CH_3}{|}}{C}}-\right]_x CH_2-\underset{\underset{CH_3}{|}}{\overset{\overset{CH_2}{||}}{C}} + BF_3\cdot H_2O \quad (35)$$

This mechanism has been substantiated [38, 39] by analytical data which showed the presence of the two end groups proposed by this mechanism, i.e.,

$$(CH_3)_3C- \qquad \text{and} \qquad -CH_2-\underset{\underset{CH_3}{|}}{C}{=}CH_2$$

Furthermore, the use of the tracer complex $BF_3\cdot D_2O$ showed [39] that the polymer contained deuterium while the initiator became converted to $BF_3\cdot H_2O$.

This mechanism actually involves initiation by addition of a proton to the monomer and subsequent termination by loss of a proton to the initiator anion. The chain growth, therefore, occurs during the brief lifetime of the carbenium ion, and the initiator is constantly regenerated. It should be pointed out again, at this point, that it is doubtful, in view of the hydrocarbon media, whether the active species in these systems are in the free ionic state, so that the counterion has been depicted as being associated with the carbenium ion throughout the propagation step.

As will be seen later, there is substantial evidence that the growing carbenium ion can also disappear by donating a proton to a monomer molecule; thus

$$(CH_3)_3C-\left[CH_2-\underset{\underset{CH_3}{|}}{\overset{\overset{CH_3}{|}}{C}}-\right]_x CH_2-\underset{\underset{CH_3}{|}}{\overset{\overset{CH_3}{|}}{C^+}} + BF_3\cdot OH^- + CH_2{=}\underset{\underset{CH_3}{|}}{\overset{\overset{CH_3}{|}}{C}} \xrightarrow{k_{tr}}$$

$$(CH_3)_3-\left[CH_2-\underset{\underset{CH_3}{|}}{\overset{\overset{CH_3}{|}}{C}}-\right]_x CH_2-\underset{\underset{CH_3}{|}}{\overset{\overset{CH_2}{||}}{C}} + (CH_3)_3C^+ + BF_3\cdot OH^- \quad (36)$$

This is, of course, a *transfer step*, which, assuming that the trimethyl carbenium ion is capable of propagating with no serious loss of activity, would have no effect on the rate, but only on the chain length.

The foregoing mechanism is amenable to kinetic analysis, using the steady-state method, as in the case of the free radical mechanism. Using the notation HA to designate the acid initiator, we can write

Initiation	$HA + M \xrightarrow{k_i} HM^+ + A^-$	(37)
Propagation	$HM^+ + A^- + M \xrightarrow{k_p} \cdots \rightarrow HM_x^+ + A^-$	(38)
Termination	$HM_x^+ + A^- \xrightarrow{k_t} M_x + HA$	(39)
Transfer	$HM_x^+ + M + A^- \xrightarrow{k_{tr}} M_x + HM^+ + A^-$	(40)

Rate of initiation

$$R_i = k_i[HA][M] \tag{41}$$

Rate of termination

$$R_t = k_t[HM_x^+] \tag{42}$$

Rate of propagation

$$R_p = k_p[HM_x^+][M] \tag{43}$$

Rate of transfer to monomer

$$R_{tr} = k_{tr}[HM_x^+][M] \tag{44}$$

Here R_t has been assumed to be a first-order reaction, since the counterion A^- is considered to be specifically associated with the carbenium ion and not as a separate species. Using the steady-state assumption, we equate R_i and R_t and thus obtain for the steady-state concentration of growing chains

$$[HM_x^+] = (k_i/k_t)[HA][M] \tag{45}$$

Hence,

$$R_p = -dM/dt = (k_p k_i/k_t)[HA][M]^2 \tag{46}$$

Here again, the propagation rate is virtually the polymerization rate, since the consumption of monomer by the initiation step is negligible. Unlike the free radical case, the rate here is *first-order in initiator concentration*, obviously due to the first-order termination step.

Experimental verification of the foregoing kinetic scheme has been obtained in the case of the cationic polymerization of styrene [40] and vinyl alkyl ethers [41], where the polymerization rate was indeed found to be dependent on the first power of the initiator and on the square of the monomer concentration.

The molecular weight of the polymer can again be expressed in terms of x_n, the number-average number of units per chain, which can be defined here as the ratio of the propagation rate to the sum of the rates of all processes leading to chain termination (including transfer). Hence from Eqs. (42), (43), and (44),

$$x_n = \frac{R_p}{R_t + R_{tr}} = \frac{k_p[HM_x^+][M]}{k_t[HM_x^+] + k_{tr}[HM_x^+][M]}$$

or

$$1/x_n = k_{tr}/k_p + k_t/k_p[M] \tag{47}$$

Equation (47) provides a means of determining the relative contribution of the termination and transfer steps. Thus, if k_{tr} is large relative to k_t, the molecular weight will be virtually independent of monomer concentration, but if the reverse is true, x_n will be directly proportional to $[M]$. Hence this relation lends itself to a simple experimental test, i.e., a plot of $1/x_n$ (the reciprocal of the *initial* x_n value) against $1/[M]$ (the reciprocal of the initial monomer concentration). It has actually been found that the polymerization of styrene by $SnCl_4$ in ethylene dichloride [40], and of vinyl alkyl ethers by $SnCl_4$ in m-cresol [41], showed a dominance of termination over transfer, i.e., $x_n \propto [M]$, while isobutylene polymerization by $TiCl_4$ in n-hexane [42] yielded a polymer whose molecular weight was independent of monomer concentration, i.e., transfer appeared to predominate.

It is interesting to compare the nature of the individual steps in cationic polymerization with those of the free radical mechanism. Thus, unlike the situation in the latter case, the termination step in cationic polymerization may be expected to require a greater energy than that of propagation, since it involves σ-bond rupture compared to the low-energy attack of the growing carbenium ion on the π bond of the monomer. If, indeed, the termination step has a higher activation energy than that of propagation, then a rise in temperature should lead to an increase in termination relative to propagation and thus to a lower steady-state concentration of growing chains. The net result would thus be a *decrease* in polymerization rate and molecular weight, i.e., a "negative" apparent overall activation energy for polymerization. This might, of course, be partially or wholly offset if the activation energy of the initiation step were sufficiently high. However, in the majority of cases it appears that this is not the case, so that faster rates (and higher molecular weights) are indeed obtained at reduced temperatures (about $-100°C$).

One interesting feature which may arise in the case of the carbenium ion polymerizations is the occurrence of a rapid isomerization reaction [43] during the propagation step. These reactions occur by means of a shift of a hydrogen (or even a methyl) group near the tip of the growing chain, leading to a rearrangement toward a more stable carbenium ion structure. Thus, for example, in the cationic polymerization of 3-methyl butene-1, the following reaction takes place:

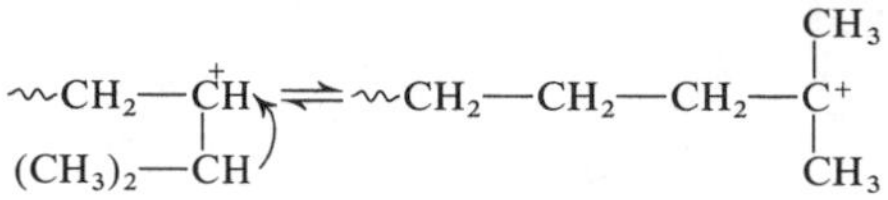

$$
\begin{array}{c}
\text{CH}_3 \\
\text{|} \\
\sim\!\!\text{CH}_2\!\!-\!\!\overset{+}{\text{CH}}\!\!\rightleftharpoons\!\!\sim\!\!\text{CH}_2\!\!-\!\!\text{CH}_2\!\!-\!\!\text{CH}_2\!\!-\!\!\overset{+}{\text{C}} \\
\text{|} \qquad\qquad\qquad\qquad\qquad\qquad \text{|} \\
(\text{CH}_3)_2\!\!-\!\!\text{CH} \qquad\qquad\qquad\qquad \text{CH}_3
\end{array}
$$

Obviously, the attainment of the more stable tertiary carbenium ion is the driving force for this isomerization. This would play an important role in controlling the chain structure, depending on the relative rates of the propagation and rearrangement reactions.

B. BUTYL RUBBER

The only important commercial elastomer prepared by a cationic polymerization is butyl rubber, i.e., a copolymer of isobutene and isoprene. The latter monomer is incorporated in relatively small proportions (~ 2–3%) in order to introduce sufficient unsaturation for sulfur vulcanization. The process [44] can be described by the accompanying "flow sheet." In this process the polymerization is almost instantaneous and extensive cooling by liquid ethylene is required to control the reaction.

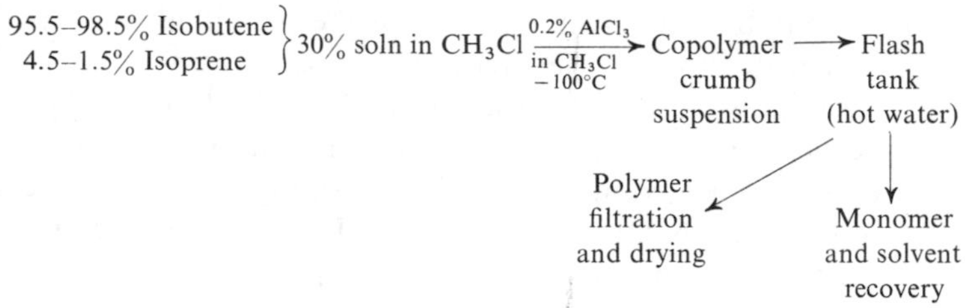

$$
\left.\begin{array}{l}
95.5\text{–}98.5\%\ \text{Isobutene} \\
4.5\text{–}1.5\%\ \text{Isoprene}
\end{array}\right\}\ 30\%\ \text{soln in CH}_3\text{Cl}\ \xrightarrow[\substack{\text{in CH}_3\text{Cl} \\ -100^\circ\text{C}}]{0.2\%\ \text{AlCl}_3}\ \underset{\substack{\text{crumb} \\ \text{suspension}}}{\text{Copolymer}}\ \longrightarrow\ \underset{\substack{\text{tank} \\ \text{(hot water)}}}{\text{Flash}}
$$

Polymer
filtration   Monomer
and drying   and solvent
              recovery

The molecular weights of butyl rubber grades are in the range of 300,000–500,000 and they are very sensitive [44] to polymerization temperature above -100°C. For example, a rise in polymerization temperature of 25°C can result in a five- or tenfold *decrease* in molecular weight, presumably due to the kinetic factors discussed previously.

Isoprene is used as the comonomer in butyl rubber because the isobutene–isoprene reactivity ratios are more favorable for inclusion of the diene than those of the isobutene–butadiene pair [44]. Thus, for the former pair, the r(isobutene) = 2.5, and r(isoprene) = 0.4. It should be noted, however, that, as discussed previously, such r values can be markedly influenced by the nature of the initiator and solvent used in the polymerization. The values just quoted are applicable to the commercial butyl rubber process, as described earlier.

C. OTHER CATIONIC POLYMERIZATIONS

Although butyl rubber is by far the most important commercial elastomer to be synthesized by cationic polymerization, there are several others, which illustrate the type of monomers susceptible to this mechanism. These are listed in Table XIII. The first two polymers have been commercially developed, while the last two have not, except for the use of the poly(tetrahydrofuran) as a low molecular weight polyglycol for the formation of urethane polymers.

TABLE XIII

NEW ELASTOMERS BY CATIONIC POLYMERIZATION[a]

Name	Monomer	Polymer unit	Ref.
Poly(propylene oxide)	CH_2—$CHCH_3$ with O bridge	—O—CH_2—CH— with CH_3	45, 46
Poly(epichlorohydrin)	CH_2—$CHCH_2Cl$ with O bridge	O—CH_2—CH— with CH_2Cl	45, 47
Poly(vinyl ethers)	CH_2=CH with OR^*	—CH_2—CH— with OR^*	48
Poly(tetrahydrofuran)	CH_2—CH_2 / CH_2 CH_2 with O bridge	O—$(CH_2)_4$—	49

[a] $R^* =$ alkyl.

The cationic polymerization of tetrahydrofuran, although not leading directly to a commercial elastomer, deserves special attention because of its kinetic features. Thus it has been shown by Dreyfuss and Dreyfuss [49] that, when an appropriate Lewis acid initiator is used, the polymerization of this monomer assumes the character of a "living" polymerization, i.e., it does not exhibit any termination step, very much like the analogous anionic polymerizations which are discussed in Section VIII. This is rather remarkable, in view of the known high reactivity of carbenium ions and their propensity to participate in various side reactions during polymerization. This remarkable stability of the growing chain in the cationic polymerization of cyclic ethers has been ascribed [50] to an "oxonium" ion mechanism, whereby the carbenium ion is actually "solvated" by coordination with the oxygen of the monomer, as follows:

Initiation

$$R^+ + O\!\!<\!\!\square \longrightarrow R\!\!-\!\!\overset{+}{O}\!\!<\!\!\square \tag{48}$$

Propagation

$$R—\overset{+}{O}\!\!\diagdown\!\!\square + O\!\!\diagdown\!\!\square \rightleftharpoons \cdots \rightleftharpoons R—[O(CH_2)_4]_x—O(CH_2)_3CH_2—\overset{+}{O}\!\!\diagdown\!\!\square \quad (49)$$

where R^+ is the initiating cation.

It is the propagation step (Eq. (49)) which has the nonterminating (or living) character. It is also shown as a reversible step, leading to an "equilibrium" polymerization, which is so common in ring-opening addition polymerization.

VIII. Addition Polymerization by Anionic Mechanism

A. Mechanism and Kinetics

An anionic mechanism is proposed for those polymerizations initiated by organometallic species, where there is good reason to assume that the metal is strongly electropositive relative to the carbon (or other) atom at the tip of the growing chain. Hence the metal becomes a cation, either in the free state or coupled with the growing carbanion. This type of mechanism probably includes the vast array of organometallic complexes known under the heading of Ziegler–Natta catalysts. However, most of these consist of heterogeneous systems which are not readily amenable to analysis. The simpler alkali metal organometallics can be used in homogeneous media and can therefore be subjected to a kinetic analysis.

Although the ability of alkali metals, such as sodium, to initiate polymerization of unsaturated organic molecules has long been known (the earliest record dating back to the work of Matthews and Strange [8] and of Harries [7], around 1910, on polymerization of dienes), the mechanism had remained largely obscure due to the heterogeneous character of the type of catalysis. The pioneering work of Higginson and Wooding [51] on the homogeneous polymerization of styrene by potassium amide in liquid ammonia, and that of Robertson and Marion [52] on butadiene polymerization by sodium in toluene, merely showed the important role of the solvent in participating in transfer reactions.

The true nature of homogeneous anionic polymerization only became apparent through studies of the soluble aromatic complexes of alkali metals, such as sodium naphthalene. These species are known [53] to be radical anions, with one unpaired electron stabilized by resonance and a high solvation energy, and are therefore equivalent to a "soluble sodium." They initiate polymerization by an "electron transfer" process [54], just as in the case of the metal itself, except that the reaction is homogeneous and therefore involves a much higher concentration of initiator. The mechanism of polymerization initiated by alkali metals (or their soluble complexes) can therefore be written as follows, using styrene as an example:

Initiation

$$Na + CH_2\!\!=\!\!CH \longrightarrow \dot{C}H_2\!\!-\!\!\ddot{C}H^-Na^+ \qquad\qquad (50)$$
$$\underset{\textstyle C_6H_5}{|} \qquad\qquad \underset{\textstyle C_6H_5}{|}$$

$$2\dot{C}H_2\!\!-\!\!\underset{\textstyle C_6H_5}{\underset{|}{\ddot{C}H}}{}^-Na^+ \longrightarrow Na^{+-}\underset{\textstyle C_6H_5}{\underset{|}{\ddot{C}H}}\!\!-\!\!CH_2\!\!-\!\!CH_2\!\!-\!\!\underset{\textstyle C_6H_5}{\underset{|}{\ddot{C}H}}{}^-Na^+ \qquad (51)$$

Propagation

$$Na^{+-}\underset{\textstyle C_6H_5}{\underset{|}{\ddot{C}H}}\!\!-\!\!CH_2\!\!-\!\!CH_2\!\!-\!\!\underset{\textstyle C_6H_5}{\underset{|}{\ddot{C}H}}{}^-Na^+ + CH_2\!\!=\!\!\underset{\textstyle C_6H_5}{\underset{|}{CH}} \longrightarrow \cdots \longrightarrow$$

$$Na^{+-}\left[\!\!\begin{array}{c}=\!\!CH\!\!-\!\!CH_2\!\!=\\[-2pt] \underset{\textstyle C_6H_5}{|}\end{array}\!\!\right]_x\left[\!\!\begin{array}{c}=\!\!CH_2\!\!-\!\!CH\!\!=\\[-2pt] \underset{\textstyle C_6H_5}{|}\end{array}\!\!\right]_x\!\!{}^-Na^+ \quad (52)$$

Thus the initiation steps (50) and (51) involve an electron transfer step and subsequent coupling of the radical anions to form di-anions which can grow a polymer chain at both ends. In the case of the soluble alkali metal aromatic complexes, this initiation step is extremely fast, due to the high concentrations of complex ($\sim 10^{-3}$ M) and monomer (~ 1 M), and so is the subsequent propagation reaction. However, in the case of the alkali metal initiators, the electron transfer step (50) is very much slower, due to the heterogeneous nature of the reaction, so that the buildup of radical anions is much slower. In fact, there is good evidence [52] that, in such cases, a second electron transfer step can occur between the metal and the radical anion, rather than coupling of the radical anions. In either case, the final result is a di-anion, i.e., a difunctional growing chain.

However, it is the homogeneous systems initiated by sodium naphthalene which demonstrate the special nature of anionic polymerization, i.e., the fact that a termination step may be avoided under certain circumstances, leading to the concept of "living" polymers [54]. Since these are homogeneous systems, the stoichiometry of the reaction becomes apparent, i.e., two molecules of sodium naphthalene generate one chain. Furthermore, since all the chains are initiated rapidly and presumably have an equal opportunity to grow, their molecular weight distribution becomes very narrow, approximating the Poisson distribution [55]. These aspects are obscured in the metal-initiated polymerizations owing to the continued slow initiation over a long period of time, leading to a great difference in "age" of the growing chains and hence in their size distribution.

Polymerization initiated by electron transfer from a metal, or metal complex, represents only one of the anionic mechanisms. It is, of course, possible to consider separately those polymerizations initiated directly by organometallic compounds. Of the latter, the organolithium compounds are probably the best examples, since they are soluble in a wider variety of solvents and are relatively stable. The mechanism of these polymerizations is somewhat simpler than in the case of sodium naphthalene, since there is no electron transfer step; thus

Initiation

$$\overset{\ominus\ \oplus}{R Li} + CH_2{=}CH \longrightarrow R{-}CH_2{-}\overset{-}{C}H\ Li^+ \qquad\qquad (53)$$
$$\underset{\ \ C_6H_5}{|}\qquad\qquad\underset{\ \ C_6H_5}{|}$$

Propagation

$$R{-}CH_2{-}\overset{-}{C}H\ Li^+ + CH_2{=}CH \longrightarrow \cdots \longrightarrow R{-}\left[CH_2{-}CH{-}\right]_x^- Li^+ \qquad (54)$$
$$\underset{C_6H_5}{|}\qquad\quad\underset{C_6H_5}{|}\qquad\qquad\qquad\qquad\underset{\quad C_6H_5}{|}$$

Termination by impurity

$$R{-}\left[\quad\right]_x^- Li^+ + HOH \longrightarrow R\left[\quad\right]_x^- H + Li^+ + OH^-$$

Hence each organolithium molecule generates one chain and there is no termination of the growing chains in the absence of adventitious impurities, such as water and acids, and if higher temperatures are avoided to prevent side reactions.

Unlike sodium naphthalene, which requires the presence of highly solvating solvents, such as H_4furan, the organolithium systems can operate in various solvents such as ethers or hydrocarbons. However, the rates are much slower in the latter than in the former solvents. Hence, if the initiation reaction (53) is very much slower than the propagation reaction, the molecular weight distribution may be considerably broadened. This does not, of course, vitiate the "living" polymer aspect of the polymerization, which has been shown [56] to operate in these systems, regardless of type of solvent, if side reactions do not intervene.

The nonterminating character of homogeneous anionic polymerization can lead to many novel synthetic routes. Thus, since each chain continues to grow as long as any monomer remains, it is possible to synthesize block polymers by sequential addition of several monomers. Another possibility is the synthesis of linear chains with various functional end groups, by allowing the polymer chain end to react with various agents, e.g., CO_2 to form COOH groups. These possibilities are, of course, of considerable industrial interest.

In view of the unusual mechanism of anionic polymerization, especially the absence of termination, the kinetics of these systems can be treated quite differently than for the other mechanisms. Thus it is possible, by suitable experimental techniques, to examine separately the initiation and propagation reaction. The latter is, of course, of main interest and can be studied by making sure that initiation is complete. In this way, the kinetics of homogeneous anionic polymerization have been extensively elucidated with special reference to the nature of counterion and role of the solvent.

It has been found universally that, in accordance with Eqs. (52) and (54), the propagation rate is always first order with respect to monomer concentration, regardless of solvent system or counterion. However, in contradiction to the foregoing equations, the rate dependency has generally been found to be *lower* than first order with respect to the concentration of growing chains, and the order was found to be strongly dependent on the nature of the solvent and counterion [57]. Strongly solvating solvents, such as ethers and amines, lead to much faster rates than nonpolar solvents and affect the kinetics of these polymerizations quite differently than the hydrocarbon media, presumably due to a different mechanism. This will be discussed first.

In ether solvents, the rate dependency on the growing chain concentration has been found to vary between half order and first order [58], whereby the rate constants increase with decreasing concentrations of growing chains. This has been explained [58, 59] most logically as being due to the simultaneous propagation by free carbanions and ion pairs,

$$
\begin{array}{ccc}
\text{RM}_j^-\text{Li}^+ & \underset{}{\overset{K_e}{\rightleftharpoons}} & \text{RM}_j^- + \text{Li}^+ \\
{\scriptstyle k_p} \big\uparrow {\scriptstyle M} & & {\scriptstyle k_p^-} \big\uparrow \\
\text{RM}_{j+1}^-\text{Li}^+ & \overset{K_e}{\rightleftharpoons} & \text{RM}_{j+1}^- + \text{Li}^+
\end{array}
\tag{55}
$$

where RM_j^-Li^+ represents an ion-pair growing chain, RM_j^- a free carbanionic growing chain, M a monomer, and K_e, k_p, and k_p^- are, respectively, the ionization constant, propagation rate constant for the ion pairs, and propagation rate constant for free ions.

This proposed mechanism was supported by measurements [58, 59] which actually showed the presence of a small degree of ionization in these ether solvents. The appropriate rate expression for the propagation step R_p then becomes

$$
R_p = k_p[\text{RM}_j^-\text{Li}^+][\text{M}] + k_p^-[\text{RM}_j^-][\text{M}]
\tag{56}
$$

Taking into account the relations of Eq. (55), this becomes

$$
R_p/[\text{M}][\text{RM}_j\text{Li}]^{1/2} = k_p^- K_e^{1/2} + k_p[\text{RM}_j\text{Li}]^{1/2} \quad \text{(if } K_e \text{ is small)}
\tag{57}
$$

where $[\text{RM}_j\text{Li}]$ is the total concentration of all growing chains.

Equation (57) can be represented by a plot of the left-hand side against $[\text{RM}_j\text{Li}]^{1/2}$, i.e., the half-power of the total lithium species (or initiator). The intercept defines $k_p^- K_e^{1/2}$, which leads to an evaluation of k_p^- if the ionization constant is first determined. The slope of this plot leads to the direct evaluation of k_p, the ion-pair propagation rate constant. Such measurements have been made [58, 59] for several monomers and counterions, and lead to the general conclusion that the free ion is much more reactive than the ion pair, with values of k_p^- being of the order of 10^4–10^5, while those of k_p are in the range of 10^2–10^3, all expressed as liters per mole per second. K_e was generally found to be about

10^{-7}. In this connection, then, it can be said that most of the propagation reaction occurs while the growing chain end is in its dissociated state.

The kinetics of organolithium polymerization in nonpolar solvents (i.e., hydrocarbons) have also been subjected to intensive study [57, 62], but the results are far less clear than those obtained in ether solvents. Thus again the rate of polymerization was found to be consistently dependent on the first power of the monomer concentration, in line with Eqs. (54) or (57). However, the dependency on organolithium concentration was found to be of a fractional order of *one half or lower*. Since it is not reasonable to expect ionic dissociation in media of such low dielectric constant or solvating power, these fractional orders must have some other explanation. In view of the strong experimental evidence for association between the organolithium species, the kinetic order ascribed to this phenomenon was postulated in a way analogous to Eq. (55), i.e.,

$$(RM_j^-Li^+)_n \overset{K_e}{\rightleftharpoons} nRM_j^-Li^+ \tag{58}$$

$$RM_j^-Li^+ + M \overset{k_p}{\rightarrow} RM_{j+1}^-Li^+ \tag{59}$$

on the assumption that *only* the dissociated chain ends are active. Actually, measurements of the state of association, i.e., the value of n in Eq. (58), have been carried out for styrene, isoprene, and butadiene, and have been found [60, 61] to be consistently very close to 2. This could explain the half-order kinetics found by most investigators [62] in the case of styrene, but cannot account for the lower orders ($\frac{1}{4}$ to $\frac{1}{6}$) found for butadiene and isoprene in various hydrocarbon solvents. Hence it appears that the propagation reaction of these associated growing chains in nonpolar media may be more complex than proposed in Eqs. (58) and (59), and probably involves a direct interaction between the monomer and the associated complex [63].

In conclusion, it should be noted that the molecular weights and their distribution follow the rules originally discussed under living polymers. This means that, regardless of the solvents and counterions used, if no termination or side reactions occur, and if the initiation reaction is fast relative to the propagation reaction, then the molecular weight distribution will approach the Poisson distribution, i.e.,

$$x_w/x_n = 1 + 1/x_n \tag{60}$$

This means that, in principle, a polymer chain of 100 units should have an x_w/x_n ratio of 1.01. This is, of course, impossible to prove experimentally, and it can be assumed that the real distribution would be somewhat broader, due for one thing to imperfect mixing in the reaction mixture. However, values of 1.05 for x_w/x_n are commonly found in these systems [62, 64].

B. CHAIN MICROSTRUCTURE OF POLYDIENES

Although the alkali metals, unlike the Ziegler–Natta systems, do not generally polymerize unconjugated olefins and are not known to lead to any tacticity, they do affect the chain microstructure of polydienes. Thus the proportion of cis-1,4 and trans-1,4 addition versus the 1,2 (or 3,4) mode can be markedly affected by the nature of the counterion as well as the solvent. Ever since the discovery that lithium polymerization of isoprene can lead to a high-cis-1,4 structure [65], close to that of natural rubber, there have been many studies of these effects [66]. Table XIV shows some of these results for isoprene polymerization. These data may be subject to some modification, since they were obtained by infrared spectroscopy, which gives a much less accurate analysis than the more recent nuclear magnetic resonance methods. However, it is obvious from these data that the stereospecific high-cis-1,4 polymer is obtained only in the case of lithium and in the absence of polar solvents. Other solvents and/or counterions exert a dramatic effect in altering the chain microstructure.

Similar effects have been observed with butadiene [66] and other dienes [66]. However, in the case of butadiene, the maximum cis-1,4 content attainable is much less than for isoprene, generally less than 50%. The effect of polar solvents, or of the more electropositive alkali metals, is to produce a high-1,2 polybutadiene.

This marked sensitivity of anionic polymerization to the nature of the counterion and solvent can be traced to the structure of the propagating chain end. The latter involves a carbon–metal bond and can have variable characteristics, ranging all the way from covalency to ionic dissociation. The presence of a more electropositive metal and/or a cation-solvating solvent, such as

TABLE XIV

MICROSTRUCTURE OF POLYISOPRENE[a]

Initiator	Solvent	Chain microstructure (mole %)			
		1,2	3,4	trans-1,4	cis-1,4
n-Butyllithium	n-Heptane	0	7	—	93
Phenylsodium	n-Heptane	8	45	47	0
Benzylpotassium	n-Heptane	10	38	52	0
Rubidium	None	8	39	47	5
Cesium	None	8	37	51	4
n-Butyllithium	H$_4$ furan	16	54	30	0
Phenylsodium	H$_4$ furan	13	49	38	0
Benzylpotassium	H$_4$ furan	17	40	43	0

[a] From Tobolsky and Rogers [67].

ethers, can lead to delocalization of the π electrons and even partial dissociation into carbanions and cations. Direct evidence for this effect, based on NMR [68] and electrolytic conductance [58] measurements, has actually been obtained.

The control of chain structure and molecular weight afforded by the organo-lithium polymerization of dienes has, of course, been of great technological interest. Such product developments have been mainly in the form of (1) poly-butadiene elastomers of various chain structures, (2) functionally terminated liquid polybutadienes (OH- and COOH-"telechelic" [69a–c]), (3) butadiene–styrene copolymers (solution SBR), and (4) styrene–diene triblock copolymers (thermoplastic elastomers).

C. Copolymers of Butadiene

The possibilities inherent in the anionic copolymerization of butadiene and styrene by means of organolithium initiators, as might have been expected, have led to many new developments. The first of these would naturally be the synthesis of a butadiene–styrene copolymer to match (or improve upon) emulsion-prepared SBR, in view of the superior molecular weight control possible in anionic polymerization. The copolymerization behavior of butadiene (or isoprene) and styrene is shown in Table XV. As indicated earlier, unlike the free radical type of polymerization, these anionic systems show a marked sensitivity of the reactivity ratios to solvent type (a similar effect is noted for different alkali metal counterions). Thus, in nonpolar solvents, butadiene (or isoprene) is preferentially polymerized, to the virtual exclusion of the styrene, while the reverse is true in polar solvents. This has been ascribed [67] to the profound effect of solvation on the structure of the carbon–lithium bond, which becomes much more ionic in such media.

The data in Table XV illustrates the problems encountered in such copolymerizations, since the use of polar solvents to assure a random styrene–diene copolymer of desired composition will, at the same time, lead to an increase in side vinyl groups (1,2 or 3,4) in the diene units (see Table XIV). This is of course quite undesirable, since such chain structures result in an increase in the glass

TABLE XV

Alkyllithium Copolymerization of Styrene and Dienes

Monomer 1	Monomer 2	Solvent	r_1	r_2	Ref.
Styrene	Butadiene	Toluene	0.1	12.5	70
Styrene	Butadiene	H_4 Furan	8	0.2	70
Styrene	Isoprene	Toluene	0.25	9.5	71
Styrene	Isoprene	Triethylamine	0.8	1.0	71
Styrene	Isoprene	H_4 Furan	9	0.1	70
Butadiene	Isoprene	n-Hexane	3.4	0.5	72

transition temperature (T_g) and therefore to a loss of good rubbery properties. Hence two methods are actually used to circumvent this problem: (1) the use of limited amounts of polar solvents to accomplish a reasonable compromise between diene chain structure and monomer sequence distribution, and (2) use of an appropriate monomer feed ratio in a continuous polymerization.

As mentioned earlier, the "living" nature of the growing chain in anionic polymerization makes this mechanism especially suitable for the synthesis of *block* copolymers, by sequential addition of *different* monomers. Since such copolymers have markedly different properties than simple copolymers, they will be discussed separately (in Section X).

D. TERMINALLY FUNCTIONAL POLYDIENES

Another characteristic of these homogeneous anionic polymerizations, as mentioned earlier, is their potential for the synthesis of polymer chains having reactive end groups. The production of liquid short-chain *difunctional* polymers by this means is of considerable technological interest and importance, and has attracted much attention in recent years, since it offers an analogous technology to that of the polyethers and polyesters used in urethane polymers. Such liquid "telechelic" polydienes could thus lead, by means of chain extension and cross-linking reactions, directly to "castable" polydiene networks.

Methods of preparing difunctional polydienes have recently been reviewed [69a]; these include both free radical and anionic mechanisms. The main difficulties that must be overcome in this area are (1) the synthesis of a highly difunctional short-chain polymer, and (2) selection of a suitable chemical reaction that can achieve a high degree of linking of functional end groups. The magnitude of these difficulties can be appreciated when one realizes that the final molecular weight of a linearly chain-extended short-chain polymer should be of the order of 10^5, so that the final concentration of functional groups would be only about $10^{-2}\ M$.

A very recent publication [69b] describes the synthesis of a highly difunctional α,ω-dihydroxypolyisoprene, by means of a dilithium initiator, and its use in the formation of a rubber network by linking the terminal hydroxy groups with a triisocyanate. In this way, a remarkably uniform polyisoprene network was obtained, due to the very narrow molecular weight distribution of the polyisoprene glycol prepared from the anionic dilithium initiator.

IX. Stereospecific Polymerization and Copolymerization by Coordination Catalysts

The term "Ziegler–Natta catalysts" refer to a wide variety of polymerization initiators generally involving mixtures of transition metal halides and alkyls of Group II or III metals. It arose from the spectacular discovery of

Ziegler *et al.* [73] that mixtures of titanium tetrachloride and aluminum alkyls polymerize ethylene at low pressures and temperatures; and from the equally spectacular discovery by Natta [74] that the Ziegler catalysts can produce stereospecific polymers from the monoolefins. As can be imagined, these systems can involve many variations of catalyst mixtures, each capable of producing various stereoisomeric polymers. However, we shall be concerned here only with polymerizations involving the commercial elastomers, principally poly-isoprene, polybutadiene, and the ethylene–propylene copolymers.

Unfortunately, it is not yet possible to discuss the mechanisms of these polymerizations in fully satisfactory terms, since they are not yet properly understood. In general, these polymerizations are considered to proceed by a "coordinate anionic" mechanism [75], where the growing chain end is attached to the transition metal, and the aluminum alkyl acts mainly as an alkylating agent, as well as a transfer agent. One of the few rational mechanisms proposed for these systems is that of Cossee [76, 77], who suggested that the active site for the propagation step was a transition metal atom having an octahedral configuration in which one position is vacant due to a missing ligand, as shown in Fig. 2, which also shows the sequence of steps involved in the addition of each monomer unit to the growing chain. It can be seen that this mechanism involves a coordination of the olefin to the transition metal atom at the vacant octahedral position through π bonding, followed by insertion of the monomer into the weakened metal–carbon bond, and an exchange of positions between the growing alkyl group and the orbital vacancy. This model was also used to explain the stereospecific polymerization of propylene [78].

The Cossee mechanism, however, although very plausible and attractive, is a "monometallic" mechanism, on a relatively simple basis. No similar in-terpretations of the more complex systems involving two (or more) organo-

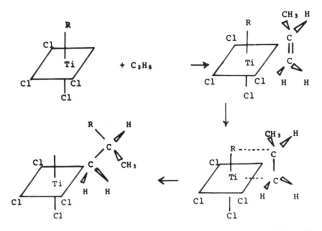

Fig. 2. Mechanism of stereospecific polymerization of propylene. (After Cossee [77].)

metallic species have so far been developed. Hence the bulk of the research studies in this field have been based on empirical results, which have not been amenable to the development of an overall theory. These investigations have led to the accumulation of a vast array of empirical data concerning the relation between type of catalyst and the resulting polymeric structure.

Some selected data which show the types of catalysts used to obtain various stereospecific polymerizations of butadiene and isoprene are shown in Tables XVI and XVII. Similarly, Table XVIII shows the results obtained with various catalyst systems in the copolymerization of ethylene and propylene. Thus

TABLE XVI

STEREOSPECIFIC POLYMERIZATION OF BUTADIENE

Catalyst system			Polymer structure (%)			
Metal alkyl (M)	Transition metal salt (T)	M/T ratio	cis-1,4	trans-1,4	1,2	Ref.
$Al(C_2H_5)_3$	TiI_4	5	92	4	5	79
$Al(C_2H_5)_2Cl$ or $Al(C_2H_5)Cl_2$	Cobalt salts or compounds	~100	95–98	—	—	80, 81
$Al(C_2H_5)_3$	$TiCl_4$	1	49	49	2	82
$Al(C_2H_5)_3$	VCl_4 or $VOCl_3$	2	—	95–100	—	83
$Al(C_2H_5)Cl_2$	β-$TiCl_3$	1–5	—	~100	—	84
	$RhCl_3$[a]	—	—	~100	—	85
$Al(C_2H_5)_3$	$Ti(OC_4H_9)_4$	3	10	—	90	86
$Al(C_2H_5)_3$	Cr compounds	10	—	—	Isotactic	87
$Al(C_2H_5)_3$	Cr compounds	3	—	—	Syndiotactic	88

[a] In presence of ligands such as alcohols or sulfate esters.

TABLE XVII

STEREOSPECIFIC POLYMERIZATION OF ISOPRENE

Catalyst system			Polymer structure (%)			
Metal alkyl (M)	Transition metal salt (T)	M/T ratio	cis-1,4	trans-1,4	3,4	Ref.
$Al(i\text{-}C_4H_9)_3$	$TiCl_4$	1	95	—	—	82
$Al(C_2H_5)_3$	$TiCl_4$	>1	95	—	4	88
$Al(C_2H_5)_3$	β-$TiCl_3$	—	85	—	—	89
$Al(C_2H_5)_3$	$TiCl_4$	>1	—	95–100	5	88
$Al(C_2H_5)_3$	VCl_3, $VOCl_3$	2	—	95–100	—	83
$Al(C_2H_5)_3$	$Ti(OR)_4$	5	—	—	95	86
$Al(C_2H_5)_3$	$VOAc_3$	—	—	—	90	90

TABLE XVIII

REACTIVITY RATIOS IN ETHYLENE–PROPYLENE COPOLYMERIZATION

Catalyst system	$r_1(C_2H_4)$	$r_2(C_3H_5)$	Ref.
$TiCl_4 + Al(C_6H_{13})_3$	33.4	0.032	91
$TiCl_3 + Al(C_6H_{13})_3$	15.7	0.11	91
$TiCl_2 + Al(C_6H_{13})_3$	15.7	0.11	91
$VCl_4 + Al(C_6H_{13})_3$	7.1	0.088	92
$VCl_3 + Al(C_6H_{13})_3$	5.6	0.145	93
$VOCl_3 + Al(C_6H_{13})_3$	18.0	0.065	94
$ZrCl_4 + Al(i\text{-}C_4H_9)_3$	61	—	95
$HfCl_4 + Al(i\text{-}C_4H_9)_3$	76	—	95

Tables XVI and XVII show the profound changes in chain microstructure of polybutadiene and polyisoprene, respectively, which can occur with the use of different combinations of metal compounds, and even with different *ratios* of such compounds. These changes have, of course, not been rationalized by any proposed mechanism, and thus remain in the realm of empirical knowledge, and speculation.

From Table XVIII it can be seen that ethylene enters the copolymer chain at a very much faster rate than does propylene. The effect of the catalyst system, although noticeable, is not profound, with the possible exceptions of the hafnium and zirconium systems, which seem to exaggerate this trend even more. Although it is not shown in Table XVIII, the reactivity ratios are not noticeably affected by temperature [88], within the range of -20 to $75°C$. It should also be noted that the product r_1r_2 is fairly close to 1, as would be expected in ionic polymerization, where the nature of the terminal unit on the growing chain has little effect on its relative reactivity toward the two monomers.

The EPDM terpolymers contain a small proportion of unsaturation, resulting from the inclusion of a small amount of a third monomer, generally a non-conjugated diene. This permits the use of sulfur vulcanization instead of peroxides. The inclusion of the third monomer must be taken into consideration when selecting a suitable catalyst system, but it does not significantly alter the relations shown in Table XVIII.

A unique type of polymerization by coordination catalysts is that involving "olefin metathesis" (Fig. 3). All the polymerizations described so far were

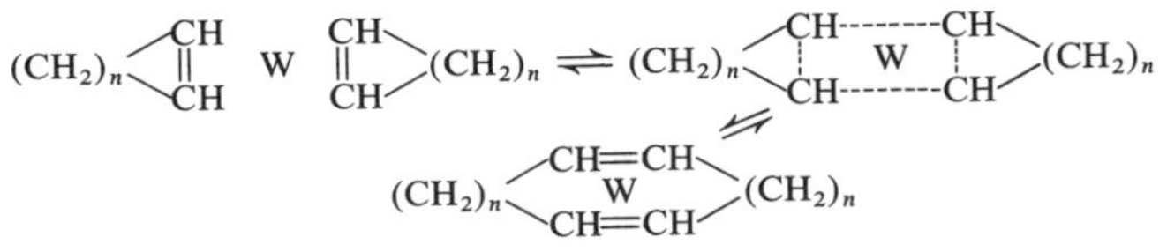

Fig. 3. Polymerization of a cycloalkene by olefin metathesis.

based on either the opening of the π bond of unsaturated compounds, or of the σ bond of cyclic compounds. The transition metal catalysts were also found to be capable of ring-opening polymerization of *unsaturated* cyclic monomers of the cycloalkene type, e.g., cyclopentene [96], leading to an *unsaturated* polymer.

Although it was originally assumed that this type of polymerization involved the usual σ-bond rupture, it was later convincingly shown by Calderon and his co-workers [97, 98] that this reaction proceeds by double-bond interchange, as shown in Fig. 3. This is apparently a special feature of the soluble tungsten or molybdenum complexes, such as (WCl_6 + AlR_3), and is visualized as the growth of a macrocyclic polymer by means of a reversible low-energy (~ 5 kg cal mole^{-1}) bond interchange. It requires only very low concentrations ($\sim 0.01-$ 0.02 mole %) of catalyst and is very rapid (~ 30 min at $0°C$). These catalysts can also be stereospecific, e.g., tungsten yields a trans polymer from cyclopentene, while molybdenum yields a cis polymer.

This type of system has good commercial potentialities. The "trans-polypentenamer" obtained from cyclopentene, for example, exhibits very good properties as a vulcanizable elastomer, combining some of the advantageous properties of *cis*-polybutadiene and *cis*-polyisoprene, e.g., strain-induced crystallization like Hevea, with a T_g of $-90°C$.

X. Graft and Block Copolymerization*

The idea of graft or block copolymerization probably first arose as a means of modifying naturally occurring polymers, such as cellulose (cotton), rubber, or wool. Graft copolymerization, by analogy to the botanical term, refers to the growth of a "branch" of different chemical composition on the "backbone" of a linear macromolecule. In contrast, the related term of block copolymerization refers to the specific case of growth of a polymer chain from the *end* of a linear macromolecule, thus leading to a composite linear macromolecule consisting of two or more "blocks" of different chemical structure. The importance of these types of polymer structures is basically due to the fact that polymer chains of different chemical structure, which are normally incompatible (because the small entropy of mixing is insufficient to overcome the mostly positive free energy of mixing), are chemically bonded to each other. This leads to the formation of microheterogeneities which can have a profound effect on the mechanical properties of the composite material.

As one might expect, graft and block copolymerization can be accomplished by means of each of the three known mechanisms, i.e., radical, cationic, and anionic, each of which shows its own special characteristics. Hence these mechanisms have been used wherever appropriate for the polymer and mono-

*See also Chapters 12 and 13.

mer involved. The examples quoted in the following discussion will deal primarily with elastomers.

A. Free Radical Mechanism

This has been the most widely applied system for the formation of graft copolymers, since it provides the simplest method and can be used with a wide variety of polymers and monomers. It has not been very useful in the synthesis of block copolymers, as will become obvious from an examination of the methods used. These can be listed as follows.

a. Chemical Initiation

This is still the most popular method for graft copolymerization via free radicals. It is mainly based on transfer reactions between the radicals and the polymer chain, which can occur in two ways:

(i)   *Direct Initiator Attack*

$$RO^\cdot + PH \rightarrow ROH + P^\cdot$$

$$P^\cdot + M \rightarrow \cdots \rightarrow \text{graft copolymer}$$

Here $RO^\cdot$ represents the initiating free radical formed from, for example, the decomposition of a peroxide, and PH is the polymer chain, containing sufficiently labile hydrogen atoms which can be abstracted by the initiator radical.

(ii)   *Chain Transfer with Polymer*

$$ROM_x^\cdot + PH \rightarrow ROM_xH + P^\cdot$$

$$P^\cdot + M \rightarrow \cdots \rightarrow \text{graft copolymer}$$

In this case, the growing chain $ROM_x^\cdot$ undergoes chain transfer with the polymer, again by hydrogen abstraction, leading to the initiation of a graft.

In both of these cases, there is always the possibility of formation of *homopolymer* during the grafting reaction, via chain transfer of the growing chain with species *other* than the polymer backbone (e.g., monomer, solvent, initiator). Obviously, direct initiator attack on the polymer chain is likely to lead to a higher proportion of graft copolymer, all things being equal, so that initiators which react preferentially with the polymer chain rather than with the monomer are chosen whenever possible.

This type of graft copolymerization has been applied to the grafting of monomers like styrene and methyl methacrylate to natural rubber, directly in the latex [99, 100]. In this case, for example, oxy radicals of the $RO^\cdot$ type, such as result from peroxides, were found to be much more active [100] in direct attack on the polymer than hydrocarbon radicals, such as are obtained from decomposition of azo compounds, e.g., azobisisobutyronitrile. Similar methods have been developed for grafting the foregoing monomers, and many other vinyl monomers, to synthetic rubbers like SBR, leading to a variety of plastic-reinforced elastomers and rubber-reinforced high-impact plastics [101]. In this case, grafting can also occur by the "copolymerization" of the monomer with the unsaturated bonds (mainly vinyl) in the polymer; thus

$$RM_x^{\cdot} + \text{~~CH}_2\text{—CH~~} \longrightarrow \text{~~CH}_2\text{—CH~~} \longrightarrow \cdots \longrightarrow$$

$$CH_2\text{=}CH \qquad\qquad RM_x\text{—CH}_2\text{—CH}^{\cdot}$$

This reaction can, of course, also lead to cross-linking of the polymer chains, and this must be controlled.

b. Other Methods

Other methods of generating free radicals can also be used to initiate graft polymerization with elastomers, both natural and synthetic. These include irradiation of polymer–monomer mixtures by ultraviolet light [102], high-energy radiation [103], and mechanical shear. The latter is of particular interest because of its unique mechanism, and has been extensively investigated.*

Thus it has been convincingly demonstrated that elastomers, when subjected to severe mechanical shearing forces, undergo homolytic bond scission to form free chain radicals. The latter, when in the presence of oxygen, may then undergo various reactions, either becoming stabilized ruptured chains, or reacting with other chains to form branched or cross-linked species [104, 105]. When blends of different elastomers are masticated, "interpolymers" are formed by the interaction of the radicals formed from the copolymers [106]. A further extension of such mechanochemical processes occurs when elastomers are masticated in the presence of polymerizable monomers, the chain radicals initiating polymerization and leading to formation of block and graft structures. This was clearly demonstrated in the case of natural rubber [106] and of other elastomers [107, 108].

B. CATIONIC MECHANISM

Graft and block copolymerization can also be accomplished by means of cationic polymerization, although this method is not used as widely as the free radical mechanism. The most common approach is to utilize the chain transfer reactions of the carbenium ions on a suitable substrate, especially if the latter contains unsaturated double bonds, which are readily attacked. Many of these systems, however, lead to mixtures with homopolymers formed by the strong tendency of cationic polymerization to undergo transfer to monomer, as previously illustrated in Eq. (36).

Very recently, a novel method of cationic graft and block copolymerization was developed [109] which leads to a much higher efficiency. It is based on the alkylation reaction between tertiary alkyl halides (or allyl halides) and alu-

---

* See also Chapter 11.

minum alkyls. Thus, for example, aluminum trimethyl alkylates t-butylchloride to neopentane [110], as follows:

$$(CH_3)_3CCl + Al(CH_3)_3 \longrightarrow (CH_3)_3\overset{+}{C}[(CH_3)_3AlCl]^- \qquad (61)$$
$$\downarrow$$
$$(CH_3)_4C + (CH_3)_2AlCl$$

In the presence of a cationically susceptible monomer, the intermediate carbenium ion is presumably capable of initiating polymerization, before undergoing the final alkylation (termination) step. Hence, if the tertiary alkyl halide is part of a polymer chain, a graft copolymerization may be initiated upon addition of a suitable aluminum alkyl compound.

This method is novel in that the polymerization is initiated *solely* at the halide substituents in the polymer chain, since the aluminum alkyl is only a coinitiator and cannot, by itself, initiate the polymerization of the monomer. Hence only graft copolymerization will occur and not any homopolymerization. However, this is unfortunately not entirely true, due to the propensity of most cationic polymerizations to exhibit a substantial amount of transfer to monomer, which would, of course, lead to homopolymer formation. By this means it has been found possible to graft polymerize isobutylene onto a poly-(vinyl chloride) backbone [111], or styrene onto a halogenated EPDM backbone [109].

Although these special methods are suitable for graft copolymerization by the cationic mechanism, they are not particularly useful for block polymerization, which requires reactive *end groups* to initiate new blocks. In this connection, one type of cationic polymerization is particularly suitable, i.e., that based on the oxonium mechanism of polymerization of cyclic oxides, which has been previously discussed [49]. Since the oxonium ion chain end can be made to retain its reactivity without termination, it is possible to utilize this type of "living" system for the sequential polymerization of different cyclic oxides to form block copolymers.

C. ANIONIC MECHANISM

It is, of course, the anionic mechanism which is most suitable for the synthesis of block copolymers, since many of these systems are of the "living" polymer type, as described previously. Thus it is possible to use organoalkali initiators to prepare block copolymers in homogeneous solution, where each block has a prescribed molecular weight, based on monomer–initiator stoichiometry, as well as a very narrow molecular weight distribution (Poisson). As would be expected, such block copolymers are very "pure," due to the absence of any side reactions during the polymerization (e.g., termination, monomer transfer, branching).

Organolithium initiators have been particularly useful in this regard, since they are soluble in a variety of solvents and since they can initiate the polymerization of a variety of monomers, such as styrene and its homologs, the 1,3-dienes, cyclic oxides and sulfides, and cyclic siloxanes. Various block copolymers of these monomers have been synthesized, some commercially, but the outstanding development in this area has been in the case of the ABA type of triblock copolymers, and these deserve special mention.

The ABA triblocks which have been most exploited commercially are of the styrene–diene–styrene type, prepared by sequential polymerization initiated by alkyllithium compounds. The behavior of these block copolymers illustrates the special characteristics of block (and graft) copolymers, which are based on the general *incompatibility* of the different blocks. Thus the polystyrene end blocks ($\sim$ 15,000–20,000 MW) aggregate into a separate phase, which forms a microdispersion within the matrix composed of the polydiene chains (50,000–70,000 MW). A schematic representation of this morphology is shown in Fig. 4. This phase separation, which occurs in the melt (or swollen) state, results, at ambient temperatures, in a network of elastic polydiene chains held together by glassy polystyrene microdomains. Hence these materials behave as virtually cross-linked elastomers at ambient temperatures, but are completely thermoplastic and fluid at elevated temperatures.

Such "thermoplastic elastomers" are very attractive technologically, since they can be heat molded like thermoplastics, yet exhibit the behavior of rubber vulcanizates. As would be expected, their structure, morphology, and mechanical properties have been studied extensively [112–114]. An electron photo-

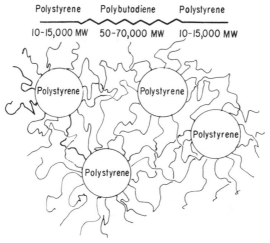

Fig. 4. Structure of thermoplastic elastomers from ABA triblock copolymers. (After Morton [112].)

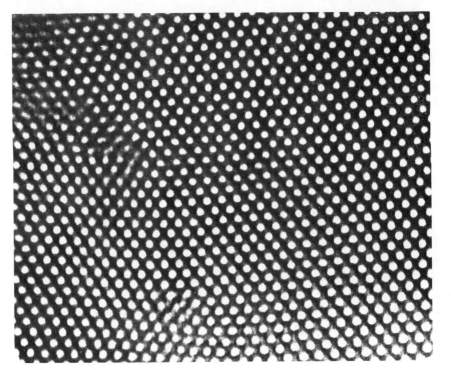

Fig. 5. Transmission electron photomicrograph of an ultrathin film of a styrene–isoprene–styrene triblock copolymer (MW 16,200–75,600–16,200) ($\times$ 100,000).

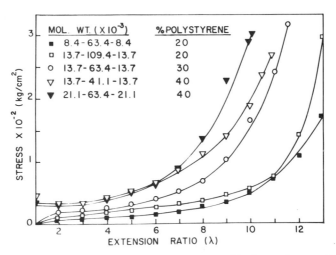

Fig. 6. Tensile properties of styrene–isoprene–styrene triblock copolymers. (After Morton [112].)

micrograph of a typical styrene–isoprene–styrene (SIS) triblock film is shown in Fig. 5, while the tensile properties of a series of such triblock copolymers are shown in Fig. 6. The unusually high tensile strength of these elastomers, better than that of conventional vulcanizates, is ascribed both to the remarkable regularity of the network, as illustrated in Fig. 5, and to the energy-absorbing characteristics of the polystyrene domains, which yield and distort under high stress [112].

This interesting behavior of the ABA triblock copolymers is not a unique feature of the styrene–diene structure, but can be found in the case of other analogous chemical structures. Thus thermoplastic elastomers have recently been obtained from other triblock copolymers, where the dienes have been replaced by cyclic sulfides [115] or cyclic siloxanes [116].

Although they do not possess the elegant regularity of structure of the block copolymers, graft copolymers can also be prepared by the anionic mechanism. These can most conveniently be synthesized by first metallating the desired polymer backbone and then initiating grafts of the desired monomer at the metallated sites. This has been done, for example, in grafting polystyrene branches onto a polybutadiene backbone [117], as illustrated in reaction (62).

$$\sim CH_2-CH=CH-CH_2\sim + RLi \longrightarrow RH + \sim CH_2-CH=CH-CH\sim \underset{\overset{|}{Li}}{} \Bigg] \underset{\overset{|}{C_6H_5}}{CH_2{=}CH}$$

$$\sim CH_2-CH=CH-\underset{\overset{|}{\left[\underset{\overset{|}{C_6H_5}}{CH_2-CH-} \right]_x}}{CH}\sim \longleftarrow \qquad Li^+ \qquad (62)$$

Although the spacing of such graft branches will be random, the length of each branch should be very similar (Poisson). Hence it is not surprising to find that if each polybutadiene chain has *at least two* grafts, the graft copolymer shows somewhat similar behavior to the thermoplastic elastomers obtained from the triblocks.

REFERENCES

1. G. Williams, *Proc. R. Soc. London* **10,** 516 (1860).
2. G. Bouchardat, *C. R. H. Acad. Sci.* **89,** 1117 (1879).
3. W. A. Tilden, *J. Chem. Soc.* **45,** 411 (1884).
4. I. Kondakow, *J. Prakt. Chem.* **62,** 66 (1900).
5. J. Thiele, *Justus Liebigs Ann. Chem.* **319,** 226 (1901).
6. S. V. Lebedev, *Zh. Russ. Fiz.-Khim. Ova., Chast* **42,** 949 (1910).
7. C. D. Harries, *Justus Liebigs Ann. Chem.* **383,** 190 (1911).
8. F. E. Matthews and E. H. Strange, Br. Patent 24,790 (1910).
9. P. J. Flory, "Principles of Polymer Chemistry," Ch. 1. Cornell Univ. Press, Ithaca, New York, 1953.
10. M. Morton, *Rubber Plast. Age* **42,** 397 (1961).

11. M. S. Matheson, E. E. Auer, E. B. Bevilacqua, and E. J. Hart, *J. Am. Chem. Soc.* **71**, 497, 2610 (1949); **73**, 1700, 5395 (1951).
12. M. Morton, P. Salatiello, and H. Landfield, *J. Polym. Sci.* **8**, 215 (1952).
13. M. Morton and S. D. Gadkary, *130th Meet., Am. Chem. Soc., Atlantic City, N.J., 1956;* S. D. Gadkary, M.S. Thesis, Univ. of Akron, Akron, Ohio, 1956.
14. G. V. Schulz, *Z. Phys. Chem., Abt. B* **43**, 25 (1939).
15. G. S. Whitby and R. N. Crozier, *Can. J. Res.* **6**, 203 (1932).
16. W. D. Harkins, *J. Am. Chem. Soc.* **69**, 1428 (1947).
17. M. Morton, S. Kaizerman, and M. W. Altier, *J. Colloid Sci.* **9**, 300 (1954).
18. W. V. Smith and R. H. Ewart, *J. Chem. Phys.* **16**, 592 (1948).
19. M. Morton and W. E. Gibbs, *J. Polym. Sci., Part A* **1**, 2679 (1963).
20. M. Morton and P. P. Salatiello, *J. Polym. Sci.* **6**, 225 (1951).
21. W. S. Richardson and A. Sacher, *J. Polym. Sci.* **10**, 353 (1953).
22. K. E. Beu, W. B. Reynolds, C. F. Fryling, and H. L. McMurry, *J. Polym. Sci.* **3**, 465 (1948).
23. A. W. Meyer, *Ind. Eng. Chem.* **41**, 1570 (1949).
24. M. Morton, J. A. Cala, and M. W. Altier, *J. Polym. Sci.* **19**, 547 (1956).
25. M. Morton and I. Piirma, *J. Polym. Sci.* **19**, 563 (1956).
26. A. M. Neal and L. R. Mayo, *in* "Synthetic Rubber" (G. S. Whitby, ed.), p. 770. Wiley, New York, 1954.
27. W. E. Mochel and J. H. Peterson, *J. Am. Chem. Soc.* **71**, 1426 (1949).
28. C. A. Hargreaves, *in* "Polymer Chemistry of Synthetic Elastomers" (J. P. Kennedy and E. Tornqvist, eds.), p. 233. Wiley (Interscience), New York, 1968.
29. J. T. Maynard and W. E. Mochel, *J. Polym. Sci.* **13**, 227 (1954).
30. F. R. Mayo and F. M. Lewis, *J. Am. Chem. Soc.* **66**, 1594 (1944).
31. G. E. Ham, ed., "Copolymerization." Wiley (Interscience), New York, 1964.
32. C. A. Uraneck, *in* "Polymer Chemistry of Synthetic Elastomers" (J. P. Kennedy and E. Tornqvist, eds.), p. 157. Wiley (Interscience), New York, 1958.
33. F. A. Bovey, I. M. Kolthoff, A. I. Medalia, and E. J. Meehan, "Emulsion Polymerization," p. 257. Wiley (Interscience), New York, 1955.
34. C. F. Fryling, *in* "Synthetic Rubber" (G. S. Whitby, ed.), p. 248. Wiley, New York, 1954.
35. W. Hofmann, *Rubber Chem. Technol.* **37**(2), 1 (1964).
36. R. G. W. Norrish and K. E. Russell, *Trans. Faraday Soc.* **48**, 91 (1952).
37. A. G. Evans and M. Polanyi, *J. Chem. Soc.* p. 252 (1947).
38. D. C. Pepper, *Sci. Proc. R. Dublin Soc.* **25**, 131 (1950).
39. F. S. Dainton and G. B. B. M. Sutherland, *J. Polym. Soc.* **4**, 37 (1949).
40. D. C. Pepper, *Trans. Faraday Soc.* **45**, 404 (1949).
41. D. D. Eley and A. W. Richards, *Trans. Faraday Soc.* **45**, 425, 436 (1949).
42. P. H. Plesch, *J. Chem. Soc.* p. 543 (1950).
43. J. P. Kennedy and A. W. Langer, *Fortschr. Hochpolym.-Forsch.* **3**, 508 (1964).
44. J. P. Kennedy, *in* "Polymer Chemistry of Synthetic Elastomers" (J. P. Kennedy and E. Tornqvist, eds.), p. 291. Wiley (Interscience), New York, 1968.
45. E. J. Vandenberg, *J. Polym. Sci.* **47**, 486 (1960).
46. E. E. Gruber, D. A. Meyer, G. H. Swart, and K. V. Weinstock, *Ind. Eng. Chem., Prod. Res. Dev.* **3**, 194 (1964).
47. E. J. Vandenberg, *Rubber Plast. Age* **46**, 1139 (1965).
48. J. Lal, E. F. Devlin, and G. Trick, *J. Polym. Sci.* **44**, 523 (1960).
49. P. Dreyfuss and M. P. Dreyfuss, *Fortschr. Hochpolym.-Forsch.* **4**, 457 (1967).
50. H. Meerwein, *Angew. Chem.* **59**, 168 (1947); H. Meerwein, D. Delfs, and H. Morschel, *Angew. Chem.* **72**, 927 (1960).
51. W. C. E. Higginson and N. S. Wooding, *J. Chem. Soc.* p. 760 (1952).
52. R. E. Robertson and L. Marion, *Can. J. Res., Sect. B* **26**, 657 (1948).

53. D. E. Paul, D. Lipkin, and S. I. Weissman, *J. Am. Chem. Soc.* **78,** 116 (1956).
54. M. Szwarc, M. Levy, and R. Milkovich, *J. Am. Chem. Soc.* **78,** 2656 (1956).
55. P. J. Flory, *J. Am. Chem. Soc.* **62,** 1561 (1940); see also Ref. 9, p. 336.
56. M. Morton, A. Rembaum, and J. L. Hall, *J. Polym. Sci., Part A* **1,** 461 (1963).
57. M. Morton, *in* "Vinyl Polymerization," Part II (G. Ham, ed.), p. 211. Dekker, New York, 1969.
58. S. K. Bhattachariyya, C. L. Lee, J. Smid, and M. Szwarc, *J. Phys. Chem.* **69,** 612, 624 (1965).
59. H. Hostalka and G. V. Schulz, *Z. Phys. Chem.* (*Frankfurt am Main*) **45,** 286 (1965).
60. M. Morton, L. J. Fetters, R. A. Pett, and J. F. Meier, *Macromolecules* **3,** 327 (1970).
61. L. J. Fetters and M. Morton, *Macromolecules* **7,** 552 (1974).
62. M. Morton and L. J. Fetters, *Rubber Chem. Technol.* **48,** 359 (1975).
63. J. B. Smart, R. Hogan, P. A. Scherr, M. T. Emerson, and J. P. Oliver, *J. Organomet. Chem.* **64,** 1 (1974).
64. M. Morton and L. J. Fetters, *Macromol. Rev.* **2,** 71 (1967).
65. F. W. Stavely and co-workers, *Ind. Eng. Chem.* **48,** 778 (1956).
66. L. E. Foreman, *in* "Polymer Chemistry of Synthetic Elastomers" (J. P. Kennedy and E. Tornqvist, eds.), p. 491. Wiley (Interscience), New York, 1968.
67. A. V. Tobolsky and C. E. Rogers, *J. Polym. Sci.* **40,** 73 (1959).
68. M. Morton, R. D. Sanderson, and R. Sakata, *J. Polym. Sci., Part B* **9,** 61 (1971).
69a. D. M. French, *Rubber Chem. Technol.* **42,** 71 (1969).
69b. M. Morton and D. C. Rubio, *Am. Chem. Soc., Meet. Rubber Div., Cleveland, Ohio 1975* Pap. No. 43. See also "Proceedings of the International Rubber Conference, Brighton, England, May 16–20, 1977," Vol. 1, pp. 15.1–15.14. The Plastics and Rubber Institute, London (1977).
69c. M. Morton, L. J. Fetters, J. Inomata, D. C. Rubio, and R. N. Young, *Rubber Chem. Technol.* **49,** 303 (1976).
70. Y. L. Spirin, A. A. Arest-Yakubovich, D. K. Polyakov, A. R. Gantmakher, and S. S. Medvedev, *J. Polym. Sci.* **58,** 1181 (1962).
71. Y. L. Spirin, D. K. Polyakov, A. R. Gantmakher, and S. S. Medvedev, *J. Polym. Sci.* **53,** 233 (1961).
72. G. V. Rakova and A. A. Korotkov, *Dokl. Akad. Nauk SSSR* **119**(5), 982 (1958); *Rubber Chem. Technol.* **33,** 623 (1960).
73. K. Ziegler, E. Holzkamp, H. Breil, and H. Martin, *Angew. Chem.* **67,** 541 (1955).
74. G. Natta, *J. Polym. Sci.* **16,** 143 (1955).
75. J. Boor, *Macromol. Rev.* **2,** 115 (1967).
76. P. Cossee, *Tetrahedron Lett.* **17,** 12, 17 (1960).
77. P. Cossee, *J. Catal.* **3,** 80 (1964).
78. E. J. Arlman and P. Cossee, *J. Catal.* **3,** 99 (1964).
79. R. P. Zelinski, D. R. Smith, G. Nowlin, and H. D. Lyons, Belg. Patent 551,851 (1957).
80. M. Gippin, *Rubber Chem. Technol.* **39,** 508 (1966).
81. A. Takahashi and S. Kambara, *J. Polym. Sci., Part B* **3,** 279 (1965).
82. Goodrich-Gulf Chemicals Co., Br. Patent 916,383 (1963); *Chem. Eng. News* **32,** 4913 (1954).
83. G. Natta, L. Porri, P. Corradini, and D. Morero, *Chim. Ind.* (*Milan*) **40,** 362 (1958).
84. G. J. van Amerongen, *Polym. Prepr., Am. Chem. Soc., Div. Polym. Chem.* **5,** 1156 (1964); see also *Adv. Chem. Ser.* **52,** 140 (1966).
85. M. Morton, I. Piirma, and B. Das, *Rubber Plast. Age* **46,** 404 (1965).
86. G. Natta, Belg. Patent 549,554 (1957).
87. G. Natta, *Chim. Ind.* (*Milan*) **42,** 1207 (1960).
88. S. E. Horne, Jr., C. F. Gibbs, V. L. Folt, and E. J. Carlson, Belg. Patent 543,292 (1956); Goodrich-Gulf Chemicals Co., Br. Patent 827,365 (1960).
89. G. Natta, L. Porri, and L. Fiore, *Gazz. Chim. Ital.* **89,** 761 (1959).

90. G. Wilke, *Angew. Chem.* **68,** 306 (1956).
91. G. Natta, A. Valvassori, G. Mazzanti, and G. Sartori, *Chim. Ind. (Milan)* **40,** 896 (1958).
92. G. Mazzanti, A. Valvassori, and G. Pajaro, *Chim. Ind. (Milan)* **39,** 825 (1957).
93. G. Natta, G. Mazzanti, A. Valvasorri, and G. Sartori, *Chim. Ind. (Milan)* **40,** 717 (1958).
94. G. Mazzanti, A. Valvassori, and G. Pajaro, *Chim. Ind. (Milan)* **39,** 743 (1957).
95. F. J. Karol and W. L. Carrick, *J. Am. Chem. Soc.* **83,** 2654 (1961).
96. G. Natta, G. Dall'Asta, and G. Mazzanti, *Angew. Chem.* **76,** 765 (1964).
97. N. Calderon, E. A. Ofstead, J. P. Ward, W. A. Judy, and K. W. Scott, *J. Am. Chem. Soc.* **90,** 4133 (1968).
98. N. Calderon, *Acc. Chem. Res.* **5,** 127 (1972).
99. G. F. Bloomfield and P. McL. Swift, *J. Appl. Chem.* **5,** 609 (1955).
100. P. W. Allen, *in* "Chemistry and Physics of Rubberlike Substances" (L. Bateman, ed.), p. 97. MacLaren, London, 1963.
101. H. A. J. Battaerd and G. W. Tregear, "Graft Copolymers." Wiley (Interscience), New York, 1967.
102. W. Cooper, P. R. Sewell, and G. Vaughan, *J. Polym. Sci.* **41,** 167 (1959).
103. E. G. Cockbain, T. D. Pendle, and D. J. Turner, *J. Polym. Sci.* **39,** 419 (1959).
104. R. J. Ceresa, "Block and Graft Copolymers," p. 65. Butterworth, London, 1962.
105. M. Pike and W. F. Watson, *J. Polym. Sci.* **9,** 229 (1952).
106. D. J. Angier and W. F. Watson, *J. Polym. Sci.* **18,** 129 (1955); *Trans. Inst. Rubber Ind.* **33,** 22 (1957).
107. D. J. Angier, R. J. Ceresa, and W. F. Watson, *J. Polym. Sci.* **34,** 699 (1959).
108. R. J. Ceresa and W. F. Watson, *J. Appl. Polym. Sci.* **1,** 101 (1959).
109. J. P. Kennedy and R. R. Smith, *in* "Recent Advances in Blends, Grafts and Blocks" (L. Sperling, ed.), Vol. 4, p. 303. Plenum, New York, 1974.
110. J. P. Kennedy, N. V. Desai, and S. Sivaram, *J. Am. Chem. Soc.* **95,** 6386 (1973).
111. J. P. Kennedy, J. J. Charles, and D. L. Davidson, *Polym. Prepr., Am. Chem. Soc., Div. Polym. Chem.* **14,** 973 (1973).
112. M. Morton, "Encyclopedia of Polymer Science and Technology," Vol. 15, p. 508. Wiley, New York, 1971.
113. G. Holden, E. T. Bishop, and N. R. Legge, *J. Polym. Sci., Part C* **26,** 37 (1969).
114. T. L. Smith and R. A. Dickie, *J. Polym. Sci., Part C* **26,** 163 (1969).
115. M. Morton, R. F. Kammereck, and L. J. Fetters, *Br. Polym. J.* **3,** 120 (1971).
116. M. Morton, Y. Kesten, and L. J. Fetters, *Polym. Prepr., Am. Chem. Soc., Div. Polym. Chem.* **15,** 175 (1974).
117. D. P. Tate, A. F. Halasa, F. J. Webb, R. W. Koch, and A. E. Oberster, *J. Polym. Sci., Part A-1* **9,** 139 (1971).

Chapter 3

Structure Characterization in the Science and Technology of Elastomers

G. VER STRATE

ELASTOMERS TECHNOLOGY DIVISION
EXXON CHEMICAL COMPANY
LINDEN, NEW JERSEY

Introduction

There is an abundance of literature on the structural characterization of polymers. The fundamental relationships between solution properties and molecular weights were worked out in the 1940s, and have been reviewed repeatedly [1–5]. Spectroscopic and chromatographic techniques have been developed since then, but these too have been incorporated in reviews and recent texts [1, 2, 7, 9, 10, 12, 14, 15, 20].* Students and researchers who intend to use a particular technique should consult these comprehensive texts and the original references cited. Thus, the primary objectives of this chapter are twofold: first, to present in tabular form a listing of available characterization techniques, the information to be gleaned by a given technique, the principle of its operation, and in the text to make some comment about a technique's

* Numbers refer to general references; if a Roman numeral appears, it refers to a reference with the tables.

Copyright © 1978 by Academic Press, Inc.
All rights of reproduction in any form reserved.
ISBN 0-12-234360-3

utility and associated pitfalls; second, to discuss the application of certain techniques in the particular area of elastomers.

Techniques at times are sophisticated and there may be a tendency to lose sight of the principle while executing measurements. Just to take a number provided and attempt to correlate some property with it is poor practice, since the details of the execution of a characterization experiment can often provide as much information as the result itself. In some cases it becomes apparent that the results must be suspect.

Tables I–VI,* which constitute the substance of the chapter, are reasonably complete and should be consulted for reference to a particular technique and for a general outline of what there is to be learned about a particular polymer.

The parameters which can be used to describe a particular polymer have been grouped in these tables in a somewhat arbitrary manner. Thus, one could dispute that chemical composition, crystallinity, and the rubber–glass transition temperature (T_g) have sufficient points in common to be tabulated in the same way. After determining all of the parameters in Tables I–III, one would hope to be able to calculate, or at least explain, what is measured in Tables IV–VI, and what is measured there *should* be related to mechanical properties (Chapters 4, 5, 10), rheology (Chapters 6, 14), vulcanization (Chapter 7), interactions with fillers, etc. (Chapter 8), and chemical reactivity (Chapter 11). Alas, today's knowledge of chemical composition and repeat unit, of compositional and sequence distribution, as well as of molecular weight, its distribution, the degree of branching, and gel content and structure, does not permit prediction of the degree of crystallinity or the position of the rubber–glass transition; again, knowledge of these dependent parameters does not permit prediction of specific behavior. Recognizing that the parameters of Tables I–III, which are really properties of individual molecules, are barely adequate to give even a qualitative understanding of the mechanical and rheological properties which are important in practical applications, we have provided the parameters of Tables IV–VI, which involve the cooperative behavior of many molecules, as also being fundamental properties to be measured.

The discussion that follows may seem somewhat unbalanced. Thus, molecular weight methods are discussed at much greater length than the spectroscopic methods. This bias derives from the decision that within the limitations of space those areas where the most common problems would arise should receive most consideration.

I. Chemical Composition and Repeat Unit Structure (Table I)

The chemical elemental analysis of polymers can often be carried out by the methods used for low molecular weight organic compounds [I-1, 2, 4, 5, 7].

---

* Tables I–VI are located at the end of the chapter, beginning on p. 128.

This is particularly true when combustion of the sample is involved. Thus, C, H, and N can be determined on milligram samples by complete combustion followed by gas chromatographic analysis of the gases evolved. Accuracy is about 0.3%. Sulfur and halogens are also easily determined after combustion, by titration of sulfate or SO_2 for S, and by potentiometric titration with $AgNO_3$ for halogens after treatment of the gases with NaOH and hydrazine sulfate, for example. Interference by nitrogen on sulfur tests must be watched for. (These kinds of problems can be handled by a good analytical laboratory, especially if the proper questions are asked; see advertisements in *ASTM Standardization News* or *Analytical Chemistry* if your laboratory is not so equipped.)

Assuming that the question of what elements the polymer comprises is solved by these destructive techniques, the next question is "How are the elements put together?" This is often answered best by a spectroscopic method. The various atoms in the polymer will absorb and emit radiation at characteristic frequencies depending on the structure. Skeletal bond transitions can be detected in the infrared and Raman spectra; electronic transitions typical of unsaturated bonds can be detected at ultraviolet and visible wavelengths; atoms with magnetic moments can be detected and their positions found by magnetic resonance experiments.

For basic information, infrared is often the most accessible technique. Elastomers are particularly suited to being handled as films. Simple pressing of 20–30 mg of polymer between Mylar® or aluminum foil, or Teflon®-coated Al foil, at an elevated temperature (150°C) for 60 sec or so can yield a film which is easily mounted on a sample holder. Scanning from 4000 to 600 cm^{-1} can provide a wealth of information through comparison with compilations of characteristic absorptions (see references in Table I) for individual groups or polymers themselves. When a polymer cannot be pressed into a wrinkle-free film, then the polymer is either cross-linked (covalent or ionic links) or it contains some crystalline or glassy domains which were not rendered fluid at the pressing temperature. Higher temperature should relieve the latter problem. If the elastomer contains covalent cross-links, the useful techniques are more restricted. (Care must be taken that oxidatively formed cross-links are not formed during pressing.) Infrared can still be used, but with microtomed specimens or in reflection or emission, which require more equipment and finesse [I-10, 15]. Use of these techniques should be preceded by consultation of the references and with someone experienced in the area.

Raman spectra can be used in much the same manner as infrared. Raman is quite supplementary to infrared in that many vibrations which are infrared inactive are Raman active. In particular, carbon–sulfur bonds are easily detectable in Raman [I-19b]. These spectra can be determined on solid specimens.

Whereas Raman and infrared are useful for "fingerprinting," extension of the absorption spectra into the visible and ultraviolet (uv) regions is done most often for quantitative analyses. Extinction coefficients for conjugated unsaturated structures are very large. Details of the use of these techniques can be found in the cited references.

Although the spectral absorption methods just discussed can be used in a quantitative fashion, calibration is required. With nuclear magnetic resonance (NMR), particularly proton NMR, the absorption intensity is often directly proportional to the number of atoms present, so that ratios of absorption intensities can be used directly to find the amounts of different kinds of protons in a sample. Protons in different environments resonate at different frequencies and thus can be identified by characteristic positions, δ (measured in parts per million, ppm, of the magnetic field strength), called chemical shifts.

Carbon-13 (^{13}C) NMR involves use of the magnetic moment of the carbon-13 nucleus. Thus, the different types of C in a polymer can be evaluated. Whereas proton NMR is most often done in solution to sharpen up lines, ^{13}C spectra can be obtained on solid amorphous rubbery specimens. Chemical shifts in ^{13}C are much larger than in proton NMR, permitting better resolution at lower fields. Due to the nuclear Overhauser effect, absorption is not directly proportional to the numbers of nuclei if the nuclei are in different environments. Thus, ^{13}C NMR is not as easily rendered quantitative as is proton NMR.

Given elemental analyses, infrared, and NMR data, one should be able to evaluate repeat unit structures in a polymer. In some cases where concentrations of particular units or groups are low, quantitative estimation may be difficult, and analysis based on the chemical reactivity of the particular group may be useful. For example, acid groups can be titrated accurately with base; olefins will add ozone and halogens quantitatively in some cases; hydroxyl groups can be esterified with anhydrides, with back titrations to evaluate hydroxyl content. In many cases, experiments can follow established procedures for small molecules. Sometimes, problems are posed by reaction rates, due to low concentration or the problem of finding appropriate solvents in which both reagents and polymers are soluble. Brief studies of the results as a function of time or temperature can usually establish the reliability of a technique.

Pyrolysis of samples in inert atmospheres can lead to the production of characteristic fragments which may be analyzed by gas chromatography (GC) or mass spectrometry. Since the relationship between fragments and the original polymer may be complex, this technique is used more as a last resort, for insoluble polymers or samples not amenable to characterization by other methods, or as a fingerprinting technique for routine analyses.

Electron paramagnetic resonance can be used to detect amounts and types of free radicals which may be present in unequilibrated rubber specimens, and is finding considerable use in the detection of the chemistry occurring during deformation processes.

II. Compositional and Sequence Distribution of Copolymers; Additives (Tables IIA, IIB)

Quite often the composition of an elastomer is only an average of a wide distribution of structures. It can be a composite of polymer, additives, and impurities, and if the polymer itself contains more than one type of monomer unit, a variety of polymer compositions is possible.

In order to determine whether the elastomer contains additives or impurities (or both), or is heterogeneous itself, the components of the mixture must be separated, i.e., fractionated. Many methods to separate components are available, which then may be identified by methods discussed in Section I. These methods utilize differences in solubility of the components (precipitation, dissolution), differences in hydrodynamic volume (gel permeation, also called size exclusion), differences in rates of diffusion (thermal and isothermal diffusion and sedimentation), and diffusion–adsorption phenomena (thin layer chromatography). These are cited in Table IIA.

The methods in Table III are intimately related to these; thus further discussion will be presented later where molecular weight and compositional distributions are considered together.

Heterogeneity within a sample may result from differences in the arrangement of monomer units in a particular chain. Chemically different monomer units may be placed in sequences of varying length. Chemically identical monomer units, by virtue of their geometrical placement, may present different structures, with different properties. This geometrical, or "stereoregular," state of the polymer can often be determined by the same techniques used to solve the problems of the sequence distribution of chemically different units. The most fruitful techniques are NMR and ir (see Table IIA).

Different arrangements of monomer units give rise to different chemical shifts and splitting in the NMR. Using selection rules and empirical knowledge of shifts in model compounds, appropriate structures can be assigned. Carbon-13 NMR has made an enormous contribution in this area. Since chemical shifts in ^{13}C spectroscopy are larger than in proton spectra, it is possible to see the influence of carbon atoms separated by at least five bonds, making this a very useful method for describing the distribution of monomer units in a polymer.

As a concrete example of what is discussed in Sections I and II, let us assume that a polymer "fraction," uniform in composition and molecular weight, has been obtained. (The criteria for establishing uniformity will be discussed later.) Pressing a film and recording an infrared spectrum produces the results in Fig. 1. Major bands at ~ 2900 and at 1500–1300 cm^{-1} are recognized as carbon–hydrogen stretching and bending bands from a spectra–

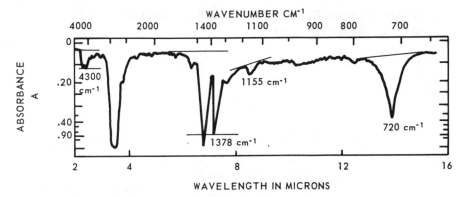

Fig. 1. Infrared spectrum of a film of the elastomer used in the discussion in the text. Possible baseline constructions for quantitative analyses are shown.

structure correlation chart [I-11, 12, 15–18]. The absorption at ~ 720 cm^{-1} could be due to sequences of methylene groups, cis unsaturation

$$\overset{\frown}{(C=C)}$$

carbon chlorine bonds, or perhaps some amine or other structure containing groups other than carbon and hydrogen. The absorption at ~ 1155 cm^{-1} falls in a region typical of isopropyl groups, ethers, sulfates, and absorptions involving atoms other than carbon and hydrogen. However, elemental analysis shows carbon and hydrogen to be the only elements present. Thus the sample is a polymer which is probably composed of the units

$$\overset{\displaystyle C}{\underset{\displaystyle |}{C}}\!\!-\!\!C, \quad C\!\!-\!\!C\!\!-\!\!C\!\!-\!\!C, \quad \text{and} \quad \overset{\frown}{(C=C)}$$

Dissolution of the polymer in CCl_4 and obtaining a 60-MH$_z$ proton NMR spectrum at ambient temperature produces a spectrum like that in Fig. 2. Resonances in the 0.6–2 ppm region are consistent with the methyl and methylene groups detected in the ir spectrum, but there is no resonance in the 4.5–6.0 ppm region, indicating an absence, or low concentration, of the olefin structure. Reaction of the polymer with ICl and O_3 under mild conditions reveals no olefin present. Thus, the polymer is basically composed of the units

$$(C\!-\!\overset{C}{\underset{|}{C}}\!-\!C), \quad (C\!-\!C\!-\!C\!-\!C) \quad \text{or} \quad C\!-\!C_y\!-\!\overset{\xi}{\underset{\displaystyle \underset{|}{\overset{|}{C_x}}}{C}}\!-\!C(C\!-\!C\!-\!C\!-\!C)_z$$

How are the units put together? Since the sample is rubbery, transparent, and flows upon deformation by hand (psycho-rheology) at ambient temperatures, one can conclude that the CH$_2$ sequences cannot be too long, or must be

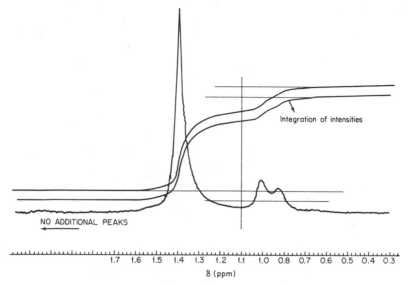

Fig. 2. Proton NMR spectrum (60 MHz) of the elastomer discussed in the text (10% elastomer in CCl_4 run at ~23°C). The chemical shift δ, obtained by dividing the chemical shift (in Hz × 10^6) by the spectrometer frequency, is independent of the spectrometer frequency. Resolution can be improved by running spectra at higher temperature or field strengths (e.g., 220 MHz). [Peak assignments from R. Ferguson, *Macromolecules* **4**, 324 (1971).]

in low concentration, else they would crystallize into a polyethylenelike lattice. For similar reasons, the side groups are either not stereoregular, or stereoregular in low concentration.

Inquiring further into the ir spectrum, it is found [I-11] that the geminal dimethyl (gem) structure should yield a strong absorption doublet at 1250 and 1200 cm^{-1}. Thus, the presence of a structure with $x = y = 0$ is ruled out. Polyisobutylene has the structure with $x = y = z = 0$, with no 720 cm^{-1} absorption and a gemdimethyl absorption at ~1220 cm^{-1}. To get much more information, two alternatives are available. One could compare these ir and NMR spectra with spectra of other known rubbery compounds. Identical ir spectra should indicate identical polymer composition. Having established the monomer structures, one could use the NMR spectrum to calculate the ratio of structures by counting methyl to methylene proton ratios, etc. The other alternative is to go to more sophisticated spectrometers. Of the alternatives presenting themselves, either higher-resolution proton NMR (say 220 MH$_z$) or ^{13}C NMR data can be obtained. The latter is considerably more informative. Using ^{13}C NMR, spectra such as Fig. 3 can be obtained, in which the peaks resulting from individual carbon atoms can be assigned for groups of up to about five adjacent bonds. The shifts are determined empirically. The structures which appear in this spectrum are consistent with the sample's being an ethene–

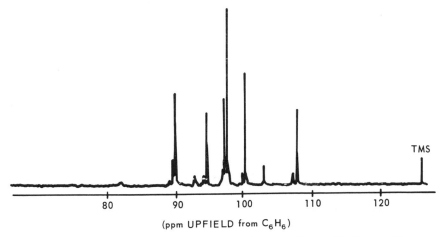

Fig. 3. Carbon-13 NMR spectrum (30 MHz) of elastomer discussed in the text. Note wealth of information as compared to Fig. 2 [C. Wilkes, C. Carman, and R. Harrington, *J. Polym. Sci., Polym. Symp.* **43**, 237 (1973); reprinted by permission of John Wiley & Sons, Inc.].

propene copolymer. This is deduced by consulting published data on chemical shifts of branched alkanes and α-olefin copolymers, which are obvious choices, considering both the ir and proton NMR spectra. Such structures could result from either direct polymerization or polymerization–modification processes, but the point is that given the structure, one begins to know what properties to expect, in this case of ethene–propene copolymers, which have been reviewed [16, 17].

Suppose that the polymer was known to be a copolymer of ethene and propene and that a knowledge of the sequence distribution of units in the polymer is desired. This problem has been tackled most fruitfully by NMR, both proton and ^{13}C. A combination of empirical and theoretical principles is used to obtain self-consistent peak assignments for the various possible arrangements of the monomer units. These sequence distribution data can then be correlated with models for polymerization or resultant properties of the polymer [16, IIB-13].

III. Molecular Weight and Its Distribution (MWD); Branching; Gel (Tables III)

There is no question that the chemical and stereochemical nature of the repeat unit has a profound effect on the polymer properties, but for a polymer to be a useful elastomer, the repeat unit will have to have been chosen so that the glass transition temperature $T_g \lesssim -30°C$, and the degree of crystallinity f_c stays below a volume fraction of ~ 0.3. Possible exceptions to this are urethane or other block copolymers. At processing temperatures, f_c is usually zero and thus the viscoelastic response is basically determined by the weight average molecular weight $\overline{M}_w$ and MWD of the polymer [1, 8]. Defects in the vulcanized network are caused by the number of chain ends present and thus depend on the number average molecular weight $\overline{M}_n$ of the polymer. Other causes of network defects are closed chain loops and nonuniform distributions of cross-link density [4, 18]. Frequently observed MWD's in elastomers are shown in Fig. 4.

The distribution is usually presented as a plot of the mole fraction, or of the percentage (number) of molecules of a given molecular weight $N(M)$ versus the molecular weight M, or as the weight of molecules of a given molecular weight $W(M)$ versus M. The distributions shown are continuous, although they are discrete, on a molecular scale, since the molecular weight must increase in jumps of monomer unit molecular weight. For large M, the distinction between sums and integrals becomes small and since the mathematics is generally simpler for integrals, the (discrete versus continuous summation) Euler–MacLaurin sum formula is used to convert the sums to integrals. The average molecular weights are defined as

$$\overline{M}_j = \frac{\sum_{i=1}^{\infty} N_i M_i^j}{\sum_{i=1}^{\infty} N_i M_i^{j-1}} = \frac{\int_0^{\infty} N(M) M^j \, dM}{\int_0^{\infty} N(M) M^{j-1} \, dM} = \frac{\int_0^{\infty} W(M) M^{j-1} \, dM}{\int_0^{\infty} W(M) M^{j-2} \, dM} \tag{1}$$

Setting $j = 1, 2,$ or 3 yields $\overline{M}_1 =$ number average $\overline{M}_n$, $\overline{M}_2 =$ weight average $\overline{M}_w$, $\overline{M}_3 = z$ average $\overline{M}_z$, etc. Thus, the molecular weight averages, which are related to the polymer's behavior in various physical phenomena, are ratios of moments of the distributions.

Consider first the number average molecular weight which determines the colligative properties of the polymer solutions; i.e., measurements of freezing point depression (cryoscopy) or boiling point elevation (ebulliometry) will yield $\overline{M}_n$ for macromolecules as for small molecules. However, the number of moles of molecules present at mass concentrations low enough to obtain ideal solutions is very small for macromolecules; therefore, the anticipated changes ΔT in boiling or freezing points are correspondingly small; for $\overline{M}_n = 10^5$, $\Delta T \sim 10^{-5}$ to 10^{-4} °C. This is so close to the limits of detection of ΔT that these classical techniques are not used above $\overline{M}_n \sim 10^4$. Even below that they have been replaced by vapor pressure osmometry (VPO). This technique records the ΔT between two thermistors (in a thermostated chamber saturated with solvent vapor, one coated with solvent and one coated with solution) caused by differences in solvent condensation rate resulting from the vapor pressure lowering in the presence of the solute. VPO can be pushed to $2 \times 10^4 \overline{M}_n$. The problem with this method which is not absolute, is related to solute dependence of the calibration constant. At the very least, instruments should be calibrated over a range of molecular weights with polymeric standards

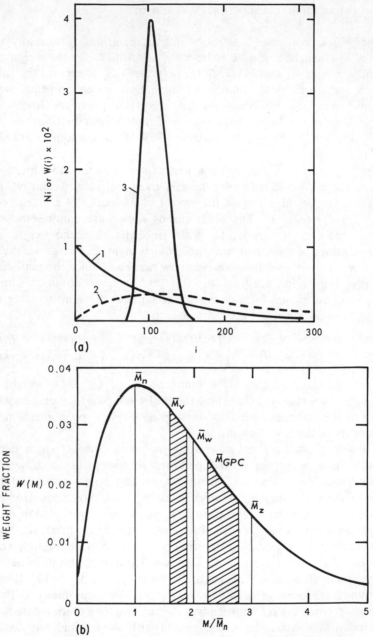

Fig. 4. Molecular weight distributions. (a) Differential number and weight distribution curves for a polymer with a number average of 100 units per chain. Curve 1, most probable number distribution; curve 2, most probable weight distribution; curve 3, Poisson number distribution. [From "Polymers: Structure and Bulk Properties" ©1965 by Patrick Meares. Reprinted by permission of Van Nostrand Reinhold Company, a division of Litton Educational Publishing, Inc.]. (b) Distribution of molecular weights in a typical polymer, showing the positions of important averages [7]. $\bar{M}_{GPC}$ is not the gel permeation chromatography peak molecular weight but is defined in reference [7].

and calibration "constants" tested for dependence on M. Errors of 25% in $\overline{M}_n$ are easily encountered at $\overline{M}_n = 10^4$ if calibration is performed only at $M \sim 2 \times 10^2$ (IIIA-8a–e).

Ebulliometry and cryoscopy are absolute techniques for which calibration is unnecessary. The molecular weights (Table IIIA) from membrane osmometry, light scattering, and equilibrium centrifugation (the latter two determine $\overline{M}_w$) are also absolute in the sense that instrument calibration with polymers of known $\overline{M}_n$ or $\overline{M}_w$ is in principle unnecessary though often practically convenient. The numbers obtained are independent of the dimensions of the molecule in solution. Intrinsic viscosity, gel permeation chromatography (GPC), sedimentation, and related measurements (to be discussed later) require calibration, and the results are a function of polymer geometry, and thus, e.g., of branching. Since branching is ubiquitous with elastomers, absolute techniques are very important. However, before discussing these techniques further, a discussion of molecular weight distributions in more depth is appropriate.

Polymerization processes generally form polymers with smooth distributions of molecular weights. For example, any stepwise monomer addition process (uncomplicated by branching) that is characterized by a single propagation rate constant, be it free radical (not terminated by coupling), cationic, or anionic, will yield a most probable distribution of molecular weights in a well-mixed continuous flow tank reactor.

For that distribution

$$N(M/M_0) = p^{(M/M_0-1)}(1 - p) = p^{(M/M_0-1)}(1 - e^{-(1-p)}) \qquad (2)$$

with $p = 1 - \overline{M}_0/\overline{M}_n$, $M/M_0 \gg 1$, and M_0 the monomer unit molecular weight. [Note that $\overline{M}_w/\overline{M}_n = (1 + p) \approx 2$ for this distribution (curves 1 and 2, Fig. 4a). M/M_0 is the degree of polymerization, designated as i in the figure and in formula (1).] As elastomers go, this is a relatively narrow molecular weight distribution. For a polymer of $\overline{M}_n = 5 \times 10^4$ and $M_0 = 50$, there is essentially no weight or mole fraction present at $M > 5 \times 10^5$, but there is some 10 mole % of polymer of $M < 5 \times 10^3$. This low molecular weight material is an important source of error in membrane osmometry, a widely used absolute method.

Membrane osmometry operates on the principle that dissolution of a solute in a solvent lowers the activity (free energy) of the solvent. If solvent containing polymer is brought into contact with pure solvent, the concentration gradient will induce mixing by diffusion (slow—days or weeks) of the solvent into the solution and vice versa, until the system is homogeneous and the free energy of the system is minimized. If a membrane with fine pores is placed between the pure solvent and solution, so that only solvent can pass through, only the solvent can now diffuse. The tendency of the solvent to pass through the membrane can be measured as a pressure π which will depend only on the number of polymer molecules present, as long as polymer–polymer interactions are absent, at very low concentration. The equation relating these quantities is

$$\pi/RTC = (1/\overline{M}_n) + A_2 C + A_3 C^2 + \cdots \tag{3}$$

where π is the osmotic pressure, R the gas constant, T the absolute temperature, C the polymer concentration, and A_i the virial coefficients. Thus measurements at several concentrations with extrapolation of (π/C) to $C = 0$ yield
$\overline{M}_n$. Even for high molecular weight materials, the osmotic pressure is significant. For a 1% solution of a material of $\overline{M}_n = 10^5$, the pressure is 2.5 cm
H_2O at 25°C (approx).

Concerning frequent difficulties, (π/C) may not be linear in C for experiments
at higher concentrations. This may be due to the fact that agglomeration is
occurring, or that higher-order terms in the series (3) are important. Solvent
or temperature changes can sometimes eliminate such problems. To get measur-
able osmotic pressures, C must be increased with $\overline{M}_n$. A_2 decreases with $\overline{M}_n$
in good solvents $(A \sim M^{-0.2})$, but not as fast as C must be increased. Therefore, nonlinear plots are often encountered when studying high $\overline{M}_n$ samples.
There are theories which relate A_2 to A_3 and which suggest means for linearizing
plots; often $(\pi/C)^{1/2}$ versus C is plotted. The greatest difficulties, though, relate
to imperfections in the membrane. Sometimes the polymer can permeate in
some sort of wormlike (segmentwise) mode of motion. The question is, how
long does it take the polymer to permeate, and thus whether a "steady state"
can be set up. Steady-state values have been obtained which indicate that the
membranes are permeable to molecules of $M < 5 \times 10^3$ (good) and in some
special cases only $M < 5 \times 10^2$. Recall that a polymer with the $\overline{M}_n$ and distribution discussed above has 10 mole % of the distribution below 5×10^3,
so a systematic error of $+10\%$ would occur in $\overline{M}_n$ even in a satisfactory measurement. For distributions skewed to low M species, the errors can be much
larger. The $\overline{M}_n$ of an elastomer is often of interest as a measure of cross-linked
network defects. Networks have sol fractions which are usually those of the
low molecular weight species. To account for network defects properly, one
should use the $\overline{M}_n$ of the material in the gel. Thus, a high value of $\overline{M}_n$ due to
membrane permeation from osmometry may luckily turn out to approximate
a correct $\overline{M}_{n(gel)} \approx \overline{M}_{n(orig)}$. In order to check for problems, it is useful to run
some standards with the particular elastomer and membrane in question. The
standard could be one of the commercially available low molecular weight
polystyrenes of narrow distribution, or polyethers, or perhaps a low molecular
weight extract from the particular sample in question, which extract has been
characterized by VPO or GPC (see the following discussion).

Light scattering, another absolute and essential technique, sometimes
presents even more problems. Light passing through a layer of material is
scattered from regions of different refractive index (polarizibility). Polymer
molecules dissolved in a solvent which has a refractive index different from
the polymer will scatter light. The refractive index n of the whole solution

will change with the polymer concentration in a manner approximated by

$$\frac{dn}{dc} = \frac{n_{pol} - n_{solv}}{\rho_{pol}} = \frac{cc}{g} \tag{4}$$

where ρ is the polymer density. For the normal range of polymer $\overline{M}_w$ (e.g., $<10^7$) and dn/dc (e.g., <0.2), too little light is scattered from a solution for it to be detected by the eye. The turbidity of the solution, which is defined as the fractional decrease in the irradiance (intensity of the incident light) per centimeter of medium, is

$$\tau = \frac{1}{l} \ln\left(\frac{I_0}{I}\right) = 10^{-4} \text{ to } 10^{-3} \tag{5}$$

where τ is the turbidity, l the length of solution, and I_0/I the initial/final irradiance. Therefore, we are talking about only a 1% decrease in intensity upon passage through 10 cm of solution, which thus appears transparent.

For polymer molecules, the experimental relationship is

$$\frac{KC}{R_\theta} = \frac{1}{\overline{M}_w P(\theta)} + 2A_2 CQ(\theta) + \cdots \tag{6}$$

where

$$K = \frac{2\pi^2}{\lambda_0 N} \left[n \frac{dn}{dc}\right]^2$$

$$R_\theta = \frac{r^2 i_\theta}{I_0(1 + \cos^2 \theta)} = \text{Raleigh ratio} = \text{instrument}$$
$$\text{characteristic, usually evaluated}$$
$$\text{by scattering from known materials}$$

N is Avogadro's number, λ_0 the wavelength of radiation, dn/dc the refractive index increment in cc/g, r the distance from scattering center to detector, i_θ the intensity of light scattered at angle θ, n the refractive index of solution, $Q(\theta)$ depends on intramolecular scattering and becomes equal to unity at $\theta = 1$, $P(\theta)$ the intramolecular scattering factor is given by

$$P(\theta) = 1 - \frac{\langle S_z^2 \rangle}{2} \left(\frac{4\pi}{\lambda}\right)^2 \sin^2\left(\frac{\theta}{2}\right) + \cdots$$

C is the concentration in g/cc, and $\langle S_z^2 \rangle$ the radius of gyration of the scattering particle as the mean square z average. The particle scattering factor is unity at $\theta = 0$ and decreases from unity as θ increases due to destructive interference, caused by light scattered from different parts of the same molecule. $P(\theta)$ can be calculated theoretically; however, it varies, depending on the shape of the molecule and the MWD. To avoid this uncertainty, one extrapolates to $\theta = 0$.

To avoid the effects of higher-order terms in C, one must also extrapolate to $C = 0$. This can be done simultaneously in a Zimm plot. There are suggested methods by which to linearize the concentration extrapolations, as there were in osmometry. $P(\theta)$ will appear again in other scattering phenomena.

A proper Zimm plot should yield a radius of gyration and virial coefficient consistent with the $\overline{M}_w$ obtained for the polymer (see Fig. 5a–c). Curvature in $P(\theta)$ should decrease as one goes to low angles, and if it does not, the reason should be determined (e.g., aggregates, gel). Solutions must be cleared of adventitious scattering particles (dust, etc.) before analysis is attempted. With

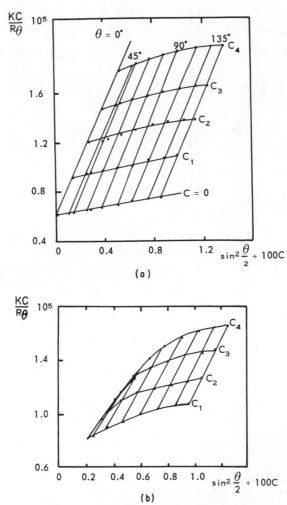

Fig. 5a,b. Zimm plots. (a) Proper appearance. (b) Showing aggregation at higher concentration [J. Vavra, J. Lapcik, and J. Sabados, *J. Polym. Sci., Part A-2* **5**, 1305 (1967)].

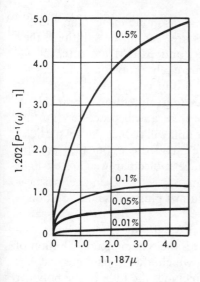
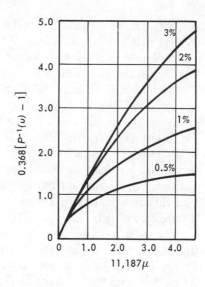

(c)

Fig. 5c. Calculated particle scattering factors (i.e., $C = 0$ in Zimm plot) for model systems containing microgel. Percent gel is indicated on curves. $\mu = \frac{8}{3}(\pi n C/\lambda)^2 \sin^2(\theta/2)$, $S = (1ZC/6)^{1/2}$, $\lambda = 5460$ Å, $l = 2.8$ Å, $n = 1.5$, $C = 10$ Å, polymer $Z = 10^3$, microgel $Z = 10^6$. Therefore, the radius of gyration of the microgel is $\sim 0.4\lambda$. For $11,187\mu = 1, 2$, $\theta = 84°$ and $140°$, respectively [111A-25].

elastomers, clarification problems are serious due to the possibility of removing some of the polymer. One ought to be able to get the same $\overline{M}_w$ regardless of the type of solution clarification procedure, and also regardless of whether the solutions of different concentration are clarified individually, or made up from a more concentrated stock solution. It is interesting to note that cross-linking of a polymer to a point short of gelation does not produce strong low-angle scattering [IIIA-31a]. A very informative experiment prior to $\overline{M}_w$ determination is to try to make a concentrated solution of the polymer. For example, dissolve two separate 10-g portions of polybutadiene (BR), natural rubber (NR), poly(styrene-*co*-butadiene) (SBR), poly(ethene-*co*-propene-diene) (EPDM), and Pressure Chemical Lot 14b polystyrene (PS), in 90 cc heptane and in 90 cc toluene. Let them dissolve over several days by gently rotating on a wheel, not by vigorous stirring. The solutions range from clear to opaque in a variety of tints of brown and yellow. The color is probably from antioxidant, or due to conjugated aromatic structures, in some cases complexed with catalyst residues left from the polymerizations. Although the quantitative results depend on the particular "grade" of polymer type chosen, the qualitative results are as follows: SBR, NR, and EPDM "solutions" contain clumps of material. BR contains some more finely divided material which is more evident in heptane than in toluene solution (due to larger dn/dc). The PC Lot 14b (weight average molecular weight $\sim 2 \times 10^6$) is insoluble in heptane and the toluene solution is quite clear. Where particles are seen in the solutions, they are probably cross-linked polymer. SBR, NR, and EPDM are often cross-linked past the gel point as they are sold commercially; polybutadiene less so. This might not be expected for anionically produced polybutadienes, but "processible" grades have added complexities (e.g., intentional branching), and they are readily cross-linked by oxidative free radical processes. Putting the material in a Waring blender or on a rubber mill will reduce the gel particles to a size which can no longer be seen with the unaided eye, but they are still there. Gel particles have a huge weight average molecular weight; by light scattering or by equilibrium ultracentrifugation, $\overline{M}_w = \infty$. Using gel permeation chromatography (discussed later), the columns plug unless the solution is filtered free of gel.

The most probable distribution of molecular weight in Fig. 4a, when cross-linked to a point slightly past the gel point, gives the schematic distribution of Fig. 6. There is some macroscopic gel fraction which, due to its finite size, has $M \lesssim 10^{22}$, and the sol fraction (95%), which includes the bulk of the polymer with $\overline{M}_n$ not significantly increased over the original (say 30%). There is also a small portion of an enormous $\overline{M}_w$, comprising the branched species which exist in low concentration just short of gellation; this portion causes the overall turbidity at low concentrations. "Solutions" such as these may present no difficulty in obtaining $\overline{M}_n$ data by membrane osmometry; however, the particles are so large that it becomes impossible to make light scattering measurements at low enough angles to permit linear extrapolation of $P(\theta)$ to $\theta = 0$ [IIIA-27, 31b]. (Scattering photometers using lasers are becoming commercially

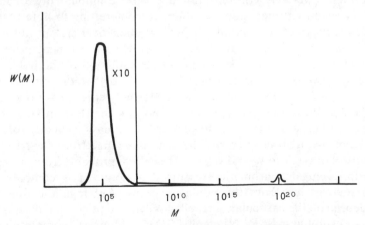

Fig. 6. Schematic representation of MWD of a polymer cross-linked past the gel point. A cross-linked gel with swell ratio of ~ 100 and 1 mm³ dimension (visible clump) must have $M \sim 10^{20}$. This is enormous compared to the microgel of Fig. 5c.

available. These permit low angle measurements, thus eliminating some of the uncertainty in extrapolation to zero angle.) Including the gel, $\overline{M}_w \approx 10^{20}$ for the sample in Fig. 6. Considering the sol fraction alone, it would be $\overline{M}_w \approx 3 \times 10^6$ and $\overline{M}_n \approx 6 \times 10^4$. This example was calculated using relationships from the cited references [4, IIIA-98] assuming tetrafunctional cross-linking and a most probable distribution.

At this point it is worth inquiring what the $\overline{M}_w$ of such a sample relates to. If $\overline{M}_w$ were going to be used to analyze the polymerization kinetics, one becomes aware that there is a chemical process which is prone to lead to gelation, so that the experiments and kinetic schemes should be set up accordingly. If the $\overline{M}_w$ value is to be used to correlate with a rheological property, such as the zero shear rate viscosity, the conclusion would be that there is no zero shear rate viscosity which is reproducible, since the gel network breaks irreversibly before the sample is able to flow. If the gel were already ruptured, or was originally formed as discrete particles, the flow process should be considered as that of a two-phase system (e.g., [IIIC-41]).

If one should need to characterize the polymer minus the gel component, and to treat the properties of the sample as a composite of gel and sol, the question arises how to separate the sol from the gel. We have seen that a gentle dissolution process leads to a dispersion which requires some special separation technique. Filtration is an obvious possibility. Presumably, a solvent-swollen three-dimensional object such as a piece of gel can be separated from the sol with reasonable efficiency with a filter in the 10^4-Å (1-μm) porosity range. However, there may be considerable amount of "nongel" material present which is of large M; as a result one finds that portions of the sol fraction are also removed upon filtration. The $\overline{M}_w$ value thus becomes a function of the technique of clarification. Although little work has been reported, it appears that ultracentrifugation is probably the only practical method available for well-defined reproducible solution clarification.

If, on the other hand, the original solution has the appearance of the polystyrene 14b, if one can obtain $\overline{M}_w$ values which are independent of filter porosity in the 0.2- to 1-μm range, and if the different solutions can be filtered separately to produce an undistorted Zimm plot, one is justified in assuming that there are no unseen problems, and in using the $\overline{M}_w$ data on their face value.

Equilibrium ultracentrifugation has been referred to as an absolute method of obtaining $\overline{M}_w$. Actually, there are several ways to use the ultracentrifuge, some of which are not absolute and some of which even provide absolute molecular weight averages other than $\overline{M}_w$ (e.g., M_n and M_z). In principle, one can obtain the whole distribution of molecular weights. Unfortunately, the performance of an ultracentrifuge experiment is time consuming and involves obtaining considerably more data than are involved in either osmometry or light scattering; however, the technique yields results which cannot be obtained by any other technique.

The ultracentrifuge operates on the principle that a polymer molecule in a solvent of differing density, when placed in a large centrifugal field, encounters gravitational potential energy differences associated with settling in the rotating cell. These differences are of the same magnitude as the free energy changes associated with minor concentration differences. For macromolecules, fields of the order of 10^3–10^4 g produce gradients of molecular numbers which are a function of molecular weight. Equilibration between settling and diffusion in the centrifuge takes times of the order of days. For molecular weights of $\sim 10^7$, gravitational effects alone can set up measurable gradients [IIIA-51]. Depending on how the concentration gradient measurement (usually done optically) is manipulated mathematically, $\overline{M}_w$, $\overline{M}_n$, $\overline{M}_z$, etc., can be obtained, and of these $\overline{M}_w$ with the greatest accuracy. The experiments should be performed in Θ solvents, and extrapolation of M_{app} to $C = 0$ is again necessary.

Two variations of ultracentrifugation are worth mentioning. One is performed in a mixed solvent system, chosen so that the gradient set up by the demixing solvents leads to a region of density equal to that of the dissolved polymer somewhere in the center of the cell. This sort of experiment is valuable for the determination of gel fractions [IIIC-9] which would equilibrate to the end of the cell in the standard experiment.

In the other variation, the changes in polymer concentration are studied as a function of time and centrifuge speed. A boundary of the sedimenting species is set up in the field and the velocity of this boundary is evaluated by an optical method. Diffusion due to the concentration gradient is superposed on the sedimentation rate. To eliminate the diffusion effect, the velocity of the boundary is extrapolated to $t = \infty$. Diffusion broadening is eliminated by this procedure, since it is proportional to $t^{1/2}$, whereas the sedimentation is proportional to t. By studying the sedimentation as a function of time, qualitative aspects of the distribution can be brought out. In particular, if gel fractions are present, they sediment rapidly. If there are more or less distinct modes of the MWD, these can be seen to separate with time. Such qualitative information alone is very useful. To convert sedimentation data to absolute molecular weights, the diffusion coefficient, or a relationship between the sedimentation coefficients and molecular weight, must have been previously established, except in the case of Archibald's method (see Table IIIA, references [IIIA-44, 45]).

Although ultracentrifugation is disadvantageous from the standpoint of expense of equipment, length of experiment, and data workup, it appears to be the only technique which permits the study of unfiltered (e.g., gel-containing) solutions.

Another commonly measured parameter of polymer solutions which can be related to molecular weight is the ratio of the viscosity of the solution η to the viscosity of the solvent η_0. These viscosities are determined by determining the flow times of solution and solvent through capillary tubes, usually with gravity as the driving force for flow. Rotational and oscillatory type viscometers are used where a uniform, well-defined or low shear rate is required. The ratio of the two viscosities, η/η_0, is called the relative viscosity, and the difference between that number and unity is the "specific" viscosity. Extrapolation of the specific viscosity, divided by the solution concentration, to $C = 0$ yields the intrinsic viscosity $[\eta]$. Thus

$$\frac{\eta_{\text{soln}}}{\eta_{\text{solv}}} = \eta_{\text{rel}}, \qquad \eta_{\text{rel}} - 1 = \eta_{\text{sp}} \tag{7}$$

$$\lim_{C \to 0} \frac{\eta_{\text{sp}}}{C} = [\eta] \tag{8}$$

Also

$$\lim_{C \to 0} \frac{\ln \eta_{\text{rel}}}{C} = [\eta] \tag{9}$$

Units for this quantity are usually cc/g, or deciliter/gram $= 100$ cc/g, of polymer. The preferred metric unit will be m^3/kg, which reduces the latter number by 10^{-1}.

Although the intrinsic viscosity cannot be related directly to a molecular weight average of a dissolved polymer on a rigorous thermodynamic basis (as can $\overline{M}_n$ to osmotic pressure and $\overline{M}_w$ to light scattering), it is still possible to make calculations using realistic kinematic models on the basis of which it is possible to derive [4, 5]

$$[\eta] = \Phi_0 \frac{\langle R_0^2 \rangle^{3/2}}{M} = 6^{3/2} \Phi_0 \frac{\langle S_0^2 \rangle^{3/2}}{M} = K_\theta M^{1/2} \tag{10}$$

where $\langle R_0^2 \rangle$ and $\langle S_0^2 \rangle$ are, respectively, the mean square end-to-end distance and radius of gyration of the polymer in question, and Φ_0 is a constant which is the same for all flexible linear macromolecules at theta (Θ) conditions. These conditions occur when the polymer–solvent interactions are repulsive, such that they balance the tendency for a polymer chain to be expanded over the configuration it would have if there were no intramolecular interactions. The intramolecular interactions, called the excluded volume effect, tend to expand the polymer beyond a random coil configuration in a good solvent. Both Φ_0 and $\langle R_0^2 \rangle^{3/2}/M$ can be calculated, the former from a variety of theories, and the latter using the rotational isomeric models [19] for short-range interactions along a polymer chain. Even in Θ solvents (i.e., Θ conditions) it turns out that Φ_0 is a function of a quantity h, called the draining parameter, which relates to the nature of the flow of the solvent through the polymer. For $h = 0$ (called free draining) the solvent passes through the polymer (or polymer through solvent) as though one portion of the macromolecule did not influence the flow of solvent in the region of another portion of the macromolecule. This is presumed not to be the case for dilute solutions of macromolecules based on the fact that $[\eta]$ becomes proportional to M rather than $M^{1/2}$ if the equations are solved with $h = 0$. That is not the experimental result, although data for low molecular weight polymers (IIIA-38b) or stiff molecules [5] tend to that direction. For finite values of h, the flow of solvent at a given polymer unit is perturbed by the other units. If the mathematics are done correctly [IIIA-33, 35] one finds that Φ_0 is a constant for $h = h^* = \infty$ or $h = 0.25N^{1/2}$, $h^* = .025$ (note that Osaki's $h^* = h/N^{1/2}$). Otherwise Φ is a function of the molecular weight (N = number of submolecules into which the molecule is divided in the theory $N \propto$ molecular weight). The case $h = \infty$ is called dominant hydrodynamic interaction and the intrinsic viscosity behaves as though the polymer were a hard sphere of Stokes radius $R_e = ((3/4\pi)V_e)^{1/3}$ where V_e is the effective volume of the sphere and

$$V_e = \frac{\Phi_0 \langle S_0^2 \rangle^{3/2}}{0.025N_A}$$

where N_A is Avogadro's number [4].

Inside the sphere polymer molecules are only about 1% dense in their own segments when in dilute solutions, so to do more than observe that the mathematical form is that of a hard sphere is, perhaps, misleading. It seems, at this point, that experimentally $h \neq \infty$ in Θ solvents but Φ_0 may be nearly constant in practical regions of molecular weight [IIIA-43b].

The hydrodynamic parameter h^* may be related to the solvent quality; experimentally $([\eta]_{\text{good solv}}/[\eta]_\theta)^{1/3}h^* \approx 0.2$, indicating that as the solvent becomes better and the intrinsic viscosity increases, the hydrodynamic interaction decreases. In any event, at Θ conditions Φ_0 will be near a value of 2.7×10^{23} ($[\eta]$ in dl/g). Since $\langle S_0^2 \rangle/M$ can be calculated, given $[\eta]$ one can calculate M. That is quite an accomplishment for polymer science.

Frequently, working at Θ conditions is inconvenient, since the required temperature control is exacting, and the polymer tends to agglomerate and precipitate due to the limiting solubility. Thus, for practical work under any desired conditions one applies the Mark–Houwink expression:

$$[\eta] = K\overline{M}_v^a \tag{11}$$

where a ranges from 0.5 to about 1, and even higher for stiff chains (see also [IIIA-43a]). K and a take on characteristic values for a given polymer, solvent, and temperature. For a polymer with a distribution of molecular weights, the viscosity average molecular weight is

$$\overline{M}_v = (\sum N_i M_i^{(1+a)}/\sum N_i M_i)^{1/a} \tag{12}$$

However, exact calculations of K and a are not possible at present, so that one usually establishes these constants empirically from (11) with the help of polymers of known molecular weight.

It is interesting, and convenient, that for all good solvents for a given polymer sample $[\eta]$ varies only $\pm 30\%$; and for a variety of polymers in good solvents $[\eta]$ can be estimated with reasonable accuracy on an empirical basis [6]. Elastomer molecules must be flexible and will nearly always have a between 0.6 and 0.8.

Solution viscosities are often reported as inherent viscosities, i.e., $\ln \eta_{rel}/C$ at low, but finite, C. If one works with solutions sufficiently dilute so that $[\eta]C \ll 1$, then $[\eta] = 1/C[2(\eta_{sp} - \ln \eta_{rel})]^{1/2}$. This requires that the plot of η_{sp}/C versus C be linear. If $[\eta] \approx 5$ dl/g, then C should be less than 0.3 mg/ml (0.03 g/dl). For accurate determination of $[\eta]$ in capillaries, the flow times should be long enough (> 100 sec) to make kinetic energy corrections negligible, or else the corrections must be made. Also, the shear rates in the experiments should be low enough to avoid non-Newtonian viscosity effects. An approximate criterion for this is that $\beta_0{}^2 = ([\eta]_0 M\tau/RT)^2 = ([\eta]_0 M\eta_0 G_0/RT)^2 \ll 1$, where τ is the shear stress, G_0 the shear rate at the wall in the tube, and η_0 the solvent viscosity. Typically, one works with $G_0 \approx 10^3$ sec^{-1} and $[\eta]_0 < 5$ as safe regions in a standard capillary. Low concentrations of a very large molecule will not be detected due to the high shear rate; therefore some other type viscometer should be employed if large molecules are suspected to be present. The full contribution to $[\eta]$ from the large molecules is decreased due to the non-Newtonian viscosity. Since solutions are commonly filtered prior to running capillary viscosities, careful work requires determining what is filtered out.

The concept of polymer coils in solution acting like Stokes spheres of radius R_e may also be used to discuss the principle of gel permeation chromatography (GPC). In GPC, a dilute solution of the polymer to be analyzed is passed through a column packed with small porous particles. Their pore diameters are of the order of the size of the region occupied by a polymer molecule. As the polymer solution passes through such a column at a flow rate slow enough so that molecular diffusion permits the polymer to move into and out of all accessible pores, small molecules, which can fit into more pores than larger ones, take longer to be eluted from the column. If a pulse of polymer with a distribution of molecular weights is pumped through a column containing particles with a range of pore sizes, and the polymer concentration and molecular weights are determined in the effluent, it is found that the polymer stream has been separated into a range of elution volumes V_e on the basis of the molecular sizes. A typical chromatogram is shown in Fig. 7. Based on experiments on the same columns under identical conditions, a reproducible relationship of elution volume to molecular weight like that in Fig. 8 can be set up.

Using such a relationship one can assign molecular weights to the polymer. It has been observed empirically that if a variety of polymer types is eluted from the same column, then at a given elution volume $([\eta]M)_{pol\ 1} = ([\eta]M)_{pol\ 2}$, where $[\eta]$ is determined under the same conditions (T, solvent) as in the GPC

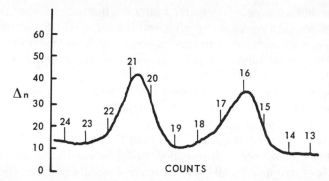

Fig. 7. GPC chromatogram. Lower molecular weight polymer is eluted last due to its accessibility to more pores. Concentration axis is often measured by refractive index increment Δn but many other techniques are useful also; spectroscopic techniques can give compositional distributions.

experiment. Thus, one can set up rather reliable calibration of molecular weights for a given polymer using known standards of other flexible polymers which are commercially available (e.g., polystyrenes).

Looking at the calibration curve for the given set of columns, we see that the resolution becomes poor at very high and low molecular weights; i.e., large

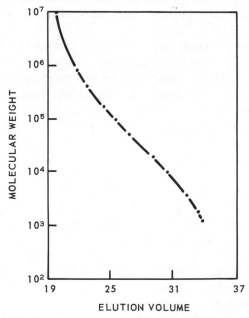

Fig. 8. GPC calibration curve. Note how resolution falls off near the ends (i.e., small V_e change is large M change) and how separation efficiency decreases with increasing molecular weight even in the linear region (• are peak positions for polystyrene standards) [III-69].

differences in molecular weight are represented by small changes in elution volume. Thus, in a particular sample, material of $M = 10^8$ may be indistinguishable from $M = 10^{6.5}$. It is also useful to know at what elution volume the antioxidant or other additives appear, so these materials are not confused with and counted as low molecular weight polymer. Parameters such as concentration and flow rate should be studied to determine their effects on elution volume. Changing concentration or flow rate can displace calibration curves by factors of 2 in molecular weight.

GPC is one of the routine characterization techniques where it is necessary to present the raw data. Such qualitative features as multimodal distributions are easily discerned visually but are not reducible to "numbers" without additional assumptions. If the data are converted to molecular weight averages or a plot of weight fraction versus M, such as Fig. 4, certain manipulations are necessary. Since the plot of M versus V_e is not linear (nor is log M versus V_e), it is apparent that the column has not resolved the polymer equally on the basis of molecular weight increments. For example, it is seen from Fig. 8 that ± 5 units of elution volume about $M = 10^5$ include all of the polymer from $M = 10^6$ to 10^4, that is, differing in molecular weight by 990,000, whereas ± 5 counts about $M = 10^4$ include only a 99,000 molecular weight range. Therefore, if one wants to compare a theoretical MWD with the experimental one, or compare experimental data from different calibrations and columns so that equal areas represent equal amounts of polymer, the chromatogram must be modified. The appropriate modification of the $W(V_e)$ curve is related to the slope of the elution volume–molecular weight curve. To compare the GPC distribution with a plot of $W(M)$ versus M, one must multiply $W(M)$, the weight fraction of given molecular weight M from the GPC, by the slope of the elution volume versus molecular weight relationship, i.e.,

$$W(M)_{\text{lin } M \text{ plot}} = W(M)_{\text{GPC}} \; dV_e/dM$$

If V_e is changing rapidly with M, a given height $W(M)_{\text{GPC}}$ represents proportionately more polymer in a given linear M range, and vice versa.

If the GPC and the theoretical distribution are to be plotted on logarithmic paper so that equal areas on the plot represent equal amounts of polymer, the theoretical expression and GPC should be modified as follows. For the theoretical expression:

$$W(M)_{\text{log plot}} = W(M)_{\text{lin}} \; dM/d \log M = W(\log M) \tag{13}$$

This preserves the integral

$$\int_{\text{all } M} W(M) \, dM = 1 = \int_{\text{all log } M} W(\log M) \, d \log M$$

For the GPC data

$$W(M)_{\text{log plot}} = W(M)_{\text{GPC trace}}\, dV_e/d \log M \qquad (14)$$

The GPC data should be normalized, so that $\int W(V_e)\, dV_e = 1$ prior to comparison. For linear calibration plots, $dV_e/d \log M = $ constant, and $dV_e/dM = $ constant $\cdot M$.

Scheme A: Steps Involved in the Interpretation of a GPC Chromatogram

---

1. Construct baseline.
2. Abstract chromatogram heights at equal increments (usually in counts of 5 ml or 2.5 ml).
3. Normalize to give fraction of each species per increment of retention volume $dw/dv = h_i$ normalized $= W(M)_{\text{GPC trace}}$.
4. Make any corrections for peak broadening with respect to retention volume.
5. Transpose retention volume scale to molecular weight scale via relevant calibration curve.
6. Make any corrections for change in detector response with molecular weight. This gives data for dw/dv versus $\log M$ (or M).
7. Apply correction from dw/dv to $dw/d \log M$. That is,

$$\frac{dw}{d \log M} = \frac{dw}{dv}\left[\frac{dv}{d \log M}\right]$$

   This gives coordinates for the differential molecular weight distribution curve.
8. Calculate M_n, M_w, M_z, etc. (from Ref. [IIIA-65]).[a]

---

[a] Note sums must be over equal intervals in $\log M$ not over V_e as in computer printout.

The usual steps in working up GPC data are enumerated in Scheme A and are discussed next. Although the construction of a baseline seems innocuous enough, it is a very important step. The baselines before and after the polymer is eluted should be smooth linear continuations of one another, i.e., there should be no shifts. If a chromatogram cannot be obtained without shifting baselines, one should ask why. Is the column plugging? What was filtered out prior to injecting the polymer into the column? The latter question should be answered for all GPC experiments on elastomers.

The pips on the chromatogram indicate 5-ml increments of elution volume on most instruments.* Normalization involves forming the sum $N = \sum_{i=1}^{n} h_i$ and then dividing each h_i by N. The sum is performed over all elution volume

---

* Newer high speed chromatographs rely on accurate pumps and the pips are merely time intervals presumably representing approximately milliliter increments which are no longer measured.

counts to be included in the analysis. This procedure leads to the representation of the distribution as a histogram with intervals equal to the count width chosen, and height h_i/N. The width of the intervals determines the accuracy of the approximation; 0.5 count intervals are usually adequate.

Corrections for peak broadening all contain assumptions and unknown parameters which generally are determined by running samples of known MWD, seeing how they behave, fitting the parameters in the assumptions to make the standards correct, and finally using the same parameters on the unknowns. There are a variety of procedures suggested which give results in reasonable agreement [IIIA-56b, 60, 64, 65, 74b]. Broadening in the GPC trace arises from a number of factors, many of which can be eliminated by proper column design. It appears that going to small columns packed with small-diameter (5 μm) particles often obviates the need for broadening corrections. In the absence of this type of column, the behavior of standard polymers must be studied to assess performance.

The "variations of detector response" with molecular weight include refractive index changes at low molecular weight if the refractive index is used as concentration measure. For copolymers, refractive index will also be a function of the amounts of the various monomers. Unless the polymer is homogeneous with respect to composition, some correction will have to be made for that. Because of these problems, other methods (e.g., ir, uv) are often employed to characterize the effluent and are very useful detectors of compositional heterogeneity. Once all the corrections are made, plots can be compared with theoretical ones, or if only molecular weight averages are desired they can be calculated according to formulas (15) and (16).

$$\overline{M}_{n} = \sum_{i} h_{i} \Big/ \sum_{i} \frac{h_{i}}{M} \qquad = \sum_{i} h_{i} \frac{dV_{e}}{d \ln M} \Big/ \sum_{i} h_{i} \left(\frac{dV_{e}}{d \ln M} \right) \left(\frac{1}{M} \right) \qquad (15)$$

$\qquad\qquad$ equal intervals of V_{e} $\qquad$ equal intervals of $\ln M$

$$\overline{M}_{w} = \sum_{i} h_{i} M \Big/ \sum_{i} h_{i} \qquad = \sum_{i} h_{i} \frac{dV_{e}}{d \ln M} M \Big/ \sum_{i} h_{i} \frac{dV_{e}}{d \ln M} \qquad (16)$$

It is possible to write expressions for all $\overline{M}_{i}$; however, they become meaningless when the sums have major contributions from h_i, which are small fractions of the maximum chromatogram height, or for M where the column resolution tails off, as shown in the calibration curve. $\overline{M}_v$ should then be calculated and compared with the experimental value of the whole polymer; see Eq. (12).

On a final note, which leads into the next characterization technique of Table IIIA, GPC separates on the basis of size, so that material of great compositional heterogeneity can emerge from a column at the same elution volume, i.e., the compositional heterogeneity is not resolved. Since the molecular dimensions of compositionally different molecules may be the same, but their solubility characteristics are usually not, fractionation based on solubility characteristics may be resorted to to separate polymeric components. However, both molecular weight and composition affect the solubility.

Precipitation of polymer from a solution can be brought about by addition of a nonsolvent or by raising or lowering the temperature. Some solvents are so good that temperature reduction may never produce precipitation before solvent crystallization, or vitrification, intercedes (e.g., polystyrene in toluene). Similarly, temperature increases to reach lower critical solution temperatures (see Fig. 9) may be inconvenient due to the need for pressure vessels. Thus, the most common technique is the use of some nonsolvent to cause phase separation of the polymer. Analytical description of the precipitation behavior of polymers in such systems is difficult.

Let us assume that some nonsolvent miscible with the GPC solvent is known. Take, e.g., a sample containing long styrene and butadiene sequences as measured spectroscopically, which could be either a blend of homopolymers or some type of block copolymer. A GPC trace of the polymer dissolved in tetrahydrofuran shows a relatively narrow band of polymer sizes. Addition of acetone to the GPC effluent produces precipitation which causes a dramatic increase in turbidity (see IIIA-75c) over a narrow range of nonsolvent addition, and a second region of precipitation after much more solvent is added. The polymer, although uniform in elution volume, is not of uniform composition. Some semiquantitative information on the relative amounts of material can be made by turbidimetric measurements. Further characterization of the material is called for, perhaps a larger-scale solvent–nonsolvent approach, with characterization of the fractions by a spectroscopic technique. The use of two methods which fractionate on a different basis is called "cross fractionation."

As stated, the classical methods of fractionation according to MWD and compositional distribution depend on the solubility differences among molecularly and chemically heterogeneous species. These methods, whether used as removing successive "cuts" of the whole polymer (a) by addition of successive batches of nonsolvent with removal of the precipitate, (b) by precipitation of nearly all the polymer with removal of the supernatant as a "cut" with dissolution of the precipitated phase, or (c) by deposition of a thin film of the polymer on some substrate and removal by elution with successively better solvent mixes, or by some other method, are empirical in nature and involve characterization of the cuts as they are recovered. The assumption is made that if analyses of the cuts (each of which represents, say, $<10\%$ by weight of the original) revealed no molecular weight or compositional heterogeneity, then the polymer would be homogeneous, i.e., monodisperse. If heterogeneity is revealed, as is usually the case, it can be described qualitatively by plots of M versus $W(M)$ as before for GPC data, or various molecular weight averages can be calculated. To be reasonably sure that the whole extent of heterogeneity has been revealed, one has to obtain the same MWD, etc., using a variety of fractionation conditions or, better, by refractionating the fractions. The need for all this experimentation arises because the theory of solutions is not adequate to permit calculation of the results of fractionation procedures. Many references are cited in Table IIIA [IIIA-76–78] which demonstrate, for particular model systems and using a particular solution theory, how well one might expect to resolve the molecular weight species present. The results more often than not show how difficult it is to show subtle, yet essential, differences.

The phase diagrams of Figs. 9–11 are presented to show typical polymeric behavior. Figure 9, which was referred to before, points out how polymer may be precipitated by either raising or lowering the temperature. The cloud points, or precipitation points, depend strongly on molecular weight. Obviously, these differences can be used to fractionate the components of a polydisperse system. Unfortunately, the presence of each component of a polydisperse mixture affects the cloud point of the other components. Figure 10 shows, for

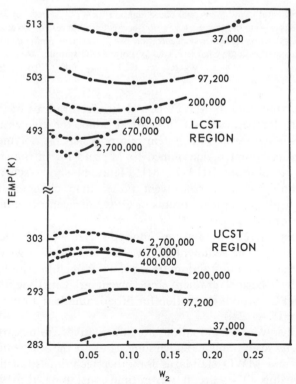

Fig. 9. Critical temperatures in polystyrene solutions: the (temperature–weight fraction polymer W_2) phase diagram for the polystyrene–cyclohexane system for samples of indicated molecular weight. [S. Saeki, N. Kuwakara, S. Komno, and M. Kaneko, *Macromolecules* **6**, 246 (1973); Copyright by the American Chemical Society. Reprinted with permission.]

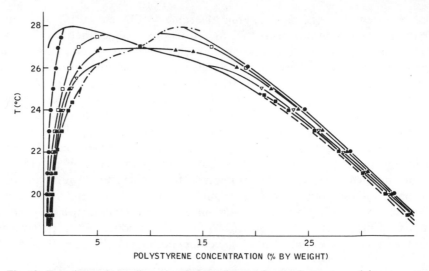

Fig. 10. Experimental two-dimensional phase diagram for a polystyrene–cyclohexane system. Cloud point curve, ———; shadow curve, —•—; coexistence curves for 2 (●); 6(□), 10 (▲), 15 (▽) and 20% (■) whole polymer concentration (wt %); critical point, determined with the phase–volume ratio method, ■. Characteristics of polystyrene sample, $M_n = 2.1 \times 10^5$; $M_w = 3.46 \times 10^5$; $M_z = 5.5 \times 10^5$. (Data of G. Rehage, G. Möller, and O. Ernst, *Makromol. Chem.* **88,** 232 (1965).)

a particular polymer and its MWD, how the concentrations of the coexisting phase vary with temperature. At the critical point, the concentration of coexisting phases is the same. In a system of a polydisperse polymer in a single solvent, the critical point is determined by $\overline{M}_w$ and $\overline{M}_z$ in the context of the lattice theory of solutions [IIIA-77, 81]. Figure 11 shows schematically the situation for addition of a nonsolvent rather than changing temperature. Figure 12a–c displays "typical" results of GPC, ultracentrifugation, and fractionation data for comparison.

The polymers investigated were "monodisperse" poly(α-methylstyrene), and blends of these. The sedimentation velocity method shows double peaks slightly better than column fractionation, but column elution and precipitation chromatography (thermal gradient and solvent gradient) yielded about the same results. GPC without correction for broadening had a much lower resolution (see also [IIIA-124]).

Solvent–nonsolvent fractionations are informative, can be carried out with little or no expensive equipment, and in some cases are the only source of narrow compositional and MWD standards for other techniques, such as GPC or $[\eta]$–$\overline{M}_w$ relationships. They are, however, time consuming. It may take a week for phases to separate, or the polymer must be eluted from a column slowly. Efforts have therefore been made to automate the temperature- (or nonsolvent-)

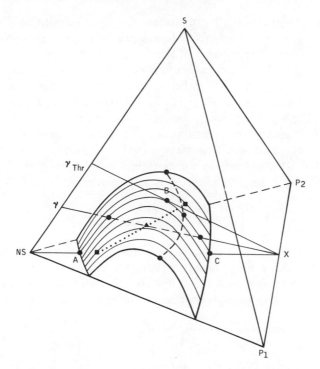

Fig. 11. Miscibility gap in a quaternary system containing two macromolecular homologs P_1 and P_2, a solvent S, and a nonsolvent NS. The chain lengths of P_1 and P_2 differ $(M_2 > M_1)$. ABC, cloud point curve of polymer mixture X; –•–•–, critical line; ■, coexisting-phase compositions relating to system ▲; · · ·, tie line. γ is the nonsolvent fraction and γ_{thr} is the concentration at the precipitation threshold.

induced precipitation method. The most effective technique for this has been found to be turbidimetry. An instrument which is capable of measuring light transmission is used to construct plots of turbidity versus concentration of the precipitant. If the turbidity can be related to molecular weight or heterogeneity on a quantitative basis, then data can be obtained rapidly. Because of the complex interrelationships among M, MWD, concentration, solubility, and particle size as they affect light scattering, this technique is really useful only if calibrated with known standard blends of polymers, or if only qualitative information is desired.

All of the methods discussed heretofore in Table IIIA involved measurements on polymer solutions because, however imperfectly solutions are understood, pure polymers present even greater difficulties. Table IIIA contains two techniques which involve measurements on pure polymeric samples, i.e., bulk viscosity and gel–sol analyses.

It is now reasonably well established that the viscosity of a pure linear polymer at a given temperature and pressure, at shear rates low enough so that the viscosity is independent of shear rate, depends on molecular weight as depicted in Fig. 13. Thus, calibration of a viscosity–$\overline{M}_w$ relationship at a particular temperature permits rather accurate $\overline{M}_w$ determination by viscosity measurement. Since elastomers are of high molecular weight, the main difficulty rests in the high viscosity η, and its correspondingly high variability with shear rate. At ambient temperatures, $\eta > 10^8$ poises. Pieces of unvulcanized SBR, EPDM, etc., may not flow for weeks (10^6 sec) under their own weight; i.e., a stress of 10^4 dyn/cm^2 on the bottom for a 10-cm-thick piece. At this shear rate of $<10^{-6}$ sec, the viscosity is approximately

$$\eta = \frac{\tau}{\dot{\gamma}}\left(\frac{\text{dyn/cm}^2}{\text{sec}^{-1}}\right) > \frac{10^4}{10^{-6}} > 10^{10}\,(\text{poises}) \tag{17}$$

To lower the viscosity, measurements are made at elevated temperatures, but even at 200°C shear rates of less than 10^{-3} sec must be used for η to remain constant with shear rate, and times of $>10^3$ sec are needed to have the flow equilibrate, often long enough so that chemical processes can intercede and

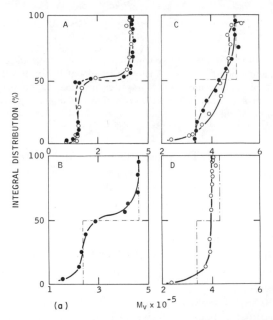

Fig. 12a. Comparison of the integral MWD fractionation results determined by precipitation chromatography (○) and elution method (●). (–•–) denotes the MWD of the mixture calculated from those of the constituent components, assuming ideal monodispersity of the components. An integral distribution is simply the sum of the weight of the polymer that has M less than the M in question, i.e., $\int_0^M W(M)\,dM$. A–D are different blends.

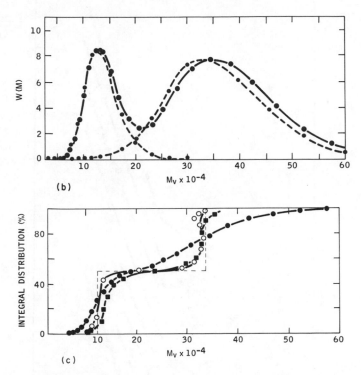

Fig. 12b,c. (b) MWD of mixture No. 9 determined by GPC. Solid line, MWD of sample; broken lines, MWD's of the components. Mixture 9 is a blend of poly(α-methylstyrene) of $\overline{M}_n$ 14 and 34 × 10³. (c) MWD's of mixture No. 9 determined by various methods (notation as in Fig. 12b): precipitation chromatography, ○; GPC, ●; sedimentation velocity, ■. Note the lower resolution of GPC. (A. Yamamoto, J. Noda, M. Nagasawa, *Polym. J.* **1**, 304 (1970).)

confuse the results. If microgel is present, equilibration times may extend far beyond 10^3 sec.

If one employs standard polymers, or uses existing universal relationships between polymer molecular weight and viscosity, it is reasonable to combine such work with solution property measurement and gel–sol studies. Where the presence of gel is suspected, the latter studies can be particularly revealing. If it can be established by some means that a given number of cross-links is introduced by a given amount of peroxide, radiation dose, or some other curative, one can obtain the weight average molecular weight of a polymer by observing the relationship of dose to gelation. Thus,

$$\overline{M}_w = M_0/\rho_{gel} \tag{18}$$

where M_0 is the molecular weight of a monomer unit, and ρ_{gel} is the probability that any given monomer unit is coupled into a cross-link at the gel point, i.e., the fraction of units in cross-links at the gel point. This relation is independent

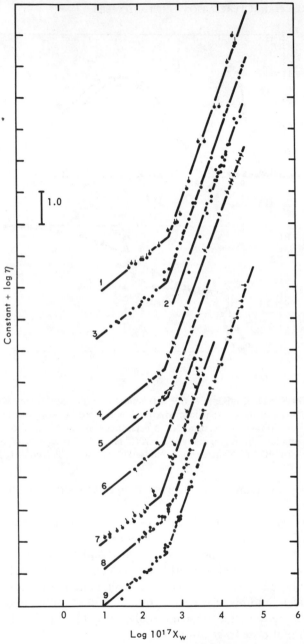

Fig. 13. Log η: the viscosity corrected to constant friction factor ζ versus log $X_w = \log[(\langle S^2{}_0 \rangle/ M)Z_w/v_2]$ (where Z_w is the weight average degree of polymerization, v_2 the specific volume) for the linear polymers: 1, poly(dimethyl siloxane); 2, poly(iso-butylene); 3, polyethylene; 4, polybutadiene; 5, poly(tetramethyl-p-diphenylsiloxane); 6, poly(methyl methacrylate); 7, poly(ethylene glycol); 8, poly(vinyl acetate); 9, polystyrene. The straight lines have slope 1.0 or 3.4. The curves have been shifted arbitrarily along the ordinate [IIIA-107].

of the MWD of the sample. However, as more cross-links are added past the gel point, the partition between gel and sol is governed by the MWD of the sample. This relation has been employed to get a measure of MWD, but one has to be sure of the chemistry so that the functionality of the cross-links is known. Allowances must be made for scission reactions. If the scission/cross-linking ratio is constant, methods are available for evaluating the molecular weight average in the absence of prior knowledge of the effective yield of cross-links produced per dose [IIIA-98, 104].

Gels possessing cross-link densities just past the gel point are very fragile. Dissolution of soluble material must be achieved with little more than the slow rotation of the specimen in a good solvent. One finds often that the data extrapolate to what appear to be negative doses of cross-linking agent for gelation, i.e., the original material was gelled, although this may not have been readily apparent. A lightly cross-linked polymer can be milled, mixed in a Banbury, and have a measurable Mooney viscosity, but if the polymer is gelled, solution property measurements must be treated with caution, and properly interpreted. Much of the work done in connection with networks has been recently reviewed in an extensive article on entanglements [IIIA-104].

BRANCHING (TABLE IIIB)

Elastomers, except for the thermoplastic variety, are generally made with unsaturation in the polymer backbone so that they can be cross-linked. Once there, unsaturation is difficult to prevent from reacting and producing branched or gelled polymer. Thus, on chemical grounds, one should always suspect branching in synthetic elastomers.

Branching is of concern since it introduces extra chain ends for a given $\overline{M}_n$ and thus more network defects. It affects, moreover, the transport properties of the molecule, and this affords some means by which the degree of branching can be measured and the corresponding changes in properties be determined. The most obvious mechanism by which branching changes transport properties is by its effect on the geometry and/or size of the molecule. Figure 14 shows how, for the same molecular weight, branching reduces the radius of gyration. The latter can be calculated, for a given molecular weight, as a unique function

Fig. 14. Schematic representation of how branching causes the overall extent of the molecule in space to be decreased at a given molecular weight. The dimensions can be calculated as a function of branch type and density at Θ conditions.

of the branch type, number, and distribution. Intuitively one would expect the intrinsic viscosity (cf. Eq. (10)) to be reduced by branching. This is found, and the reduction in intrinsic viscosity is reported as a function of g, the ratio of the mean square radii of gyration of the branched and unbranched polymer of the same molecular weight ($\overline{M}_w$).

$$g = \langle S^2 \rangle_{br} / \langle S^2 \rangle_{lin}, \qquad [\eta]_{br} = f(g)[\eta]_{lin} \tag{19}$$

Calculation of $f(g)$ involves dilute solution theory, as discussed following Eq. (10), with the type of branching explicitly considered. Alternatively, g may be determined by working with model compounds. Calculations have been performed on a limited number of branched structures with results such as those shown in Fig. 15 [IIIB-35]. Agreement with experiment (i.e., using model branched structures) is approximately correct, but highly branched structures appear to have $f(g)$ smaller than that calculated.

There is little question that g values of less than unity indicate branching and that relative values of g can be used to indicate relative degree of branching if the branching is of the same type, but assignment of absolute levels of branching on the basis of intrinsic viscosity measurements alone is not yet possible. Together with light scattering measurement of the radius of gyration (see Eq. (6)), a direct comparison of g values can be made. In Θ solvents, the experimental g values agree quite satisfactorily (with some notable exceptions) [IIIB-31].

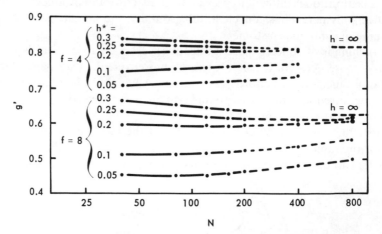

Fig. 15. Intrinsic viscosity ratio $g' = [\eta]_f / [\eta]_{lin}$ plotted against N for various values of h^*. Thick dashed lines are the nonfree draining limit obtained by Zimm and Kilb; f is the number of "arms" extending from a center branch point; h^* is discussed in the text. [From K. Osaki and J. Shrag, *J. Polym. Sci., Polym. Phys. Ed.* **11**, 549 (1973); reprinted by permission of John Wiley & Sons, Inc.]

The reduction in molecular size by branching, which is directly measurable in light scattering and which reduces the intrinsic viscosity, also affects the sedimentation coefficient, making it larger for a given molecular weight. This affords another means for measuring branching [IIIB-26] in conjunction with GPC and [η], and data have been reported for some model systems.

In any case, the effect of branching on many forms of behavior can be based on the experimental GPC result, observed previously, that $([\eta] M)_{\text{pol 1}} = ([\eta] M)_{\text{pol 2}}$ holds also for branched molecules. Therefore, if one runs a GPC state shear compliance (see Chapter 5) to be decreased by 30% by a single branch. This approach has been utilized to study EPDM elastomers [IIIB-36].

The most widely used, though least rigorous, method to detect branching is based on the experimental GPC result, observed previously, that $([\eta] M)_{\text{pol 1}} = ([\eta] M)_{\text{pol 2}}$ holds also for branched molecules. Therefore, if one runs a GPC scan on a homogeneous fraction of polymer, the intrinsic viscosity and viscosity–molecular weight relationship of which are known, the product $[\eta]M$, and thus M, can be evaluated. Knowing M, we can calculate $[\eta]_{\text{lin}}$ and thus g can be found, as discussed above. The postulate of equivalent hydrodynamic volumes should be regarded as an approximation [IIIB-12a, b]. More often than not, homogeneous fractions are not available and some assumptions have to be made to describe the branching and its distribution. To substantiate the assumptions, chromatographs have been fitted with viscometers so that the intrinsic viscosity at a particular elution volume can be measured. This approach has dramatized one of the often neglected difficulties, that of GPC peak broadening, in that at the extremes of the elution volume range for a given broad distribution sample a given elution volume may have been shifted by as much as one count (e.g., 50% in molecular weight). That is, if the material eluted at a particular count were rerun through the instrument, it would now be shifted, its original position having been influenced due to diffusion and concentration effects, and perhaps in part to hydrodynamic volume dependence on concentration. In any event, when intrinsic viscosities are measured and compared with the expected intrinsic viscosities for the given elution volume, the nonsensical result that $[\eta]_{\text{br}}/[\eta]_{\text{lin}} = f(g) > 1$ may be found. Thus dispersion corrections must be made.

As an example of what can be done, consider the data in Fig. 16, from the class of the second entry of Table IIIB. These results were obtained by a preparative fractionation of an emulsion SBR and analyzed on the GPC for g values, which were then used to correct the MWD of the polymer for branching. Thus in Fig. 16 the data points (upper curve) refer to the g values from the fractions, while the lower curve, $W(\log M)$, is the weight distribution from GPC regardless of degree of branching. It is seen that the random branching processes encountered in free radical SBR polymerization lead to increasing degrees of branching with increasing molecular weight. Even if these results are not

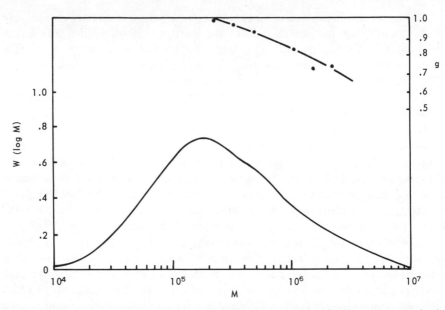

Fig. 16. Representation of branching in an SBR which had been fractionated to obtain the $[\eta]_{\text{br}}/[\eta]_{\text{lin}} = f(g) \approx g^{1/2}$ for fractions (upper curve), these data then being used to correct GPC data for the whole polymer to give the solid curve, $\overline{M}_{\text{w}} = 5.2 \times 10^5$, $\overline{M}_{\text{n}} = 8.6 \times 10^4$. Note how branching increases with molecular weight. [From G. Kraus and C. Stacy, *J. Polym. Sci. Part A-2* **10,** 657 (1972); reprinted by permission of John Wiley & Sons, Inc.]

rigorously correct, a nice qualitative picture is presented. However, one should keep in mind that although no mention is made in the paper, most likely some gelled material was removed during solution preparation.

If well-characterized samples of a material are available, one may also try to "calibrate" properties of undiluted bulk polymers to detect branching. Unfortunately, the most common measurement, that of bulk viscosity, is influenced by branching in ways as yet not very well defined [IIIB-38b]. Below a certain molecular weight, branching reduces the bulk viscosity, but as the molecular weight increases, so does the sensitivity of the viscosity to branching. Eventually, the branched polymers become more viscous than their linear counterparts of the same $\overline{M}_{\text{w}}$. Similar difficulties exist with compliances, moduli, and other bulk property measurements. Polymers should thus again be characterized by several techniques.

Recently laser light scattering photometers have become available which have scattering cells small enough to permit on-line detection when coupled to a GPC. Direct detection of $\overline{M}_{\text{w}}$ coupled with knowledge of $[\eta]M$ from elution volume is the most attractive (and adaptable to high speed GPC) system for branching detection currently available. Similarly, analysis of the quasielastic scattering components of the same laser light scattering signal can provide diffusion coefficients [22] which include the same branching information as [η]. This technique is yet to be developed but is under study.

GEL (TABLE IIIC)

For the purpose of this discussion, "gel" is best defined by referring to its formation by the combination of macromolecules into chemically connected groups which possess properties "significantly different" from simple branched structures. The significantly different properties will be manifested in ways depending on the macroscopic state of the gel. Unavoidably, the distinction between an assembly of very high molecular weight branched structures and an actual gel is chemically and mechanically difficult. If the number of chemical bonds needed to form a gel is n, and $n - x$ and $n + y$ links give a pre-gel and a definitely gelled structure, respectively, x and y are $\ll n$, so that studies right at the gel point are rare. There are, though, discrete changes in the light scattering behavior (cf. [IIIA-25, 31a]) on passing into the gelled state. Gels also should support small enough stresses (no chemical bond breakage) indefinitely after some finite equilibrium deformation. Gelled structures will imbibe finite amounts of good solvent, and swollen gels will attain their equilibrium strain under an imposed stress relatively rapidly as opposed to the indefinite relaxation of uncross-linked material. Gel, however, may be present not as a macroscopic piece, but rather as "microgel," by which one means finely divided gel, say of particles of the order of 1 μm (10^4 Å). It may have formed by the original cross-linking process, by some mechanical process which broke up the original gel (e.g., by milling a cross-linked piece of gelled rubber), or during a high shear solution polymerization where loosely inter- and intramolecularly linked structures are torn apart.

Examination of a polymer containing microgel by membrane osmometry will yield a proper average molecular weight. By light scattering, if data are taken at low scattering angles, $\overline{M}_{\rm w}$ appears infinite, and the depolarization measurements will be unusual. During ultracentrifugation, with appropriate manipulation of rotor speeds and sedimentation patterns, the gel can be readily distinguished. Microgel may be filtered from solutions, but filtration of polymer solutions is always tricky because nongelled as well as gelled polymer tends to plug filters. If the gel is macroscopic, it can be characterized by standard volume-swelling techniques as to its cross-link density; these matters are treated in Chapter 1; attendant problems are discussed elsewhere also [IIIA-104]. Even after the amount of gel particles and their average molecular weight have been evaluated, one is still faced with the task of finding how they will affect other polymer properties. If, for example, the gel were heavily cross-linked, it might behave much like a low-modulus filler [IIIC-41], whereas if the gel particles were just near the gel point, they might participate in the rheological behavior of the sample nearly the same as high molecular weight branched structures with extremely long equilibration times at low stresses [IIIC-40].

On the commercial scale, the presence of gel affects primarily the compounding phases, and then only if the amount of gel is unduly large. Most of the rheological processes to which rubbers are subjected, e.g., putting a piece on a mill or into a Banbury, are such that the gel structures are partially destroyed by the enormous stresses. Cold flow problems with polybutadienes are mostly solved by introducing small doses of high molecular weight or gelled polymer, which is a practical example of the importance of light gelation.

As a final note, polymer solutions typically readily undergo molecular association and agglomeration and/or require long times for complete molecular dissolution. At concentrations barely adequate to give reasonable light scattering measurement in even reasonably good solvents, rather permanent aggregates of molecules may remain or form. This phenomenon is most common with polymers which are semicrystalline (e.g., EPDM, PVC, or polyurethanes). One must thus be careful not to confuse gel with reversible association, which may change with solvent or temperature changes.

IV. Rubber–Glass and Other Secondary Transitions (Table IV)

Given the molecular weight and chemical structure of a polymer, which are fundamental properties, one might anticipate being able to predict the rheological characteristics of the polymer. Experimentally and empirically it is a fact that most viscoelastic behavior can be correlated in terms of models which in effect assign the material a set of characteristic relaxation times. The memory (elasticity) of the material and the dissipative (viscous) properties it possesses will depend on the time scale of the experiment in relation to these relaxation times. The relaxation times are functions of MW, MWD, cross-linking (if any), temperature, and pressure. The temperature dependence is empirically correlated by the WLF equation (see Chapter 5). When polymers are compared at equal temperature differentials from a reference temperature T_g, which appears in the WLF equation, they all behave similarly. It turns out that elastomers generally have their T_g in the range of -30 to $-100°C$. That is because at ambient temperatures (say 75°C above T_g) where elastomers are rubbery, the relaxation times of the chain segments must be short enough to let the material deform under relatively low stresses (e.g., 10^7 dynes/cm^2) in reasonably short periods. If ambient temperatures were $\sim -60°C$ most elastomers would be plastics. As one gets farther above T_g, the relaxation rate of the material gets faster.

If T_g has such importance, it seems that one should set about calculating it in terms of fundamental polymer structures. Such attempts have been made, and a WLF-type equation can be derived from statistical mechanics and from concepts such as chain stiffness and its effect on the entropy of a polymer molecule, and the volume of the system and its effect on the freedom of chain segments to rearrange. Rigorous theories with parameters which can be fixed by means external to the theory do not exist, however, and one is left with a measurement of T_g rather than its calculation; thus we regard it as a fundamental property to be measured.

Obviously, if mechanical properties are controlled by T_g, they can be used to measure it. That discussion is presented in Chapter 5.

Other properties undergo changes at T_g as well. In a thermodynamic sense T_g is not a first-order transition like melting or boiling but is second order. There is not a discontinuous change in volume at T_g but there is a discontinuous change in the rate of change of volume with temperature (Fig. 17). Even this second-order thermodynamic sense of T_g must be regarded with caution. The actual position of the discrete change in the second-order quantity is a function of the rate of measurement. The experimentally measured T_g is a function of the relaxation time of the variable under observation (IV-56).

Since V and T are fundamental thermodynamic variables which are adequate to describe the equilibrium properties of a closed system at constant pressure, given the behavior shown in Fig. 17, one can anticipate that most physical properties will undergo some discontinuity in their rate of change with temperature at T_g. Thus, a myriad of experimental possibilities suggest themselves as means to detect T_g. Optical properties such as light, or X-ray scattering from density fluctuations, or refractive index change with temperature undergo discrete changes. Thermal propertirs such as heat capacity and even associated changes in the absorption spectra can be monitored. Other examples are cited in Table IV.

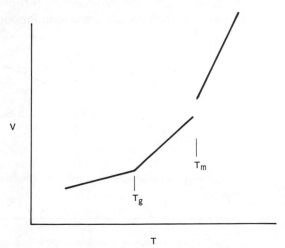

Fig. 17. Schematic representation of dependence of V on T suggesting how T_g (glass transition) may be detected. At T_g there is dV/dT discontinuity; at the melting point there is both dV/dT and ΔV discontinuity. A completely amorphous polymer would exhibit no T_m.

For all measurements, the exact location found for T_g will depend on the thermal history of the sample, the rate at which the temperature is changed during the measurement, and the characteristic time scale of the measuring probe. If one measuring technique senses molecular motions at higher frequencies than another, the apparent discontinuity in behavior appears at higher temperatures for the higher-frequency technique. Thus changes in volume measured over a period of days (e.g., lowering of T by $1°C/day$) will give a lower T_g than measuring the modulus at 1 Hz, or than NMR relaxation (T_1) times at 10^6 Hz even when the cooling rates are the same. This is presented schematically in Fig. 18. This dependence of T_g on measurement frequency can be correlated with the WLF equation.

Consider the special case of heat capacity (C_p) measurement. This is an easily detected thermal property. It is usually not measured statically in an adiabatic calorimeter, but in commercially available differential scanning calorimeters (DSC). These instruments heat (or cool) a sample of the material and an inert reference material of similar total heat capacity at a programmed rate. Either the temperature differential (ΔT) which develops between the reference and sample (in the du Pont instrument) or the difference in amount of energy needed to keep the reference and sample at the same temperature (in the Perkin-Elmer instrument) is recorded. When the sample undergoes a discrete change in heat capacity this shows up as a discontinuity in the smooth baseline. The instruments can be calibrated to give quantitative C_p data. The ΔC_p of most elastomers is in the 0.1 cal/g range at T_g and C_p is of the order of 0.5 cal/g °C above T_g. This is not a zero (apparent) frequency technique. The T_g recorded is a function of heating rate. A typical trace is shown in Fig. 19a where we have also included first-order melting transition for comparison.

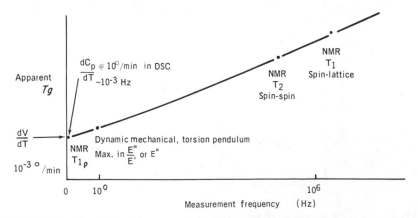

Fig. 18. Schematic of how the apparent measured T_g depends on the frequency of the technique used to measure it. NMR, dielectric, creep, and stress relaxation measurements may be made over decades of rate.

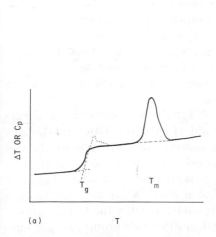

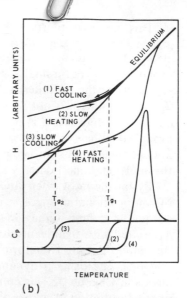

Fig. 19. (a) Schematic of a DSC trace showing departure from baseline at T_g and a first-order transition T_m. Dashed line indicates behavior that can be induced by annealing the sample slightly below T_g for long periods compared to the heating rate of the DSC. (b) Schematic of changes of enthalpy and heat capacity employing unequal cooling and heating rates. [From S. Wolpert, A. Weitz, and B. Wunderlich, *J. Polym. Sci., Part A-2* **9**, 1887 (1971); reprinted by permission of John Wiley & Sons, Inc.]

The portion of the DSC trace to identify with T_g (or T_m) is a somewhat subjective matter. The best procedure is to be consistent and to run "control" samples of the materials to which a comparison is to be made. A possible line construction is indicated in Fig. 19a. DSC scans often appear to be a bunch of wiggly lines. Often that is all they are, and there is no sense in trying to interpret scans until they are reproducible. Punching out a 10-mg specimen from an 0.030 in. pad of the rubber often produces a bizarre initial DSC scan. Rerunning the same specimen gets rid of most of the apparent transitions. The transitions seen on the first scan may or may not have been real. Unless a sample of rubber is fabricated at temperatures high enough and times long enough, (e.g., 30 min at 150°C) to relax most of the orientation, there is a tendency for the specimen to change dimensions in the DSC when it gets to a high enough temperature to relax. This causes apparent heat capacity changes which do not reappear on a second scan.

There are also real effects which do not appear on a second scan, however. Although all but the vibrational motions of the polymer chains in their lattice have essentially ceased below T_g, there is some slow relaxation occurring. If a sample is held just below T_g for a long period compared to the scanning time of the DSC, an apparent endotherm is observed upon heating (Fig. 19).

This has been interpreted as due to the heating rate of the sample being greater than the rate at which the glassy lattice can rearrange to an equilibrium state at the particular temperature just above T_g. When the rearrangement or expansion does occur finally, a certain amount of heat is immediately absorbed, thus causing the endotherm. If the scan is repeated with no annealing below T_g, the endotherm is not reproduced. Also, if the initial scanning rate is slow enough, no endotherm is produced, and in certain cases an exotherm can be produced. Whatever other property is chosen for T_g detection much the same kind of time-dependent behavior can be found so care should be exercised in the interpretation of data.

By far the most often studied aspects of T_g with elastomers are the effects of compositional changes in copolymers and of additives. Often the DSC is adequate for these studies. In certain cases, however, the heat capacity changes associated with significant mechanical property changes are so small as to be difficult to detect. Therefore, if possible, it is wise to use a combination of methods, one being a dynamic mechanical method. Such methods are discussed in Chapter 5. This concern is further reinforced by the fact that other molecular motion processes such as methyl or other side group motions may exhibit "sudden" onset as temperature is raised, thus causing apparent T_g's. These occur predominantly below the T_g associated with the rubber–glass transition. A mechanical property measurement establishes definitely whether the modulus is dropping from 10^{10} dynes/cm^2 by 2 or 3 decades, which is what is of primary interest if one is trying to determine a lower bound for service temperature of an elastomer.

V. Crystallinity (Table V)

Crystallinity in "plastics," i.e., in hard solid polymers, frequently determines their moduli and usually upgrades their mechanical performance and strength. Rubberiness, e.g., recoverable extensibility produced by low stresses, is a property of amorphous polymers above T_g, but many rubbers contain small to moderate amounts of crystallinity embedded in the elastomeric matrix. Crystallites act as reinforcing filler particles and as "physical" cross-links which permit an unvulcanized semi-crystalline rubber to support stress indefinitely. Most importantly crystallization can be induced by stressing producing a state in which the stress-induced degree of crystallinity raises the strength of the rubber. Rapid melting of the strain-induced crystallites permits rapid strain and stress recovery. Thus, in contrast to the importance of the degree of crystallinity for plastics, it is the crystallizability which engenders the best properties of an elastomer.

Polybutadienes, some ethene–propene, and many thermoplastic elastomers possess significant degrees of crystallinity at ambient temperatures, while

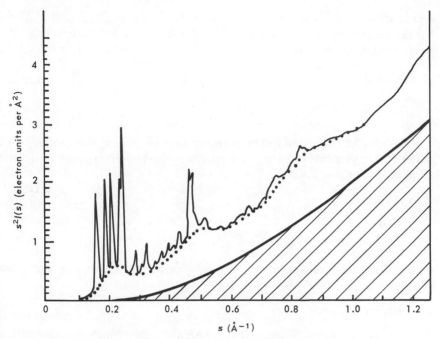

Fig. 20. Curve of $s^2 I(s)$ versus s for polypropylene sample. Hatched area, incoherent scatter; coherent amorphous scatter above that; coherent crystalline scatter above dotted line; see Eq. (20). [From L. Alexander, "X-Ray Diffraction Methods in Polymer Sciences," copyright ©1969; reprinted by permission of John Wiley & Sons, Inc.]

natural and butyl rubbers, among others, crystallize significantly when stressed. In view of the profound effects of crystallinity on all properties, the fraction present and detailed morphology must be characterized as quantitatively as possible.

The nature of crystallinity in polymers is a field of active research, and a number of models have been proposed for the wide variety of states of semi-crystalline polymers. The simplest description for the polymer is a two-phase system of crystalline and amorphous domains, each phase possessing the characteristics of the pure phase unperturbed by the other. However, there are also interphase regions which are far from negligible.

The degree of crystallinity is described as a weight (W_c) or a volume fraction (f_c). Classically, the structure of crystal lattices is studied by X rays (Fig. 20), but more often than not the degree of crystallinity in polymers is so low, and the long-range order in the crystal so poor, that only a few diffraction maxima can be observed. This makes the determination of the crystal structure, i.e., of the positions of the atoms in the unit cell, very difficult. In certain cases, only the most intense maxima from a particular geometry can be observed, and the crystal structure is assumed to be that of polymer of the same repeat unit which has crystallized more completely under other conditions. Concerning the two-phase model, one may define the weight fraction W_c by a variety of methods.

Specific volume

$$W_c = \frac{v_a - v}{v_a - v_c} \tag{20}$$

v_a is the specific volume of the pure amorphous phase a, v_c the specific volume of the pure crystalline phase c, v the specific volume of the sample in question, all in cc/g, and $\rho = 1/v$.

Specific heat

$$W_c = \frac{(C_p)_a - C_p}{(C_p)_a - (C_p)_c} \tag{21}$$

$(C_p)_i$ are the heat capacities of the pure phases, and C_p that of the sample, in cal/deg g.

Enthalpy of fusion

$$W_c = \frac{\Delta H_f}{(\Delta H_f)_c} \tag{22}$$

ΔH_f is for the sample and $(\Delta H_f)_c$ for the pure crystalline phase, in cal/g.

Infrared extinction coefficient

$$W_c = \frac{\varepsilon_{\lambda c}}{(\varepsilon_{\lambda c})_c} = 1 - \frac{\varepsilon_{\lambda a}}{(\varepsilon_{\lambda a})_a} \tag{23}$$

$\varepsilon_{\lambda i}$ are the absorbance of amorphous and crystalline phases; parentheses denote pure phases.

$$W_c = \left(1 + \frac{Q\varepsilon_{\lambda a}}{P\varepsilon_{\lambda c}}\right)^{-1} \tag{24}$$

Q and P are proportionality constants between absorbance and weight of phase present

X-ray scattering

$$W_c = \frac{\int I_c}{\int I_c + I_a} \tag{25}$$

I_i are scattering intensities from the amorphous (a) and crystalline (c) phases.

$$W_c = 1 - \frac{\int I_a}{\int (I_a)_a} \tag{26}$$

$$W_c = \left(1 + \frac{qI_a}{pI_c}\right)^{-1} \tag{27}$$

p,q are proportionality constants between weight of phase present and intensity of scattering.

$$W_c = \frac{\int_0^\infty S^2 I_c\, ds}{\int_0^\infty S^2 I\, ds} \frac{\int_0^\infty S^2 \bar{f}^2\, ds}{\int_0^\infty S^2 \bar{f}^2 D\, ds} \tag{28}$$

$S = (2 \sin \theta)/\lambda$, $I(s)$ is the scattering intensity, D a lattice imperfection factor, and $\bar{f}^2$ the weighted mean square atomic scattering factor.

NMR

$$\frac{W_c}{1 - W_c} = \frac{\text{area of broad line component}}{\text{area of narrow line component}} \tag{29}$$

Specific volume (i.e., density) measurements are easily made by a variety of techniques, the simplest of which is a density gradient column. The gradient is formed by filling a vertical column several feet in length with a solution of two liquids which differ in density. The liquids are added in batches of differing proportions, so as to produce a concentration gradient and thus a density gradient from one end to another. If good temperature control is maintained, the gradient is stable up to several months, since diffusive mixing rates are low. Calibrated glass beads are dropped into the column and the levels attained are plotted versus height, yielding a calibration curve. The polymer of unknown density is then dropped into the column and the density found from the height to which it settles. An accuracy of 0.0001 g/cc is easily attained. Obviously the polymer must be unaffected by the liquids and voids must be absent. If the densities of the amorphous and crystalline phases are known, W_c can be calculated from formula (20). Often the amorphous density is obtained at a particular temperature by extrapolation of data from above the melting point. The density of the crystal can be evaluated from unit cell dimensions and the atomic weights of the elements. However, the crystallinity cannot be calculated unambiguously by this method, since the interfacial regions and poorly formed small crystallites lack the assumed perfect unit cell density, and the amorphous material that is strained enough to prevent further crystallization will not have the relaxed amorphous density.

Specific heats are determined by adiabatic calorimetry or differential scanning calorimetry (DSC). Statements like those made above about densities can also be made about ambiguities in C_p. When the heats of fusion are used as measures of crystallinity, DSC is the easiest means for obtaining such data. (An example of a DSC trace was shown in Fig. 19a.) The experimental apparatus is as described in the section on glass transition detection (Section IV). Endothermic processes (melting) show up as negative (cooling) ΔT values or extra power consumption departures from the baseline. For exothermic processes (crystallizations), the opposite departures are found. After the transition occurs, the system should return toward the original ΔT or C_p, but the heat capacity of the system is now different and what would have been the baseline in the absence of the transition becomes difficult to determine. Usually, a smooth transition is drawn. The area above this baseline (in Fig. 19a) is proportional to ΔH_f, so that these instruments must be calibrated with standards for ΔH_f. If melting occurs over a very broad region, even the low-temperature end of the transition becomes hard to establish. Since crystallization is often quite slow below the main melting region of a polymer, it is usually possible to cool the sample for a DSC run without causing significant crystallization. One should, therefore, also study the annealing process at a variety of temperatures—in particular, just below a region where a baseline is to be drawn. In view of the crystallite imperfections and large surface areas, all determinations of degree of crystallinity by heat of fusion yield low values compared to the amount measured by density or X ray (see below).

If particular spectroscopic absorptions can be assigned to crystalline and amorphous regions of the polymer, a degree of crystallinity can be found through the use of Eqs. (23) and (24). Equation (24) has been obtained by assuming only that a constant proportionality between the absorbance and concentration of the phase exists over the range of W_c. The ratio Q/P is obtained by plotting ε_a versus ε_c over a range of samples with different W_c. The intercepts at ε_c and $\varepsilon_a = 0$ yield the proportionality constants. Inadequacies of the method arise from changing proportionalities and the neglect of contributions from interfacial material.

The X-ray methods of Eqs. (25)–(28) represent various degrees of approximation or sophistication. When X rays (or electrons, or neutrons, etc.) impinge upon a sample, a fraction of the radiation is scattered. The intensity of the X rays scattered over all angles depends on the number of the scattering centers (whose patterns form the diffraction grating) and their electron densities. The close intervals of interatomic distances give rise to scattering and interference patterns over wide angles (of the order of 10–20°), whereas the larger patterns of the arrangements of crystallites or other domains give rise to small-angle scattering (SAXS). In any case, the total scattering intensity is independent of the state of order of the atoms. Therefore, if crystalline scattering can be separated from the amorphous one, the crystalline fraction is given by the ratio of the integrated crystalline intensity to the total integrated intensity (Fig. 20). A variety of procedures exist for doing this, each with its drawbacks. Scattering from different atoms on an ordered crystal lattice bears a distinct phase relationship, producing reinforcement at particular scattering angles. This gives rise to characteristic diffraction maxima which are easily distinguished from the diffuse scatter from the armophous regions. There are defects in the crystals, however, plus the effects of thermal vibrations, so that part of the scattering from the crystal appears to come from the amorphous part. This problem has been treated by the method of Ruland, represented by Eq. (28). A disorder factor is employed to evaluate how much of the crystal scattering appears as "amorphous scatter."

Equation (27) is less rigorous, but a rather good approximation is achieved in practice if one decomposes the scattering into amorphous and crystalline contributions and simply assumes the scattering to be proportional to the amount of each phase present. Since the proportionality constants may be different, only a portion of the whole angular region need be studied as long as it contains contributions from both amorphous and crystalline phases. Study of samples with varying W_c, or of a completely amorphous (quenched) sample and a crystalline one, permits evaluation of the proportionality constants, as in the method of Hermans and Weidinger [V-2]. The analogy to the ir method mentioned earlier is obvious. Unfortunately, the proportionality constants may change with degree of crystallinity, reflecting a change in the nature (and thus scattering) of the crystalline or amorphous phase [V-18].

Equation (26) is based on the use of amorphous scattering only. Here again, the intensity of the amorphous scatter must be separable from the background and from the crystalline scattering, and a pure amorphous specimen of the polymer must be available for comparison; a knowledge of the sample thickness is important here.

Equation (25) may be used without correction for disorder. This will lead to lower W_c than does Eq. (28). In some cases Eq. (25) is used when only a portion of the total coherent scattering is evaluated. This will cause errors depending on the relative proportions of the neglected scattering, which then can be corrected by using the Hermans–Weidinger method.

All these methods can be employed to study crystallinity in stressed specimens. They will, though, exhibit an orientation of the crystalline (and some strain birefringence in the amorphous) material which will cause the intensity of the scattering to depend on the geometrical orientation of the specimen with reference to the beam and detector. One should then study the intensity as a function of sample orientation to ascertain the degree of orientation of molecules and crystals. The orientation may be presented in the form of pole figures, or as mathematical averages over specific functions of interest, and used to compute the degree of crystallinity by utilizing the intensities to evaluate relationships (25)–(28). For comparison, the sample may be mechanically random-ized relative to the X-ray beam when the experiment is run, so that the diffraction pattern becomes that of an unoriented specimen. Actually, even unstressed specimens should be examined to see whether oriented diffraction intensities are present or not. Birefringence (and thus orientation) may also be measured in a common Abbé refractometer. Important information about crystal size and perfection can be obtained also from X-ray line width studies.

Among other techniques useful for estimates of the degree of crystallinity, Eq. (29) is concerned with NMR line widths. To a reasonable approximation, the line widths for protons (for example) from crystalline phases are broad, since the spin–spin relaxation times T_2 become coupled and mutually interfere when they are in rigid relative positions in a lattice. In the amorphous phase, the signals (line widths) sharpen due to the lack of correlation among the protons. Analyzing the NMR spectrum for the broad and narrow components yields fractions of protons, and thus of the phases. This same line width phenomenon is used to detect the onset of molecular motion in the amorphous phase corresponding to glass–rubber transitions. Lastly, relationships which rely on the refractive indices of amorphous and crystalline phases which differ can be set up on either an absolute or an empirical basis by correlation with other methods.

VI. Morphology (Table VI)

The morphology of a material is its organization on a supramolecular scale, i.e., the form, size, and orientation of its crystallites, domains, the structure of groups of molecules in the specimen and of their boundaries, and the degree of crystallinity. For crystals, one is concerned with the details of habit and perfection, how the polymer chains are incorporated and may form links between the crystals. The amorphous or noncrystalline material present is characterized with respect to its density and density fluctuations, orientation, and degree of disorder on a microscale. In the presence of filler particles or of finely divided domains of other components, their size, dispersion, and distribution, too, can have a profound effect upon the physical properties of the total material. For example, as seen in Fig. 21, the anisotropy (e.g., length/diameter) of crystals alone can give vast differences in the modulus of a material. For specimens which undergo crystallization when stressed, and thus exhibit high strength even in an unreinforced state (e.g., natural rubber, butyl) the nature of the strain-induced anisotropies and/or of the corresponding domains is of particular importance. Many of the thermoplastic elastomers have crystallizable blocks or segments which associate to form a hard phase; noncrystallizable segments of high T_g also may form islands of a hard, glassy character. Polymers containing polar (e.g., ionic) groups which tend to form separate phases are another example where morphological studies are important. Even pure

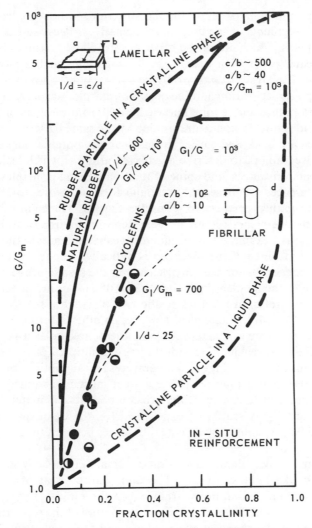

Fig. 21. Experimental relationships between normalized shear stiffness G and volume fraction of the crystalline reinforcing phase for natural rubber, crystalline polymers, and model in situ grown composites of crystallized acetanilide in rubber (cricles; ◕, $T = 0°$; ◑. $T = 25°$; ●, $T = 40°$). Different crystallization temperatures imply different morphologies; a, b, c, d, and l denote the dimensions of the embedded crystallites. (J. Halpin and J. Kardos, *J. Appl. Physics* **43**, 2235 (1972).)

amorphous homopolymers may contain structures which are considerably larger than the segments of the molecules themselves. These so-called nodules or denser domains, which are of the order of 50–150 Å, presumably represent regions of greater order than that in the homogeneous matrix (see Table VI for references). The structures may only be surface effects however. Thus, the nature of the phase separation and the details of morphology have considerable effect on the properties and require knowledge of the morphology of the system.

All the various morphologies are studied by the same battery of techniques, many of which are discussed earlier in this chapter and elsewhere in the text. At this point we attempt to describe a general outline of which techniques are used in particular instances, and some of their underlying principles. The techniques and references are presented in tabular form as before.

In order to gather perspective, consider Fig. 22. Depending on what is known about the chemical structure of the system and how many similar systems have been previously examined, some judgment can be made as to the techniques to be used. For instance, if the specimen is optically clear to the unaided eye, either it is homogeneous, or the particle sizes in the system are less than $\frac{1}{10}$ of the wavelength of the incident light (less than 0.05 μm) or are in very low concentration, or the refractive indices of the two phases are so close that light scattering is nil [VI-14b]. If chemical analyses or other information leaves no indication that the system is heterogeneous, then one can let it go at that. If, however, DSC data reveal melting endotherms or multiple T_g's, or if mechanical property data taken over a range of temperatures reveal multiple moduli or loss tangent maxima, then a multiple-phase system must be assumed. Since it is unlikely that the electron density of the two components will match, one should run a small-angle X-ray scattering experiment (on the clean specimen) where scattering will be determined by density differences between the morphological units.

For systems of low volume concentration, small-angle X-ray scattering can be treated just like light scattering. The same particle scattering factor will apply (under certain assumptions) from which the radius of gyration of the scattering centers can be determined. From the nature of the scattering pattern, information about the shape of the particles and their polydispersity can be determined. Since X rays are of higher energy and thus shorter wavelength than light, information about smaller particles can be obtained than from light scattering. For scattering centers of high concentration, periodic maxima may occur in the scattering function which may be related to inter- or intraparticle dimensions. Similar behavior is found with other forms of radiation. The interpretation of scattering from concentrated systems is always less rigorous. Even homogeneous polymers will scatter some light and some X rays. This scattering arises from normal fluctuations whose range and average can be calculated by the methods of statistical mechanics from a knowledge of the overall material compressibility.

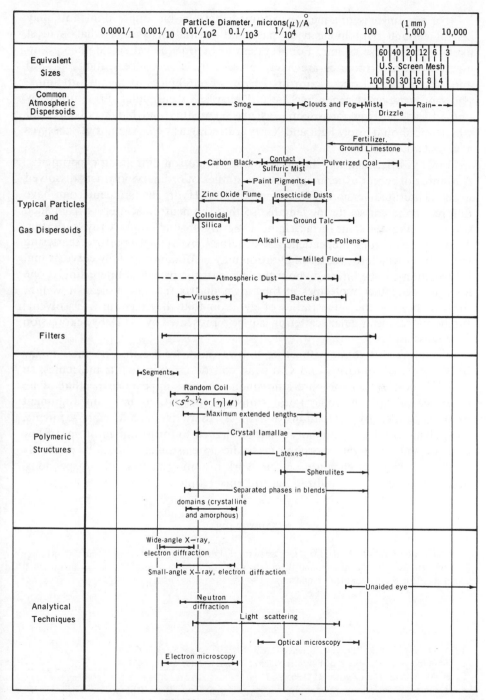

Fig. 22. Characteristics of particles.

If a specimen is opaque, microtomed sections or simply thin films may permit enough light intensity to pass and permit analysis of what is most likely a multiphase system. For a system to be opaque yet single phase, conjugated olefine structures are likely present in large concentration. Optical and phase-contrast microscopy are obvious first choices to examine the morphology however. Carbon blacks and blends, in particular, are studied by optical and electron microscopy, as are segmented and block polymers, to which one should apply light and X-ray scattering as well. Again, Fig. 22 serves as a first guide.

Finally, brief mention should be made of neutron diffraction experiments. A number of recent experiments involve studies of ordinary polymers dissolved in the deuterated version of the same polymer. Hydrogen and deuterium have different coherent scattering lengths, so that the neutron scattering from such solutions takes the same form as light scattering or small-angle X-ray scattering. That is, the angular dependence is governed by a single particle scattering factor from which the radius of gyration may be obtained and, by curve fitting, information can be inferred about the shape of the particles. Interestingly, one finds that polymer molecules in their own matrix (no low molecular weight diluent) possess the same radius of gyration that they have in a Θ solvent; furthermore, the angular scattering can be represented by the Debye expression for Gaussian coils.

Coupling this result with other known data, we see that polymer in the bulk normally conforms to a Gaussian coil description for its molecules, so that the best model for pure amorphous rubbers appears to be that of an ensemble of random coils. Local transient correlations in chain alignment exist, as do density fluctuations, but unless there are crystallizable segments, the polymer molecules are not folded or aligned to form configurations of low entropy. It follows that morphology studies in elastomers should be primarily directed at signs of chemical and network inhomogeneities, filler dispersions, and strain-induced crystallizations and anisotropies.

Acknowledgment

Thanks are due to F. P. Baldwin, C. Cozewith, I. Duvdevani, I. J. Gardner, and J. J. Maurer for suggestions, helpful criticism, and their help with organizing the text. Thanks go also to Mrs. Carol Baus for typing a difficult manuscript.

General References

1. A. D. Jenkins, ed., "Polymer Science." Am. Elsevier, New York, 1972.
2. F. Billmeyer, "Textbook of Polymer Science." Wiley, New York, 1971.
3. H. Morawetz, *High Polym.* **21** (1965).
4. P. Flory, "Principles of Polymer Chemistry." Cornell Univ. Press, Ithaca, New York, 1953.

5. H. Yamakawa, "Modern Theory of Polymer Solutions." Harper, New York, 1971.
6. D. Van Krevelen, "Properties of Polymers." Elsevier, Amsterdam, 1972.
7. E. Collins, J. Bares, and F. Billmeyer, "Experiments in Polymer Science." Wiley, New York, 1973.
8. A. Tobolsky and H. Mark, eds., "Polymeric Science and Materials." Wiley, New York, 1971.
9. C. Wadelin and M. Morris, *Anal. Chem.* **49,** 133R (1977).
10. J. Cobbler and C. Chow, *Anal. Chem.* **49,** 159R (1977).
11. F. Sliemers and K. Boni, eds., *J. Polym. Sci., Part C* **43** (1974).
12. H. Cantow, "Polymer Fractionation." Academic Press, New York, 1967.
13. D. Allport and W. James, "Block Copolymers." Wiley, New York, 1973.
14. M. Ezrin, ed., *Adv. Chem. Ser.* **125** (1973).
15. A. Weissberger and B. Rossiter, eds., "Physical Methods of Chemistry." Wiley (Interscience), New York, 1971. (This is a comprehensive series with additional volumes being added; it should always be consulted for basic techniques).
16. F. P. Baldwin, G. Ver Strate, *Rubber Rev.* **44,** 709 (1972).
17. S. Cesca, *J. Polym. Sci. Macromol. Rev.* **10,** 1 (1975).
18. W. Graessley, *Adv. Pol. Sci.* **16,** 1 (1974).
19. P. J. Flory, "Statistical Mechanics of Chain Molecules." Wiley, New York, 1971.
20. H. G. Elias, "Macromolecules" Vols. 1, 2. Plenum, New York, 1977.
21. J. Brandrup and E. Immergut, eds. "Polymer Handbook" 2nd ed. Wiley, New York, 1975.
22. J. Green and R. Dietz, eds., "Industrial Polymers: Characterization by Molecular Weight," Transcrypta, London, 1973.

TABLE 1 STRUCTURE: CHEMICAL COMPOSITION

Characterization technique[a]	Variables measured	Symbol (and unit)	Principle of operation	References
Elemental analysis	Fraction of particular element in a polymer	Number of atoms/ base unit (weight fraction, or percent)	Varied, usually involves destruction of sample into fragments which are analyzed	1–8
Absorption spectroscopy				
Infrared (ir), $1-16\mu m^c$	Wave number Intensity of absorption	n (cm^{-1})^b or v A (none)	Measures characteristic energies of bond deformations by their location in the spectrum	8–18
Raman	Wave number Intensity of emission	n (cm^{-1})	Same as infrared	13, 19a, 19b, 32
Nuclear magnetic resonance H (NMR) $5 \times 10^7 \ \mu m$	Chemical shift, coupling constant Intensity of absorption	δ (ppm), J (Hz) (Arbitrary)	Measures characteristic energies of magnetic moment transitions	8a, 9, 20–22, 25
^{13}C	Chemical shift Intensity of absorption		Same as H except ^{13}C nucleus transitions	23, 24
Ultraviolet (uv) $0.2-0.35\mu m$	Wave number Intensity of absorption	n (cm^{-1}) A	Measures characteristic energies of electronic transitions	8a, 9, 26, 27
Visible $0.35-0.8 \ \mu m$	Wave number Intensity of absorption	n (cm^{-1}) A	Same as ultraviolet	8a, 9, 27, 28
Functional group analysis	Acid, hydroxyl, olefins, amine, isocyanate, etc. concentration	Fraction, percent, or concentration	Depends on known reactions of the particular type of functionality	5, 29
Pyrolysis				
Gas chromatography	Small molecules and oxidation products indicative of macromolecular composition		Sample is pyrolyzed with chromatographic or mass spectrographic identification of characteristic fragments.	30
Mass spectroscopy				8, 9, 32, 33
Electron paramagnetic resonance (EPR) $10^5 \ \mu m$	Position of resonance Intensity of absorption Hyperfine splitting constants	g a_x (G)	Measures characteristic energies of unpaired electron spin transitions which are a function of the chemical environment	8a, 9, 31, 34

[a] Other techniques discussed in general references [9, 10] and Table I [32].
[b] Absorption location can be wavelength λ (μm), wave number $n = 1/\lambda$ (cm^{-1}), or frequency Hz = (c/λ).   [c] Range of wavelength applicable.

REFERENCES FOR TABLE I

1. T. S. Ma and M. Gutterson, *Anal. Chem.* **44,** 445R (1972).
2. D. J. Pasto and C. Johnson, "Organic Structure Determination," pp. 221ff. Prentice-Hall, Englewood Cliffs, New Jersey, 1969.
3. General references 2, 7, 9, 10.
4. D. Hercules and J. Carrer, *Anal. Chem.* **46,** 133R (1974); see also Gen. ref. 7.
5. F. Weiss, *Chem. Anal. Ser. Monogr. Anal. Chem. Appl.* **32,** (1970).
6. Annual Book of ASTM Standards (see particular tests).
7. J. Winefordner and T. Vickers, *Anal. Chem.* **46,** 192R (1974); see also general reference 9.
8a. W. Tyler, *Rubber Rev.* **40,** 238 (1967).
8b. P. Griffiths, "Chemical Infrared Fourier Transform Spectroscopy." Wiley, New York, 1975.
9. General references 1, 2, 7, 9, 10.
10a. R. S. McDonald, *Anal. Chem.* **46,** 521R (1974).
10b. E. Knözinger, *Angew. Chem. Int. Ed. Engl.* **15,** 25 (1976).
11. L. J. Bellamy, "Infrared Spectra of Complex Molecules." Wiley, New York, 1958.
12a. L. J. Bellamy, "Advances in Infrared Group Frequencies." Barnes & Noble, New York, 1968.
12b. E. Crandall and A. Jagtap, *J. Appl. Polym. Sci.* **21,** 449 (1977).
13. F. J. Boerio and J. Koenig, *J. Macromol. Sci., Rev. Macromol. Chem.* **7,** 209 (1972).
14a. R. Hampton, *Rubber Rev.* **45,** 546 (1973).
14b. W. Hart, C. Painter, J. Koenig, and M. Coleman, *Appl. Spectros.* **31,** 220 (1977).
15. J. Henniker, "Infrared Spectrometry of Industrial Polymers." Academic Press, New York, 1967.
16. D. Hummel, "Infrared Spectrometry of Polymers," Wiley (Interscience), New York, 1966.
17. R. Zbinden, "Infrared Spectroscopy of High Polymers." Academic Press, New York, 1964.
18. D. Hummel and F. Sholl, "Infrared Analysis of Polymers." Halsted, New York, 1969.
19a. B. Schrader, *Angew. Chem., Int. Ed. Engl.* **12,** 884 (1973).
19b. M. Coleman, J. Shelton, and J. Koenig, *Ind. Eng. Chem., Prod. Res. Dev.* **13,** 154 (1974).
20. L. M. Jackman and S. Sternhell, "Applications of Nuclear Magnetic Resonance Spectroscopy in Organic Chemistry." Pergamon, Oxford, 1969.
21. P. Corio, S. Smith, and J. Wasson, *Anal. Chem.* **46,** 314R (1974).
22. F. Bovey, *Prog. Polym. Sci.* **3,** 3 (1971).
23. V. D. Mochel, *J. Macromol. Sci., Rev. Macromol. Chem.* **8,** 289 (1972).
24. G. C. Levy and G. L. Nelson, "Carbon-13 NMR for Organic Chemists." Wiley, New York, 1972.
25. F. Bovey, "High Resolution NMR of Macromolecules." Academic Press, New York, 1972.
26. R. Hummel and D. Kaufman, *Anal. Chem.* **46,** 354R (1974).
27. F. Grum, *in* "Visible and Ultraviolet Spectrophotometry" (A. Weissberger and B. Rossiter, eds.), Physical Methods of Chemistry, Part IIIB. Vol. 1, p. 207. Wiley (Interscience). New York, 1972.
28. D. Boltz and M. Mellon, *Anal. Chem.* **46,** 227R (1974).
29. W. T. Smith and J. Patterson, *Anal. Chem.* **46,** 394R (1974).
30. General references 9, 10 (see references therein).
31. E. Janzen, *Anal. Chem.* **46,** 378R (1974).
32. J. Petrazzi, ed., *Anal. Chem. Fundamental Revs.* **40,** (1976).
33. D. Braun and E. Canji, *Angew. Makromol. Chem.* **36,** 75 (1974).
34. W. Potter and G. Scott, *Eur. Polym. J.* **7,** 489 (1971).

TABLE IIA STRUCTURE: COMPOSITIONAL DISTRIBUTION

Characterization technique	Variables measured	Symbol (and unit)	Principle of operation	References[a]
Fractionation:				
Precipitation	Composition, molecular weight of fractions		Different composition and molecular weight polymer components have different solubility.	1–4, 7
Solution	Composition, molecular weight of fractions	Composition of components must be measured after separation by these techniques.	Different composition and molecular weight polymer components have different solubility.	1–4, 7
Gel permeation chromatography (GPC)	Composition, molecular weight of fractions		Different size molecules are separated on the basis of exclusion from pores; uv, ir, etc. detectors on column show compositional distribution.	5, 12, 15, IIIA-75c
Thermal diffusion	Composition, molecular weight of fractions		A gradient in concentration and molecular weight is set up by a thermal gradient.	10, 11
Isothermal diffusion	Diffusion coefficients of components	D (cm^2/sec)	Diffusion coefficients depend on molecular weight and composition.	14, 15
Sedimentation	Sedimentation and diffusion coefficients, density gradient equilibrium	S (sec) D (cm^2/sec)	Sedimentation and diffusion coefficients depend on molecular weight and composition.	5a, 5b
Thin layer chromatography (TLC)	Ratio of distance moved by polymer component relative to solvent front, must be calibrated.	R_f (0 to 1)	Separations based on differences in diffusion and adsorption through a particular matrix	6, 8, 9, 12
Light scattering (LS)	Apparent molecular weights Heterogeneity parameter	$\overline{M}_w$ Q	Apparent $\overline{M}_w$ depends on the solvent refractive index, which is related to compositional heterogeneity of the polymer.	1, 13

[a] Additional references for these techniques appear in Table III.

REFERENCES FOR TABLE IIA

1. General references 1, 2, 9, 10, 12, 13, 21, 22.
2. L. H. Tung, *J. Macromol. Sci., Rev. Macromol. Chem.* **7,** 51 (1971).
3. W. V. Smith, *Rubber Rev.* **45,** 667 (1972).
4. G. Kegeles *et al.*, eds., "Characterization of Macromolecular Structure." Natl. Acad. Sci., Washington, D.C., 1968.
5a. General references 1, 2, 7, 9–13, 15, 21, 22.
5b. C. Stacy, *J. Appl. Polym. Sci.* **27,** 2231 (1977).
6. J. G. Kreiner, *Rubber Rev.* **44,** 381 (1971).
7. S. Teramachi and Y. Kato, *Macromolecules* **4,** 55 (1971).
8. T. Kotaka and J. White, *Macromolecules* **7,** 106 (1974).
9. E. P. Otocka, *Adv. Chem. Ser.* **125,** 55 (1973).
10. F. Gaeta and A. Di Chiara, *J. Polym. Sci.* **13,** 163, 177, 203 (1975).
11a. J. Giddings, Y. H. Yoon, and M. Myers, *Anal. Chem.* **47,** 126 (1975).
11b. J. Giddings, L. Smith, and M. Myers, *Anal. Chem.* **48,** 1587 (1976).
12. J. Profivova, J. Pospisil, and J. Holcik, *J. Chromatogr.* **92,** 361 (1974).
13. T. Chau and A. Rudin, *Polym. J.* **15,** 593 (1973).
14. General references 1, 12, 15.
15. J. Anderson, *J. Appl. Polym. Sci.* **18,** 2819 (1974).

TABLE IIB

STRUCTURE: MONOMER SEQUENCE DISTRIBUTION, TACTICITY, AND CONFORMATIONS

Characterization technique	Variables measured	Symbol (and unit)	Principle of operation	References
Infrared (ir)	Intensity of absorption	A	Arrangement and geometry of units	1–3
	Frequency	n (cm^{-1})	affects spectra.	
Nuclear magnetic resonance	Chemical shift	δ (ppm)	Arrangement and geometry of units	1–3, 13, 14
(NMR)	Intensity of absorption	Area (arbitrary units)	affects spectra.	
	Coupling constants	J (Hz)		
Pyrolysis–gas chromatography	Size of fragments	g/mole	Arrangement and chemical structure	4–8
(GC)			of units affects types of fragments	
Infrared	Structure of fragments		produced.	
Mass spectroscopy (MS)	Size of fragments			
Dipole moments	Mean square dipole moment	$\langle u^2 \rangle$ (D^2)	Geometry and sequence distribution	9
			affects dipole moments.	
Light scattering	Mean square radius of gyration	$\langle S^2 \rangle$ (cm^2)	Geometry and sequence distribution	10–12
	Characteristic ratio	C_∞	affects polymer dimensions.	
Viscometry	Characteristic ratio	C_∞	Geometry and sequence distribution	10–12
			affects polymer dimensions.	

REFERENCES FOR TABLE IIB

1. General references 1, 2, 7, 9, 10.
2. T. Shimanouchi, "NMR Basic Principles and Progress" (P. Diehl, ed.), Vol. 4, p. 287. Springer-Verlag, 1971.
3. F. A. Bovey, "High Resolution NMR of Macromolecules." Academic Press, New York, 1972.
4. General references 7, 9, 10.
5. L. Michajlov, P. Zugenmaier, and H. Cantow, *Polymer* **9**, 6325 (1968).
6. P. E. Slade and L. T. Jenkins, "Thermal Characterization Techniques," Ch. 2. Dekker, New York, 1970.
7. H. Kuo, H. Pfeffer, and J. Gillham, *Am. Chem. Soc., Div. Org. Coat. Plast. Chem., Pap.* **35,** 434 (1975).
8. R. Beimer, *Am. Chem. Soc., Div. Org. Coat. Plast. Chem., Pap.* **35,** 428 (1975).
9. J. E. Mark, *Acc. Chem. Res.* **7,** 218 (1974).
10. General references 1–5, 13.
11. P. Flory, "Statistical Mechanics of Chain Molecules." Wiley (Interscience), New York, 1969.
12. S. Bruckner, G. Allegra, G. Gianotti, and G. Moroglio, *Eur. Polym. J.* **10,** 347 (1974).
13. C. Carman, R. Garrington, and C. Wilkes, *Macromolecules* **10,** 536 (1977).
14. J. Randall, "Polymer Sequence Determination." Academic Press, New York, 1977.

TABLE IIIA

STRUCTURE: MOLECULAR WEIGHT AND ITS DISTRIBUTION

Characterization technique	Variables measured	Symbol (and unit)	Principle of operation	References
Membrane osmometry	Number average molecular weight Virial coefficient	$\overline{M}_n$ (g/mole) A_i (mole cc/g^2)^a	Thermodynamic potential for mixing is measured by separating solvent and solution by membrane "impermeable" to polymer, a colligative property.	1–5b
Vapor pressure osmometry (VPO)	Number average molecular weight	$\overline{M}_n$ (g/mole)	Vapor pressure lowering measured by a dynamic evaporation technique, needs calibration	6–10
Cryoscopy	Number average molecular weight	$\overline{M}_n$ (g/mole)	Freezing point depression colligative property	11
Ebullometry	Number average molecular weight	$\overline{M}_n$ (g/mole)	Boiling point elevation, colligative property	12–14
Light scattering (LS)	Weight average molecular weight Virial coefficient Radius of gyration	$\overline{M}_w$ (g/mole) A_i (mole cc/g^2)^a $\langle S^2 \rangle_z$ (cm^2) $\langle R_G \rangle_z$ (cm^2)	Light scattered by a polymer solution is related to the molecular weight of the solute. Angular dependence of scattering is related to particle size (0.05–0.2 μm).	15–31c
Intrinsic viscosity	Intrinsic viscosity Viscosity average molecular weight Estimates of MWD breadth	$[\eta]$ (dl/g) $\overline{M}_v$ (g/mole) $\overline{M}_w$ (g/mole)	Ability of a polymer to increase the viscosity of a solution depends on polymer molecular weight.	32–43b
Ultracentrifugation Equilibrium	$\overline{M}_w$ and other molecular weight averages	$\overline{M}_w$, $\overline{M}_z$ (g/mole)	Concentration gradient in a large gravitational field is related to molecular weights.	44–50

Equilibrium in density gradient	General heterogeneity	$\overline{M}_w$, $\overline{M}_z$	Concentration gradient in a large gravitational field is related to molecular weights.	49
	Molecular weight averages	$\overline{M}_w$, $\overline{M}_z$		
Sedimentation	Molecular weights		Rates of diffusion in a large gravitational field are related to molecular weights.	44–50
	Sedimentation coefficient	S (sec)		
Terrestrial sedimentation	Molecular weight		Polymer settles under the influence of gravity near the critical point, or if of large enough size in normal solutions.	51, 90, 91
Gel permeation chromatography	Average molecular weights	$\overline{M}_v$, $\overline{M}_w$, $\overline{M}_n$	Permeation of polymer into a porous structure depends on molecular weight.	52–75c
	Shape of distribution			
Fractionation by solubility	Molecular weight averages if measurements are made on fractions	$\overline{M}_w$, $\overline{M}_n$ (g/mole)	Solubility of polymer depends on molecular weight and composition.	76–84
	Shape of distribution			
Turbidimetry or cloud point measurement	Cloud point temperature	T_c (°C)	Phase separation temperatures depend on molecular weight.	85–91
	Spinodals, concentration of precipitated material			
	Molecular weight	$\overline{M}_w$		
Diffusion	Diffusion coefficients	D (cm^2/sec)	Diffusion coefficients depend on molecular weight.	92–95
Gel–Sol analysis of cross-linked sample	Gel fraction	W_g gel fraction	Gel point and partition between gel and sol depends on molecular weight and MWD.	96–104
	Sol fraction	S sol fraction		
Bulk viscosity	Viscosity	η (poise)	Empirical correlation of viscosity with molecular weight	104–112
Filtration	Gel fraction	W_g		17, 113–115
Comparisons of methods				116–124

---

[a] Changes depending on form of equation; this unit corresponds to the equation in the text.

REFERENCES FOR TABLE IIIA

1. General references 1–5, 7, 9, 10, 12, 13, 15, 21, 22.
2. J. Brown and P. Verdier, *J. Res. Natl. Bur. Stand., Sect. A* **76,** 161 (1972).
3a. H. Vink, *Eur. Polym. J.* **10,** 149 (1974).
3b. A. Sikora, *Makromol. Chem.* **176,** 3501 (1975).
4. H. Coll and F. Stross, *in* "Characterization of Macromolecular Structure," p. 10. Natl. Acad. Sci., Washington, D.C., 1968.
5a. H. Elias, *in* "Characterization of Macromolecular Structure," p. 28. Natl. Acad. Sci., Washington, D.C., 1968.
5b. C. Strazielle and R. Duck, *Makromol. Chem.* **142,** 241 (1971).
6. General references 1, 2, 7, 10, 15.
7. J. Brzezinski, H. Glowala, and A. Karnas-Calka, *Eur. Polym. J.* **9,** 1251 (1973).
8a. B. Bersted, *J. Appl. Polym. Sci.* **17,** 1415 (1973).
8b. B. Bersted, *J. Appl. Polym. Sci.* **18,** 2399 (1974).
8c. G. Ver Strate, unpublished results (1972).
8d. K. Kamide, T. Terakawa, and H. Uchiki, *Makromol. Chem.* **177,** 1447 (1976).
8e. C. Morris, *J. Appl. Polym. Sci.* **21,** 435 (1977).
9. B. Hudson, *Polym. Prepr., Am. Chem. Soc., Div. Polym. Chem.* **12,** 259 (1971).
10. J. Van Dan, *in* "Characterization of Macromolecular Structure," p. 336. Natl. Acad. Sci., Washington, D.C., 1968.
11. General references 1–5, 7, 9, 10, 14, 15, 21, 22.
12. P. Parrini and M. Vacanti, *Makromol. Chem.* **175,** 935 (1974).
13. M. Ezrin, *in* "Characterization of Macromolecular Structures," p. 3. Natl. Acad. Sci., Washington, D.C., 1968.
14. General references 1–5, 7, 9, 10, 14, 15.
15. General references 1–5, 7, 9, 10–13, 14, 15.
16. J. Kratohvil, *in* "Characterization of Macromolecular Structure," p. 59. Natl. Acad. Sci., Washington, D.C., 1968.
17. M. Kerker, "The Scattering of Light and Other Electromagnetic Radiation." Academic Press, New York, 1969.
18. E. Casassa, *Polym. J.* **3,** 517 (1972).
19. L. Frolen, G. Ross, A. Wims, and P. Verdier, *J. Res. Natl. Bur. Stand., Sect. A* **76,** 156 (1972).
20. J. Lorimer, *Polymer* **13,** 52 (1972).
21. H. Wagner, *J. Res. Natl. Bur. Stand. Sect. A* **76,** 151 (1972).
22. W. Bahary and L. Bsharah, *J. Polym. Sci., Part A-1* **6,** 2819 (1968).
23. A. R. Shultz and W. H. Stockmayer, *Macromolecules* **2,** 178 (1969).
24. Von H. Lange, *Kolloid Z. Z. Polym.* **240,** 747 (1970).
25. G. Greschner, *Makromol. Chem.* **170,** 203 (1973).
26. T. Wallace, M. Volosin, R. Delunyea, and A. Gingello, *J. Polym. Sci., Part A-2* **10,** 193 (1972).
27. P. Mijnlieff and D. Coumori, *J. Colloid Interface Sci.* **27,** 553 (1968).
28. I. Serdyuk and S. Grenader, *J. Polym. Sci., Part B* **10,** 241 (1972).
29. J. Prudhomme and Y. Sicotte, *J. Colloid Interface Sci.* **27,** 547 (1968).
30. G. Miller, F. Filippo, and D. Carpenter, *Macromolecules* **3,** 125 (1970).
31a. K. Kajiwara and M. Gordon, *J. Chem. Phys.* **59,** 3623 (1973).
31b. R. Arnett, *J. Phys. Chem.* **79,** 85 (1975).
31c. V. Morris, H. Coles, and B. Jennings, *Nature (London)* **249,** 340 (1974).
32. General references 1–12, 14, 15.
33. K. Osaki, *Macromolecules* **5,** 141 (1972).
34. K. Osaki, *Fortschr. Hochpolym.-Forsch.* **12,** 1 (1973).
35. N. Tschoegl, *Polym. Prepr., Am. Chem. Soc., Div. Polym. Chem.* **15,** 7 (1974).

36. W. R. Moore, *Prog. Polym. Sci.* **1**, 1 (1967).
37. R. Christensen, *J. Res. Natl. Bur. Stand., Sect. A* **76**, 147 (1972).
38a. A. Rudin, G. Strathdee, and W. Edey, *J. Appl. Polym. Sci.* **17**, 3085 (1973).
38b. J. Freire, A. Horta, I. Katime, and J. Figueruelo, *J. Chem. Phys.* **65**, 2867 (1976).
39. N. Yamaguchi, Y. Sugiura, K. Okano, and E. Wada, *J. Phys. Chem.* **75**, 1141 (1971).
40. A. Rudin, G. Bennett, and J. McLaren, *J. Appl. Polym. Sci.* **13**, 2371 (1971).
41. R. Koningsveld and C. Tuihnman, *Makromol. Chem.* **38**, 39 (1960).
42. R. Harrington and B. H. Zimm, *J. Phys. Chem.* **69**, 161 (1964).
43a. D. McIntyre, *Polym. Prepr., Am. Chem. Soc., Div. Polym. Chem.* **16**, 22 (1975).
43b. F. Wang, *Polym. Prepr., Am. Chem. Soc., Div. Polym. Chem.* **16**, 701 (1975).
44. General references 1–5, 9, 10, 12, 14, 15, 21, 22.
45a. H. Fujita, *J. Phys. Chem.* **73**, 1759 (1969).
45b. D. Soucek and E. Adams, *J. Colloid Interface Sci.* **55**, 571 (1976).
46a. H. Utiyama, N. Tazata, and M. Kurata, *J. Phys. Chem.* **73**, 1448 (1969).
46b. V. Vosicky and M. Bohdanecky, *J. Polym. Sci. Polym. Phys. Ed.* **15**, 757 (1977).
47. W. Bengough and G. Grant, *Eur. Polym. J.* **7**, 203 (1971).
48. J. Henderson, J. Hulme, R. Small, and H. Williams, *Rubber Chem. Technol.* **38**, 817 (1965).
49. J. J. Hermans and H. Ende, *in* "Newer Methods of Polymer Characterization" (B. Ke, ed.), p. 525. Wiley (Interscience), New York, 1964.
50. E. Adams, *in* "Characterization of Macromolecular Structure," p. 84. Natl. Acad. Sci., Washington, D.C., 1968.
51. B. Wolf, *Makromol. Chem.* **161**, 277 (1972).
52. General references 1, 2, 7, 9–11, 13–15, 21, 22.
53. J. F. Johnson and R. Porter, *Prog. Polym. Sci.* **2**, 201 (1970).
54. S. Abbott, *Amer. Lab.* **9**, 41 (1977).
55a. D. Bly, *J. Polym. Sci., Part B* **9**, 401 (1971).
55b. E. F. Casassa, *Macromolecules* **9**, 182 (1976).
56a. E. F. Casassa and Y. Tagami, *Macromolecules* **2**, 14 (1969).
56b. R. Kelley and F. Billmeyer, *Anal. Chem.* **42**, 399 (1970).
57a. C. DeLigny and W. Hammers, *J. Chromatogr.* **141**, 91 (1977).
57b. J. V. Dawkins and G. Taylor, *Polymer* **15**, 687 (1974).
57c. A. Cooper and A. Bruzzone, *J. Polym. Sci., Part A-2* **11**, 1423 (1973).
58. P. James and A. Ouano, *J. Appl. Polym. Sci.* **17**, 1455 (1973).
59. M. Ambler, *J. Polym. Sci., Part A-1* **11**, 191 (1973).
60a. G. Ross and L. Frolen, *J. Res. Natl. Bur. Stand., Sect. A* **76**, 163 (1972).
60b. L. Fetters, *J. Appl. Polym. Sci.* **20**, 343 (1976).
61. A. Spatorico and B. Coulter, *J. Polym. Sci., Part A-2* **11**, 1139 (1973).
62. J. Chuang and J. Johnson, *J. Appl. Polym. Sci.* **17**, 2123 (1973).
63. H. E. Adams *et al.*, *J. Appl. Polym. Sci.* **17**, 269 (1973).
64. A. Cooper, J. Johnson, and A. Brazzone, *Eur. Polym. J.* **9**, 1381 (1973).
65. J. Evans, *Polym. Eng. Sci.* **13**, 401 (1973).
66. Y. Kato, S. Kido, and T. Hashimoto, *J. Polym. Sci., Part A-2* **11**, 2329 (1973).
67. A. Rudin and H. Hoegy, *J. Polym. Sci., Part A-1* **10**, 217 (1972).
68a. B. A. Whitehouse, *Macromolecules* **4**, 463 (1971).
68b. J. Dawkins and M. Hemming, *J. Appl. Polym. Sci.* **19**, 3107 (1975).
69a. J. V. Dawkins, *Br. Polym. J.* **4**, 87 (1972).
69b. J. V. Dawkins and M. Hemming, *Makromol. Chem.* **176**, 1777 (1975).
70. R. C. Williams, J. A. Schmit, and H. Suchan, *J. Polym. Sci., Part B* **9**, 413 (1971).
71. A. Ouano, *J. Polym. Sci., Part A-1* **10**, 2169 (1972).
72. J.-Y. Chuang, A. Cooper, and J. Johnson, *J. Polym. Sci., Polym. Symp.* **43**, 291 (1973).
73. A. Ouano, *J. Polym. Sci., Polym. Symp.* **43**, 299 (1973).

74a. K. Unger, R. Keru, M. Ninou, and K. Krebs, *J. Chromatogr.* **99**, 435 (1974).

74b. W. Smith, *J. Appl. Polym. Sci.* **18**, 925 (1974).

74c. A. Ram and J. Miltz, *Polym. Plast. Technol. Eng.* **4**, 23 (1971).

74d. G. Longman, G. Wignall, M. Hemming, and J. Dawkins, *Colloid Polym. Sci.* **252**, 298 (1974).

74e. S. Yozka and M. Kubin, *J. Chromatogr.* **139**, 225 (1977).

74f. W. You, H. Stoklosa, and D. Bly, *J. Appl. Polym. Sci.* **21**, 1911 (1977).

74g. L. Marais, Z. Gallot, and H. Benoit, *J. Appl. Polym. Sci.* **21**, 1955 (1977).

74h. S. Mori, *J. Appl. Polym. Sci.* **20**, 2157 (1976).

74i. S. Mori, *J. Appl. Polym. Sci.* **21**, 1921 (1977).

74j. J. Janca and S. Pokorny, *134*, 263, 273 (1977).

74k. D. Freeman and I. Poinescu, *Anal. Chem.* **49**, 1183 (1977).

74l. S. Mori, R. Porter, and J. Johnson, *Anal. Chem.* **46**, 1599 (1974).

74m. A. Krishen and R. Tucker, *Anal. Chem.* **49**, 898 (1977).

74n. W. Dark, R. Limpert, and J. Carter, *Polym. Eng. Sci.* **15**, 831 (1974).

74o. J. Giddings, L. Bowman, and M. Myers, *Anal. Chem.* **49**, 243 (1977).

74p. P. Szewczyk, *Polymer* **17**, 90 (1976).

75a. G. Kegeles *et al.*, *in* "Characterization of Macromolecular Structure," pp. 259ff. Natl. Acad. Sci., Washington, D.C., 1968.

75b. A. Ouano and W. Kaye, *J. Polym. Sci., Polym. Chem. Ed.* **12**, 1151 (1974).

75c. M. Hoffman and H. Urban, *Makromol. Chem.* **178**, 2683 (1977).

76. General references 1–4, 6, 9–13, 15.

77. R. Koningsveld, *Fortschr. Hochpolym.-Forsch.* **7**, 1 (1970).

78a. J. Welch and U. Bloomfield, *J. Polym. Sci., Part A-2* **11**, 1855 (1973).

78b. K. Kamide, Y. Miyazaki, and T. Abe, *Polym. J.* **9**, 395 (1977).

79a. K. Kamide and K. Yamaguchi, *Makromol. Chem.* **167**, 287 (1973).

79b. Y. Migazaki and K. Kàmide, *Polym. J.* **9**, 61 (1977).

80. S. Teramachi, *J. Macromol. Sci., Chem.* **6**, 403 (1972).

81. R. Koningsveld, W. Stockmayer, J. Kennedy, and L. Kleintjens, *Macromolecules* **7**, 73 (1974).

82. G. Allen, R. Koningsveld, and G. Molau, *in* "Characterization of Macromolecular Structure," pp. 155ff. Natl. Acad. Sci., Washington, D.C., 1968.

83. S. Teramachi and T. Fukao, *Polym. J.* **6**, 532 (1974).

84. E. Schroeder and C. Hannemann, *Wiss. Z. Tech. Hochsch. Chem.* "Carl Schorlemmer" Leuna-Merseburg, **16**, 364 (1974).

85. General references 7, 15.

86. M. Hosono, S. Sugii, O. Kusudo, and W. Tsuji, *Bull. Inst. Chem. Res., Kyoto Univ.* **51**, 104 (1973).

87. F. Peaker and M. Rayner, *Eur. Polym. J.* **6**, 107 (1970).

88. J. Springer, K. Veberreiter, and W. Weinle, *Eur. Polym. J.* **6**, 87 (1970).

89. R. Dobbins and G. Jismagian, *J. Opt. Soc. Am.* **56**, 1345 (1966).

90. V. Klenin and S. Shchyogolev, *J. Polym. Sci., Part C* **42**, 965 (1973).

91. V. Klenin, S. Shchyogolev, and E. Fein, *J. Polym. Sci., Polym. Symp.* **44**, 181 (1974).

92. General references 1–6, 9, 10, 12, 15.

93a. M. Kubin and B. Porsch, *Eur. Polym. J.* **6**, 97 (1970).

93b. B. Chu, *Pure Appl. Chem.* **49**, 944 (1977).

94. B. Porsch and M. Kubin, *Collect. Czech. Chem. Commun.* **39**, 3494 (1974).

95. A. Rudin and H. Johnston, *J. Polym. Sci., Part B* **9**, 55 (1971).

96. General reference 4.

97. D. Kobelt and H. Stemmer, *Kautsch. Gummi, Kunstst., Asbest* **195**, 197 (1963).

98. N. R. Langley, *Macromolecules* **1**, 348 (1968).

99. J. Rehner, *J. Appl. Polym. Sci.* **4**, 95 (1960).

100. F. Bueche and S. Harding, *J. Appl. Polym. Sci.* **2**, 273 (1959).

101. M. Gordon, S. Kucharik, and T. Ward, *Collect. Czech. Chem. Commun.* **35**, 3252 (1970).
102. K. Dusek and W. Prins, *Fortschr. Hochpolym.-Forsch.* **6**, 1 (1969).
103. M. Falk and R. Thomas, *Can. J. Chem.* **52**, 3285 (1974).
104. W. Graessley, *Fortschr. Hochpolym.-Forsch.* **16**, 100 (1974).
105. General references 1, 12.
106. R. Rudin and K. Chee, *Macromolecules* **6**, 613 (1973).
107. G. Berry and T. Fox, *Fortschr. Hochpolym.-Forsch.* **5**, 261 (1968).
108. W. Graessley, *Fortschr. Hochpolym.-Forsch.* **16**, 1 (1974).
109. G. Locati, L. Gargani, and A. De Chirico, *Rheol. Acta* **13**, 278 (1974).
110. W. Graessley and E. Shinbach, *J. Polym. Sci., Polym. Phys. Ed.* **12**, 2047 (1974).
111. R. Penwell and W. Graessley, *J. Polym. Sci., Polym. Phys. Ed.* **12**, 213 (1974).
112. A. Rudin and K. Chee, *Macromolecules* **6**, 613 (1973).
113. General reference 15.
114. R. W. Baker, *J. Appl. Polym. Sci.* **13**, 369 (1969).
115. See also references 11–15 in Table IIIC.
116. General references 1–3, 12, 14.
117. T. G. Scholte, *Eur. Polym. J.* **6**, 51 (1970).
118. L. Tung and J. Runyon, *J. Appl. Polym. Sci.* **17**, 1589 (1973).
119. A. Yamamoto, I. Nada, and M. Nagasawa, *Polym. J.* **1**, 304 (1970).
120. M. Veda, *Polym. J.* **3**, 431 (1972).
121. M. Ambler and R. Mate, *J. Polym. Sci., Part A-1* **10**, 2677 (1972).
122. C. Uraneck, M. Barker, and W. Johnson, *Rubber Chem. Technol.* **38**, 802 (1965).
123. C. Strazielle and H. Benoit, *Pure Appl. Chem.* **26**, 451 (1971).
124. J. Polacek, V. Bohackova, Z. Pokorna, and E. Sinkulova, *Coll. Czeh. Chem. Comm.* **41**, 2510 (1976).

TABLE IIIB

STRUCTURE: BRANCHING

Characterization technique	Variables measured	Symbol (and unit)	Principle of operation	References
Fractionation with $[\eta]$ and $\overline{M}_w$ measurements or D and $\overline{M}_w$ measurement	$\overline{M}_w$ and $[\eta]$ on fractions	$g = \dfrac{[\eta]_{br}}{[\eta]_{lin}}$	Branching reduces intrinsic viscosity by amounts which can be correlated with theoretical calculations or model polymers. Need $[\eta] = KM^a$ relation for linear polymer.	1–13, 28, 34
GPC with $[\eta]$ measurement on fractions, or on line	GPC of fractions for which $[\eta]$ is known, or $[\eta]$ of GPC effluent	g	Same as above except the molecular weight of the polymer is determined by postulating $(M[\eta])_{lin} = (M[\eta])_{br}$ at a given elution volume. $\overline{M}_n$ appears to be the correct MW average.	12, 14–16c
GPC with $[\eta]$ on whole polymer	GPC, $[\eta]$ on whole polymer	g	Assumptions are made about branching distribution and the degree of branching is chosen to make GPC and whole polymer $[\eta]$ agree, or $[\eta]$ measured as GPC cuts elute.	17–25

GPC–sedimentation and [η]	GPC Sedimentation coefficients [η]	S_i (g)	Branching increases sedimentation coefficient.	4, 26, 27, IIIA–45b
Huggins constant	Concentration dependence of intrinsic viscosity	k	Branching causes "systematic" variations in these quantities which have been recommended as measures of branching.	28, 29, 30, 32
Virial coefficients	Concentration dependence of osmotic pressure	A_2 (mole cm^3/g^2)		
	Theta temperatures	Θ (°C)		
Relaxation times	Time dependence of mechanical properties	τ_i (sec)	Either by comparison with model polymers or through use of theoretical calculations on models such as the Zimm theory, changes in these quantities at given molecular weights in dilute solution are related to degrees of branching.	33–36
Compliances	Strain response to imposed stress is measured	J', J_e'' (area/force)		
Moduli	Stress response to imposed strain is measured	G', G'' (force/area)		
Concentrated polymer viscosity	Viscosity is measured as a function of shear rate and temperature.	η (poise)	Changes in this quantity are related to branching empirically by comparison with model compounds; least rigorous of the methods, because no theory equivalent to that for dilute solutions is involved.	8, 10, 33, 37, 38a,b

REFERENCES FOR TABLE IIIB

1a. General references 1, 2, 9–13.
1b. P. A. Small, *Adv. Polym. Sci.* **18,** 1 (1975).
 2. W. Graessley, *in* "Characterization of Macromolecular Structure," p. 371. Natl. Acad. Sci., Washington, D.C., 1968.
 3. L. Mrkvickova-Vacvlova and P. Kratochvi, *Collect. Czech. Chem. Commun.* **37,** 2015, 2029 (1972).
 4. J. Roovers and S. Bywater, *Macromolecules* **5,** 384 (1972).
 5. M. Kurata, M. Abe, M. Iwama, and M. Matsushima, *Polym. J.* **3,** 729 (1972).
 6. L. H. Tung, *J. Polym. Sci., Part A-2* **11,** 1247 (1973).
 7. G. Kraus and C. Stacy, *J. Polym. Sci., Part A-2* **10,** 657 (1972).
 8. T. Fujimoto, H. Narukawa, and M. Magasawa, *Macromolecules* **3,** 57 (1970).
 9. I. Noda, T. Horikawa, T. Kato, T. Fujimoto, and M. Nagasawa, *Macromolecules* **3,** 795 (1970).
10. T. Masuda, Y. Ohta, and S. Onogi, *Macromolecules* **4,** 763 (1971).
11. K. Kamada and H. Sato, *Polym. J.* **2,** 489 (1971).
12a. T. Kato, A. Itsubo, Y. Yamamoto, T. Fujimoto, and M. Nagasawa, *Polym. J.* **7,** 123 (1975).
12b. M. Ambler and D. McIntyre, *J. Polym. Sci. Polym. Lett. Ed.* **13,** 589 (1975).
12c. J. Dawkins and M. Hemming, *Polymer* **16,** 554 (1975).
13. S. Krozer, *Makromol. Chem.* **175,** 1893, 1905 (1974).
14. General references 9, 10.
15a. W. Park and W. Graessley, *J. Polym. Sci. Polym. Phys. Ed.* **15,** 71, 85 (1977).
15b. G. Kraus and C. Stacy, *J. Polym. Sci., Polym. Symp.* **43,** 329 (1973).
15c. J. Janca and M. Kolinsky, *J. Chromatogr.* **132,** 187 (1977).
16a. A. Servotte and R. DeBrulle, *Makromol. Chem.* **176,** 203 (1975).
16b. A. Hamelic and A. Ouano, *J. Chromat. Sci.* **1,** (1978).
16c. S. Nakano and Y. Goto, *J. Appl. Polym. Sci.* **20,** 3313 (1976).
17. General references 9, 10.
18a. M. Kurata, H. Okamoto, M. Iwama, M. Abe, and T. Homma, *Polym. J.* **3,** 739 (1972).
18b. M. Ambler, *J. Appl. Polym. Sci.* **21,** 1655 (1977).
19. A. Shultz, *J. Polym. Sci., Part A-2* **10,** 983 (1972).
20. G. R. Williamson and A. Cervenka, *Eur. Polym. J.* **8,** 1009 (1972).
21. J. Pannell, *Br. Polym. J.* **13,** 277 (1972).
22. A. Cervenka and T. W. Bates, *J. Chromatogr.* **53,** 85 (1970).
23. M. Ambler, R. Mate, and J. Pardon, *J. Polym. Sci., Polym. Chem. Ed.* **12,** 1759, 1771 (1974).
24. R. Bartosiewicz, C. Booth, and A. Marshall, *Eur. Polym. J.* **10,** 783 (1974).
25. G. Williamson and A. Cervenka, *Eur. Polym. J.* **10,** 295 (1974).
26a. L. H. Tung, *J. Polym. Sci., Part A-2* **9,** 759 (1971).
26b. H. Matsuda, I. Yamada, and S. Kurolwa, *Polymer J.* **8,** 415 (1976).
27. W. Bengough and G. Grant, *Eur. Polym. J.* **7,** 203 (1971).
28. N. Hadjichristidis and J. Roovers, *J. Polym. Sci., Polym. Phys. Ed.* **12,** 2521 (1974).
29. No theoretical justification for this method has been derived at present.
30. F. Candau, P. Rempp, and H. Benoit, *Macromolecules* **5,** 627 (1972).
31. G. Berry, *J. Polym. Sci., Part A-2* **9,** 687 (1971).
32. J. Roovers and S. Bywater, *Macromolecules* **7,** 443 (1974) and references therein.
33. T. Fujimoto, H. Kajiura, M. Hirose, and M. Nagasawa, *Polym. J.* **3,** 181 (1972).
34. Y. Mitsuda, J. Schrag, and L. D. Ferry, *Polym. J.* **4,** 668 (1973).
35. K. Osaki and J. Schrag, *J. Polym. Sci., Polym. Phys. Ed.* **11,** 549 (1973).
36. Y. Mitsuda, J. Schrag, and J. Ferry, *J. Appl. Polym. Sci.* **18,** 193 (1974).
37. W. Graessley and E. Shinbach, *J. Polym. Sci., Polym. Phys. Ed.* **12,** 2047 (1974).
38a. A. Ghijesels and H. Mievas, *J. Polym. Sci., Polym. Phys. Ed.* **11,** 1849 (1973).
38b. W. Graessley, *Accounts Chem. Res.* **10,** 332 (1977).

TABLE IIIC

STRUCTURE: GEL-AGGREGATES

Characterization technique	Variables measured	Symbol (and unit)	Principle of operation	References
Centrifugation Sedimentation	Gel or aggregate fraction Sedimentation coefficient	W_g S_i (sec)	High molecular weight material can be sedimented and separated mechanically from the "solution," or identified as sedimenting species.	1–7
Density gradient	Gel fraction	W_g	High molecular weight material can be concentrated in a particular density region and identified as such.	8–10
Filtration	Gel fraction	W_g	High molecular weight material can be removed from solutions by filtration.	11–15
Light scattering	Molecular weight Gel fraction Aggregation	$\overline{M}_w$ (g/mole) W_g	Microgel contents can be estimated by decomposition of scattering diagram into low- and high-angle portions.	16–24
Extraction	Gel fraction	W_g	Gelled material can be separated from "soluble" fractions by extraction.	25–28
Microscopy	Particle sizes, shapes		Gelled material may have distinctive appearance.	26, 29–31
GPC	Elution volume Aggregate fraction	V_e (cc or "counts")	Aggregated polymer can be studied through effects of solution treatment on elution volume.	32–34
General particle size	Particle size	—	Various techniques including light scattering.	35–37
Model systems	Various properties			38–43

REFERENCES FOR TABLE IIIC

1. General reference 15.
2. H. Suzuki and C. Leonis, *Br. Polym. J.* **5**, 485 (1973).
3. H. Suzuki, C. Leonis, and M. Gordon, *Makromol. Chem.* **172**, 227 (1973).
4. R. Skerrett, *J. Chem. Soc.* p. 1328 (1974).
5. C. G. Overberger and B. Sedlacek, eds., *J. Polym. Sci., Polym. Symp.* **44** (1974).
6. H. Lange and W. Scholtan, *Angew. Makromol. Chem.* **26**, 59 (1972).
7. See Table IIIA, references 25, 49.
8. General reference 15.
9. R. Buchdahl, H. Ende, and L. Peebles, *J. Polym. Sci., Part C* **1**, 143 (1963).
10. R. Trautman and M. Hamilton, *in* "Princ. Tech. Plant Viral" (C. Kdao, ed.). Van Nostrand, New York, 1972.
11a. J. Purdon and R. Mate, *J. Polym. Sci., Part A-1* **8**, 1306 (1970).
11b. M. Ambler, *J. Appl. Polym. Sci.* **20**, 2259 (1976).
12. M. D. Maijal, *Anal. Chem.* **44**, 1337 (1972).
13. R. Harrington and B. H. Zimm, *J. Polym. Sci., Part A-2* **6**, (1968).
14. R. Baker and H. Strathman, *J. Appl. Polym. Sci.* **14**, 1197 (1970).
15. General reference 15.
16. J. Lynguae-Jorgersen, *Makromol. Chem.* **167**, 311 (1973).
17. B. Rietveld and T. G. Scholte, *Macromolecules* **6**, 468 (1973).
18. See Table IIIA, references 24, 25.
19. H. Lange, *Kolloid. Z. Z. Polym.* **250**, 775 (1972).
20. R. A. Isaksen, C. Williams, J. Heaps, and R. Clark, *Ind. Eng. Chem., Prod. Res. Dev.* **10**, 298 (1971).
21. P. Kratochvil, *Collect. Czech. Chem. Commun.* **30**, 1119 (1965).
22. H. Lange, *Kolloid Z. Z. Polym.* **240**, 747 (1970).
23a. See Table IIIB, reference 22.
23b. D. Carpenter, G. Santiago, and A. Hunt, *J. Polym. Sci., Polym. Symp.* **44**, 75 (1974).
24. W. Burchard, K. Kajiwara, M. Gordon, J. Ralel, and J. Kennedy, *Macromolecules* **6**, 642 (1973).
25. See Table IIIA, reference 104, Chap. 7. Details of network extractions can be found in the references of this chapter.
26. M. Morimoto and Y. Okamoto, *J. Appl. Polym. Sci.* **17**, 2801 (1973).
27. W. Wake, "The Analysis of Rubber and Rubber Like Polymers." Wiley, New York, 1968.
28. General reference 6, Chapter 17.
29. General reference 15.
30. W. F. Gemeyer, *J. Polym. Sci., Polym. Symp.* **44**, 25 (1974).
31. R. Tal'Roze, U. Shibaev, and N. Pate, *J. Polym. Sci., Polym. Symp.* **44**, 35 (1974).
32. A. Abdel-Alim and A. Hamielec, *J. Appl. Polym. Sci.* **16**, 1093 (1972).
33. V. Gaylor, H. James, and J. Herdering, *J. Polym. Sci., Polym. Chem. Ed.* **13**, 1575 (1975).
34. C. Price, J. Forget, and C. Booth, *Polymer* **18**, 526 (1977).
35. General references 1, 2, 7, 9, 13, 15.
36. R. Davies, *Ind. Eng. Chem.* **62**, 87 (1970).
37. R. Davies, *Am. Lab.* Jan. (1974). "Rapid Response Instrumentation for Particle Size Analysis."
38. H. Eschwey, M. Hallensleben, and W. Burchard, *Makromol. Chem.* **173**, 235 (1975).
39. M. Hoffman, *Makromol. Chem.* **175**, 613 (1974).
40. R. Valentine, J. Ferry, T. Homma, and K. Ninomiya, *J. Polym. Sci., Part A-2* **6**, 479 (1968).
41. S. Rosen, *Appl. Polym. Symp.* **7**, 127 (1968).
42. D. Plazek, *Am. Chem. Soc., Div. Org. Coat. Plast. Chem., Pap.* **35**, 389 (1975).
43. P. Thirton and R. Chasset, *Pure Appl. Chem.* **23**, 183 (1970).

TABLE IV STRUCTURE: GLASS

Characterization technique	Variables measured	Symbol (and unit)	Principle of operation	References
Differential thermal analysis	Heat capacity as a function of temperature Thermooptical (TOA), electrothermal (ETA), thermomechanical (TMA)	ΔC_p (cal/g deg)	At T_g, a discontinuity in C_p occurs. Optical, electrothermal, mechanical properties also undergo abrupt changes at T_g.	1–13
Dilatometry	Volume as a function of temperature	α (deg^{-1}) $\Delta\alpha$ (deg^{-1})	At T_g, a discontinuity in the expansion coefficient occurs.	14–18
Mechanical spectroscopy	Moduli, compliances as a function of time and temperature	G', G'' (dyn/cm^2)	G' increases by $\sim 10^3$, there is a maximum in G'', and $G''/G' = \tan\delta$	19–25
Dielectric spectroscopy	Complex dielectric constant	ε', ε''	Segmental motion associated with dipoles leads to maxima in ε'' and $\varepsilon''/\varepsilon'$ about T_g.	26–29
Nuclear magnetic resonance	Relaxation times and line widths as a function of temperature	τ_1, τ_2, τ_p (sec) M_2, M_4 (Hz)	Motions of the chain segments determine relaxation times and line widths.	30–37
Small molecule diffusion	Diffusion coefficient	D (cm^2/sec)	Mobility in polymer matrix is determined by polymer segmental motion.	38–40
Neutron diffraction	Segment diffusion coefficient	D (cm^2/sec)	Direct measure of segment mobility.	41–43
Birefringence relaxation	Stress-induced birefringence as a function of time and temperature	Δn	Adequate molecular mobility permits relaxation of birefringence.	44–45
Infrared	Infrared absorption wave number versus temperature	(cm^{-1})	Changes in coefficient of expansion and molecular mobility lead to changes in relationships of ir absorption wave numbers and T.	46
Theoretical or predictive estimates of T_g, compilations				47–55
Miscellaneous				56–59

REFERENCES FOR TABLE IV

1. General references 1, 2, 7–10, 13, 15.
2. P. Slade and L. Jenkins, "Techniques and Methods of Polymer Evaluation, Vol. 1, Thermal Analysis." Dekker, New York, 1966.
3. P. Slade and L. Jenkins, "Techniques and Methods of Polymer Evaluation, Vol. 2, Thermal Characterization Techniques." Dekker, New York, 1970.
4. J. J. Maurer, *Rubber Rev.* **42,** 110 (1969).
5. B. Wunderlich, *J. Therm. Anal.* **5,** 117 (1973).
6. R. Haward, "The Physics of Glassy Polymers." Halsted, New York, 1972.
7. S. Wolpert, A. Weitz, and B. Wunderlich, *J. Polym. Sci., Part A-2* **9,** 1887 (1971).
8. J. Chiu, *J. Macromol. Sci., Chem.* **8,** 3 (1974).
9. J. Flynn, *Thermochim. Acta* **8,** 69 (1974).
10. G. Gee, *Contemp. Phys.* **11,** 313 (1970).
11. B. Wunderlich and H. Baur, *Fortschr. Hochpolym.-Forsch.* **7,** 151 (1970).
12. A. Wertz and B. Wunderlich, *J. Polym. Sci., Polym. Phys. Ed.* **12,** 2473 (1974).
13. D. Leary and M. Williams, *J. Polym. Sci., Polym. Phys. Ed.* **12,** 265 (1974).
14. General references 1, 2, 7, 8, 10, 13, 15.
15. J. McKinney and M. Goldstein, *J. Res. Natl. Bur. Stand., Sect. A* **78,** 331 (1974).
16. D. Kaelble, *in* "Rheology: Theory and Applications" (F. R. Eirich, ed.), Vol. 5, p. 223. Academic Press, New York, 1969.
17. G. Kraus and J. Gruver, *J. Polym. Sci., Part A-2* **8,** 571 (1970).
18. K. Chee and A. Rudin, *Ind. Eng. Chem., Fundam.* **9,** 177 (1970).
19. General references 1, 2, 7–10, 13, 15.
20. J. D. Ferry, "Viscoelastic Properties of Polymers," 2nd Ed. Wiley, New York, 1970.
21a. D. Massa, *J. Appl. Phys.* **44,** 2595 (1973).
21b. T. Murayama and A. Armstrong, *J. Polym. Sci., Polym. Phys. Ed.* **12,** 1211 (1974).
22. A. Yim, R. Chanal, and L. St. Pierre, *J. Colloid Interface Sci.* **43,** 583 (1973).
23. D. Masa, J. Flick, and S. Petrie, *Am. Chem. Soc., Div. Org. Coat. Plast. Chem., Pap.* **35,** 371 (1975).
24. T. L. Smith, *in* "Rheology: Theory and Applications" (F. R. Eirich, ed.), Vol. 5. Academic Press, New York, 1969.
25a. L. Nielson, "Mechanical Properties of Polymers and Composites." Dekker, New York, 1974.
25b. B. Shah and R. Darby, *Polym. Eng. Sci.* **16,** 46 (1976).
26. General references 1, 2.
27. N. McCrum, B. Read, and G. Williams, "Anelastic and Dielectric Effects in Polymeric Solids." Wiley, New York, 1967.
28. M. E. Baird, *Prog. Polym. Sci.* **1,** 161 (1967).
29. W. MacKnight, *Am. Chem. Soc., Div. Org. Coat. Plast. Chem., Pap.* **35,** 398 (1975).
30. G. E. Johnson, A. Anderson, and G. Link, *Am. Chem. Soc., Div. Org. Coat. Plast. Chem., Pap.* **35,** 404 (1975).
31. G. Allen, *Rev. Pure Appl. Chem.* **17,** 67 (1967).
32. D. McCall, *Accounts Chem. Res.* **7,** 223 (1971).
33. General references 1, 2.
34. K. Liu and J. Anderson, *Rev. Macromol. Chem.* **6,** 1 (1971).
35. J. E. Anderson, P. Davis, and W. Slichter, *Macromolecules* **2,** 166 (1969).
36. G. Wardell, V. McBrievty, and D. Douglass, *J. Appl. Phys.* **45,** 3441 (1974).
37. V. McBrierty and I. McDonald, *Polymer* **16,** 125 (1975).
38. General reference 1.
39. G. Rabold, *J. Polym. Sci., Part A-1* **7,** 1203 (1969).
40. P. Lumler and R. Boyer, *Polym. Prepr., Am. Chem. Soc., Div. Polym. Chem.* **16,** 572 (1975).

41. General reference 1.
42. G. Allen, *Pure Appl. Chem.* **39**, 151 (1974).
43. R. Ober, J. Cotton, B. Farnoux, and J. Higgins, *Macromolecules* **5**, 634 (1974).
44. General references 1, 9, 10.
45. T. Hammack and R. Andrews, *J. Appl. Phys.* **38**, 5182 (1967).
46. N. Brockmeir, *J. Appl. Polym. Sci.* **12**, 2128 (1968).
47. General references 1, 2, 8.
48. A. Tonelli, *Macromolecules* **7**, 632 (1974).
49. M. Shen and A. Eisenberg, *Rubber Rev.* **43**, 95, 156 (1970).
50. R. Boyer, *Macromolecules* **7**, 142 (1974).
51. T. Nose, *Polym. J.* **4**, 217 (1973).
52. W. Lee and G. Knight, *Br. Polym. J.* **2**, 73 (1970).
53. M. Goldstein, *J. Phys. Chem.* **77**, 667 (1973).
54. V. Privalko and Y. Lipatov, *J. Macromol. Sci., Phys.* **9**, 551 (1974).
55. D. Morley, *J. Mater. Sci.* **9**, 619 (1974).
56. G. D. Patterson, *Polym. Prepr., Am. Chem. Soc., Div. Polym. Chem.* **16**, 747 (1975).
57. M. Naoki and T. Nose, *Polym. J.* **6**, 45 (1974).
58. J. Stevens and R. Rowe, *J. Appl. Phys.* **44**, 4328 (1973).
59. D. Gray and J. Guillet, *J. Polym. Sci., Polym. Lett. Ed.* **12**, 231 (1974).

TABLE V

STRUCTURE: CRYSTALLINITY

Characterization technique	Variables measured	Symbol (and unit)	Principle of operation	References
Density–specific volume	Can be calibrated to get degree of crystallinity	W_c, f_c (weight, volume fraction)	Knowing the densities of pure crystal and amorphous phases, one can interpolate a degree of crystallinity.	1–10
Differential scanning calorimetry or adiabatic calorimeter	Heat of fusion Crystalline fraction Heat capacities	ΔH_f (cal/g) f_c, W_c C_p (cal/g °C)	Heats associated with transitions are compared with those of the pure crystal.	12–18
Infrared	Degree of crystallinity	W_c, f_c	Peaks assigned to crystalline and amorphous regions are compared to determine the relative amounts of each phase.	11, 19–23
Wide-angle X-ray	Degree of crystallinity	W_c, f_c (λ)	Obtained by various rigorous methods involving separation at crystalline and amorphous scatterings	11, 24–34
	Unit cell dimensions	a, b, c (Å)	Obtained from diffraction maxima	
	Crystallite dimensions	(Å)	Obtained from line broadening	
	Orientation	Presented in a variety of ways	Obtained from studies of scattering intensity as a function of diffraction direction	

Small-angle X-ray	Crystallite dimensions	(Å)		35–38
	Radius of gyration	$\langle S^2 \rangle$		
	Degree of crystallinity	W_c, f_c		
	Density	ρ_a, ρ_c (g/cc)		
NMR	Absorption intensities	f_c, W_c	Numbers of atoms exhibiting a particular line width are counted as crystalline or amorphous	39–40
Miscellaneous				
Refractive index	Refractive index	n	Crystalline and amorphous regions have different refractive indexes.	22
Diffusion	Rates of diffusion of small molecules	D (cm^2/sec)	Diffusion in amorphous regions is much faster than in crystals.	22, 42
Chemical reactivity	Amount of polymer susceptible to chemical attack		Crystalline material is less susceptible to chemical reaction due to lower accessibility of crystals to diffusing species.	22
Modulus, compliance, hardness	Modulus		Mechanical properties of crystalline and amorphous phases differ; therefore, the composite has some average property depending on the crystalline fraction.	41

150　　　　　　　　　　　　　　　　　　　　　　　G. VER STRATE

REFERENCES FOR TABLE V

1. General references 1, 2, 7–10, 13, 15.
2. L. Alexander, "X-Ray Diffraction Methods in Polymer Science." Wiley, New York, 1969.
3. J. Fortuin, *J. Polym. Sci.* **44,** 505 (1960).
4. L. Mandelkern, "Crystallization of Polymers." McGraw-Hill, New York, 1964.
5. W. Statton, *J. Polym. Sci., Part C* **18,** 33 (1967).
6. S. Kavesh and J. Schultz, *Polym. Eng. Sci.* **9,** 452 (1969).
7. S. Kavesh and J. Schultz, *J. Polym. Sci., Part A-2* **9,** 85 (1971).
8. V. Grmela and L. Megarskaja, *J. Sci. Instrum.* **2,** 2 (1969).
9. H. G. Zachman, *Angew. Chem., Int. Ed. Engl.* **13,** 244 (1974).
10. A. Gent, *J. Polym. Sci., Part A-2* **4,** 447 (1966).
11. R. J. Samuels, "Structured Polymer Properties." Wiley, New York, 1974.
12. General references 7, 9, 10, 15.
13. M. Richardson, *J. Polym. Sci., Part C* **38,** 251 (1972).
14. V. Bares and B. Wunderlich, *J. Polym. Sci., Part A-2* **11,** 1301 (1973).
15. S. Hobbs and G. Menkin, *J. Polym. Sci., Part A-2* **9,** 1907 (1971).
16. B. Wunderlich, *Thermochim. Acta* **5,** 369 (1973).
17. S. Clough, *J. Macromol. Sci., Phys.* **4,** 199 (1970).
18. G. Ver Strate and Z. Wilchinsky, *J. Polym. Sci., Part A-2* **9,** 127 (1971).
19. General references 1, 15.
20. W. Glenz and A. Peterlin, *J. Macromol. Sci., Phys.* **4,** 473 (1970).
21. H. Hendus and G. Schnell, *Kunststoffe* **51,** 69 (1961).
22. B. Wunderlich, "Macromolecular Physics," Vol. 1. Academic Press, New York, 1973.
23. C. Sung and N. Schneider, *Macromolecules* **8,** 68 (1975).
24. General references 1, 2, 7, 8–10, 13, 14.
25. R. Hoseman. *Crit. Rev. Macromol. Sci.* **1,** 351 (1972).
26. W. Statton, *J. Polym. Sci., Part C* **18,** 33 (1967).
27. C. Vonk, *J. Appl. Crystallogr.* **6,** 148 (1973).
28. A. Kulshreshtha, N. Deweltz, and T. Radhakrishnan, *J. Appl. Crystallogr.* **4,** 116 (1971).
29. O. Yoda, K. Doi, N. Tamura, and I. Kuriyama, *J. Appl. Phys.* **44,** 2211 (1973).
30. H. Hill and A. Keller, *J. Macromol. Sci., Phys.* **3,** 153 (1969).
31. W. Krigbaum, J. Dowkins. G. Via, and Y. Bulta, *J. Polym. Sci., Part A-2* **4,** 475 (1966).
32. W. Krigbaum, "Stressed Crystallinity References," AD 690,870. Clearinghouse Fed. Sci. Tech. Inf., Springfield, Virginia, 1969.
33. A. Ghaffar, I. Goodman, and I. Hall, *Br. Polym. J.* **5,** 315 (1973).
34. Y. Chang and G. Wilkes, *J. Polym. Sci., Polym. Phys. Ed.* **13,** 455 (1975).
35. General references 1, 2, 7–10, 13, 14.
36. G. Strobl and N. Muller, *J. Polym. Sci., Polym. Phys. Ed.* **11,** 1219 (1973).
37. D. Brown. K. Fulcher, and R. Wetton, *Polymer* **14,** 379 (1973).
38. H. Brunberger, ed.. "Small Angle X-Ray Scattering." Gordon & Breach, New York, 1967.
39. General reference 1.
40. V. McBrierty, *Polymer* **15,** 503 (1974).
41. S. Davidson and G. Taylor, *Br. Polym. J.* **4,** 65 (1972).
42. A. Peterlin, *IUPAC Pure Appl. Chem.* **39,** 239 (1974).

TABLE VI

STRUCTURE: MORPHOLOGY

Characterization technique	Variables measured	Symbol (and unit)	Principle of operation	References
Microscopy				
Electron	Size and shape of particles		Magnetic or glass lens systems are used to enlarge image of reflected or transmitted radiation.	1–10
Optical	Size and shape of particles			
Light scattering	Particle scattering factor	$P(\theta)$	Light is scattered from regions of varying refractive index. Radius of gyration determines initial terms of scattering function for dilute systems.	11–15
	Radius of gyration	$\langle S^2 \rangle_z^{1/2}$		
	For concentrated systems, correlation functions, lengths, and volumes of heterogeneities can be evaluated.		Characteristic maxima occur in scattering intensity for concentrated systrms.	
X-ray	Particle scattering factor	$P(\theta)$	X rays are scattered from regions of varying electron density; same analysis as for light scattering.	16–22b
Small angle	Radius of gyration	$\langle S^2 \rangle_z^{1/2}$		
	Long period		Characteristic maxima occur in scattering intensity for concentrated systems. Correlation lengths, surface area parameters can be evaluated.	

TABLE VI (*continued*)

Characterization technique	Variables measured	Symbol (and unit)	Principle of operation	References
Wide angle	Line width		Small crystallites broaden diffraction maxima.	
	Orientation		Intensity of maxima depend on scattering direction from oriented specimens.	
Neutron diffraction	Particle scattering factor	$P(\theta)$ $\langle S^2 \rangle_z^{1/2}$	Neutrons are scattered from regions of varying scattering length; same particle scattering factor as for light scattering.	23–28
Infrared dichroism	Orientation		Infrared absorption depends on the polarization of the radiation when the absorbing matrix is oriented.	57
Birefringence	Orientation	Δn	Refractive index parallel and perpendicular to chain axis differs, therefore birefringence increases with orientation.	58
Combinations of techniques directed to particular morphology				
Amorphous homopolymers				29–31
Segmented polymers				32–35
Block polymers				36–44
Ionomers				45–54
Blends				55–57

REFERENCES FOR TABLE VI

1. General references 1, 2, 7–15.
2. J. Kruse, *Rubber Rev.* **46,** 653 (1973).
3. R. Seward, *Rubber Rev.* **43,** 1 (1970).
4. D. Luch and G. Yeh, *J. Appl. Phys.* **43,** 4326 (1972).
5. G. Yeh, *J. Macromol. Sci., Phys.* **6,** 451 (1972).
6. G. Yeh, *J. Macromol. Sci., Phys.* **6,** 465 (1972).
7. A. Burmiester and P. Geil, *in* "Advances in Polymer Science and Engineering," Vol. 12, p. 43. Plenum, New York, 1972.
8. G. L. Wilkes, *J. Macromol. Sci., Rev. Macromol. Chem.* **10,** 149 (1974).
9. G. L. Wilkes, *Am. Chem. Soc., Div. Org. Coat. Plast. Chem., Pap.* **35,** 426 (1973).
10. G. Cocks, *Anal. Chem.* **46,** 420R (1974).
11a. General references 1, 2, 7–15.
11b. K. Wun and W. Prins, *J. Polym. Sci., Polym. Phys. Ed.* **12,** 533 (1974).
12a. F. Robillard, A. Patitsas, and B. Kaye, *Powder Technol.* **10,** 307 (1974).
12b. S. Visconti and R. Marchessault, *Macromolecules* **7,** 913 (1974).
13. M. Moritani *et al., in* "Colloidal and Morphological Behavior of Block and Graft Copolymers" (G. Molau, ed.), Plenum. New York, 1971.
14a. W. Kaye, *J. Colloid Interface Sci.* **44,** 384 (1973).
14b. B. Conaghan and S. Rosen, *Polym. Eng. Sci.* **12,** 134 (1972).
15. R. Koningsveld, L. Kleintjens, and H. Schoffeleers, *IUPAC Pure Appl. Chem.* **39,** 1 (1974).
16. General references 1, 2, 7–15.
17. A. Keller and M. Mackley, *Pure Appl. Chem.* **39,** 195 (1974).
18. M. Myers and D. McIntyre, *Polym. Eng. Sci.* **13,** 176 (1973).
19. C. Wilkes and M. Lehr, *J. Macromol. Sci.., Phys.* **7,** 225 (1973).
20. J. Wendorff and E. Fischer, *Kolloid Z. Z. Polym.* **251,** 876 (1973).
21. J. Wendorff and E. Fischer, *Kolloid Z. Z. Polym.* **251,** 884 (1973).
22a. C. F. Pfluger, *Anal. Chem.* **46,** 469R (1974).
22b. G. Nelson and S. Jabarin, *J. Appl. Phys.* **46,** 1175 (1975).
23. General reference 1.
24. J. Cotton *et al., Macromolecules* **7,** 863 (1974).
25. R. Kirste, W. Kruse, and K. Ibel, *Polymer* **16,** 121 (1975).
26. W. Schmatz, T. Springer, J. Schelten, and K. Ibel, *J. Appl. Crystallogr.* **7,** 96 (1974).
27a. G. Wignall, D. Ballard, and J. Schelten, *Eur. Polym. J.* **10,** 861 (1974).
27b. A. Hamada, *Kobunshi* **25,** 743 (1976).
28. D. Ballard, G. Wignall, and J. Schelten, *Eur. Polym. J.* **9,** 965 (1973).
29. General references 9, 10.
30. P. Lindenmeyer, *J. Macromol. Sci., Phys.* **8,** 361 (1973).
31. E. Helfand and Y. Tagani, *J. Chem. Phys.* **56,** 3592 (1972).
32. General reference 13.
33. C. Wilkes and C. Yusek, *J. Macromol. Sci., Phys.* **7,** 157 (1973).
34. G. Estes, S. Cooper, and A. Tobolsky, *Rev. Macromol. Chem.* **5,** 167 (1970).
35. H. Ng, A. Allegrezza, R. Seymour, and S. Cooper, *Polymer* **14,** 255 (1973).
36. General references 1, 3, 10, 13.
37. G. E. Molau, ed., "Colloidal and Morphological Behavior of Block and Graft Copolymers." Plenum, New York, 1971.
38. R. Ceresa, ed., "Block and Graft Copolymerization." Wiley, New York, 1973.
39. A. Jenkins, ed., *Prog. Polym. Sci.* **3** (1971).
40. D. Leary and M. Williams, *J. Polym. Sci., Polym. Phys. Ed.* **11,** 345 (1973).
41. M. Kojima and J. Magill, *J. Appl. Phys.* **45,** 4159 (1974).

42. M. Shen, U. Mehra, M. Niinomi, J. Koberstein, and S. Cooper, *J. Appl. Phys.* **45,** 4182 (1974).
43. R. Mayer, *Polymer* **15,** 137 (1974).
44. T. Smith, *J. Polym. Sci., Polym. Phys. Ed.* **12,** 1825 (1974).
45. General references 9, 10.
46. E. P. Otocka, *J. Macromol. Sci., Rev. Macromol. Chem.* **5,** 275 (1971).
47. C. Marx, D. Caufield, and S. Cooper, *Macromolecules* **6,** 344 (1973).
48. M. Pineri, C. Meyer, A. Levelut, and M. Lambert, *J. Polym. Sci., Polym. Phys. Ed.* **12,** 115 (1974).
49. M. Navratil and A. Eisenberg, *Macromolecules* **7,** 84 (1974).
50. A. Eisenberg and M. Navratil, *Macromolecules* **7,** 90 (1974).
51. R. J. Roe, *J. Phys. Chem.* **76,** 1311 (1973).
52. J. Kao, R. Stein, W. MacKnight, W. Taggart, and G. Cargill, *Macromolecules* **7,** 95 (1974).
53. A. Eisenberg, *J. Polym. Sci., Polym. Symp.* **45,** 99 (1974).
54. W. MacKnight, W. Taggart, and R. Stein, *J. Polym. Sci., Polym. Symp.* **45,** 113 (1974).
55. General references 9, 10.
56. R. Deanin, *in* "Recent Advances in Polymer Blends, Grafts and Blocks" (L. Sperling, ed.), p. 63. Plenum, New York, 1974.
57. S. Onogi and T. Asada, *Prog. Polym. Sci. Jpn.* **2,** 261 (1971).
58. R. Morgan and L. Treloar, *J. Polym. Sci., Part A-2* **10,** 51 (1972).

Chapter 4

The Molecular and Phenomenological Basis of Rubberlike Elasticity

MITCHEL SHEN

DEPARTMENT OF CHEMICAL ENGINEERING
UNIVERSITY OF CALIFORNIA
BERKELEY, CALIFORNIA

I. Introduction

As a class of materials, elastomers are certainly unique in their mechanical behavior. No other solid can be stretched up to 1000% and still return to its original length upon release. The elastic moduli of elastomers, on the other hand, are only of the order of 10^7 dyn/cm^2, which is lower than that of ordinary solids by a factor of more than 10^4.

In ordinary solids it is generally accepted that the atoms or molecules are situated at equilibrium positions with respect to each other in a regular array, as in the case of crystalline solids, or in a quasi-lattice, as in glassy solids. Upon application of an external force, these atoms or molecules are displaced from

155

Copyright © 1978 by Academic Press, Inc.
All rights of reproduction in any form reserved.
ISBN 0-12-234360-3

their equilibrium positions. The resistance offered by the interatomic or inter-molecular forces to this displacement gives rise to the observed rigidity of the material. Since such interatomic or intermolecular forces are short range in nature, and decrease rapidly with increasing interatomic or intermolecular distances, the elastic limit of such solids is very low ($\sim 1\%$). At larger deforma-tions these interatomic or intermolecular energies can no longer sustain the applied stress, so the solid fractures. The fact that elastomers have elastic limits some 10^3 times greater immediately implies that the mechanism just described must be inoperative, and a different interpretation for their elastic behavior should be sought.

Numerous early theories were proposed to account for the remarkable behavior of rubberlike elasticity [1, 2]. Ostwald and Fikentscher and Mark observed that in a solvent only part of natural rubber is soluble. They postulated that the insoluble part has an open network structure whose spiral configura-tion gives rise to the elastic restoring force. The soluble part is supposed to act as a liquid medium that fills the open space in the network. Mack envisioned the rubber molecules as possessing a folded conformation maintained by forces between adjacent hydrogen atoms. When extended, the molecule not only unfolds, but also turns in the direction of the applied force. Griffith hypothe-sized that rubber molecules can undergo thermal motions like a skipping rope, which tends to pull its ends toward each other and thus exerts an elastic restor-ing force.

The fundamental concept for the currently accepted molecular mechanism for rubberlike elasticity was first proposed by Meyer *et al.* [3]. These authors noted that intramolecular forces along the chain must be much greater than the intermolecular ones in the lateral directions. Thus thermal motions of segments of a long-chain molecule must be greater in the latter sense, which results in a "repulsive pressure" between parallel chains. This repulsive pressure causes the stretched rubber to retract until an "irregular, statistically determined ar-rangement of the molecules" is reached. This relation between the statistical configuration and elasticity is the basis of modern theories of rubberlike elasticity.

An understanding of the molecular mechanism of rubberlike elasticity is important not only in explaining the observed unique extensibility of elastomers, but also in providing a basis for elucidating a host of other physical properties of polymers. In this chapter, we shall first discuss the elasticity of a single polymer chain. This treatment will then be generalized to a network of chains. The molecular and thermodynamic significance of rubberlike elasticity will be explored in the light of these treatments. Next we abandon the molecular concept and discuss elasticity strictly on a phenomenological basis. These theories will then be applied to explain the elastic behavior of swollen and filled elastomers as well as elastomers undergoing chemical reaction. Our discussions will thus be limited to the equilibrium state; the transient viscoelastic effects belong to the realm of Chapter 5. These discussions are further confined to deformations before the elastic limit is reached. The ultimate mechanical properties of elastomers will be the subject of Chapter 10.

II. Statistics of a Polymer Chain

A. FREELY ORIENTING RANDOM CHAIN

Our model for polymer molecules is that they are long, flexible chains. Like thin threads, if unconstrained they are likely to assume many different shapes (conformations). Because of Brownian motion, these macromolecules are constantly changing their conformations. The number of conformations of about the same internal energy which a given chain can assume is very large; therefore, we can employ statistical methods to study their average properties.

For a polymer chain with n links each of which has a length l, the fully extended length of the entire chain would thus be nl. However, because the fully extended conformations is only one of myriad conformations that the chain can assume, it is more meaningful to consider the number of conformations available to the chain for each specific end-to-end separation. In the simplest case of randomly coiled one-dimensional chains, the number of conformations for a given separation of chain ends x has been shown to be [1, 4–8]:

$$w(x) = (3/2\pi nl)^{1/2} \exp(-3x^2/2nl^2) \tag{1}$$

Equation (1) is the well-known Gaussian error function. An alternative interpretation of Eq. (1) is that it represents the probability of the ends of an n-linked chain to be a distance x apart. This equation, however, applies only to integral values of x. For a long chain (large n), we substitute for this discontinuous function a continuous distribution function, so that $w(x)\,dx$ expresses the probability of finding a value of x between x and $x + dx$:

$$w(x)\,dx = (3/2\pi nl^2)^{1/2}[\exp(-3x^2/2nl^2)]\,dx \tag{2}$$

Equation (2) is known as the Gaussian distribution function. In three dimensions, the distribution function becomes

$$w(x, y, z)\,dx\,dy\,dz = (b/\pi^{1/2})^3[\exp(-b^2r^2)]\,dx\,dy\,dz \tag{3}$$

where

$$b^2 = 3/2nl^2$$

and

$$r^2 = x^2 + y^2 + z^2$$

Equation (3) gives the probability that, if one end of a random chain is located at the origin of a Cartesian coordinate system, the other end can be found in

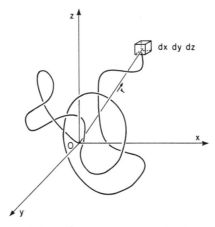

Fig. 1. Schematic representation of a random chain in a Cartesian coordinate system.

the volume element $dx\,dy\,dz$ a distance r away (Fig. 1). If instead of the vectorial distance $\vec{r}$ we desire to know the probability of finding the free chain end anywhere a distance r from the origin (where r is a scalar quantity), then it can be expressed by the radial distribution function

$$w(r)\,dr = (b/\pi^{1/2})^3[\exp(-b^2r^2)]4\pi r^2\,dr \tag{4}$$

The quantity $4\pi r^2\,dr$ is the volume of a spherical shell of thickness dr located a distance r from the origin. Figure 2 shows a plot of this function for a chain consisting of 2000 links, when we set the length of each link equal to 1.

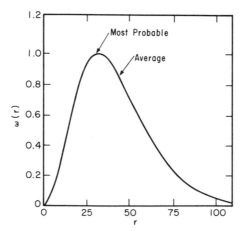

Fig. 2. Radial distribution function for the end-to-end distances of a Gaussian chain consisting of 2000 links of unit lengths.

From Eq. (4) we can find the mean square end-to-end distance of an unperturbed random chain by taking the average

$$\langle r_0{}^2 \rangle = \int_0^\infty r^2 w(r)\, dr \Big/ \int_0^\infty w(r)\, dr = 3/2b^2 = nl^2$$
$$b^2 = 3/2\langle r_0{}^2 \rangle \tag{5}$$

Thus the average end-to-end separation is

$$\langle r_0{}^2 \rangle^{1/2} = n^{1/2} l \tag{5a}$$

which is indicated in Fig. 2 for our hypothetical chain. It is not the same as the most probable value of r, which can be obtained by setting the derivative of Eq. (4), with respect to r, equal to zero:

$$\frac{dw(r)}{dr} = 8\pi r (b/\pi^{1/2})^3 (1 - b^2 r^2) \exp(-b^2 r^2) = 0 \tag{6}$$

Solution of Eq. (6) gives the most probable value of r as $1/b$ or $(2nl^{2/3})^{1/2}$.

Knowing the number of conformations available to a random chain is useful in that we are now in a position to calculate the elasticity of the chain. To accomplish this we first recall the entropy S is related to the number of conformations through the Boltzmann relations:

$$S = k \ln w \tag{7}$$

where k is the Boltzmann constant. The first law of thermodynamics states that

$$dU = dQ - dW \tag{8}$$

where U is internal energy, Q is heat, and W is work. If the only work is the stress–strain work of the chain, then Eq. (8) becomes

$$dU = T\, dS + f\, dr \tag{9}$$

where f is the elastic force. Transforming to the Helmholtz free energy function A, we get (with dL as the change in length)

$$dA = dU - T\, dS = -S\, dT + f\, dL \tag{10}$$

Therefore, the force can be easily obtained from

$$f = (\partial A / \partial r)_T \tag{11}$$

In our model random chain, the internal energy is not affected by the conformational changes, and the Helmholtz free energy will be determined by the entropy term alone. Thus, from Eqs. (4) and (7) we can write

$$A = A' + kTb^2 r^2 \tag{12}$$

where A' is that part of the free energy unaffected by conformations. Differentiation of Eq. (12) according to Eq. (11) immediately yields [9]

$$f = 2kTb^2r \tag{13}$$

Equation (13) predicts that the elastic force of a chain is directly proportional to the end-to-end length of that chain. The proportionality constant is $2kTb^2$. Thus according to this simple theory, a single random chain will obey Hooke's law in that stress is proportional to strain.

B. STATISTICALLY EQUIVALENT RANDOM CHAIN

The model chain we discussed in Section II.A is a freely orienting random chain. Each link in this chain can orient in any direction, independent of the other links. It is also supposed to be volumeless in that two or more links can presumably occupy the same space. In addition, the internal energy of the chain is not dependent on the conformation. Obviously, this "phantom" model chain is highly idealized. The following effects must be taken into account for a more realistic model [1, 2].

1. The excluded volume effect. Each link must occupy a finite amount of space, and the space occupied by one link is excluded from being occupied by another.

2. The effect of bond angle restrictions. In real chains, bonds along the chain backbone cannot be oriented in any direction, but must be restricted by the bond angles. For instance, in hydrocarbon type chains, neighboring C–C bonds must form tetrahedral angles.

3. Hindered rotation effects. Rotations of neighboring bonds in a real macromolecule are not free, but are hindered by steric interferences. Figure 3 shows a schematic illustration of the rotation of two monomeric segments of a polyethylene chain (four CH_2 units). Because the neighboring nonbonded atoms exert a repulsive force against each other, the chain possesses a minimum potential energy when the atoms are situated as far away from each other as possible. This is achieved in (a), where the chain assumes a staggered conformation. Rotation of the segments by $60°$ clockwise causes the nonbonded atoms to be in close proximity to each other; the potential energy then reaches a maximum, as in (b). A further rotation of $60°$ results in another staggered conformation, but now the two bulky nonbonded carbon atoms are closer to each other (c), and the energy is not as low as in (a). The conformation of maximal energy is (d), where the two nonbonded carbon atoms are in close proximity to each other.

An elegant technique, which is based on the rotational isomeric model, takes into account the detailed molecular characteristics of real polymer chains when calculating their conformational properties. A detailed discussion of this

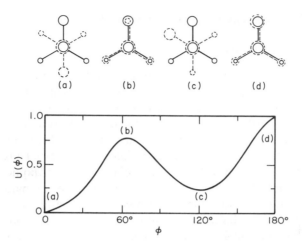

Fig. 3. Potential energy diagrams for four consecutive units of a hydrocarbon chain.

model is beyond the scope of this chapter. Interested readers are referred to the literature for further information [10–12]. The point we wish to make here is that although the hypothetical phantom random chain that we discussed in Section II.A is not realistic, this model is still widely used, because of its simplicity, in many calculations involving the physical properties of polymers. The use of this simple model can be justified by the concept of the "statistically equivalent" random chain [9, 13]. If a real chain is very long, we can represent its average properties by a hypothetical random chain of a different number of links, the lengths of which may be determined by the following two conditions:

$$\langle r_0^2 \rangle = \langle r_e^2 \rangle = n_e l_e^2 \tag{14}$$

$$R = R_e = n_e l_e \tag{15}$$

where R is the fully extended length of the real chain and the subscript e refers to the equivalent random chain. Simultaneous solution of Eqs. (14) and (15) gives

$$n_e = R^2/\langle r_0^2 \rangle, \qquad l_e = \langle r_0^2 \rangle/R \tag{16}$$

This means that if the real macromolecule is sufficiently long, it will still obey Gaussian statistics. Its average properties are represented by a statistically equivalent random chain with n_e links of lengths l_e. For instance, for a high molecular weight polyethylene chain, the equivalent chain contains one third more links whose lengths are about two and a half times longer than those in the real chain.

III. Rubberlike Elasticity of a Polymer Network

A. GAUSSIAN THEORY OF RUBBERLIKE ELASTICITY

Polymer chains above the glass transition temperature are capable, as mentioned in the last section, of undergoing Brownian motion. However, besides the translational (diffusion) and vibrational modes of thermal motion that are found in ordinary liquids, thermal segmental rotation is the most important form of diffusion in polymers. Even so, many of the properties of elastomers resemble those of liquids. For instance, the internal pressure and the cohesive energy density of rubbers are similar to those of liquids. The thermal expansion coefficient and bulk modulus of rubbers are comparable to those of many liquids, implying that their intermolecular forces are of the same order of magnitude. Uncross-linked elastomers, being liquid, flow under stress, except that because of their high molecular weight, the viscosities are much higher. Thus, in order to exhibit long-range elasticity, the polymer molecules must be cross-linked; i.e., their motions are now constrained so that they cannot flow past each other.

Our model for the calculation of rubberlike elasticity is thus a network of long chains connected with each other at one, or several points through cross-links. The Gaussian theory of rubber elasticity is formulated on the basis of the following assumptions:

1. The internal energy of the network is independent of deformation.
2. The individual network chains can be described by Gaussian statistics.
3. The free energy of the network is the sum of the free energies of the individual chains.
4. Cross-link sites are fixed at their mean position, separated by distances r_f, which upon deformation are homogeneously displaced in the same way as the bulk sample (affine deformation).

We first write the thermodynamic equation of state for a bulk elastomer as

$$dA = -S\,dT + f\,dL - P\,dV \tag{17}$$

which differs from that for a single chain [Eq. (10)] only by $P\,dV$ in the work term. If we now make the assumption that the volume is unchanged during the deformation,* then

$$f = (\partial A/\partial L)_{T,V} \tag{18}$$

---

* It has been amply demonstrated that the volume of an elastomer dilates by a few hundredths of 1% upon application of a uniaxial extension [2]. We make the constant volume assumption for the sake of simplicity. However, this constraint can be subsequently removed without undue difficulty.

Now, because of assumptions 1–3, we can write the free energy of a network consisting of N chains from Eq. (12) as

$$A_u = A' + kT \sum_{i=1}^{N} b^2(x_i^2 + y_i^2 + z_i^2) \tag{19}$$

where A_u is the free energy of the undeformed network. In the deformed state, we use assumption 4 and write

$$A_d = A' + kT \sum_{i=1}^{N} b^2(\alpha_1^2 x_i^2 + \alpha_2^2 y_i^2 + \alpha_3^2 z_i^2) \tag{20}$$

where α_1, α_2, and α_3 are the deformation ratios in the x, y, and z directions (corresponding to λ in Chapter 1). Since by definition [(Eq. 3)] for the ith chain

$$r_i^2 = x_i^2 + y_i^2 + z_i^2 \tag{21}$$

and for a random isotropic network all directions are equally probable:

$$x_i^2 = y_i^2 = z_i^2 = r_i^2/3 \tag{22}$$

we can therefore rewrite Eq. (20) as

$$\Delta A = \tfrac{1}{3}kT \sum_{i=1}^{N} b^2 r_i^2(\alpha_1^2 + \alpha_2^2 + \alpha_3^2 - 3)$$

$$= \tfrac{1}{3}NkT\langle b^2 r^2\rangle(\alpha_1^2 + \alpha_2^2 + \alpha_3^2 - 3) \tag{23}$$

where $\langle b^2 r^2\rangle = \sum_{i=1}^{N} b^2 r_i^2/N$. Using Eq. (5) and defining $\sum_{i=1}^{N} r_i^2/N \equiv \langle r_f^2\rangle$, we get

$$\langle b^2 r^2\rangle = \tfrac{3}{2}(\langle r_f^2\rangle/\langle r_0^2\rangle) \tag{24}$$

Equation (23) can now be written as

$$\Delta A = \tfrac{1}{2}NkT(\langle r_f^2\rangle/\langle r_0^2\rangle)(\alpha_1^2 + \alpha_2^2 + \alpha_3^2 - 3) \tag{25}$$

In uniaxial extension, if we select the x axis as the direction of stretch, then

$$\alpha_1 = L/L_0 \equiv \alpha \tag{26}$$

where L_0 is the initial and L the final sample length. Since we have already assumed that the volume of the sample is not changed upon deformation,

$$\alpha_1 \alpha_2 \alpha_3 = 1 \tag{27}$$

Combining Eqs. (26) and (27), we find that the contractions of the sample in the lateral directions are

$$\alpha_2 = \alpha_3 = 1/\alpha^{1/2} \tag{28}$$

Equation (25) now becomes

$$\Delta A = \tfrac{1}{2}NkT(\langle r_f{}^2\rangle/\langle r_0{}^2\rangle)(\alpha^2 + 2/\alpha - 3) \tag{29}$$

To find the elastic force, we differentiate according to Eq. (18) as follows:

$$f = \left(\frac{\partial A}{\partial L}\right)_{T,V} = \left(\frac{\partial A}{\partial \alpha}\right)_{T,V}\left(\frac{\partial \alpha}{\partial L}\right)_{T,V} = \frac{1}{L_0}\left(\frac{\partial A}{\partial \alpha}\right)_{T,V} \tag{30}$$

and thereby obtain the Gaussian equation of state for rubber elasticity [14–18]:

$$f = \frac{NkT}{L_0}\frac{\langle r_f{}^2\rangle}{\langle r_0{}^2\rangle}\left(\alpha - \frac{1}{\alpha^2}\right) \tag{31}$$

In Eq. (31) the quantity $\langle r_0{}^2\rangle/\langle r_f{}^2\rangle$ is often referred to as the front factor.

We have noted earlier that the volume of the elastomer in uniaxial tension increases by hundredths of a percent. To take this into account, we substitute the following transformation into Eq. (31):

$$\alpha = \lambda'(V_0/V)^{1/3} \tag{32}$$

where V_0 and V are the rubber volumes before and after stretching. In addition, the unperturbed dimension of a chain in the network should also increase by a factor of $(V/V_0)^{2/3}$. Equation (31) now becomes

$$f = GA_0(\lambda' - V/V_0\lambda'^2) \tag{33}$$

where A_0 is the cross-sectional area of unstrained rubber, $N_0 = N/V_0$, and

$$G = N_0kT\langle r_f{}^2\rangle/\langle r_0{}^2\rangle \tag{34}$$

for elastomers that undergo strain-induced volume dilation.*

To clarify the meaning of G, we rewrite the extension ratio as

$$\alpha = 1 + \varepsilon \tag{35}$$

where $\varepsilon = (L - L_0)/L_0$ is the tensile strain. Now, inserting Eq. (35) into Eq. (33), and expanding by the binomial theorem to first order, we get

$$f/A_0 = \sigma \approx 3G\varepsilon \tag{36}$$

where σ is the engineering stress. By definition the tensile modulus is $E = \sigma/\varepsilon$. For elastomers, with their very small volume change, it is permissible to set $E = 3G$, so that G is the shear modulus of the rubber.

The Gaussian equation of state predicts, therefore, that the stress in an elastomer is linearly proportional to strain [Eq. (36)] at low strains (Hookean). At higher strains, however, the behavior becomes nonlinear, as described by Eq. (33). This is sometimes referred to as neo-Hookean behavior. In both instances the proportionality constant is the shear modulus. According to

* For a more rigorous derivation of Eq. (33) see Aklonis et al. [8].

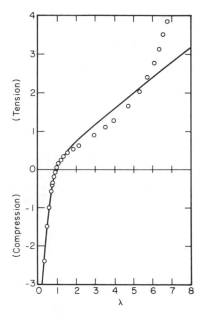

Fig. 4. Comparison of uniaxial extension and compression data of natural rubber (○) with Gaussian theory (————) [Eq. (33)]. (After Treloar [2].)

Eq. (34), the elastic force is directly proportional to absolute temperature; see also Eq. (13). It is also directly proportional to N, the number of network chains. Since, ideally, each cross-link connects four network chains, while each chain is terminated by two cross-links, the modulus is therefore also directly proportional to the degree of cross-linking. These conclusions are all in agreement with experimental findings.

Figure 4 shows that Eq. (33) provides an excellent fit to the experimental stress–strain curve in the region of low strains for both uniaxial extension and compression. The theory, however, deviates from experiment at extension ratios greater than 1.3, and fails to predict the upturn at $\lambda > 4$. Thus the validity range of the Gaussian theory for rubberlike elasticity is about 30% strain, above which it is necessary to use either non-Gaussian theories or phenomenological theories.

B. ENERGY CONTRIBUTION TO RUBBERLIKE ELASTICITY

In formulating the Gaussian theory of rubberlike elasticity, we assumed that the internal energy of the elastomer remains unaffected by conformational changes. This assumption (no. 1) is necessary to allow us to derive the free energy expression explicitly in terms of the conformational entropy alone. In real elastomers the energetic contribution obviously cannot be completely negligible. To examine this effect, we write the thermodynamic equation of state as

$$dU = +f\,dL - P\,dV + T\,dS \tag{37}$$

or, differentiated with respect to changing length,

$$f = f_e + f_s \tag{38}$$

where

$$f_e = (\partial U/\partial L)_{T,V} \tag{39a}$$

$$f_s = -T(\partial S/\partial L)_{T,V} \tag{39b}$$

We refer to f_e as the energetic component of the elastic force, and to f_s as the entropic component, while f is the total force. From the Maxwell relations, we get the thermodynamic identity

$$\left(\frac{\partial S}{\partial L}\right)_{T,V} \equiv -\left(\frac{\partial f}{\partial T}\right)_{V,L} \tag{40}$$

Combining Eq. (40) with Eqs. (38) and (39b), we get

$$f_e = f - T\left(\frac{\partial f}{\partial T}\right)_{V,L} \tag{41}$$

or

$$\frac{f_e}{f} = 1 - \left(\frac{d \ln f}{\partial \ln T}\right)_{V,L} \tag{42}$$

which is the contribution from internal energy to the total elastic force. However, the condition of constant volume is experimentally almost impossible to fulfill. Therefore, as an approximation, we use the Gaussian equation of state for rubberlike elasticity, Eq. (33), and find

$$\left(\frac{\partial \ln f}{\partial \ln T}\right)_{V,L} = \frac{d \ln G}{d \ln T} + \frac{\beta T}{3} \tag{43}$$

where β is the isobaric coefficient of thermal bulk expansion. Insertion of Eq. (43) into (42) yields [19, 20]

$$\frac{f_e}{f} = 1 - \frac{d \ln G}{d \ln T} - \frac{\beta T}{3} \tag{44}$$

Equation (44) indicates that if we know the temperature coefficient of the shear modulus of an elastomer and its thermal expansion coefficient, we can readily calculate the energy contribution to its rubberlike elasticity. For natural rubber, $f_e/f = 0.15$. In other words, 15% of the initial elastic force in natural rubber is due to internal energy and the other 85% to entropy changes; thus the foregoing approximation is not too bad.

Although the internal energy contribution to rubberlike elasticity is not negligible, Eq. (33) is still useful. The form of Eq. (44) indicates that over the range of its validity, the energy contribution is a constant fraction of the total elastic force independent of the strain. Thus the general shape of the stress–strain curve is unaffected. Neglecting the energy effects merely causes the predicted curve to shift by a constant fraction, so that the energetic component according to Eq. (38) does not affect other aspects of the Gaussian equation of state. The shear modulus G in Eq. (33) may therefore be obtained by taking the slope of a plot of f/A_0 versus $(\lambda - 1/\lambda^2)$, since it cannot be evaluated a priori from the quantities given by Eq. (34) (neither N_0 nor the front factor can be experimentally determined at present).

To clarify the mechanism by which the internal energy contributes to the elastic force, we differentiate Eq. (34) with respect to temperature and obtain

$$\frac{d \ln \langle r_f^2 \rangle}{d \ln T} = 1 - \frac{d \ln G}{\partial \ln T} - \frac{\beta T}{3} \qquad (45)$$

In performing the differentiation, we note that $N_0 = N/V_0$ and $\langle r_0^2 \rangle \propto V_0^{2/3}$. The right-hand sides of Eqs. (44) and (45) are the same; therefore

$$\frac{f_e}{f} = \frac{d \ln \langle r_f^2 \rangle}{d \ln T} \qquad (46)$$

Equation (46) indicates that the energy contribution to rubberlike elasticity is related to the temperature coefficient of the unperturbed chain dimensions, a result that has been confirmed experimentally. In other words, if the network chains are not isoenergetic, then the internal energy of the chains will change as the conformations change. The conformational changes can be induced either by stretching or by changing the supply of thermal energy.

Another very important point of Eq. (46) is that the internal energy contribution is intrachain rather than interchain in nature. This is significant because it validates the principle of free energy additivity (assumption 3) for the Gaussian theory of rubberlike elasticity. In arriving at Eqs. (19) and (20), we assumed that the free energy of the network as a whole is the sum of the free energies of the individual network chains. Thus the conformations of these network chains must be unaffected by the presence in the network of neighboring chains; i.e., they must be the same as in free space. This is possible only as long as interchain interactions are negligible. For some very polar rubbers, or at larger strains, Eq. (46) may no longer hold.

C. NON-GAUSSIAN THEORY OF RUBBERLIKE ELASTICITY

The statistical theory described in Section III.A is based on network chains that obey Gaussian statistics. This distribution function [Eq. (4)] is really valid, however, only for relatively small strains; i.e., for strains in which the end-to-end distance r is much less than nl. In the region of high strains, i.e., for $r/nl > \frac{1}{3}$ to $\frac{1}{2}$, this limit is exceeded. Here the finite extensibility of the network chains becomes important, and the appropriate non-Gaussian statistics must be employed. This distribution function was first derived by Kuhn and Grün [21] and is given by

$$w(r) = C \exp\left\{- n\left[\frac{r\mathscr{L}^{-1}(r/nl)}{nl} + \ln\frac{\mathscr{L}^{-1}(r/nl)}{\sinh \mathscr{L}^{-1}(r/nl)}\right]\right\} \tag{47}$$

where C is a normalization constant and $\mathscr{L}^{-1}$ is the inverse Langevin function. For small r/nl, it can be shown that Eq. (47) reduces to the Gaussian expression Eq. (4).

Consider a model for the rubber network that consists of three chains, each of which is parallel to the principal coordinates x, y, and z of a Cartesian system. We now assume that the volume will remain constant during a uniaxial extension, and that the deformation is affine. Using the Boltzmann relation, the entropy expression is

$$S = s_x + 2s_y = -kn\left[\frac{r\lambda}{nl}\mathscr{L}^{-1}\left(\frac{r\lambda}{nl}\right) + \ln\frac{\mathscr{L}^{-1}(r/nl)}{\sinh \mathscr{L}^{-1}(r\lambda/nl)}\right]$$
$$- 2kn\left[\frac{r\lambda^{-1/2}}{nl}\mathscr{L}^{-1}\left(\frac{r\lambda^{-1/2}}{nl}\right) + \ln\frac{\mathscr{L}^{-1}(r\lambda^{-1/2}/nl)}{\sinh \mathscr{L}^{-1}(r\lambda^{-1/2}/nl)}\right] \tag{48}$$

In Eq. (48) the total entropy S is considered to be the sum of the entropies of the three chains (s) in the three directions. For a rubber network consisting of N_0 chains, the total entropy of the network per unit volume can thus be obtained by simply multiplying each s by $N_0/3$. If we neglect the energy contribution, then from Eq. (38) it can be shown that [2]

$$f = \frac{N_0 kT}{3} n^{1/2}\left[\mathscr{L}^{-1}\left(\frac{\lambda}{n^{1/2}}\right) - \lambda^{-3/2}\mathscr{L}^{-1}\left(\frac{1}{\lambda^{1/2}n^{1/2}}\right)\right] \tag{49}$$

In Eq. (49) r has been replaced by the expression $(nl^2)^{1/2}$ for free chains. The equation reduces to the Gaussian expression for small values of λ. Figure 5

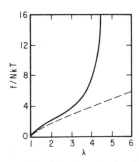

Fig. 5. Stress–strain relations predicted by Gaussian (-- --) and non-Gaussian (———) theories. (After Treloar [2].)

shows the stress–strain curve for $n = 25$. The most important feature of this theory is that it predicts an upturn at high strains, as observed in experiments (Fig. 4).

Although Eq. (49) seems to predict the sigmoidal stress–strain curve of elastomers, it is not necessarily a quantitative resprensentation of rubberlike elasticity. To formulate a complete molecular theory, other factors, such as excluded volume effects, chain entanglements, energy contributions, nonaffine deformation, and chain alignments, must also be considered. Such a theory has yet to be developed. In its absence, the Gaussian theory because of its simplicity remains a viable theory, providing at least to first approximation a basic molecular mechanism for rubberlike elasticity.

D. CONTINUUM THEORY OF FINITE DEFORMATION

The statistical theories discussed in Sections III.A and III.C were arrived at through consideration of the underlying molecular dynamics. The equation of state was obtained from considerations of their conformational properties. We shall now discuss the continuum theory of finite deformations, which concerns only the observed behavior of the rubbers and does not consider the molecular mechanism of the polymer. The central problem here is to find an expression for the elastic energy stored in the system that is analogous to the free energy expression in statistical theory [Eq. (25)].

Consider the deformation of a cube.* In order to arrive at the state of strain, a certain amount of work must be done, which is stored in the body as strain energy:

$$\overline{W}(\lambda_i) = \int_{\lambda_1 = 1} \sigma_1 \, d\lambda_1 + \int_{\lambda_2 = 1} \sigma_2 \, d\lambda_2 + \int_{\lambda_3 = 1} \sigma_3 \, d\lambda_3 \qquad (50)$$

where the λ's are the principal extension ratios, used as the α's in Eq. (20). This energy is a unique function of the state of strain, and if its amount were known as a function of strain, the elastic properties of the material would be completely defined. Although the strain energy function, expressed in terms of the principal extension ratios, is chosen without regard to the molecular mechanism, it must satisfy certain symmetry conditions in the case of isotropic solids.† First we consider a mirror reflection of the cube; i.e., we replace one of the λ's, say λ_1, by $-\lambda_1$. This is a rigid body motion in which no actual deformation is suffered by the cube. In the case of such an operation, the strain energy function must remain unaffected. It thus follows that $\overline{W}$ must depend explicitly on λ^2. Next, the assignment of λ's to each of the three principal axes is arbitrary. For instance, in the case of uniaxial extension, we assigned the λ's (α's) according to

---

* See also Chapter 1, Fig. 7.

† This theory is only valid as long as the deformed elastomers stay isotropic, i.e., up to 100–200% strain.

Eq. (26). Of course, there is no reason why we cannot assign λ_2 or λ_3 to be α. The strain energy function should not depend on such arbitrary choices, and must be invariant to any such permutations. These considerations then lead to expressions of $\overline{W}$ in terms of I, the so-called strain invariants [2, 22]:

$$\overline{W} = \overline{W}(I_i), \qquad i = 1, 2, 3 \tag{51}$$

where

$$I_1 = \lambda_1{}^2 + \lambda_2{}^2 + \lambda_3{}^2 \tag{52a}$$

$$I_2 = \lambda_1{}^2\lambda_2{}^2 + \lambda_2{}^2\lambda_3{}^2 + \lambda_3{}^2\lambda_1{}^2 \tag{52b}$$

$$I_3 = \lambda_1{}^2\lambda_2{}^2\lambda_3{}^2 \tag{52c}$$

The third strain invariant is obviously

$$I_3 = (V/V_0)^2 \tag{53}$$

which is equal to unity for an incompressible material [see Eq. (27)].

The most general form of strain energy function for an isotropic material is a power series:

$$\overline{W} = \sum_{i,j,k=0}^{\infty} C_{ijk}(I_1 - 3)^i(I_2 - 3)^j(I_3 - 1)^k \tag{54}$$

Quantities in the parentheses are so chosen that the strain energy vanishes at zero strain; see also Eq. (23). Since we cannot determine a priori the set of terms in Eq. (54), let us examine the lowest members of the series. For $i = 1, j = 0$, $k = 0$,

$$\overline{W} = C_{100}(I_1 - 3) \tag{55}$$

which is functionally identical to the Gaussian free energy of deformation [Eq. (25)]. The stress–strain relation can then be obtained from Eq. (50) by differentiation. For the case of uniaxial extension

$$\lambda_1 = \lambda \tag{56}$$

Using Eqs. (52c) and (53) we get

$$\lambda_2 = \lambda_3 = (V/V_0\lambda)^{1/2} \tag{57}$$

Equation (55) now becomes

$$\overline{W} = C_{100}(\lambda^2 + 2V/V_0\lambda) \tag{58}$$

Setting $2C_{100} = G$, we can readily show that the stress $\sigma = f/A_0$ is

$$\sigma = \partial\overline{W}/\partial\lambda = G(\lambda - V/V_0\lambda^2) \tag{59}$$

which is the neo-Hookean equation, and is identical in form to the Gaussian expression [Eq. (33)]. Suppose we retain an additional term with $i = 0, j = 1, k = 0$; then

$$\overline{W} = C_{100}(I_1 - 3) + C_{010}(I_2 - 3) \qquad (60)$$

For uniaxial extension, we obtain

$$\sigma = (2C_1 + 2C_2/\lambda)(\lambda - V/V_0\lambda^2) \qquad (61)$$

where for simplicity we have set $C_1 = C_{100}$ and $C_2 = C_{010}$. Equation (61) is known as the Mooney–Rivlin equation [22, 23].

The Mooney–Rivlin equation differs from the neo-Hookean equation, or the Gaussian equation, by the extra C_2 term. If we set $V/V_0 = 1$, then Eq. (59) predicts that $\sigma/(\lambda - 1/\lambda^2)$ is a constant (G). Equation (61), on the other hand, would predict that $\sigma/(\lambda - 1/\lambda^2)$ varies as a function of λ^{-1}. If we plot $\sigma/(\lambda - 1/\lambda^2)$ against $1/\lambda$, then according to Eq. (61) we should find a straight line whose slope is $2C_2$ and the intercept is $2C_1$. The stress–strain data of Fig. 4 were plotted in this manner in Fig. 6. It is seen that up to $1/\lambda > 0.4$ they agree indeed with the Mooney–Rivlin equation. Below this value $(\lambda > 2.5)$, the deviations from the Mooney–Rivlin theory become significant. To account for this upturn at high strains, we can retain one more term [24–26] in the strain energy function [Eq. (54)], e.g., the C_{200} term

$$\overline{W} = C_{100}(I_1 - 3) + C_{200}(I_1 - 3)^2 + C_{010}(I_2 - 3) \qquad (62)$$

For simplicity we shall show the stress–strain relation for $V/V_0 = 1$:

$$\sigma = (C + C'/\lambda + C''\lambda^2)(\lambda - 1/\lambda^2) \qquad (63)$$

where $C = 2(C_{100} - 6C_{200})$, $C' = 2(4C_{200} + C_{010})$, and $C'' = 4C_{200}$. A much better agreement with experimental data can be obtained for this three-constant equation [25], as seen in Fig. 6.

It should be restated that all these phenomenological equations of state were developed for the purpose of describing experimental stress–strain data.

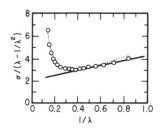

Fig. 6. Mooney–Rivlin plot for sulfur-cured natural rubber. Solid line, Eq. (61); dotted line, Eq. (63). (After Sato [25].)

No molecular model was assumed in their derivations. Although there have been some attempts in the literature to assign molecular mechanisms to the constants, for instance to the C_2 term, they have not been fruitful, since these constants were arbitrarily chosen from terms in the strain energy functions which are devoid of molecular significance.

IV. Applications of Rubberlike Elasticity Theories

A. Swollen Elastomers

When immersed in an appropriate solvent, a cross-linked elastomer will not dissolve, but becomes distended isotropically by imbibing the solvent into the network. Thus, swelling is a three-dimensional form of deformation. If the ratio of the dry volume of rubber to that of swollen rubber is V_r, then we can write the neo-Hookean strain energy function per unit volume of swollen rubber in the following manner:

$$\overline{W}_s = V_r \overline{W}_d = V_r[C_{100}(I_{1d} - 3)] \tag{64}$$

where the subscripts s and d refer to swollen and dry samples, respectively. In Eq. (64) the strain-invariant I_{1d} is defined by λ_d's, which are the strains suffered by the dry rubber through both swelling and extension:

$$I_{1d} = \lambda_{1d}^2 + \lambda_{2d}^2 + \lambda_{3d}^2 \tag{65}$$

However, since the swelling is an isotropic deformation, λ_s is just $V_r^{-1/3}$ for all three principal axes. We can relate λ_d's to λ_s's (deformation of swollen rubber by extension only) as follows:

$$\lambda_{1d} = \lambda_{1s}V_r^{-1/3}, \qquad \lambda_{2d} = \lambda_{2s}V_r^{-1/3}, \qquad \lambda_{3d} = \lambda_{3s}V_r^{-1/3} \tag{66}$$

If we assume that there is no volume change in the extension of a swollen rubber, then in analogy to Eq. (27)

$$I_{3s} = \lambda_{1s}^2\lambda_{2s}^2\lambda_{3s}^2 \tag{67}$$

For the case of uniaxial elongation, we obtain, as in Eqs. (56) and (57),

$$\lambda_{1s} = \lambda_s, \qquad \lambda_{2s} = \lambda_{3s} = \lambda_s^{-1/2} \tag{68}$$

The neo-Hookean strain energy function for swollen elastomers is thus

$$\overline{W}_s = GV_r^{1/3}(\lambda_s^2 + 2/\lambda_s - 3) \tag{69}$$

Following the method of Eq. (59), we get the stress–strain relation [2]

$$\sigma_s = GV_r^{1/3}(\lambda_s - 1/\lambda_s^2) \tag{70}$$

Equation (70) differs from Eq. (59) by the factor $V_r^{1/3}$. In other words, the shear modulus of an elastomer decreases with increasing degree of swelling in proportion to $V_r^{1/3}$. If we now consider the expression for G [Eq. (34)] in the Gaussian theory, we note that the number of network chains per unit volume must decrease with increasing swelling as $N_0 V_r$, and that the mean square end-to-end distance of the chain in a swollen network should be $\langle r_0^2 \rangle V_r^{-2/3}$; therefore the shear modulus of the swollen elastomer is

$$G_s = N_0 k T V_r^{1/3} \langle r_f^2 \rangle / \langle r_0^2 \rangle = G V_r^{1/3} \tag{71}$$

which is in agreement with the prediction of the continuum theory as well as with experiments.

B. FILLED AND SEMICRYSTALLINE ELASTOMERS

The introduction of reinforcing fillers into the elastomer generally raises the modulus of the system. The most commonly used equation for the increase in modulus as a function of the volume fraction of fillers (ϕ) is the Guth–Smallwood equation [27]:

$$G_f/G = 1 + 2.5\phi + 14.1\phi^2 \tag{72}$$

where G is the shear modulus of the unfilled elastomer and G_f is that of the filled elastomer. Equation (72) is valid for values of ϕ up to 0.3.

The physical basis of the filler effect is as follows. Since the modulus of the fillers (e.g., carbon black) is usually much greater than that of the elastomeric matrix (by about 5 orders of magnitude), application of an external force to the sample will leave the fillers essentially undeformed. All the strain suffered by the sample as a whole must therefore be borne by the elastomer alone. In other words, the actual strain sustained by the rubber phase is greater than the overall applied strain, since the fillers are not deformable. The amount by which the actual strain is greater is given by the right-hand side of Eq. (72). This quantity has in fact been referred to as the "strain amplification factor" [28]. Thus the actual elongation ratio Λ for the elastomeric matrix is related to the applied strain ε as

$$\Lambda = 1 + (1 + 2.5\phi + 14.1\phi^2)\varepsilon \tag{73}$$

Mullins and Tobin [29] have shown that Mooney–Rivlin plots for a series of carbon-black-filled natural rubber up to $\phi = 0.138$, all fall on the same curve. Their data lend strong support to the strain-amplifying role of the fillers.

An interesting case of filler effect is the elasticity of semicrystalline polymers. Crystallites in such polymers act as fillers and as cross-link sites [30, 31]. Thus, Eq. (72) becomes applicable to semicrystalline polymers if ϕ is taken as the degree of crystallinity. The shear modulus can still be defined by Eq. (34) and the Gaussian theory remains valid. Tobolsky showed that if the degree of crystallinity is not too high, Eq. (33) can be used for semicrystalline elastomers if the number of amorphous chains is expressed as [30]

$$N_0 = (1 - \phi)\rho/vm \tag{74}$$

where ρ is the density of the polymer, v the average size of the noncrystallizable sequence in the chain, and m the molecular mass of each segment.

C. Elastomers Undergoing Chemical Change

Up to this point, we have discussed rubberlike elasticity under the assumption that the chemical nature of the elastomer remains unchanged in its strained state. In practice, of course, some chemical changes can often be expected to occur [32–34]. For example, if a cross-linked elastomer is held at constant strain at elevated temperatures, it may return only partly to its unstrained position when released. Such an irreversible deformation is often called the permanent set [6].

The permanent set is caused by simultaneous reactions of chain scission and cross-linking. Chains of the original network under strain are gradually ruptured by chemical scission and thus become reactive sites. A new network is formed by cross-linking in equilibrium with the strained sample by these reactive groups. When the sample is released, the weakened original network will tend to return the sample to the unstrained length, but the new, second network will resist its return. As a consequence, the sample can only retract to a length at which these two opposing forces are equal, causing the observed permanent set.

Experimentally, the occurrence of simultaneous chain scission and cross-linking can be studied by parallel continuous and intermittent stress–relaxation. Presumably, under equilibrium conditions, the elastic force in the elastomer upon deformation can be predicted by one of the equations of state for rubber-like elasticity. If chain scission occurs, however, the stress will decay as a function of time, while the sample is held at a fixed elongation (continuous stress–relaxation) during which the number of effective cross-links changes. In an intermittent stress–relaxation experiment, the sample is left unstrained at the same elevated temperature as in the continuous stress–relaxation test. At various time intervals, the sample is rapidly stretched and its stress is quickly measured. Afterward, the sample is returned to its unstrained condition until the next measurement. If only scission takes place, then the two methods will yield an identical stress–relaxation curve. However, if cross-linking reactions proceed simultaneously, then the intermittent stress–relaxation will show a more gradual decay, since the effect of chain scission is partially compensated by the "repairing" of the original network through cross-linking in the unstrained condition. The data for a Hevea gum are shown in Fig. 7.

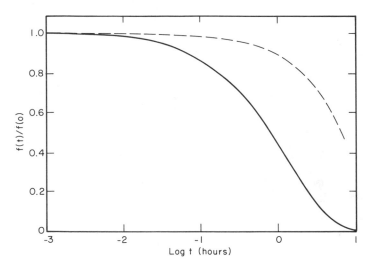

Fig. 7. Continuous and intermittent stress relaxations of Hevea gum at 130°C (50% elongation). (After Andrews *et al.* [32].)

It is possible to compute the permanent set from the continuous and intermittent stress–relaxation data by using the equation of state for rubberlike elasticity. For the sake of simplicity, we shall assume the absence of strain-induced volume dilation and use Eq. (31). The condition for permanent set is that the opposing forces from the two networks be equal. These are designated as X and U in Fig. 7. The permanent set is defined as

$$\varepsilon_p = \left(\frac{L_s - L_0}{L_x - L_0}\right) \times 100 \tag{75}$$

where L_s is the new equilibrium unstrained length of the sample at time t, L_0 the original unstrained length, and L_x the strained length. The force due to the network at equilibrium at the original length of the sample is [from Eq. (31)]

$$f_u = \frac{N_u kT}{L_0} \frac{\langle r_f^2 \rangle}{\langle r_0^2 \rangle}\left[\left(\frac{L_s}{L_0}\right) - \left(\frac{L_0}{L_s}\right)^2\right] \tag{76}$$

which can be determined from the continuous stress–relaxation curve at time t. N_u is the total number of network chains at equilibrium with the original length L_0. Since after the chemical reactions have taken place, the original sample dimensions will never be attained again, it is more sensible to express the force per cross-sectional area of the strained sample A_s. Equation (76) can now be rewritten as

$$\frac{f_u}{A_s} = \frac{N_u kT}{L_0 A_s} \frac{\langle r_f^2 \rangle}{\langle r_0^2 \rangle} \left[\left(\frac{L_s}{L_0} \right) - \left(\frac{L_0}{L_s} \right)^2 \right]$$

$$= \frac{N_u kT}{L_s A_s} \frac{\langle r_f^2 \rangle}{\langle r_0^2 \rangle} \left(\frac{L_s}{L_0} \right) \left[\left(\frac{L_s}{L_0} \right) - \left(\frac{L_0}{L_s} \right)^2 \right]$$

$$= \frac{N_u kT}{V_0} \frac{\langle r_f^2 \rangle}{\langle r_0^2 \rangle} \left[\left(\frac{L_s}{L_0} \right)^2 - \left(\frac{L_0}{L_s} \right) \right] \tag{77}$$

Since we have assumed that no volume change occurs, $L_s A_s = V_0 = L_0 A_0$. With the definition of shear modulus [Eq. (34)], we get

$$\sigma_u' = \frac{f_u}{A_s} = G \left(\frac{N_u}{N} \right) \left[\left(\frac{L_s}{L_0} \right)^2 - \left(\frac{L_0}{L_s} \right) \right] \tag{78}$$

where σ_u' is the "true stress" (force per cross-sectional area of the strained sample). Similarly, we can write the opposing force due to the second network as

$$\sigma_x' = \frac{f_x}{A_s} = G \left(\frac{N_x}{N} \right) \left[\left(\frac{L_s}{L_x} \right)^2 - \left(\frac{L_x}{L_s} \right) \right] \tag{79}$$

where N_x is the number of chains in equilibrium with the extended length of the sample. At the length of the permanent set, we have by definition

$$\sigma_u' = -\sigma_x' \tag{80}$$

The negative sign of σ_x' indicates that it is an opposing force. Inserting the expressions for σ_u' and σ_x' [Eqs. (78) and (79)] into Eq. (80), we have

$$\left(\frac{N_u}{N_x} \right) \left[\left(\frac{L_s}{L_0} \right)^2 - \frac{L_0}{L_s} \right] = \left(\frac{L_x}{L_s} \right) - \left(\frac{L_s}{L_x} \right)^2 \tag{81}$$

Equation (81) can be rearranged to read

$$\frac{L_s}{L_0} = \left[\frac{(L_x/L_0)^3 - 1}{(N_u L_x^2 / N_x L_0^2) - 1} + 1 \right]^{1/3} \tag{82}$$

We rewrite the definition of permanent set [Eq. (75)] as

$$\varepsilon_p = \frac{(L_s/L_0) - 1}{(L_x/L_0) - 1} \times 100. \tag{83}$$

Inserting Eq. (82) into (83), we obtain, after rearrangement [32],

$$\varepsilon_p = Q_3 \left\{ \left[\frac{Q_1}{Q_2(N_u/N_x) + 1} + 1 \right]^{1/3} - 1 \right\} \tag{84}$$

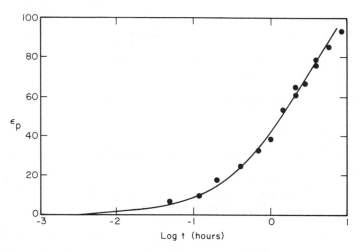

Fig. 8. Comparison of permanent set data of Hevea gum at 130°C (50% elongation) with theory [Eq. (84)]. (After Andrews *et al.* [32].)

where

$$Q = (L_x/L_0)^3 - 1$$

$$Q_2 = (L_x/L_0)^2$$

$$Q_3 = 100/[(L_x/L_0) - 1]$$

The ratio N_u/N_x can be experimentally obtained from continuous (c) and intermittent (i) stress relaxation measurements:

$$\frac{N_u}{N_x} = \frac{[\sigma'(t)/\sigma'(0)]_c}{[\sigma'(t)/\sigma'(0)]_i - [\sigma'(t)/\sigma'(0)]_c} \tag{85}$$

Figure 8 shows a comparison of permanent set data with Eq. (84). Excellent agreement between theory and experiment is evident [32].

ACKNOWLEDGMENT

The author is indebted to the Dreyfus Foundation for a Teacher-Scholar Grant.

REFERENCES

1. P. J. Flory, "Principles of Polymer Chemistry." Cornell Univ. Press, Ithaca, New York, 1953.
2. L. R. G. Treloar, "Physics of Rubber Elasticity." Oxford Univ. Press, London and New York, 1958.
3. K. H. Meyer, G. von Susich, and E. Valko, *Kolloid-Z.* **59**, 208 (1932).
4. W. Kuhn, *Kolloid-Z.* **68**, 2 (1934).

5. E. Guth and H. Mark, *Monatsh. Chem.* **65,** 93 (1934).
6. A. V. Tobolsky, "Properties and Structure of Polymers." Wiley, New York, 1960.
7. F. Bueche, "Physical Properties of Polymers." Wiley (Interscience), New York, 1962.
8. J. J. Aklonis, W. J. MacKnight, and M. Shen, "Introduction to Polymer Viscoelasticity." Wiley (Interscience), New York, 1972.
9. W. Kuhn, *Kolloid-Z.* **76,** 258 (1936).
10. M. V. Volkenstein, "Configurational Statistics of Polymeric Chains." Wiley (Interscience), New York, 1963.
11. T. M. Birshtein and O. B. Ptitsyn, "Conformations of Macromolecules." Wiley (Interscience), New York, 1966.
12. P. F. Flory, "Statistical Mechanics of Chain Molecules." Wiley (Interscience), New York, 1969.
13. W. Kuhn, *Kolloid-Z.* **87,** 3 (1939).
14. H. M. James and E. Guth, *J. Chem. Phys.* **11,** 455 (1934); *J. Polym. Sci.* **4,** 153 (1949).
15. P. J. Flory, *Chem. Rev.* **35,** 51 (1944); *J. Am. Chem. Soc.* **18,** 5232 (1956); *Trans. Faraday Soc.* **5,** 829 (1961).
16. M. S. Green and A. V. Tobolsky, *J. Chem. Phys.* **14,** 80 (1943); A. V. Tobolsky, D. W. Carlson, and N. Indictor, *J. Polym. Sci.* **54,** 175 (1961).
17. F. T. Wall, *J. Chem. Phys.* **10,** 485 (1942); **11,** 527 (1943).
18. J. J. Hermans, *Trans. Faraday Soc.* **43,** 591 (1947); *J. Polym. Sci.* **59,** 191 (1962).
19. M. Shen and P. J. Blatz, *J. Appl. Phys.* **39,** 4937 (1968).
20. M. Shen, *Macromolecules* **2,** 358 (1969).
21. W. Kuhn and F. Grün, *Kolloid-Z.* **101,** 248 (1942).
22. R. S. Rivlin, *in* "Rheology" (F. Eirich, ed.), Vol. 1, p. 351. Academic Press, New York, 1956.
23. M. J. Mooney, *J. Appl. Phys.* **11,** 582 (1940).
24. A. Signorini, *Ann. Mater. Pura Appl.* **39,** 147 (1955).
25. Y. Sato, *Rp. Prog. Polym. Phys. Jpn.* **9,** 369 (1969).
26. N. W. Tschoegl, *J. Polym. Sci., Part A-1* **9,** 1959 (1971).
27. H. M. Smallwood, *J. Appl. Phys.* **15,** 758 (1944).
28. E. Guth, *J. Appl. Phys.* **16,** 20 (1945).
29. L. Mullins and N. R. Tobin, *Trans. Inst. Rubber Ind.* **33,** 2 (1956).
30. A. V. Tobolsky, *J. Chem. Phys.* **37,** 1139 (1962).
31. L. E. Nielsen and F. D. Stockton, *J. Polym. Sci., Part A 1.* 1995 (1963).
32. R. D. Andrews, A. V. Tobolsky, and E. E. Hanson, *J. Appl. Phys.* **16,** 352 (1946).
33. J. P. Berry and W. F. Watson, *J. Polym. Sci.* **18,** 201 (1955).
34. P. Thirion and R. Chasset, *Rubber Chem. Technol.* **37,** 617 (1964); see also *in* "Physics of Non-Crystalline Solids" (J. A. Prins, ed.), p. 345. North-Holland Publ., Amsterdam, 1964.

Chapter 5

Dynamic Mechanical Properties

OLE KRAMER

UNIVERSITY OF COPENHAGEN
COPENHAGEN, DENMARK

and

JOHN D. FERRY

UNIVERSITY OF WISCONSIN
MADISON, WISCONSIN

179

Copyright © 1978 by Academic Press, Inc.
All rights of reproduction in any form reserved.
ISBN 0-12-234360-3

I. Introduction

Rubber is frequently used for applications in which it undergoes rapid cyclic deformations at a certain frequency or over a range of frequencies. Examples are the sidewall or the tread of a rotating tire; and engine mounts, where use is made of the flexibility, friction, cushioning, and damping properties of rubber. The dynamic mechanical properties are strongly dependent on temperature, frequency, the presence of fillers, and the extent of deformation if it is large. For very small deformations, the properties are found to be independent of the magnitude of the deformation (linear viscoelasticity), and temperature–frequency superposition can be applied in many cases, a feature which greatly facilitates obtaining or predicting the dynamic response over a wide frequency range.

It is important to note that the dynamic mechanical properties of different unfilled amorphous elastomers are quite similar when the proper reference state is used [1]. The glass transition temperature is often used as reference. However, at the glass transition temperature the mechanical properties are no longer elastomeric. A better reference comparison can be formulated in terms of the frequency for the onset of the transition zone at a given temperature, as will be explained in Section IV; this quantity is closely related to the local friction coefficient per monomer unit, which is an inverse measure of the molecular mobility of the polymer chains. It will be seen that a difference in chemical structure of the elastomer will change the location of a dynamic response curve on the frequency scale but it will not change the overall shape of the frequency dependence appreciably. The dependences on molecular weight, molecular weight distribution, and cross-linking are similar for different elastomers. Only the important rubber plateau modulus of an uncross-linked polymer (G_N^0), which expresses the elasticity of the entanglement network, may vary considerably from one polymer to another.

The special effects which arise due to the existence of crystallinity under certain conditions in elastomers, such as natural rubber [2] and EPDM [3], will not be covered in this chapter.

Both cgs and SI units are used here in illustrations of quantitative data, since currently the scientific community is in a process of transition between the two systems. A stress or modulus of 1 N/m² (SI) is equivalent to 10 dyn/cm² (cgs).

II. Definitions of Dynamic Mechanical Properties and Viscoelastic Functions

"Dynamic" mechanical properties refer generally to responses to periodically varying strains or stresses. They are most simply defined for a small sinusoidally varying strain or stress, for which the response is a small sinusoidally varying stress or strain, respectively, with the same frequency but generally out of phase. Most of this chapter is restricted to this simple case, although large and/or nonsinusoidal (square, triangular, pulsing, etc.) periodic stresses are of course also of great practical importance.

A. STORAGE AND LOSS MODULI AND COMPLIANCES

For small deformations in simple shear, if the strain γ varies sinusoidally as

$$\gamma(t) = \gamma^0 \sin \omega t \tag{1}$$

where γ^0 is the peak strain, ω the radian frequency (2π times the frequency in cycles per second) and t the time, the stress σ will be

$$\sigma(t) = \sigma^0 \sin(\omega t + \delta) \tag{2}$$

as illustrated in Fig. 1. Here the stress σ is the tangential force per unit area producing the deformation and the strain γ is the tangent of the angle by which the cubical element is deformed. The stress can be decomposed into a component in phase with the strain proportional to $\sin \omega t$ and another component

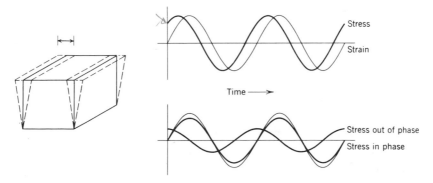

Fig. 1. Sinusoidally varying simple shear; strain and stress, and stress decomposed into in-phase and out-of-phase components [1].

90° out of phase proportional to cos ωt, as shown in the figure; the total stress can then be expressed as

$$\sigma = \gamma^\circ [G'(\omega) \sin \omega t + G''(\omega) \cos \omega t] \qquad (3)$$

thus defining two shear moduli, $G'(\omega)$ and $G''(\omega)$. The storage modulus G' is a measure of energy (elastic) stored and recovered in cyclic deformation; the loss modulus G'' is a measure of energy dissipated as heat. The ratio G''/G' is tan δ, the loss tangent [1, Chs. 1 and 3; 4].

The two moduli are functions of frequency, as illustrated in Fig. 2 for a typical gum stock rubber [5]. At low frequencies, G' approaches the equilibrium shear modulus G_e treated in Chapter 4 and G'' becomes very small; the rubber is nearly perfectly elastic in small deformations. With increasing frequency, G' and G'' increase and become roughly equal in magnitude (at room temperature around 10^6 Hz); this change is related to local molecular mobility, as will be explained later. Eventually, G' would approach a high value characteristic of

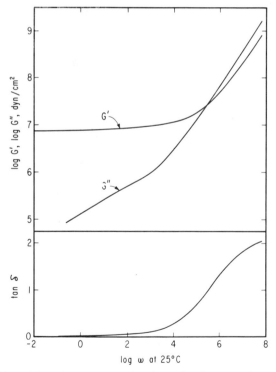

Fig. 2. Plots of log G', log G'', and tan δ against log radian frequency for a typical gum rubber: styrene–butadiene rubber [5] vulcanized with dicumyl peroxide to an equilibrium shear modulus of 7.2×10^6 dyn/cm^2.

a hard glass rather than a rubber, but at room temperature this would require exceedingly high frequencies. Here tan δ, a measure of the "lossiness" or relative dissipation versus storage of energy, would go through a maximum at a frequency somewhat to the right of Fig. 2. It is closely related to properties such as damping of vibrations and attenuation of propagated waves (see Sections VI.D and VI.F).

For small deformations in simple extension (elongations of less than a few percent, so that Hookean behavior is approached), the corresponding Young's moduli E' and E'' are defined in a manner analogous to the ratios of the in-phase and out-of-phase tensile stresses to the practical tensile strain $\varepsilon \, (= \lambda - 1$, where λ is the stretch ratio, i.e., the stretched length divided by the unstretched length). Unless the frequency is so high that the moduli approach glasslike magnitudes $(G' > 10^9 \text{ dyn/cm}^2)$, the ratios E'/G' and E''/G'' can be well approached by 3, so dynamic shear and extensional experiments give equivalent information. (For very high frequencies, E'/G' lies between 2.5 and 3.)

In an alternative formalism, the *strain* can be expressed in terms of two components in phase and 90° out of phase with the *stress*. Ratios of strain to stress are called compliances, and the storage and loss compliances $J'(\omega)$ and $J''(\omega)$ are defined by the equation

$$\gamma = \sigma^0[J'(\omega) \sin \omega t - J''(\omega) \cos \omega t] \tag{4}$$

Although J' and G' are both measures of energy storage, they differ in that comparisons of J' are comparisons at corresponding stresses, whereas G' compares at corresponding strains.

The reader should be warned that G' is not the reciprocal of J', etc. Specifically,

$$J' = G'/(G'^2 + G''^2) \tag{5}$$

$$J'' = G''/(G'^2 + G''^2) \tag{6}$$

and the same equations hold with G and J interchanged. The loss tangent, however, is the ratio of loss to storage for both modulus and compliance, i.e., $\tan \delta = G''/G' = J''/J'$. The compliances for simple extension are D' and D'', and they are smaller than J' and J'', respectively, by a factor of 3 (except at very high frequencies).

B. RELATION TO OTHER LINEAR VISCOELASTIC FUNCTIONS

The quantities G', G'', tan δ, J', and J'', all functions of frequency, are called linear viscoelastic functions because they describe the "viscoelastic" behavior of materials, like rubber, that combine both viscous and elastic character in that they both dissipate and store energy. The adjective "linear" refers to the fact that the stress/strain and strain/stress ratios, though dependent on frequency, are independent of the magnitude of peak stress as long as the latter is sufficiently small.

There are other well-known linear viscoelastic functions, such as the creep compliance $J(t)$ and the relaxation modulus $G(t)$. If a sample of material is suddenly subjected to a shear stress σ and its strain γ is followed as a function of time,

$$\gamma(t)/\sigma = J(t) \tag{7}$$

If a sample is suddenly deformed to a shear strain γ and the stress required to maintain this deformation is followed as a function of time,

$$\sigma(t)/\gamma = G(t) \tag{8}$$

For a vulcanized rubber, $J(t)$ and $G(t)$ eventually attain equilibrium values J_e and G_e ($J_e = 1/G_e$), the latter being the equilibrium shear modulus treated in Chapter 4. For an unvulcanized rubber, $G(t)$ eventually falls to zero and $J(t)$ increases without limit, corresponding to viscous flow under constant stress (Chapter 6).

The linear viscoelastic functions are all interrelated. *Very* roughly, $J'(\omega)$ approximates the value of $J(t)$ at $t = 1/\omega$, and similarly for $G'(\omega)$ and $G(t)$. In order of magnitude, $J''(\omega)$ can be approximated by the slope $dJ(t)/d \ln t$ and similarly for $G''(\omega)$ and $G(t)$. For any quantitative comparison or interconversion of these different kinds of data, however, it is necessary to use certain approximation equations [6], some of which are rather complicated but can be easily evaluated by computer.

III. Zones of Dynamic Viscoelastic Behavior and Their Qualitative Molecular Interpretation*

In Fig. 2, two characteristic zones of frequency scale are seen—at low frequencies, G' is nearly constant and G'' is quite small; at higher frequencies, they are similar in magnitude and increase with increasing frequency. These zones are part of a more extensive pattern which appears in uncross-linked (unvulcanized) polymers of high molecular weight and is illustrated in Fig. 3 for poly(n-octyl methacrylate) [7, 8]. (The general features of the pattern are much the same for all elastomers, and this one is chosen only because there are unusually extensive data for it.) Each zone will be discussed in turn, with a qualitative explanation of the viscoelastic behavior associated with it.

A. The Plateau Zone

In the plateau zone, G' changes little with frequency and G'' passes through a minimum. This behavior is usually interpreted by a somewhat vaguely defined

---

* See Ferry [1, Ch. 2] and Ward [4, Ch. 7].

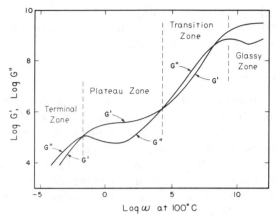

Fig. 3. Zones of viscoelastic behavior illustrated by logarithmic plots of G' and G'' against frequency for uncross-linked poly(n-octyl methacrylate) (molecular weight 3.6×10^6) at $100°C$ [7, 8].

concept, entanglement coupling. The tangled, contorted macromolecules form a network in which the molecules act in some respects as though they were rather tightly coupled at widely separated points (of the order of 200 chain atoms apart along each molecule). Within the period of oscillation in this frequency range, there is plenty of time for the network strands between coupling points to rearrange their configurations and store elastic energy through an entropy change, just as a vulcanized rubber does (as treated in Chapter 4); but there is not enough time for the much more complicated molecular rearrangements that would enable widely separated portions of a molecule to change their relative positions, since these processes would involve complicated snaking motions through the entanglements. Hence within this particular frequency range the material behaves almost as though it were cross-linked. However, there is always more energy dissipation than in a vulcanized gum rubber at low frequencies. Where the plateau zone actually falls on the frequency scale depends on the chemical nature of the polymer, the temperature, and other variables.

B. THE TRANSITION ZONE

At higher frequencies than the plateau zone, the period of oscillation becomes too short to allow all the possible configurational changes of a strand caught between two entanglements. Then the strain corresponding to a given stress is less, and the modulus increases with increasing frequency. At the same time, a lag between changing stress and changing strain causes extra dissipation of energy and G'' increases faster than G'. This behavior epitomizes the transition zone, so called because G' rises from a low magnitude, characteristic of a soft elastomer, to a high magnitude, characteristic of a hard glass. The polymer is indeed losing its elastomeric quality as the frequency increases, becoming first leathery and then hard, as felt by the dynamic measurement (though its physical state is still of course the same).

C. THE GLASSY ZONE

At very high frequencies, there are no configurational rearrangements of polymer chain backbones taking place within the period of oscillation, except for very local motions, the exact nature of which is still a matter of conjecture. The strain in response to a given stress is very small and corresponds to a storage modulus G' of the order of 10^{10} dyn/cm^2, that of a hard glasslike solid. The lag between stress and strain is less, and tan δ is of the order of 0.1, so the material is less lossy than in the transition zone (tan $\delta > 1$) but more lossy than a gum rubber (tan δ of the order of 0.01).

D. THE TERMINAL ZONE

In the other direction on the frequency scale, the plateau zone is replaced by the terminal zone at very low frequencies. Here the period of oscillation is long enough so that the molecules can snake through their entanglement constraints and completely rearrange their configurations. The boundary between plateau and terminal zones is rather abrupt if the polymer molecular weight distribution is narrow and it shifts to lower frequencies with increasing molecular weight M (the boundary frequency being approximately inversely proportional to $M^{3.5}$). For high molecular weight at room temperature, it might be of the order of one cycle per week. Viscoelastic behavior in the terminal zone is reflected in processing and molding operations and is described more fully in Chapter 6.

E. THE EQUILIBRIUM AND ENTANGLEMENT ZONES IN LIGHTLY CROSS-LINKED ELASTOMERS

In a cross-linked elastomer, at least if the number of chemical cross-links exceeds the number of entanglements originally present, the equilibrium zone replaces the plateau zone of an uncross-linked polymer, as can be seen by comparing Figs. 2 and 3; the entanglements originally present do not manifest themselves, although they may make an important contribution to the equilibrium modulus, and there is of course no terminal zone since the permanent cross-links prevent any long-range configurational changes of the original linear molecules. However, if the degree of cross-linking is very slight, an additional loss mechanism appears, as illustrated in Fig. 4 for a very lightly vulcanized styrene–butadiene rubber [5] (cf. Fig. 2). Elastic equilibrium, where G' levels off, is achieved only at extremely low frequencies. The extra loss appearing as a small maximum in G'' in the vicinity of log $\omega = 2$ and extending to lower

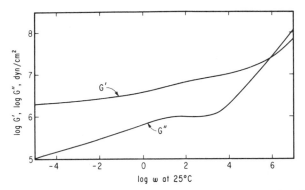

Fig. 4. Logarithmic plots of G' and G'' against radian frequency at 25°C for a very lightly cross-linked gum rubber: styrene–butadiene rubber [5] cross-linked with dicumyl peroxide to an equilibrium shear modulus of 1.5×10^6 dyn/cm².

frequencies is attributed primarily to slippage of untrapped entanglements on dangling branched structures, although rearrangement of trapped entanglements between cross-links might play a role if the entanglements greatly outnumber the cross-links.

IV. The Transition Zone*

For the rubber technologist, perhaps the greatest interest in the transition zone is to know how to avoid it, unless high loss, increased modulus, and energy dissipation as heat are specifically desired. The transition zone looks much the same whether cross-links are present or not.

A. ONSET OF THE TRANSITION ZONE; THE MONOMERIC FRICTION COEFFICIENT

The onset of the transition zone with increasing frequency might be specified with data for G' by the intersection of two straight lines, as drawn in Fig. 5 (enlarged from Fig. 3); the dashed line at the right has a slope of 1/2 because this is specified by the Rouse–Mooney theory [1, p. 259], which is sometimes applied (though with misgivings by theorists [9]) to describe viscoelastic behavior in this zone. The intersection frequency ω_{tr} is very nearly the reciprocal of the "relaxation time" τ_{tr}, which is a measure of the time required for complete configurational rearrangement of a piece of macromolecule caught between two cross-links or two entanglements. The theory cannot be discussed within the scope of this chapter, but it predicts that ω_{tr} will be proportional to $1/P_c^2\zeta_0$, where P_c is the average number of monomer units between two cross-links or

* See Ferry [1, Ch. 12; 8].

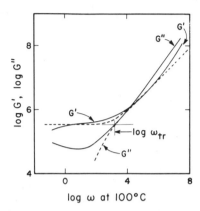

Fig. 5. Onset of transition zone from Fig. 3 (——) compared with the frequency dependence predicted by the Rouse–Mooney theory (---), with intersection of two lines to define the boundary frequency ω_{tr} between the plateau and transition zones.

two entanglements and ζ_0 is the local friction coefficient per monomer unit [1, p. 259].

The significance of the monomeric friction coefficient is shown schematically in Fig. 6: it is the frictional force per unit velocity encountered by a short segment of the macromolecule as it pushes its way through its surroundings with its thermal energy, randomly undergoing configurational rearrangements, reckoned per monomer unit. (Of course, the surroundings consist of other macromolecules, but their average resistance can be reckoned as a force per unit velocity.) It is also an inverse measure of local molecular mobility. It is the most important factor in determining ω_{tr} and many other time-dependent properties; the less frictional opposition, the higher the frequency to which oscillatory motions can go without getting into the transition zone. At a given temperature, ζ_0 depends greatly on the chemical nature of the elastomer; for a given elastomer, it changes very rapidly with temperature, pressure, or plasticization with diluent.

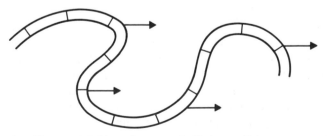

Fig. 6. Sketch to illustrate definition of monomeric friction coefficient: a group of n monomer units moving together with translational velocity v encounter a frictional force of $n\zeta_0 v$.

Another example of the changes in G' and G'' as the frequency goes through the transition zone, this time for a cross-linked elastomer (polyurethane rubber) [10], is shown in Fig. 7. Curves for the Rouse–Mooney theory are also drawn, and agree qualitatively with the experimental results. From fitting the data, ζ_0 can be obtained. Discussion of the deviations between theory and experiment at high and low frequencies is unnecessary and beyond the scope of this chapter.

The data in Fig. 7 were actually measured at several different temperatures and reduced to a reference temperature of $-42°C$ by a procedure which at the same time determines the temperature dependence of ζ_0 (and of ω_{tr}), as described in the following section.

For a given polymer, the shapes of the curves for G' and G'' in the transition zone are much the same whether cross-linked or not, and if not cross-linked, they do not depend much on molecular weight or molecular weight distribution. They do depend to some extent on the chemical nature of the polymer. However, even data for different polymers look much alike when compared in suitable reference states. This is shown in Fig. 8a where G' is plotted logarithmically against radian frequency at 25°C for four different cross-linked elastomers [5, 11–13]. The onset of the transition zone appears at lower frequencies for styrene–butadiene rubber than for natural rubber or 1,4-polybutadiene, and for butyl rubber it is lower by about two orders of magnitude. But when the

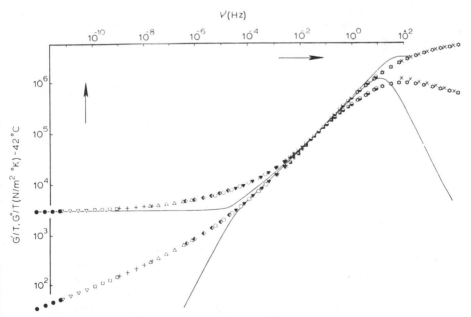

Fig. 7. Logarithmic plots of G'/T and G''/T (in N/m² deg) against frequency (in Hz) reduced to $-42°C$ for a polyurethane rubber [10] cross-linked to an equilibrium shear modulus of 7.1×10^6 dyn/cm². Curves drawn for the Rouse–Mooney theory.

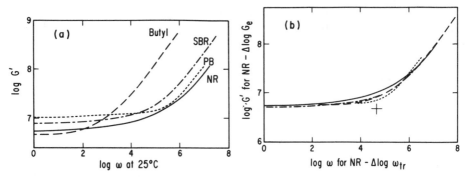

Fig. 8. Logarithmic plots of G' against radian frequency at 25°C for four elastomers. NR, natural rubber [12]; PB, 1,4-polybutadiene, about 50% trans [11]; SBR, styrene–butadiene rubber 1500 [5]; Butyl, butyl rubber [13]. All cross-linked with sulfur except SBR by dicumyl peroxide. (a) plotted directly; (b) with scales shifted to make G_e and ω_{tr} coincide with the values for natural rubber.

frequency scale is adjusted by subtracting $\Delta \log \omega_{tr}$, the difference between ω_{tr} for a given polymer and that of natural rubber, and the modulus scale is adjusted by subtracting $\Delta \log G_e$ (where G_e is the limiting low-frequency equilibrium modulus), the curves all look roughly the same. This is shown in Fig. 8b.

The differences in Fig. 8b may be partly due to details of the cross-linking process or initial molecular weight distribution, and partly to a real dependence on local chemical structure which is not well understood. It is probably more important, however, to emphasize the similarity of the dynamic behavior when compared in this manner.

At a frequency of ω_{tr}, defined as shown in Fig. 5 or Fig. 8, the loss tangent is observed to be 0.4 ± 0.1 for these various elastomers, so a substantial proportion of the energy of deformation is being dissipated (see also Section VI.C). To keep the relative dissipation very small, it is necessary to limit the frequency to values well below ω_{tr}; just how much depends on the details of the cross-linking process and other factors.

B. Temperature Dependence of the Friction Coefficient and Dynamic Viscoelastic Properties

With increasing temperature, ζ_0 diminishes rapidly. Not only the longest relaxation time of the transition zone, τ_{tr}, but also a series of shorter times reflecting other modes of configurational rearrangement are proportional to ζ_0 and all are decreased by the same factor, corresponding to more rapid motions of all types. The ratio of ζ_0 at absolute temperature T to ζ_0 at a chosen reference temperature T_0 is given by

$$a_T = \zeta_0(T)/\zeta_0(T_0) \tag{9}$$

Therefore, $a_T < 1$ for $T > T_0$, $a_T > 1$ for $T < T_0$. Since all motions are affected in the same way, a measurement of G' or G'' at T, ω is essentially equivalent to one at T_0, ωa_T, More precisely, the magnitudes of G' and G'' are expected to be proportional to the product $T\rho$ where ρ is the density (as is the equilibrium modulus G_e treated by the statistical theory of rubberlike elasticity), so $G'/T\rho$ is the quantity which should be compared at different temperatures.

On a logarithmic plot, a temperature change from T_0 to T simply shifts the entire transition zone along the $\log \omega$ axis by a displacement $-\log a_T$. This is illustrated in Fig. 9 for the polyurethane rubber of Fig. 7. (Here the ordinate is $\log(G'/T)$, the slight temperature variation of ρ having been neglected.) With increasing temperature, the onset of the transition zone is shifted to higher and higher frequencies and vice versa; at the temperature extremes it is shifted out of the experimental window of the frequency range covered, but its actual location outside the window can be predicted with confidence. The position of G'' on the logarithmic frequency scale is shifted identically. The shift factors $\log a_T$ corresponding to the various temperatures are plotted against T in Fig. 10.

Conversely, when measurements are made within a limited experimental window, as exemplified by Fig. 9, they can be plotted against $\log \omega a_T$ to provide a picture of the transition zone as it would have been measured over a very

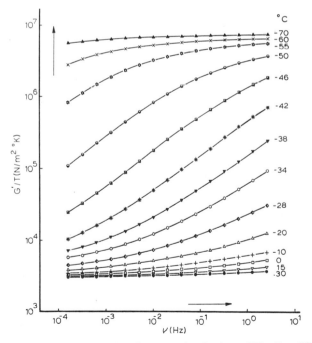

Fig. 9. Logarithmic plot of G'/T against frequency for the data of Fig. 7 at different temperatures, before reduction to a reference temperature of $-42°C$ [10].

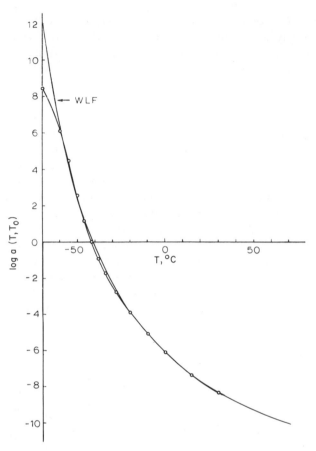

Fig. 10. Temperature shift factor $\log a_T$ for the polyurethane rubber of Figs. 7 and 9, based on $-42°C$ as reference temperature, plotted against temperature [10]. The dashed curve is drawn for the WLF equation, Eq. (10), with $c_1 = 16.7$ and $c_2 = 68.0$.

much wider frequency range at a single temperature T_0. This is, of course, the way Fig. 7 was constructed from the experimental data of Fig. 9, after choosing $T_0 = 231°K$. More commonly, the ordinates are plotted as $G'T_0\rho_0/T\rho$ and $G''T_0\rho_0/T\rho$, so that one can read off the actual magnitudes of the viscoelastic functions as they would be measured at the arbitrary reference temperature T_0.

The temperature dependence of a_T shown in Fig. 10, and for amorphous polymers generally, can be described quite closely by the Williams–Landel–Ferry (WLF) equation [1, pp. 314–318],

$$\log a_T = -c_1(T - T_0)/(c_2 + T - T_0) \tag{10}$$

in which c_1 and c_2 depend of course on the choice of T_0 as well as on the nature of the polymer. If T_0 is chosen as the glass transition T_g (the point where the thermal expansion coefficient changes abruptly), for most polymers $c_1 \cong 17.4$ and $c_2 \cong 52$. If T_0 is about $50°$ above T_g, $c_1 \cong 8.86$ and $c_2 \cong 102$. However, it is better to use values specifically determined for an individual polymer. Values of c_1 and c_2 corresponding to different choices of T_0 are easily inter-converted.

With the ordinate relabeled, Fig. 10 could serve to show the temperature dependence of the frequency ω_{tr} which corresponds to the onset of the transition zone and the beginning of the loss of elastomeric character as represented in Fig. 5.

C. Free Volume Concept*

The dependence of the local friction coefficient ζ_0 on temperature, as well as on other variables such as pressure and plasticizer content, can be usefully interpreted through the concept of fractional free volume. An amorphous polymer above T_g (also any ordinary liquid) is viewed as having a certain fraction of effective empty space f distributed around the molecules. Any configurational change of the kind to which we have repeatedly referred requires momentary concentration of this empty space to form a void into which a unit portion of the macromolecule can move. Calculations based on simplified models show that the probability of getting the necessary void is an exponential function of $-1/f$. The friction coefficient is an inverse measure of this probability, so it should depend on f as follows:

$$\log \zeta_0 = C + B/2.303 f \qquad (11)$$

where C and B are constants; it can be concluded that B must be close to unity, and it is often simply set at unity.

The magnitude of f has not yet been clearly defined but it can be specified in connection with Eq. (10) as follows. We expect f to expand linearly with temperature as follows:

$$f = f_0 + \alpha_f(T - T_0) \qquad (12)$$

Indeed, the increase in f with T is responsible for the fact that the thermal expansion of liquids is generally greater than that of solids and specifically that the expansion coefficient of a polymer above T_g (viz., α_l) is much larger than below T_g (viz., α_g). If Eq. (12) is introduced into Eq. (11), the ratio a_T is found to have exactly the form of Eq. (10) with $c_1 = 1/2.303 f_0$ and $c_2 = f_0/\alpha_f$, f_0 being the value of f at T_0 (if $B = 1$). From the usual numerical values of c_1 and c_2, it follows that f is about 0.025 at T_g and α_f is about 5×10^{-4} deg^1, which is near the difference between α_l and α_g. These values do differ somewhat, however, from one polymer to another.

* See Ferry [1, p. 313].

From these relations, the dynamic properties of an elastomer can be predicted over a wide range of temperatures from a limited amount of experimental data. It should be emphasized that the change in magnitude of G' or G'' with a change in temperature at constant frequency depends on two separate effects: (a) how far the curve slides along the $\log \omega$ axis, as in Fig. 9, determined by $\log a_T$; (b) how steep the curve is at the frequency of interest. Thus, in Fig. 9, a change from $-20°$ to $-28°$ will make relatively little difference at 10^{-4} Hz (increases G' by 30%) but much more at 1 Hz (increases G' by 120%).

D. DEPENDENCE OF THE FRICTION COEFFICIENT AND DYNAMIC PROPERTIES ON PRESSURE, PLASTICIZER, MOLECULAR WEIGHT, AND CROSS-LINKING

The effects of all the variables in this subtitle on dynamic properties in the transition zone can be described primarily as a shift along the $\log \omega$ axis, just as in the case of temperature, and the magnitude of the shift can also be related to fractional free volume through Eq. (11) [1, p. 320].

1. Pressure

With increasing pressure, the free volume is partially collapsed; decreasing f causes ζ_0 to increase. If a_P is defined as the ratio of ζ_0 at pressure P to ζ_0 at $P_0 = 1$ atm, the transition zone slides to lower frequencies by a shift $-\log a_P$ on the $\log \omega$ axis. Experimentally, it has been observed [14–16] that $\log a_P$ is approximately a linear function of P:

$$\log a_P = C'(P - P_0) \tag{13}$$

This relation can be derived from Eq. (11) if the pressure dependence of the free volume has the following plausible form:

$$f = f_0/[1 + A_P(P - P_0)] \tag{14}$$

where f_0 is the value at $P_0 = 1$ atm; $C' = A_P B/2.303 f_0$. Thus, under high confining pressure the curves of Figs. 2–5 may be shifted far to the left and an elastomer may find itself in the transition zone or even in the glassy zone.

For a more accurate description of the pressure dependence of free volume and relaxation times, the reader is referred to a recent investigation [17] which shows that some considerably more complicated relations than Eqs. (13) and (14) are required.

Under relatively small pressures (perhaps up to 100 atm) where the dependence of f on P is nearly linear instead of given by a more complicated relation such as Eq. (14), the relative effects of temperature and pressure can be described by the derivative

$$(\partial T/\partial P)_f = \beta_f/\alpha_f \tag{15}$$

where β_f is the compressibility of the fractional free volume. For many polymers, $(\partial T/\partial P)_f$ is about 0.020 ± 0.003 deg/atm; thus a temperature decrease of 1°C has an effect similar to that of a pressure increase of 50 atm.

2. Plasticizer

Except in rare instances, diluents of low molecular weight added to elastomers increase the fractional free volume. The principal effect on the dynamic properties in the transition zone is to slide curves of G' and G'' along the log ω axis to higher frequencies. In addition, the magnitudes of G' and G'' are lowered by the dilution. For an uncross-linked polymer they are diminished by a factor of v_2, the volume fraction of polymer in the mixture; for a cross-linked polymer, the reduction depends on whether the diluent is added before or after vulcanization, and in general cannot be described by a simple relation.

The increase in free volume can often be described satisfactorily by the linear equation

$$f = f_0 + \beta' v_1 \tag{16}$$

where f_0 is the value for the undiluted polymer and v_1 is the volume fraction of diluent $(= 1 - v_2)$. The coefficient β' depends on the nature of the diluent and must be determined experimentally, but it is usually in the range from 0.1 to 0.3. By substituting Eq. (16) into Eq. (11), assuming $B = 1$, one can calculate the shift factor a_v which describes the ratio of ζ_0 at diluent fraction v_1 to ζ_0 of the undiluted polymer. Thus,

$$\log a_v = -\frac{v_1}{2.303 f_0 (f_0/\beta' + v_1)} \tag{17}$$

The dynamic moduli G' and G'' will shift to higher frequencies by log a_v, while dropping slightly in magnitude. In this way, the onset of the transition zone can be pushed to higher frequencies.

The effectiveness of a small amount of plasticizer in shifting the frequency scale can be gauged by differentiating Eq. (17) and making v_1 approach zero:

$$d \log a_v/dv_1|_{v_1 \to 0} = -\beta'/2.303 f_0^2 \tag{18}$$

Near T_g where f_0 is small, the effect can be quite large; but for T far above T_g, f_0 is greater and a small amount of plasticizer is much less effective.

3. Molecular Weight and Cross-Linking

To a first approximation, the position of the transition zone on the frequency scale is independent of both molecular weight and cross-linking. However, at extremely low molecular weights it shifts somewhat to higher frequencies, because the average fractional free volume is increased by poor packing around chain ends. The relation is

$$f = f_0 + A_M/M_n \tag{19}$$

where M_n is number-average molecular weight, f_0 is the value for infinite molecular weight, and A_M is of the order of 100 g/mole. For all but very low molecular weights, the effect is trivial. For very low molecular weights, the effect is similar to that of a diluent.

Cross-linking by agents which do not alter the chemical structure greatly (e.g., natural rubber by dicumyl peroxide) may diminish the fractional free volume slightly, causing a very slight shift of the transition zone to lower frequencies, but this might be overbalanced by the diluent effect of small molecules produced as a by-product of the reaction. When rubbers are vulcanized by sulfur, a distinct shift to lower frequencies is often observed, but this is attributable to the chemical change of the combined sulfur rather than to the production of the cross-links themselves.

4. Relation of Fractional Free Volume to Glass Transition Temperature

Any variable which increases f will decrease the glass transition temperature, as can be seen qualitatively from the approximate rule that $f_g = 0.025$ at T_g; if f is larger, it will be necessary to cool to a lower temperature to reach f_g, and roughly, $\Delta T_g = -\Delta f/\alpha_f$. The effects of pressure, diluent, molecular weight, and combination with sulfur are sometimes described in the literature in terms of their influence on T_g. However, it is a more direct approach to relate them to f as we have done, since the molecular friction (or, in an opposite sense, molecular mobility) is closely related to f through Eq. (11); moreover, one is often interested in the behavior of an elastomer at temperatures far above T_g in applying the relationships summarized here.

E. EXAMPLES OF APPROXIMATE CALCULATIONS

1.   From Fig. 2, the onset of the transition zone is observed to be at $\log \omega_{tr} = 4.38$ for this styrene–butadiene rubber at 25°C. How far can the temperature be lowered if an onset frequency of 0.1 Hz is acceptable?

A frequency of 0.1 Hz corresponds to $\omega = 2\pi \times 0.1 = 0.63$, $\log \omega = -0.2$; so $\log a_T$ for the unknown temperature is 4.58. Equation (10) becomes then

$$4.58 = -\frac{c_1(T - T_0)}{c_2 + T - T_0}$$

Values of $c_1 = 4.28$ and $c_2 = 107.4$ for this particular polymer corresponding to $T_0 = 298°K$ can be found in the literature [5]. Solving for $T - T_0$, we find $-56°$. Thus the temperature can be lowered to $-31°$ without entering the transition zone at 0.1 Hz.

2.   At this temperature, how much plasticizer would be needed to push the onset of the transition zone back up to $\log \omega_{tr} = 4.38$, i.e., a frequency of 3800 Hz?

For this calculation, it is necessary to estimate f. From the literature [1, p. 316] f_g is 0.021, $T_g = 210°K$, and $\alpha_f = 8.2 \times 10^{-4}/\deg^1$. Hence, at $-31°$, $f_0 = 0.047$. We require $\log a_v = -4.58$; if β' is estimated to be 0.25, substitution into Eq. (17) and solving for v_1 gives $v_1 = 0.19$.

3.   Under a confining pressure of 100 atm, what would be the frequency for onset of the transition zone?

According to Eq. (15), a pressure increase of 100 atm is roughly equivalent to a temperature decrease of 2°. From Eq. (10), we find $\log a_T = 0.08$. Thus the frequency is decreased from 3800 Hz to 3170 Hz. At higher pressures, much larger effects would be observed, but more information is necessary to estimate their magnitudes.

V. The Plateau Zone*

The most important features of the plateau zone in amorphous uncross-linked rubbery polymers are its length, i.e., its extent along the logarithmic frequency axis; its height, i.e., the magnitude of the storage modulus G' in its flat region; and the minimum loss, measured by $\tan \delta$ or the ratio G''/G' at its lowest point. The position of the high-frequency end of the plateau on the frequency scale is already specified, since it is the same as the onset of the transition zone (ω_{tr}), which will depend on temperature, pressure, etc. as described in the preceding section. The foregoing features can be related to the average spacing between entanglements along a molecule, a quantity that will enter importantly into the discussion even though the exact physical nature of an entanglement remains somewhat uncertain, together with the molecular weight. In cross-linked polymers, especially for low degrees of cross-linking, the effects of the entanglements will contribute to the modulus, influence mechanical loss, and probably also influence the behavior in large deformations leading to rupture.

A. The Plateau Zone in Uncross-Linked Elastomers

A detailed view of the plateau zone [19] is seen in Fig. 11, where G' is plotted logarithmically against ω for nine polystyrenes with very narrow molecular weight distributions and molecular weights ranging from 47,000 to 580,000. This polymer (rubbery at a temperature well above T_g, where these data were measured) is chosen for illustration because of the extensive research which has been devoted to it.

* See Ferry [1, Chs. 13 and 14] and Graessley [18].

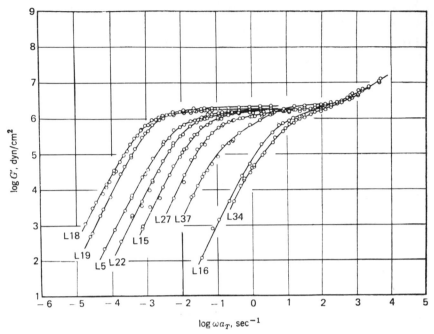

Fig. 11. Logarithmic plots of G' against radian frequency reduced to $160°C$ for nine polystyrenes with narrow molecular weight distribution and viscosity average molecular weights ranging from 47,000 (L34) to 580,000 (L18). (Reprinted with permission from S. Onogi *et al.*, *Macromolecules* **3**, 109 (1970). Copyright by the American Chemical Society.)

The height of the plateau, designated as G_N^0, the "entanglement modulus," is independent of molecular weight (although at the lowest molecular weights the plateau is not well defined). It represents the modulus of the entanglement network in a time scale where the entanglements are not slipping, and by applying the theory of rubberlike elasticity to it (as though the entanglements were cross-links), the average molecular weight between entanglement points can be estimated as $M_e = 18,000$, corresponding to 350 chain backbone atoms. When the molecular weight is less than five or six times M_e, or when molecular weight distribution is present, the plateau in G' is not very flat, but more sophisticated methods are available for estimating G_N^0 in this case. When the molecular weight is less than M_e, the plateau disappears.

Other polymers behave very similarly but with different values of G_N^0 and thus of M_e; for example, for ethylene–propylene copolymers (56 mole % ethylene), M_e is only 1700, corresponding to 100 chain backbone atoms between entanglement points. These differences are seen in the apparent elasticities of the raw rubbers; also, in block copolymers such as styrene–butadiene or styrene–isoprene, the equilibrium modulus depends primarily on the G_N^0 of the component with low T_g; and the properties of ordinary rubbers after vulcanization may be influenced by substantial contributions from entanglements trapped by the cross-linking process.

The dependence of entanglement modulus G_N^0 on temperature or pressure is usually very slight. Plasticization or addition of diluent [20] lowers it by a factor of v_2^2.

The length of the plateau on the $\log \omega$ axis may be specified by $\log \omega_{tr} - \log \omega_{te}$ where ω_{tr} is the onset of the transition zone and ω_{te} is the left end of the plateau or boundary of the terminal zone, similarly defined by the intersection of two lines. With increasing molecular weight, ω_{te} moves rapidly to lower frequencies, as seen, proportional to about the 3.5 power of M (or of the number of entanglements per molecule); ω_{tr} does not change perceptibly (Section IV.D). The frequency ω_{te} is roughly the reciprocal of the time τ_{te} required for a molecule to completely rearrange its configuration by snaking through its numerous entanglements. For most polymers, changes in temperature and pressure affect ω_{tr} and ω_{te} in the same way, so that the plateau slides along the $\log \omega$ axis without changing its length. Plasticization affects ω_{te} more than ω_{tr} (increases it by an additional factor of v_2^a where a is at least 2), so that the plateau length shrinks somewhat with added diluent.

The energy dissipation is at a minimum where the curve of G' against frequency is the flattest; if the slope $d \log G'/d \log \omega$ is sufficiently small, the loss tangent can be related to it by the approximate formula

$$\tan \delta \cong (\pi/2) \, d \log G'/d \log \omega \tag{20}$$

A minimum in $\tan \delta$ occurs at a frequency near the minimum in G'', as seen in Fig. 3, for example. For uniform molecular weight, the depth of the minimum is inversely proportional [1, p. 408] to the 0.8 power of M/M_e. Molecular weight distribution lessens the depth of the minimum.

The abrupt boundary between plateau and terminal zones seen in Fig. 11 is evident only for polymers of nearly uniform molecular weight. If there is molecular weight distribution, each species will have a different τ_{te}, and G' in the plateau zone will gradually change slope so that no specific boundary of ω_{te} on the frequency scale may be identified. This feature is associated with the shallower minimum in $\tan \delta$ as a function of frequency.

The precise mechanism of coupling molecular motions together by entanglements is not understood, but it clearly involves the topological constraints of molecules looped around each other, a feature which simply arises from their length and random configuration and has little to do with specific chemical structure [18]. The entanglement loci are not thought of as fixed, but as gradually slipping and rearranging as the molecules undergo Brownian motion. Just how the molecules snake through these entanglements at very long times, causing the rapid fall in G' at low frequencies in Fig. 11, for example, is not understood, and this problem is the subject of much current theoretical and experimental study [21–23].

It should be mentioned that in certain polymers the relaxation times in the terminal zone (τ_{te}, for example) seem to be abnormally long and their dependence on temperature and plasticization is exaggerated. This behavior suggests the presence of "hyperentanglements" in which the normal topological constraints are enhanced by specific attractive forces in appropriate intermolecular configurations [1, pp. 335, 538]. The rubber technologist would not ordinarily encounter such materials [poly(methyl methacrylate) is an example].

B. ENTANGLEMENTS IN CROSS-LINKED ELASTOMERS

When a rubber is vulcanized, some of the entanglements originally present may be trapped so that subsequent disentanglement is impossible, though slippage back and forth between the trapping cross-links may occur. The criterion for trapping is that all four molecular strands radiating from an entanglement locus are attached to the three-dimensional network structure [24]. Otherwise, eventual disentanglement can take place, though it may be very slow if the unattached strand terminates in a highly branched structure which is itself entangled. The distinction between trapped and untrapped entanglements is illustrated in Fig. 12.

In lightly cross-linked rubbers, where the entanglements outnumber the cross-links, the dynamic properties show an abnormally large loss at low frequencies, as already illustrated in Fig. 4. The effect is more prominent in the loss compliance, as shown in Fig. 13, which compares J'' for butyl rubbers vulcanized to various extents [13]. In the transition zone, on the right, all samples are nearly identical. At very low frequencies, the losses in the lightly vulcanized samples are enormously greater than those in the well-vulcanized ones (which are, however, still soft gum rubbers). These losses are rather similar for elastomers of different chemical structures if, again, they are compared in

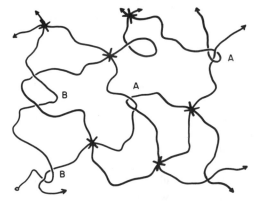

Fig. 12. Sketch illustrating trapped and untrapped entanglements in a rubber vulcanizate. Arrows denote attachment to permanent network, circle is a free end. A, trapped; B, untrapped.

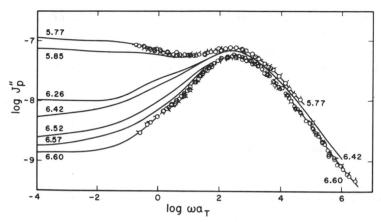

Fig. 13. Logarithmic plots of loss shear compliance J'' against radian frequency reduced to 25°C for butyl rubber vulcanized with sulfur to various extents. Numbers denote $\log G_e$ (equilibrium shear modulus in dyn/cm²). (Reprinted with permission from J. F. Sanders and J. D. Ferry, *Macromolecules* **7**, 681 (1974). Copyright by the American Chemical Society.)

suitable reference states [13], though there are some differences [23]. The extra loss has been attributed primarily to motions of dangling branched structures incompletely attached to the network (though slippage of trapped entanglements may also contribute [25]). Similar large losses are observed in networks containing a substantial proportion of unattached macromolecules which are highly entangled but free to reptate (i.e., undergo reptile-like motions) through the three-dimensional structure; an example can be prepared by curing a mixture of butyl rubber and polyisobutylene, the ingredients of which are nearly identical chemically and mix well on a molecular scale [26]. In the latter case, the extra energy dissipation and damping can be "tuned" to a desired frequency region by adjusting the molecular weight of the polyisobutylene, instead of having it spread over a broad frequency range as in Fig. 13.

VI. Practical Aspects of Dynamic Properties*

Rubber articles may have complicated geometries, and in addition to simple shear, which is the simplest type of deformation, other types of deformation, e.g., bulk compression, simple extension, and bending, may be encountered. It is beyond the scope of this chapter to discuss all these cases in detail; they have been treated in some of the references cited. In Sections VI.A and VI.B, expressions are derived for stored and dissipated energy in simple shear from the functions $G'(\omega)$ and $G''(\omega)$ or $J'(\omega)$ and $J''(\omega)$. It is hoped that these simple

---

* See Ferry [1, Ch. 19].

derivations will be helpful to readers who are not already familiar with the use of dynamic functions.

We will use the geometry shown in Fig. 1, in which the bottom surface is fastened to a base while the top surface undergoes a sinusoidal motion with a small amplitude, so that we can use linear viscoelasticity.

A. Prescribed Strain Cycle in Sinusoidal Deformation

In this case the strain is given by Eq. (1)

$$\gamma(t) = \gamma^0 \sin \omega t \tag{1}$$

while the stress is given by Eq. (3)

$$\sigma(t) = \gamma^0 [G'(\omega) \sin \omega t + G''(\omega) \cos \omega t] \tag{3}$$

We want to calculate the stored energy, which is at a maximum after a quarter of a cycle and zero after a full cycle, and the dissipated energy, which increases continuously. The work of deformation w is calculated as force times distance traveled. All quantities are per unit cubical element, so that force is given as $\sigma(t)$ while distance is given as $\gamma(t)$:

$$w = \int \sigma(t) \, d\gamma(t) \tag{21}$$

From Eq. (1)

$$d\gamma(t) = \gamma^0 \cos \omega t \, d(\omega t) \tag{22}$$

Substitution of Eqs. (3) and (22) into Eq. (21) gives

$$w = (\gamma^0)^2 G'(\omega) \int \sin \omega t \cos \omega t \, d(\omega t) + (\gamma^0)^2 G''(\omega) \int \cos^2 \omega t \, d(\omega t) \tag{23}$$

Per quarter cycle, the energies stored and dissipated are

$$\mathscr{E}_{st} = (\gamma^0)^2 G'(\omega) \int_0^{\pi/2} \sin \omega t \, d \sin \omega t = \frac{1}{2} (\gamma^0)^2 G'(\omega) \tag{24}$$

$$\mathscr{E}_d = (\gamma^0)^2 G''(\omega) \int_0^{\pi/2} \cos^2 \omega t \, d(\omega t) = \frac{\pi}{4} (\gamma^0)^2 G''(\omega) \tag{25}$$

The ratio of dissipated to stored energy is

$$\frac{\mathscr{E}_d}{\mathscr{E}_{st}} = \frac{\pi}{2} \frac{G''(\omega)}{G'(\omega)} = \frac{\pi}{2} \tan \delta \tag{26}$$

The energy dissipated per cycle is

$$(\gamma^0)^2 G''(\omega) \int_0^{2\pi} \cos^2 \omega t \, d(\omega t) = \pi (\gamma^0)^2 G''(\omega) \tag{27}$$

and per second when the frequency is $v = \omega/2\pi$ Hz:

$$\dot{\mathscr{e}} = \frac{\omega}{2} (\gamma^0)^2 G''(\omega) \tag{28}$$

B. PRESCRIBED STRESS CYCLE IN SINUSOIDAL DEFORMATION

In this case the stress is given as

$$\sigma(t) = \sigma^0 \sin \omega t \tag{29}$$

while the strain is given by Eq. (4)

$$\gamma(t) = \sigma^0 [J'(\omega) \sin \omega t - J''(\omega) \cos \omega t] \tag{4}$$

which gives

$$d\gamma(t) = \sigma^0 [J'(\omega) \cos \omega t + J''(\omega) \sin \omega t] \, d(\omega t) \tag{30}$$

Substitution of Eqs. (29) and (30) into Eq. (21) gives

$$w = (\sigma^0)^2 J'(\omega) \int \sin \omega t \cos \omega t \, d(\omega t) + (\sigma^0)^2 J''(\omega) \int \sin^2 \omega t \, d(\omega t) \tag{31}$$

Per quarter cycle

$$\mathscr{e}_{st} = (\sigma^0)^2 J'(\omega) \int_0^{\pi/2} \sin \omega t \, d \sin \omega t = \frac{1}{2} (\sigma^0)^2 J'(\omega) \tag{32}$$

$$\mathscr{e}_{d} = (\sigma^0)^2 J''(\omega) \int_0^{\pi/2} \sin^2 \omega t \, d(\omega t) = \frac{\pi}{4} (\sigma^0)^2 J''(\omega) \tag{33}$$

The ratio of dissipated to stored energy is

$$\frac{\mathscr{e}_d}{\mathscr{e}_{st}} = \frac{\pi}{2} \frac{J''(\omega)}{J'(\omega)} = \frac{\pi}{2} \tan \delta \tag{34}$$

which is the same result as we obtained for the prescribed strain cycle. The energy dissipated per cycle is

$$(\sigma^0)^2 J''(\omega) \int_0^{2\pi} \sin^2 \omega t \, d(\omega t) = \pi (\sigma^0)^2 J''(\omega) \tag{35}$$

and per second when the frequency is $v = \omega/2\pi$ Hz

$$\dot{\mathscr{E}} = \frac{\omega}{2}(\sigma^0)^2 J''(\omega) \tag{36}$$

C. HEAT GENERATION IN SINUSOIDAL DEFORMATION

The energy dissipated per second is given by Eq. (28) for the prescribed strain cycle and by Eq. (36) for the prescribed stress cycle. In both cases the energy dissipation is proportional to frequency and amplitude squared as long as we are in the linear viscoelastic range.

As an example of a numerical calculation for the case of a prescribed strain cycle, a lightly cross-linked natural rubber with [1, p. 44] $G'' = 1 \times 10^5$ N/m^2 (1×10^6 dyn/cm^2), a frequency of 10 Hz, and a peak shear strain of 10^{-2} would dissipate

$$\dot{\mathscr{E}} = 2\pi \times 10(10^{-2})^2 10^5/2 \cong 3 \times 10^2 \quad \text{J/m}^3 \text{ sec}^1 \tag{37}$$

Since the heat capacity is about 2×10^6 J/m^3 °K^1, the temperature would rise ΔT°K sec^{-1} if no heat were lost by conduction.

$$\Delta T = 3 \times 10^2/2 \times 10^6 = 1.5 \times 10^{-4} \quad \text{°K sec}^{-1} \tag{38}$$

This is a very small rate of energy dissipation. However, if we increase the strain amplitude to $\gamma^0 = 3 \times 10^{-2}$ and choose a frequency closer to the transition zone, $v = 10^3$ Hz, to which corresponds a loss modulus [1, p. 44] $G'' = 10^6$ N/m^2,

$$\Delta T \cong 3 \times 10^6/2 \times 10^6 = 1.5 \quad \text{°K/sec}^1 \tag{39}$$

again neglecting heat lost by conduction. In this case the sample would be quickly destroyed. It should be noted that it is G'' and not $\tan \delta = G''/G'$ which is important for the calculation of heat rise for the prescribed strain cycle deformation.

D. DAMPING IN SINUSOIDAL DEFORMATION

We now consider the case where a mass M is fastened to the top surface of the rubber in Fig. 1. The mass is set into motion in free oscillations and, in the absence of damping, oscillates with a constant amplitude and a characteristic frequency [1, p. 158] $\omega_c/2\pi$ given by

$$\omega_c^2 = G'(\omega)A/hM \tag{40}$$

where A is the area of the top surface, h the height of the rubber, and $G'(\omega)$ the storage modulus of the rubber. In the presence of damping the amplitude gradually decreases to zero with an exponential decay. This is illustrated in Fig. 14, though there the oscillating deformation is in torsion rather than simple

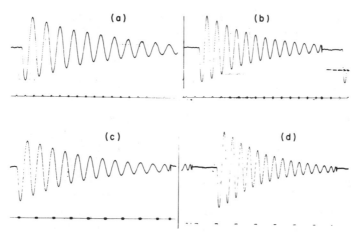

Fig. 14. Automatic recordings of angular displacement against time showing damping of free torsional oscillations of lightly vulcanized natural rubber with different moments of inertia [27]. Frequencies: (a) 0.64; (b) 0.91; (c) 1.37; (d) 2.35 Hz. Temperature 34.7°C. Over this frequency range, G' increases from 3.50 to 3.99 $\times$ 10^6 and G'' from 0.175 to 0.251 $\times$ 10^6 dyn/cm^2. Time scale shown by marks at bottom; runs (c) and (d) have expanded scale.

shear. The damping is usually given in terms of the logarithmic decrement Δ, which is the natural logarithm of the ratio between two successive displacements. The relations between ω_c and Δ and the viscoelastic functions $G'(\omega)$ and $G''(\omega)$ are quite complicated for very high degrees of damping. However, for low degrees of damping ω_c is roughly given by Eq. (40) and Δ can be estimated [1, p. 158] from Eq. (41).

$$\Delta = \pi A G''(\omega)/hM\omega_c{}^2 \quad \text{or} \quad \Delta \simeq \pi G''(\omega)/G'(\omega) = \pi \tan \delta(\omega) \quad (41)$$

for $\tan \delta(\omega) \ll 1$. This equation shows that the damping in free oscillations is determined by $\tan \delta$, i.e., the ratio of the loss to the storage modulus, in contrast to heat generation in forced oscillations, which was shown to be determined by the absolute value of the loss modulus (Section VI.C).

Machine mounts are used to isolate a vibrating machine from a building. The effectiveness of isolation is given in terms of the transmissibility T, which may be defined [28] as the ratio of the transmitted force to an externally imposed force for a mass supported by rubber mountings. The general expression [28, p. 26] for T is given by

$$T^2 = \frac{1 + [\tan \delta(\omega)]^2}{[1 - (\omega/\omega_0)^2 G'(\omega_0)/G'(\omega)]^2 + [\tan \delta(\omega)]^2} \quad (42)$$

where ω_0 is the resonance frequency, $G'(\omega_0)$ is the corresponding storage shear modulus; ω is the actual frequency of the vibrating system; $G'(\omega)$ and $\tan \delta(\omega)$

are the corresponding storage shear modulus and loss tangent. When ω/ω_0 is large, Eq. (42) simplifies to

$$T = (\omega_0/\omega)^2 [G'(\omega)/G'(\omega_0)][1 + (\tan \delta(\omega))^2]^{1/2} \tag{43}$$

for $\omega/\omega_0 \gg 1$, which shows that the transmissibility decreases by the second power of the frequency for constant $G'(\omega)$ and $\tan \delta(\omega)$. Machine mounts are therefore designed to have a resonance frequency considerably below the frequencies from which isolation is desired.

Figure 15 shows the dependence of transmissibility on the ratio of imposed frequency to resonance frequency for unfilled rubbers having various degrees of damping [29]. Rubbers with high damping have a much smaller transmissibility than rubbers with little damping at the resonance frequency. Well above the resonance frequency the situation is the opposite, i.e., a much smaller transmissibility for rubbers with little damping. The reason is that achievement of high damping in conventional rubbers usually requires getting into the transition zone where $G'(\omega)$ increases rapidly with frequency while $\tan \delta(\omega)$ goes through a maximum.

According to Eq. (43) the ideal machine mount material should have a constant or nearly constant storage shear modulus for a couple of decades in

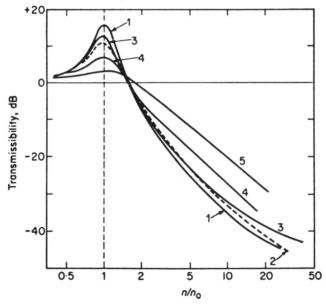

Fig. 15. Dependence of transmissibility on ratio of imposed frequency n to natural frequency n_0 for unfilled vulcanized rubber with various degrees of damping. 1, Natural rubber—least damping; 2, styrene–butadiene rubber; 3, neoprene; 4, butyl rubber; 5, nitrile rubber—highest damping. (From Payne and Whittaker [29].)

frequency, and this requires that we be well below the onset frequency of the transition zone. A loss tangent as small as possible above the resonance frequency is also beneficial, even though the effect of $\tan \delta$ is small above the resonance frequency. According to Eq. (42), the loss tangent at the resonance frequency should be as large as possible to prevent a peak in the transmissibility at resonance. Thus, the ideal machine mount material should have an extra strong loss peak at frequencies well below the transition zone and a nearly constant storage modulus for several decades above the resonance frequency. Networks containing unattached macromolecules which give rise to extra tunable loss peaks (Section V.B) come quite close to these requirements. However, these new materials have not yet been tested for machine mount applications.

E. RESILIENCE AND REBOUND

When a perfectly elastic ball rebounds from a plane surface of a viscoelastic solid, the ball will not recover its original height because of the energy dissipated in the viscoelastic solid. The resilience [30] is the ratio of the energy given up on recovery from the deformation to the energy required to produce the deformation. Since the potential energy of the ball is proportional to its height above the viscoelastic solid, the resilience can be expressed as the fraction of the original height recovered. The experiment is physically similar [1, p. 608] to half a cycle of an oscillating free vibration measurement (as in a torsion pendulum). The equivalent frequency ω_e is therefore given as

$$\omega_e \cong \pi/t \tag{44}$$

where t is the time of contact.

The rebound experiment can be used to determine viscoelastic properties [1, p. 608]. If the viscoelastic plate is thick enough so that the time required for an elastic wave to be reflected from its other side and return is longer than the time of contact, then the resilience is essentially independent of the thickness and in practice also of the velocity of impact. The logarithmic decrement can be estimated from the relation $\Delta \cong 1 - \%$ resilience/100. For a hard viscoelastic solid the deformation is a combination of shear and compression, but for the usual range of Poisson's ratio the majority of the stored energy is attributable to the shear. The rebound experiment is convenient for rapid scanning of changes in viscoelastic properties during a chemical reaction, for example [31].

F. WAVE PROPAGATION

When we oscillate the top surface of the rubber in Fig. 1 in simple shear, a transverse wave will propagate into the rubber. Under conditions where the shear wave approximates a one-dimensional disturbance in the x direction, the damping is exponential and the amplitude (shear displacement u) can be represented as [1, p. 136]

$$u = u_0 e^{-x/x_0} e^{i(\omega t - 2\pi x/\lambda)} \tag{45}$$

where ω is the angular frequency, t the time, λ the wavelength, and x_0 the distance within which the amplitude falls off by a factor of $1/e$, as shown [32] in Fig. 16. The attenuation is $\alpha = 1/x_0$ in nepers per meter; the velocity is $v = \lambda\omega/2\pi$ m/sec.

When the sample is sufficiently large so that waves are attenuated before they can experience reflection, measurements of the wavelength and attenuation make it possible to calculate the components of the complex shear modulus by the following equations [1, p. 136].

$$G'(\omega) = \rho v^2 (1 - r^2)/(1 + r^2)^2 \tag{46}$$

$$G''(\omega) = 2\rho v^2 r/(1 + r^2)^2 \tag{47}$$

where ρ is the density and $r = \lambda/2\pi x_0$. For convenience, an auxiliary parameter $\widetilde{G}(\omega) = \rho v^2$ (the "wave rigidity modulus") is sometimes calculated and then subsequently converted to $G'(\omega)$ by the correction factor $(1 - r^2)/(1 + r^2)^2$, for which only a rough estimate of the damping is needed. On the other hand, $G''(\omega)$ is directly proportional to the attenuation and requires a precise determination of the latter.

For the propagation of sound (longitudinal) waves [1, p. 189] through viscoelastic materials three different cases should be considered:

A.   The samples are thin strips or fibers for which the wavelength is large compared to the sample thickness and small compared to the length. The deformation corresponds to simple extension and the components of the complex Young's modulus $\mathbf{E}^*(\omega)$ are measured. Equations (45)–(47) are also valid in this case with $G'(\omega)$ and $G''(\omega)$ replaced by $E'(\omega)$ and $E''(\omega)$, respectively.

B.   The sample thickness is an appreciable fraction of the wavelength. This is a complicated case because the lateral contraction characteristic of

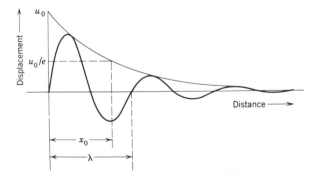

Fig. 16. Exponentially damped shear wave with definitions of characteristic parameters [32].

simple extension is opposed by the viscoelasticity and inertia of the material.

C. The sample dimensions normal to the propagating wave are large compared to the wavelength. The complex bulk (compression) modulus $K^*(\omega)$ now becomes important, and the wave propagation is governed by the complex modulus $M^*(\omega)$:

$$M^*(\omega) = K^*(\omega) + (4/3)G^*(\omega) \qquad (48)$$

Equations (45)–(47) are also valid in this case with $G'(\omega)$ and $G''(\omega)$ replaced by $M'(\omega)$ and $M''(\omega)$, respectively. The loss components $K''(\omega)$ and $G''(\omega)$ may be of similar magnitude, so Eq. (48) shows that both must be known in order to calculate the damping of sound waves by viscoelastic materials with large dimensions normal to the propagating wave.

VII. Dynamic Properties in Large Deformations

Rubber is seldom used as pure gum, since addition of fillers to elastomers improves several properties; in the case of reinforcing fillers such as carbon black the modulus is increased and the important failure properties tensile strength, tear resistance, and abrasion resistance are improved enormously. Reinforcement of rubber is discussed in Chapter 8; this section will primarily be concerned with the effect of strain on dynamic properties of filled rubbers.

A. EFFECT OF STRAIN AMPLITUDE

Addition of carbon black to elastomers increases the low-strain storage modulus strongly. A typical graph from the work of Payne and Whittaker [29] of the effects of filler content and strain on storage shear modulus is shown in Fig. 17 for butyl rubber containing HAF carbon black. The storage modulus of pure butyl rubber is essentially independent of strain, whereas it decreases by almost one order of magnitude going from small to large strain for a butyl rubber containing 23.2 vol % HAF ($\simeq 50$ phr). The transition from linear to nonlinear dynamic properties occurs .at decreasing strain amplitude with increasing filler content. At 23.2% HAF the nonlinear region begins at a double strain amplitude of approximately 0.002, which is a very small strain. The storage modulus shows a limiting value at both small strains, $G_0'(\omega)$, and high strains, $G'_\infty(\omega)$. Not only does $G'_0(\omega)$ depend on carbon black content, but it also depends on carbon black "structure" and particle size [33] and on degree of dispersion [29], while $G'_\infty(\omega)$ depends primarily on carbon black "structure" [33]. It is believed that the difference between the two limits $(G'_0(\omega) - G'_\infty(\omega))$ represents structure breakdown, i.e., separation of carbon black aggregates with strain. If the strain amplitude is quickly changed from a high value to a low value, the measured storage modulus will at first be abnormally low. However, the storage modulus will slowly recover [34] toward its original low

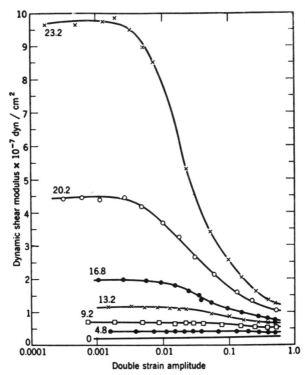

Fig. 17. Dependence of storage shear modulus on double strain amplitude at 0.1 Hz for butyl rubber containing various concentrations up to 23.2 vol % HAF carbon black. (From Payne and Whittaker [29].)

amplitude value, the rate of recovery being higher at higher temperatures [35]. This recovery indicates a reformation of structure. The reversible strain amplitude dependence of the storage modulus is of the nature of a thixotropic change.

Figure 18 shows the dependence of loss shear modulus on strain amplitude for the same series of butyl rubber compounds. The maxima in the loss modulus fall at decreasing strain amplitudes with increasing filler content in the region of strain where the storage modulus changes most rapidly. The dependence of loss tangent on strain amplitude is shown in Fig. 19. Unlike the maxima in the loss modulus curve, the maxima in the loss tangent all fall in the same region of strain amplitude, namely, a double strain amplitude of about 0.2.

Example

We want to calculate the effect of strain amplitude on heat rise for a butyl rubber pure gum and for a butyl rubber containing 28.8 vol % HAF carbon black. The heat rise is calculated from Eq. (28), assuming no heat lost by conduction. Figure 18 gives values of $G''(\omega)$ at a frequency of 0.1 Hz for a butyl

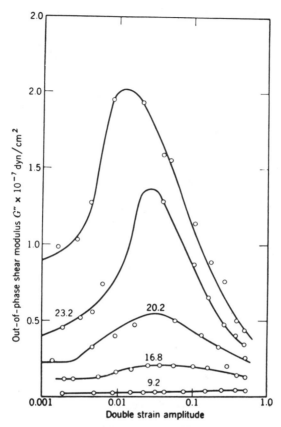

Fig. 18. Dependence of loss shear modulus on double strain amplitude at 0.1 Hz for butyl rubber containing various concentrations up to 28.8 vol % HAF carbon black. (From Payne and Whittaker [29].)

rubber containing 28.8 vol % HAF carbon black but not for butyl rubber pure gum. However, a typical value [13] is $G''(\omega) = 1 \times 10^4$ N/m² at 25°C and a frequency of 0.1 Hz. The dynamic properties are considered to be linear for butyl rubber pure gum.

Table I shows a huge effect of strain amplitude on heat rise (heat rise is proportional to the square of the amplitude). A comparison of columns 2 and 3 shows that addition of 28.8 vol % HAF may increase heat rise by approximately two orders of magnitude, while a comparison of columns 3 and 4 shows that the maximum error made by using linear dynamic properties (column 4) rather than the correct loss moduli is about a factor of 2. This is a smaller effect than is found for the storage modulus (Fig. 17), where the change with strain amplitude would change the resonance frequency of a machine mount by about one order of magnitude.

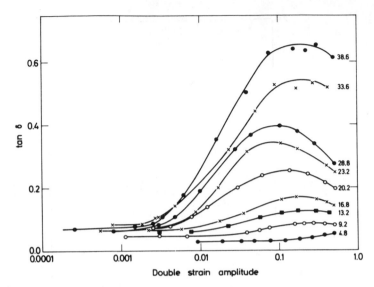

Fig. 19. Dependence of loss tangent on double strain amplitude at 0.1 Hz for butyl rubber containing various concentrations up to 38.6 vol % HAF carbon black. (Calculated from data of Payne and Whittaker [29].)

TABLE I

HEAT RISE OF FILLED AND UNFILLED BUTYL RUBBER

	Heat rise[a]		
Double strain amplitude	Butyl rubber with no HAF (°K/sec)	Butyl rubber with 28.8% HAF (°K/sec)	Butyl rubber with 28.8% HAF (°K/sec)[b]
0.002	1.5×10^{-9}	1.5×10^{-7}	1.5×10^{-7}
0.01	3.8×10^{-8}	7.6×10^{-6}	3.8×10^{-6}
0.10	3.8×10^{-6}	4.2×10^{-4}	3.8×10^{-4}
0.50	9.5×10^{-5}	4.0×10^{-3}	9.5×10^{-3}

[a] At a frequency of 0.1 Hz, assuming no heat lost by conduction.
[b] Assuming linear dynamic properties, i.e., $G''(\omega) = 1 \times 10^6$ N/m^2 at all strain amplitudes.

Medalia [33] has found that the modulus versus strain amplitude curves of vulcanizates with different carbon blacks may cross over because of the independent effects of carbon black "structure" and particle size. This is one more reason why measurements of dynamic properties of carbon-black-filled compounds should always be made at the proper amplitude for practical applications.

B. Effect of Static Strain

In addition to oscillatory motions, machine mounts are often subjected to static strains. The superposition of a sinusoidally varying strain on a static strain in compression is used in the Goodrich Flexometer, which is one of the tests for compression fatigue (ASTM D 623). Voet and Morawski [36] have recently used a modified Rheovibron to make dynamic measurements on filled rubbers under static strains [37] in simple elongation up to sample rupture. Figure 20 shows the storage modulus measured at small strain amplitudes as a function of static strain (up to 200%) for SBR pure gum and two HAF carbon-black-filled SBR vulcanizates. Unlike the SBR pure gum, the filled rubbers show an initial decrease followed by a strong increase in the storage modulus with increasing static strain. Voet and Morawski attribute the decrease in E' to a disruption of the filler network, while the subsequent increase in E' at higher elongations is attributed to the limited extensibility of the elastomer chains. These authors also observed a lower dynamic modulus for samples which had

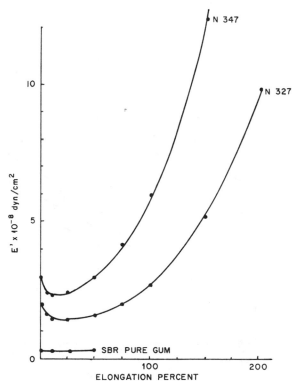

Fig. 20. Dependence of storage modulus (Young's) on elongation for SBR pure gum and two SBR vulcanizates containing HAF carbon black (50 phr), at 11 Hz, 30°C, and amplitude 1.58×10^{-3}. (From Voet and Morawski [37].)

previously undergone several extension–recovery cycles (Mullins effect—see Chapter 4, Section 6), and they followed part of the recovery curve dynamically.

C. EFFECTS OF FREQUENCY AND TEMPERATURE

Increasing frequency reduces the configurational changes with which the polymer molecules in the elastomer matrix can respond within a cycle of deformation, as discussed in Sections III.B and IV, and we should therefore expect an increase in the storage modulus. Such an increase in storage modulus has been found by Payne and Whittaker [29] for carbon-black-filled butyl rubber and by Warnaka [38] for carbon-black-filled SBR over a frequency range of about two decades. The frequency effect seems to be independent of strain amplitude. Voet and Morawski [37] have found a similar increase in storage

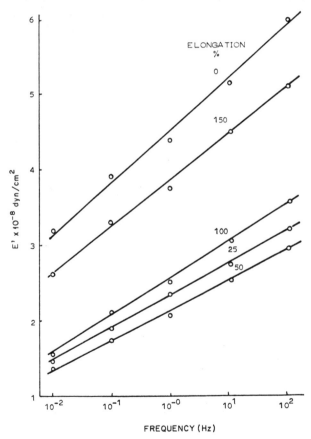

Fig. 21. Storage modulus (Young's) plotted against the logarithm of frequency at different elongations for SBR containing silica (50 phr), at 30°C, amplitude 1.58×10^{-3}. (From Voet and Morawski [37].)

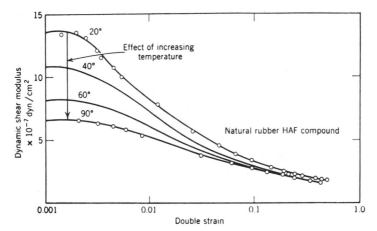

Fig. 22. Variation of storage shear modulus with double strain amplitude as a function of temperature at 0.1 Hz for natural rubber containing 32 vol % HAF carbon black. (From Payne and Whittaker [29].)

modulus over four decades in frequency for silica-filled SBR at different elongations, as shown in Fig. 21.

The effect of temperature is more complicated than the effect of frequency. A decrease in temperature means an increased resistance to molecular motions, which leads to a higher storage modulus; at the same time, the statistical theory of rubber elasticity predicts a decrease in the elastic modulus [39] which is proportional to absolute temperature (see Chapter 4). The first two effects therefore oppose each other; and finally, a change in temperature might change the strength of filler–filler and filler–matrix interactions. The effect of temperature on a natural rubber vulcanizate containing 32 vol % HAF carbon black is shown in Fig. 22. For this compound an increase in temperature from 20 to 90°C decreases the storage modulus at small strain amplitudes by about 50%, whereas the storage modulus at high strain amplitudes is practically independent of temperature.

VIII. Experimental Methods and Apparatus for Measurements and Tests

A wide variety of experimental equipment for measuring dynamic viscoelastic properties of elastomers is available commercially, and an even wider range has been developed in research laboratories for specific investigations. Many of these are described in references previously cited [1, Chs. 5–8, Appendix B; 4, Ch. 6; 30] and other reviews [40–42]. Without attempting to assess the merits of individual instruments, we discuss here some of the features that must be considered in the selection of experimental methods and equipment for measurements and tests. Some specific instruments are mentioned, but the list is not intended to be complete.

It should be noted that the dynamic moduli G' and G'' (or E' and E'') are ratios of stresses to strains, but the stresses and strains are usually not themselves measured. In many instruments, forces and displacements (or torques and angular displacements) are measured, and subsequent calculations require analysis of the spatial distribution of stresses and strains within the sample to obtain the dynamic moduli. This analysis may be quite simple, however, if the strain is homogeneous. In certain special cases, such as the Fitzgerald electromechanical transducer [43], the measurements provide force/displacement or stress/strain ratios, both in and out of phase, circumventing the need to measure these quantities separately. In others, employing forced or free oscillations, the measurements involve a resonance or characteristic frequency and some gauge of damping or breadth of resonance response, from which dynamic moduli are obtained by some of the relations in Section VI. Wave propagation methods [1, pp. 136, 189; 32] may also be employed.

1. Type of Deformation

Samples may be strained in essentially simple shear, as pictured in Fig. 1, with measurement of G' and G''. There are oscillatory rotational instruments, such as the Weissenberg rheogoniometer [44, 45], in which the mode of deformation is also shear. Alternatively, samples may be strained in simple extension, as in the Rheovibron dynamic viscoelastometer [46, 47] and the Kawai viscoelastic spectrometer [48], with measurement of E' and E''. For elastomeric materials, in small oscillatory strains and in the absence of superimposed static strains, these measurements are equivalent, as mentioned in Section II.A, differing only by a factor of 3. Some commercial devices have several interchangeable attachments for different deformation geometries [46, 49].

In the Weissenberg rheogoniometer with cone and plate geometry and the simple extension devices quoted, the strain is essentially homogeneous throughout the sample. In other types of deformation, such as torsion of a rod (shear, but inhomogeneous) or bending of a bar (extension–compression, but inhomogeneous), the sample is strained to very different extents at different points. There are also devices in which the apparatus performs a steady-state rotation but the sample undergoes periodic deformation which varies greatly from point to point; examples are the Maxwell orthogonal rheometer [50, 51] and the Contraves balance rheometer [52]. For linear viscoelastic behavior, it is possible to extract G' and G'', but otherwise analysis of the measurements may be very complicated.

The storage and loss moduli G' and G'' can be obtained either by decomposing the periodic stress into its two vector components [Eq. (3)] or by measuring the loss tangent and the ratio of peak stress to peak strain, $\sigma^0/\gamma^0 = (G'^2 + G''^2)^{1/2}$. In the latter case, $G' = (\sigma^0/\gamma^0) \cos \delta$ and $G'' = (\sigma^0/\gamma^0) \sin \delta$.

In some cases, it may be desirable to impose a static strain in addition to the oscillating strain described by the dynamic viscoelastic properties. Certain devices are designed to provide this option [36, 48] for simple extension.

2. Magnitude of Strain

The peak cyclic strain, γ_0 in Eq. (1), may be only a fraction of a percent or in some cases much larger, as in the Yerzley mechanical oscillograph [53]. In the latter case, storage and loss moduli are not exactly defined from the measurements, but "effective dynamic moduli" can be calculated. If the cyclic strain is very large, a sinusoidally prescribed strain is associated with a stress that is not sinusoidal (contains harmonics), or vice versa.

3. Size and Attachment of Samples

The samples may be quite massive, as in the Firestone resonance apparatus [54], or tiny and delicate, as in the Rheovibron [46]. They may be attached by clamps or cementing, or in certain geometries simply by self-adhesion. Vulcanized rubber samples are sometimes attached to metal mounts in the course of the vulcanization process.

Care must be taken that the measurements are not complicated by incipient propagation of an elastic wave in the sample (unless, of course, a wave propagation method is being employed with measurement of propagation velocity and attenuation). This means that the wavelength of the appropriate disturbance (shear, elongational, or flexural wave) must be much larger than the sample dimension in the appropriate direction (perpendicular to slide direction in shear, direction of stretch in simple elongation) [1, p. 121].

4. Frequency Range

There are many devices in the range from 0.01 to 100 Hz, although it is rare for a single instrument to cover much more than two logarithmic decades. Some provide a continuous range, others measure only at specific value of frequency. At higher frequencies, measurements are more difficult and require more complicated instrumentation, with very careful attention to possible spurious effects arising in the apparatus itself (see Section VIII.6).

5. Temperature

Control and measurement of temperature are very important in view of the rapid temperature dependence of friction coefficient and relaxation times which is observed frequently as illustrated in Fig. 10, and the concomitant shift of curves for G' and G'' along the frequency axis. It is particularly important that the temperature of the *sample* be known, not just that of the air near the sample, especially if there is any significant heat generation in the oscillatory deformation (Section VI.C). For measurements at any other than ambient temperature, attention must be paid to possible conduction of heat toward or away from the sample by metal parts of the apparatus which are in contact with it. It does little good to control the temperature of the sample chamber if the sample is attached to a piece of metal whose other end is at a temperature 100° different. If possible, contacts with the sample should be insulated by sections of material with low heat conductivity.

6. Accuracy

The accuracy of measurements depends on many features which must be analyzed for each specific instrument. Examination of the calculations will often reveal that in certain ranges of frequency and/or sample stiffness, subtractions of one large number from another are being performed with an alarming loss of precision. In certain circumstances, the contribution of the sample to the measured force may be largely masked by the inertia of part of the apparatus (at high frequencies) or the contribution of the sample to displacement may be masked by yielding of some part of the apparatus that appears superficially to be quite rigid. Force transducers are supposed to measure forces with very little accompanying displacement, but in some cases that displacement may be significant.

Sometimes knowledge of the sample dimensions, or uniformity of dimensions, is the limiting factor in absolute accuracy, especially in bending or torsion, where a linear dimension enters the calculation to a high power. Another source of error may be the presence of a more complicated stress distribution in the sample than that assumed in the calculations, which in an extreme case would be evident in bulging, necking, or other irregularities in the sample shape.

Careful scrutiny of possible sources of error will be rewarding in the reliability and usefulness of the measurements.

LIST OF SYMBOLS

A	area, *or* empirical constant
a_T	ratio of relaxation times at two different temperatures
a_v	ratio of relaxation times at two different concentrations (volume fractions)
a_P	ratio of relaxation times at two different pressures
B	constant in free volume equation
C, C'	empirical constants
c_1, c_2	coefficients in WLF equation referred to T_0 as reference
$c_1{}^g, c_2{}^g$	coefficients in WLF equation referred to T_g as reference
E	Young's (tensile) modulus
$\mathbf{E^*}$	complex dynamic tensile modulus
E'	tensile storage modulus
E''	tensile loss modulus
E_e	equilibrium or pseudo-equilibrium tensile modulus
$\mathscr{E}_{st}$	stored energy per cubic centimeter
$\mathscr{E}_d$	dissipated energy per cubic centimeter (in a quarter cycle)
$\dot{\mathscr{E}}$	rate of dissipation of energy per second
e	base of natural logarithms

f	fractional free volume
f_g	fractional free volume at the glass transition temperature
G	shear modulus
$G(t)$	shear relaxation modulus .
$\mathbf{G^*}$	complex dynamic shear modulus
G'	shear storage modulus
G''	shear loss modulus
G_e	equilibrium shear modulus
$G_N{}^0$	shear modulus associated with entanglement network
$\tilde{G}$	shear wave rigidity modulus
h	height
i	$\sqrt{-1}$
J	shear compliance
$J(t)$	shear creep compliance
$\mathbf{J^*}$	complex dynamic shear compliance
J'	shear storage compliance
J''	shear loss compliance
J_e	equilibrium shear compliance
$J_N{}^0$	shear compliance associated with entanglement network (in uncross-linked systems)
K	bulk modulus
$\mathbf{K^*}$	complex dynamic bulk modulus
K'	bulk storage modulus
K''	bulk loss modulus
M	mass, *or* molecular weight
$\mathbf{M^*}$	complex dynamic longitudinal bulk modulus
M'	bulk longitudinal storage modulus
M''	bulk longitudinal loss modulus
$\bar{M}_n$	number average molecular weight
$\bar{M}_w$	weight average molecular weight
$\bar{M}_z$	z average molecular weight
M_0	molecular weight per monomer unit
M_e	average molecular weight between entanglement coupling points
M_c	average molecular weight of a network strand
P	pressure, *or* degree of polymerization
P_c	average degree of polymerization of a network strand
r	wave damping parameter
T	absolute temperature, *or* transmissibility
T_0	reference temperature for reduced variables
T_g	glass transition temperature
t	time
u	displacement
v	velocity
w	work
x	linear displacement
x_0	critical damping distance
α	thermal expansion coefficient, *or* attenuation of propagated wave
α_f	thermal expansion of free volume relative to total volume
α_l	thermal expansion coefficient above T_g
α_g	thermal expansion coefficient below T_g
β	coefficient of compressibility

β' parameter relating free volume to volume concentration of diluent

γ strain

Δ logarithmic decrement

δ phase angle between stress and strain

ε practical tensile strain

ζ_0 translation friction coefficient per monomer unit

λ wavelength, *or* relative length (in large deformations)

v frequency in hertz

ρ density

σ stress

τ relaxation time

τ_{te} relaxation time associated with terminal zone

τ_{tr} longest relaxation time associated with transition zone

ω frequency in radians/second

ω_0 resonance frequency (forced oscillations)

ω_c characteristic frequency (free oscillations)

ω_e equivalent radian frequency in rebound test

ω_{te} radian frequency corresponding to high-frequency end of terminal zone

ω_{tr} radian frequency corresponding to low-frequency end of transition zone

REFERENCES

1. J. D. Ferry, "Viscoelastic Properties of Polymers," p. 375. Wiley, New York, 1970.
2. L. R. G. Treloar, "The Physics of Rubber Elasticity," p. 270. Oxford Univ. Press, London and New York, 1958.
3. F. P. Baldwin and G. Ver Strate, *Rubber Chem. Technol.* **45,** 709 (1972).
4. I. M. Ward, "Mechanical Properties of Solid Polymers," Ch. 5. Wiley, New York, 1971.
5. R. G. Mancke and J. D. Ferry, *Trans. Soc. Rheol.* **12,** 335 (1968).
6. F. R. Schwarzl, *Rheol. Acta* **8,** 6 (1969); **9,** 382 (1970); **10,** 166 (1971); N. W. Tschoegl, *Rheol. Acta* **10,** 582, 595 (1971).
7. W. Dannhauser, W. C. Child, Jr., and J. D. Ferry, *J. Colloid Sci.* **13,** 103 (1958); J. W. Berge, P. R. Saunders, and J. D. Ferry, *J. Colloid Sci.* **14,** 135 (1959).
8. J. D. Ferry, *in* "Deformation and Fracture of High Polymers" (H. H. Kausch, J. A. Hassell, and R. E. Jaffee, eds.), p. 27. Plenum, New York, 1974.
9. A. Ziabicki, *Macromolecules* **7,** 501 (1974).
10. F. Schwarzl, C. W. van der Wal, and H. W. Bree, *Chim. Ind. (Milan)* **54,** 51 (1972).
11. E. Maekawa, R. C. Mancke, and J. D. Ferry, *J. Phys. Chem.* **69,** 2811 (1965).
12. A. R. Payne, *in* "Rheology of Elastomers" (P. Mason and N. Wookey, eds.), p. 86. Pergamon, Oxford, 1958.
13. J. F. Sanders and J. D. Ferry, *Macromolecules* **7,** 681 (1974).
14. J. M. O'Reilly, *J. Polym. Sci.* **57,** 429 (1962).
15. K. H. Hellwege, W. Knappe, F. Paul, and V. Semjonov, *Rheol. Acta* **6,** 165 (1967).
16. S. Tokiura, S. Ogihara, Y. Yamazaki, K. Tabata, and H. Sasaki, *Zairyo* **17,** 365 (1968).
17. R. W. Fillers, Ph.D. Thesis, California Inst. of Technol., Pasadena, 1975; R. W. Fillers and N. W. Tschoegl, *Trans. Soc. Rheol.* **21,** 51 (1977).
18. W. W. Graessley, *Fortschr. Hochpolym.-Forsch.* **16,** 1 (1974).
19. S. Onogi, T. Masuda, and K. Kitagawa, *Macromolecules* **3,** 109 (1970).
20. N. Nemoto, T. Ogawa, H. Odani, and M. Kurata, *Macromolecules* **5,** 641 (1972).
21. P. G. de Gennes, *J. Chem. Phys.* **55,** 572 (1971).
22. S. F. Edwards and J. W. V. Grant, *J. Phys. A: Gen. Phys.* **6,** 1169, 1186 (1973).

23. P. Thirion, *Eur. Polym. J.* **10,** 1093 (1974).
24. N. R. Langley, *Macromolecules* **1,** 348 (1968).
25. R. E. Cohen and N. W. Tschoegl, *Int. J. Polym. Mater.* **2,** 49 (1972).
26. O. Kramer, R. Greco, R. A. Neira, and J. D. Ferry, *J. Polym. Sci., Polym. Phys. Ed.* **12,** 2361 (1974).
27. N. R. Langley, Ph.D. Thesis, Univ. of Wisconsin, Madison, 1968.
28. J. C. Snowdon, "Vibration and Shock in Damped Mechanical Systems," p. 22. Wiley, New York, 1968.
29. A. R. Payne and R. E. Whittaker, *Rubber Chem. Technol.* **44,** 340 (1971).
30. F. S. Conant, *in* "Rubber Technology" (M. Morton, ed.), 2nd Ed., p. 134. Van Nostrand-Reinhold, New York, 1973.
31. M. Gordon and B. M. Grieveson, *J. Polym. Sci.* **29,** 9 (1958).
32. J. D. Ferry, W. M. Sawyer, and J. N. Ashworth, *J. Polym. Sci.* **2,** 593 (1947).
33. A. I. Medalia, *Rubber World* **168**(5), 49 (1973).
34. A. R. Payne, *in* "Reinforcement of Elastomers" (G. Kraus, ed.), p. 76. Wiley (Interscience), New York, 1965.
35. W. P. Fletcher and A. N. Gent, *Trans. Inst. Rubber Ind.* **29,** 266 (1953).
36. A. Voet and J. C. Morawski, *Rubber Chem. Technol.* **47,** 758 (1974).
37. A. Voet and J. C. Morawski, *Rubber Chem. Technol.* **47,** 765 (1974).
38. G. E. Warnaka, *Rubber Chem. Technol.* **36,** 407 (1963).
39. L. R. G. Treloar, *Rubber Chem. Technol.* **47,** 625 (1974).
40. L. E. Nielsen, "Mechanical Properties of Polymers and Composites," Ch. 4. Dekker, New York, 1974.
41. A. C. Edwards and G. N. S. Ferrand, *in* "The Applied Science of Rubber" (W. J. S. Naunton, ed.), Ch. 8. Arnold, London, 1961.
42. J. R. Scott, "Physical Testing of Rubbers," p. 181. MacLaren, London, 1965.
43. E. R. Fitzgerald, *Phys. Rev.* **108,** 690 (1957).
44. K. Weissenberg, "The Testing of Materials by Means of the Rheogoniometer." Farol Res. Eng., Bognor Regis, Sussex, England, 1964.
45. J. Meissner, *J. Appl. Polym. Sci.* **16,** 2877 (1972).
46. M. Takayanagi, *Proc. Int. Congr. Rheol. 4th,* Part 1, p. 161 (1965).
47. D. J. Massa, *J. Appl. Phys.* **44,** 2595 (1973).
48. K. Fujino, I. Furuta, S. Kawabata, and H. Kawai, *Zairyo* **13,** 404 (1964).
49. C. W. Macosko and F. C. Weissert, *Am. Soc. Test. Mater., Spec. Tech. Publ.* **553,** 127 (1974).
50. B. Maxwell, *J. Polym. Sci.* **20,** 551 (1956).
51. R. B. Bird and E. K. Harris, *A.I.C.E. J.* **14,** 758 (1968).
52. T. E. R. Jones and K. Walters, *Br. J. Appl. Phys.* **2,** 815 (1969).
53. E. H. Yerzley, *Rubber Chem. Technol.* **13,** 149 (1940).
54. J. H. Dillon, I. B. Prettyman, and G. L. Hall, *J. Appl. Phys.* **15,** 309 (1941).

Chapter **6**

Rheological Behavior of Unvulcanized Rubber

JAMES LINDSAY WHITE

THE UNIVERSITY OF TENNESSEE
KNOXVILLE, TENNESSEE

Copyright © 1978 by Academic Press, Inc.
All rights of reproduction in any form reserved.
ISBN 0-12-234360-3

I. Introduction and Historical Background†

The fabrication of rubber parts generally involves the mixing and processing of bulk unvulcanized compounds and sometimes solutions and emulsions through complex equipment. The ease or difficulty of fabrication depends upon how these rubber systems respond to applied stresses and deformation, their *rheological* (from *rhein*: to flow, *logos*: science of) properties. It is the purpose of this chapter to describe these rheological properties of unvulcanized elastomers, their solutions, and their compounds. We will also consider some of the implications of rheology for processing.

The study of the rheological properties and processing of elastomers and their solutions and compounds dates to the origins of the industry in the 1820s. The patent literature, memoirs, and reviews of the nineteenth century contain numerous discussions of the flow and fabrication of natural rubber and gutta-percha [45, 104, 119, 166, 270, 303, 316]. The fundamental properties and methods of processing rubber are associated with men such as Thomas Hancock, Charles Macintosh, Edwin Chaffee, Charles Goodyear, Richard Brooman, Henry Bewley, and others, many long forgotten. It was not, however, until the development of three-dimensional linear viscoelasticity by Boltzmann [37] in 1874 that understanding of the rheological properties of rubbery materials became sophisticated enough to allow rational study. Furthermore, it was another half century before Bruno Marzetti (1884–1953) of Pirelli [79, 175] in Italy and (in the 1930s) John H. Dillon of Firestone [73, 75] and Melvin Mooney (1893–1968) of US Rubber [195, 196, 202, 301] undertook the study of the deformation and flow of unvulcanized rubber. In each case the motivation seems to have been the development of quality control instrumentation to ensure satisfactory processibility. Fortunately each of the three was a careful, observant, and thoughtful scientist. Marzetti interpreted the extrusion of rubber through a cylindrical die in terms of the flow of a fluid with a shear-rate-dependent viscosity. This view was confirmed by Dillon [73, 75] and Mooney [195, 196]. Mooney was able to obtain the first quantitative viscosity–shear rate data on rubber and the first measurements of elastic recoil in rubber compounds, and Dillon and Cooper [74] reported the first investigations of stress transients at the beginning of flow.

The studies of Mooney and Dillon in the 1930s established the prevailing view of the rheological behavior of unvulcanized rubber. The 1940s saw the coming of World War II and the Rubber Reserve synthetic rubber program in the United States. Rheology played a relatively small role in this program but that role was dominated in large part by Mooney [197, 198, 200, 202, 278], utilizing and expanding the concepts devised during the previous decade. Mooney's 1934 shearing disk viscometer [195] was established by the Rubber

† More detailed historical surveys are given by the author elsewhere [299, 301, 303].

Reserve as a quality control instrument, which it remains to this day. Mooney investigated a wide range of problems that were however reported only incompletely. These included the relationship between elastic shear recovery (measured in a disk viscometer) and shrinkage of calendered and extruded parts, surface roughness developed during calendering and extrusion, and screw extrusion behavior as a function of non-Newtonian viscosity. In succeeding years little of this research was followed up by rubber companies (though some exceptions may be noted [78, 80, 200, 203, 230]) and the interaction between polymer rheology and processing was developed in large part by plastics companies.

Post-World War II studies of the rheological properties of unvulcanized elastomers have been dominated by the idea that these materials are viscoelastic. Research along these lines was initiated by Leaderman [156, 157], who rediscovered the work of Boltzmann and his contemporaries. In 1948, Alfrey [4] published a monograph on the mechanical behavior of polymers which emphasized viscoelastic behavior and had a major influence in succeeding years. In the late 1940s, Tobolsky and Andrews [7, 8, 279] made extensive stress relaxation measurements on polyisobutylene. In succeeding years linear viscoelastic measurements were performed on a wide variety of polymers in temperature regions where they exhibited rubbery behavior. Tobolsky and Andrews devised a program to relate the viscoelastic behavior to molecular parameters, such as molecular weight and glass transition temperature. This work is reviewed in a more recent monograph by Tobolsky [280].

The 1940s and 1950s were also the period of the pioneering studies of the normal stress, or Weissenberg, effects [96, 247, 250, 251, 294, 295, 296], which were being realized to be an aspect of nonlinear viscoelastic behavior. Mooney [199] was among the early investigators and indeed had actually observed but not recognized the significance of the effect some years earlier [196, 301]. Many of the measurements were carried out on solutions of elastomers, especially polyisobutylene [44, 109, 172]. However, there was at first no explicit attempt to relate this effect to the rheological behavior of bulk elastomers. It is fair to say, though, that because similar effects had been observed for vulcanized rubber [244] it was implicitly presumed that such materials would respond as nonlinear viscoelastic fluids. Pollett [236] made early normal stress measurements on molten polyethylene. In the early 1960s, Bernstein, Kearsley, and Zapas [24, 25, 327, 328] made extensive studies of the nonlinear viscoelastic properties of polyisobutylene and proposed a tensor constitutive equation to represent its properties. Later in the decade, nonlinear transient experiments on polyisobutylene were interpreted similarly by Middleman [190].

During the 1960s rheological investigations of the processing behavior of elastomers began to receive attention again. White and Tokita [282, 299, 300, 308–310, 312] and Ninomiya et al. [212, 213] have carried out research programs in which various processing characteristics were interpreted in terms of the theory of viscoelasticity. Kraus and his colleagues [143–147] have made extensive studies of the influence of molecular structure on the rheological properties of elastomers. Since about 1970 research programs on rheological properties as related to processing of rubber have been reported by Collins and Nakajima [40, 65, 66, 208, 209], Folt [89–91] and Weissert [169, 297] and their co-workers, as well as by Tokita [234, 281] and White [98, 138, 304a, 305] independently.

There are numerous useful monographs on rheology and its application to polymers which represent a variety of viewpoints and emphases. We refer the reader to those of Bogue and White [34], Ferry [84], Lenk [161], Lodge [164], McKelvey [167], Middleman [189], Pearson [227], Reiner [241], Skelland [261], Tadmor and Klein [274], and Tobolsky [280]. Turning to rubber, Mooney [200] published a review in the same area as this article in 1958. The author's views are expanded in an earlier review [299] and a more recent paper [304a].

II. Basic Concepts

A. STRESS TENSOR AND EQUATIONS OF MOTION

In this section we will try to develop the basic ideas of classical rheological thought. We presume elastomers to deform as continuous media and to be subject to the formalism of continuum mechanics [286]. We will begin by developing the idea of the nature of applied forces and the stress tensor.

The idea of the stress tensor in a material arises from the necessity of representing the influence of applied forces upon deformation. The applied forces $\mathbf{F}$ acting on a body may be represented as the sum of contact forces acting on the surface and body forces $\mathbf{f}$, such as gravitation, which act directly on the elements of mass. We may write [286] (Fig. 1)

$$\mathbf{F} = \sum_i \mathbf{t}_i \, \Delta a_i + \sum_i \mathbf{f}_j \, \Delta m_j = \oint \mathbf{t} \, da + \int \rho \mathbf{f} \, dV \tag{1}$$

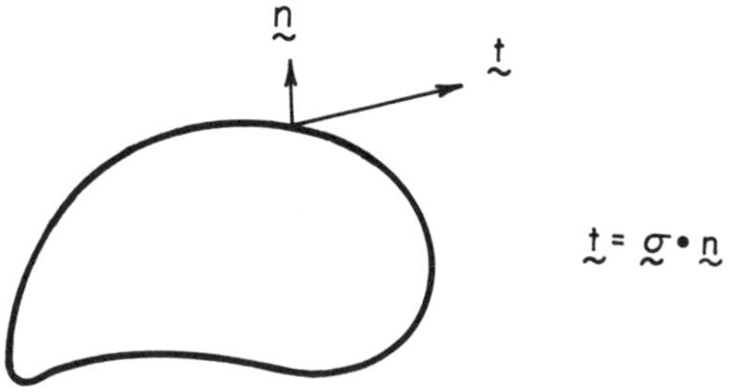

Fig. 1. Stress vector and stress tensor. Wavy underscores indicate vector quantities.

where **t** is the force per unit area (stress vector) acting on the surface area elements Δa_i, and the $\oint$ indicates that the integration exists over the entire surface.

The idea of the stress tensor comes from relating **t** to the unit normal vector **n** to the surface through

$$\mathbf{t} = \boldsymbol{\sigma} \cdot \mathbf{n} \tag{2}$$

where $\boldsymbol{\sigma}$ is the array of nine quantities

$$\boldsymbol{\sigma} = \begin{vmatrix} \sigma_{11} & \sigma_{12} & \sigma_{13} \\ \sigma_{21} & \sigma_{22} & \sigma_{23} \\ \sigma_{31} & \sigma_{32} & \sigma_{33} \end{vmatrix} \tag{3}$$

known as the stress tensor or matrix. $\boldsymbol{\sigma}$ may be considered as a second-order tensor or a Gibbs dyadic

$$\boldsymbol{\sigma} = \sum_i \sum_j \sigma_{ij} \mathbf{e}_i \mathbf{e}_j \tag{4}$$

The concepts of the stress vector and stress tensor were developed during the 1820s by Cauchy [57]. The σ_{ij} stresses represent stresses in the i direction acting on a plane normal to the j axis. The diagonal components of the stress tensor σ_{ii} represent stress components acting normal to planes. They are thus called *normal stresses*. The σ_{ij} $(i \neq j)$ are tangential to plane j and are called *shear stresses*.

Applying the divergence theorem [100, 286] to Eq. (1) gives

$$\oint \mathbf{t}\, da = \oint \boldsymbol{\sigma} \cdot \mathbf{n}\, da = \int \nabla \cdot \boldsymbol{\sigma}\, dV \tag{5}$$

$$\mathbf{F} = \int [\nabla \cdot \boldsymbol{\sigma} + \rho \mathbf{f}]\, dV \tag{6}$$

where ∇ is the del operator

$$\nabla = \mathbf{e}_1 \frac{\partial}{\partial x_i} + \mathbf{e}_2 \frac{\partial}{\partial x_2} + \mathbf{e}_3 \frac{\partial}{\partial x_3} \tag{7}$$

The complete dynamics of a deforming body requires relating the contact and body forces with inertial forces. For a macroscopic mass M

$$\mathbf{F} = \frac{d}{dt} \int \rho \mathbf{v}\, dV = \int [\nabla \cdot \boldsymbol{\sigma} + \rho \mathbf{f}]\, dV \tag{8}$$

while for a macroscopic fixed-space "control volume" through which the mass may move [286]

$$\int \frac{\partial}{\partial t} (\rho \mathbf{v}) \, dV + \oint \rho \mathbf{v}(\mathbf{v} \cdot \mathbf{n}) \, da = \int [\nabla \cdot \boldsymbol{\sigma} + \rho \mathbf{f}] \, dV \qquad (9)$$

It follows that at a point within the body [286]

$$\rho \left[\frac{\partial \mathbf{v}}{\partial t} + (\mathbf{v} \cdot \nabla) \mathbf{v} \right] = \nabla \cdot \boldsymbol{\sigma} + \rho \mathbf{f} \qquad (10)$$

Equation (10), which establishes the balance of forces at a point within the body, is known as Cauchy's law of motion.

The components of the stress tensor are not independent of each other. By a balance of torques and angular moments similar to that leading to Eq. (10) it may be shown that the stress tensor is symmetric [286]; i.e.,

$$\boldsymbol{\sigma} = \boldsymbol{\sigma}^{\mathrm{T}} \qquad \text{or} \qquad \sigma_{ij} = \sigma_{ji} \qquad (11)$$

Thus the off-diagonal components of Eq. (7), which are the shear stresses, are related:

$$\sigma_{12} = \sigma_{21}, \qquad \sigma_{13} = \sigma_{31}, \qquad \text{and} \qquad \sigma_{23} = \sigma_{32}$$

The basic problem of rheology is the development of expressions for $\boldsymbol{\sigma}$ in terms of the deformation and kinematics of materials. The deformation behavior of continuous materials may then be determined through solutions of Eq. (10).

If a body is not subjected to applied forces, the stress components reduce to equal normal hydrostatic pressure components

$$\boldsymbol{\sigma} = -p\mathbf{I}, \qquad \mathbf{I} = \begin{vmatrix} 1 & 0 & 0 \\ 0 & 1 & 0 \\ 0 & 0 & 1 \end{vmatrix} \qquad (12)$$

where p is the pressure. More generally, when forces are applied we may express the stress tensor in terms of the pressure and an extra stress tensor $\mathbf{P}$ through the relation

$$\boldsymbol{\sigma} = -p\mathbf{I} + \mathbf{P} \qquad (13a)$$

Clearly

$$\sigma_{ii} = -p + P_{ii} \qquad (13b)$$

$$\sigma_{ij} = P_{ij} \qquad (i \neq j) \qquad (13c)$$

In shearing deformations we need not distinguish between σ_{ij} and P_{ij}.

B. LINEAR VISCOELASTICITY

Traditionally one divides materials into elastic solids, where stress depends upon strain, or viscous fluids, where the stress depends solely on the rate of strain. Unvulcanized rubber, however, responds to stresses as if it were a fluid with memory. The material deforms indefinitely under the application of shear stresses but exhibits elastic recoil upon release of the stresses and stress relaxation if the flow is halted. The idea of materials with incomplete memory of their deformation history seems to have first been expressed in analytical form in 1866 by Maxwell [180], who suggested that for such materials the rate of increase of elastic modulus times rate of strain should exceed the rate of increase of stress by an amount proportional to the stress. By strain, i.e., shear strain, we mean (see Fig. 2a) the ratio of the displacement to the distance of shear

$$\gamma = [l(t) - l^*]/H \tag{14}$$

where l^* represents a reference state. For a one-dimensional shearing deformation

$$\frac{d\sigma}{dt} = G\frac{d\gamma}{dt} - \frac{1}{\tau} = G\dot{\gamma} - \frac{1}{\tau}\sigma \tag{15a}$$

where σ is a shear stress, $d\gamma/dt$ a shear rate (see Fig. 2b), G a modulus, and τ an inverse decay constant. We may solve this differential equation for an arbitrary deformation history $d\gamma/dt$ to give

$$\sigma = \int_{\infty}^{t} Ge^{-(t-s)/\tau}\frac{d\gamma}{ds}(s)\,ds \tag{15b}$$

τ plays the role of a *relaxation time*, which is the name by which it is generally designated.

A mechanical analog of Maxwell's theory of materials with incomplete memory, commonly referred to as *viscoelastic*, is sketched in Fig. 3 [5, 241]. This is a spring with constant G and a dashpot (a piston–cylinder apparatus containing a viscous liquid with resistance η) in series. Increase of the length of the model correspond to strain γ, applied forces to stress σ. The validity of

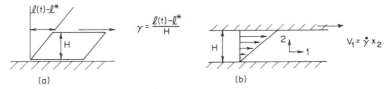

Fig. 2. (a) Simple shear strain. (b) Simple shear flow.

Fig. 3. Spring–dashpot mechanical analog for a Maxwell fluid [Eqs. (15)–(16)].

the model may be perceived by noting that the displacement of the total model is the sum of the displacement of the spring and the displacement of the dashpot.

$$\gamma = (\sigma/G) + \int (\sigma/\eta)\, dt \qquad (16)$$

$$\underbrace{\hphantom{\gamma}}_{\substack{\text{total}\\\text{for mode}}} \quad \underbrace{\hphantom{(\sigma/G)}}_{\text{spring}} \quad \underbrace{\hphantom{\int(\sigma/\eta)dt}}_{\text{dashpot}}$$

Differentiating gives

$$\frac{d\gamma}{dt} = \frac{1}{G}\frac{d\sigma}{dt} + \frac{\sigma}{\eta} \qquad (17)$$

which is equivalent to Eq. (15) with the relaxation time τ given by

$$\tau = \eta/G \qquad (18)$$

Boltzmann [37] in 1874 gave a superior formulation of the mechanical behavior of materials with incomplete memory. Consider a material which when subjected to strain ε at time zero exhibits a stress

$$\sigma = G(t)\gamma \qquad (19)$$

where $G(t)$ is a modulus function that decays with time. If the response of this material is linear, the stress σ developed for a series of sequential deformations $\gamma_1, \gamma_2, \gamma_3, \ldots$ at times $t_1, t_2, t_3, \ldots$ will be at time t

$$\sigma = G(t - t_1)\gamma_1 + G(t - t_2)\gamma_2 + G(t - t_3)\gamma_3 + \cdots \qquad (20a)$$

$$= \int_{\infty}^{t} G(t - s)\, d\gamma(s) = \int_{\infty}^{t} G(t - s)\frac{d\gamma}{ds}\, ds \qquad (20b)$$

where the integral refers to the limit for which the applied deformation is continuous. Equation (20b) is known as Boltzmann's superposition integral.

The superiority and generality of Boltzmann's work was later pointed out by Maxwell himself [181].

From comparison of Eqs. (15) and (20b), it may be seen that Maxwell's theory is the special case of Boltzmann's for which

$$G(t) = Ge^{-t/\tau} \tag{21}$$

Generally it is found that although Eqs. (15) and (21) exhibit the proper qualitative features of elastomers and other polymer fluids, the representation is not quantitative. For long times $\log G(t)$ does decay linearly with time [281]. This suggests representation of $G(t)$ in the form of a series of exponentials or in terms of a distribution function $H(\tau)$, i.e.,

$$G(t) = \sum_i G_i e^{-t/\tau_i} \tag{22a}$$

$$= \int_\infty^\infty H(\tau) e^{-t/\tau} \, d\ln \tau \tag{22b}$$

Such distribution functions, which are common in linear analysis, were introduced in viscoelasticity by Wiechert [314] and were later used by Kuhn et al. [153], Alfrey [4], Tobolsky [7, 8, 280], and later authors. The rheological characterization of a linear viscoelastic material thus consists of determining the specific $H(\tau)$.

Often, following Boltzmann [37], it is of interest to consider Eq. (20b) in terms of strain rather than strain rate. Integration of Eq. (20b) by parts yields (see White [301])

$$\sigma(t) = G(0)\gamma(t) - \int_{-\infty}^t \Phi(t-s)\gamma(s) \, ds \tag{23}$$

where we have taken $G(\infty)$ as being zero for a fluid and

$$\Phi(t) = -\frac{dG(t)}{dt} \tag{24}$$

$$= \sum \frac{G_i}{\tau_i} e^{-t/\tau_i} = \int_0^\infty \frac{H(\tau)}{\tau^2} e^{-t/\tau} \, d\tau \tag{25}$$

To proceed we must look more closely at our strain quantity γ, defined in Eq. (15). Specifically, we ask ourselves what the reference state is. Since a fluid has no preferred configuration, the only choice that is not artificial is the instantaneous state, i.e., l^* equal to $l(t)$. If we accept this definition, Eq. (23) reduces to

$$\sigma(t) = -\int_{-\infty}^t \Phi(t-s)\gamma(s) \, ds = \int_{-\infty}^t \Phi(t-s) \frac{l(t) - l(s)}{H} \, ds \tag{26}$$

The formulation just given is one dimensional. More generally, three-dimensional formulations are required. To convert Eqs. (20) and (26) to three-dimensional forms, we note that if we express the direction of flow by 1 and the direction of shear by 2, then

$$\frac{d\gamma}{ds} = \frac{1}{H}\frac{dl(s)}{ds} = \frac{V}{H} = \frac{\partial v_1}{\partial x_2} \tag{27}$$

Thus the stress tensor is to be related to the history of the velocity gradient. However, we must be careful here for it turns out that it is not the velocity gradient tensor we are concerned with, but rather the deformation rate tensor d_{ij}, defined by

$$\frac{\partial v_i}{\partial x_j} = d_{ij} + \omega_{ij} \tag{28a}$$

where

$$d_{ij} = \frac{1}{2}\left(\frac{\partial v_i}{\partial x_j} + \frac{\partial v_j}{\partial x_i}\right) = \begin{array}{l}\text{deformation} \\ \text{rate tensor}\end{array} \tag{28b}$$

$$\omega_{ij} = \frac{1}{2}\left(\frac{\partial v_i}{\partial x_j} - \frac{\partial v_j}{\partial x_i}\right) = \text{vorticity tensor} \tag{28c}$$

As pointed out by Stokes [271, 286], d_{ij} represents true material deformation, while ω_{ij} represents local rigid rotation. From these comments and Eqs. (14) and (20) it follows that we may write the stress σ as

$$\sigma = -p\mathbf{I} + 2\int_{-\infty}^{t} G(t - s)\,\mathbf{d}(s)\,ds = -p\mathbf{I} + 2\int_{0}^{\infty} G(z)\,\mathbf{d}(z)\,dz \tag{29}$$

where $\mathbf{d}(s)$ is a matrix with components d_{ij}. Note that $\mathbf{d}$, $\mathbf{I}$, and σ are all symmetric.

To express Eq. (26) in three-dimensional form, we may (for small strains only) write

$$\frac{d\gamma}{dt} = \mathbf{d} \tag{30a}$$

$$\gamma_{ij} = \frac{1}{2}\left(\frac{\partial u_i}{\partial x_j} + \frac{\partial u_j}{\partial x_i}\right) \tag{30b}$$

where u_i are displacements. It follows that

$$\sigma = -p\mathbf{I} + 2\int_{-\infty}^{t} \Phi(t - s)\gamma(s)\,ds = -p\mathbf{I} + 2\int_{0}^{\infty} \Phi(z)\gamma(z)\,dz \tag{31}$$

where γ measures strain from time s to time t.

C. Nonlinear Viscoelasticity

The problem of developing a theory of viscoelasticity valid for large strains is a rather difficult one. It was not until the work of Zaremba [329], who published in a little-known journal in the opening years of the present century, that the essentials of the problem were realized. With the exception of this paper and the later work of Hencky [121], there was little understanding of nonlinear viscoelasticity until the 1950s. A new era began with the work of Oldroyd [217, 218, 219], whose studies were followed up in succeeding years by Lodge [162, 163], Rivlin and Ericksen [246], Noll [214, 215], Green and Rivlin [107], Rivlin [245], and others. The essential point made by Zaremba, Oldroyd, and later authors is that a constitutive equation, to be valid for large deformations, must be formulated in a coordinate system embedded in the medium rather than fixed space. Furthermore, there is no unique generalization of a constitutive equation to large strains. The tensor γ, for example, may be generalized to

$$\gamma \to E = \tfrac{1}{2}(\mathbf{I} - \mathbf{c}) \tag{32a}$$

or

$$\gamma \to e = \tfrac{1}{2}(\mathbf{c}^{-1} - \mathbf{I}) \tag{32b}$$

where

$$c_{ij} = \sum_a \frac{\partial \bar{x}_a}{\partial x_i} \frac{\partial \bar{x}_a}{\partial x_j}, \qquad c_{ij}^{-1} = \sum_a \frac{\partial x_i}{\partial \bar{x}_a} \frac{\partial x_j}{\partial \bar{x}_a} \tag{33}$$

and x_i is the position of a material point at time t and $\bar{x}_a$ its position at time s. E follows from considering the strain to arise from the deformation of a differential material line segment within a body [286]. e may be interpreted in terms of the deformation of an area element [286].

The generalization of the Maxwell and similar differential equation models of viscoelastic behavior to large deformation was the concern of Zaremba [329] and many other early authors. Use of embedded or convected coordinate systems led to complex nonlinear time derivatives of stress [217, 219, 307, 329] which were not unique. Of the various formulations, that of White and Metzner [307] has proven the most useful. We shall not, however, develop these studies here. We shall rather look to the generalization of Boltzmann's viscoelasticity theory to large deformations. Determining which generalization would be appropriate to polymer systems is a difficult task that requires an extensive experimental program. Fortunately, some hints of a reasonable form can be obtained from considering molecular theories of deformation behavior. From the statistical mechanics of cross-linked networks of flexible chains [88, 285], i.e., the kinetic theory of rubber elasticity, it may be shown that

$$\sigma = -p\mathbf{I} + vkT\mathbf{c}^{-1} \tag{34}$$

where k is Boltzmann's constant, and T the thermodynamic temperature; v is proportional to the cross-link density; and $\bar{x}_a$ of $\mathbf{c}^{-1}$ refers to the virgin state. Arguments by Green and Tobolsky [108], Lodge [162, 163], and Yamamoto [319, 320] showed that if the permanent cross-links are replaced with temporary junctions (entanglements), Eq. (34) is modified to

$$\boldsymbol{\sigma} = -p\mathbf{I} + kT \int_0^\infty v(z)\mathbf{c}^{-1}\, dz \qquad (35)$$

where $v(t)$ is the junction density at time t. Yamamoto also gave a more general development of Green and Tobolsky's ideas and came up with rather more complex expressions. Since 1963, Bernstein et al. [24] and later authors [28, 33, 34, 59, 115, 189, 267, 311, 327, 328] have seriously attempted to determine quantitative constitutive equations for unvulcanized elastomers and polymer solutions and melts by generalizing Eq. (35). They have used expressions of form

$$\boldsymbol{\sigma} = -p\mathbf{I} + \int_{-\infty}^t [m_1(z)\mathbf{c}^{-1} - m_2(z)\mathbf{c}]\, dz \qquad (36)$$

where $m_1(z)$ is the dominant term. Generally $m_1(z)$ and $m_2(z)$ are taken to depend upon the deformation process through the invariants of the strain or the roles of deformation. Bernstein et al. [24] took $m_1(z)$ and $m_2(z)$ to be functions of the invariants of $\mathbf{c}$, while Spriggs et al. [267] made use of the invariants of the rate of deformation $\mathbf{d}$ and Bogue [33] considered time-averaged integrals of invariants of $\mathbf{d}$. Bogue and White [34] and the experiments of Chen and Bogue [61] suggest these forms of $m_1(z)$ and $m_2(z)$:

$$m_1(z) = \left(1 + \frac{\varepsilon}{2}\right) \sum_i \frac{G_i}{\tau_{i\text{eff}}} e^{-z/\tau_{i\text{eff}}} \qquad (37a)$$

$$m_2(z) = -\frac{\varepsilon}{2} \sum \frac{G_i}{\tau_{i\text{eff}}} e^{-z/\tau_{i\text{eff}}} \qquad (37b)$$

Here the $\tau_{i\text{eff}}$ are effective relaxation times dependent upon invariants. For small strains and deformation rates the $\tau_{i\text{eff}}$ must become constants τ_i and Eqs. (36) and (37) reduce to Eqs. (31) and (25). If we use the Bogue time-averaged invariants of the rate of deformation, a reasonable form of $\tau_{i\text{eff}}$ turns out to be

$$\tau_{i\text{eff}} = \frac{\tau_i}{1 + b\tau_i \overline{\Pi_\mathbf{d}^{1/2}}} \qquad (38a)$$

$$\overline{\Pi_\mathbf{d}^{1/2}} = \frac{1}{t} \int_0^t \Pi_\mathbf{d}^{1/2}\, dz \qquad (38b)$$

$$\Pi_\mathbf{d} = 2tr\mathbf{d}^2 = 2d_{ij}d_{ij} \qquad (38c)$$

Similar forms of $m_1(z)$ and $m_2(z)$ are suggested by Carreau [55], who takes different forms for the $\tau_{i\text{eff}}$ in the coefficients and exponentials of Eqs. (37a,b). This introduces additional parameters and greater flexibility in fitting data. The Bogue–White theory has been extensively compared with experiment on polymer melts and elastomers [61, 98, 128a, 190]. This has recently been reviewed by White [304a].

A more detailed review of early studies of constitutive equations is given by Bogue and White [34]. A critical discussion has been given by Yamamoto [322] and a recent review by Han [115].

A word of warning on the applications of the above theories should be given. They generally work well with homogeneous solutions and melts. However they cannot be applied to highly filled melts.

A summary of many of the theories is given in Table I.

D. Special Types of Deformations

1. Shearing Flows

Perhaps the most important class of deformations in which we must study polymer fluids are laminar shear flows. These motions derive from simple shear flow

$$v_1 = \dot{\gamma} v_2, \qquad v_2 = v_3 = 0, \qquad \mathbf{d} = \tfrac{1}{2} \begin{vmatrix} 0 & \dot{\gamma} & 0 \\ \dot{\gamma} & 0 & 0 \\ 0 & 0 & 0 \end{vmatrix} \tag{39}$$

(see Fig. 2) and include Couette flow between coaxial cylinders, Poiseuille flow in a tube, torsional flow between rotating disks, and flow between a cone and a plate. The rheological response of a Newtonian fluid to this type of motion is a constant shear stress proportional to the rate of shear.

$$\sigma_{12} = \eta \dot{\gamma} \tag{40}$$

The constant of proportionality η is the viscosity. For polymer fluids, the response includes not only a shearing stress, but unequal normal stresses in the direction and perpendicular to the direction of flow.

$$\begin{aligned} \sigma_{12} &= \eta \dot{\gamma} \\ \sigma_{11} &= -p + \beta_1 \dot{\gamma}^2 \\ \sigma_{22} &= -p + \beta_2 \dot{\gamma}^2 \\ \sigma_{33} &= -p \end{aligned} \tag{41}$$

These normal stresses give rise to the stirring-rod-climbing normal stress effects

TABLE I

INTEGRAL CONSTITUTIVE EQUATIONS FOR NONLINEAR
VISCOELASTIC FLUIDS IN TERMS OF EQ. (36)

Memory functions	References
$m_1(t) = \sum_i \dfrac{G_i}{\tau_i} e^{-t/\tau_i}$ $m_2(t) = 0$	Lodge [163, 164]
$m_1(t) = 2\dfrac{\partial U}{\partial I_1}$ where U is a function of strain invariants I_1 and I_2 $m_2(t) = 2\dfrac{\partial U}{\partial I_2}$	Bernstein *et al.* [24, 25, 327, 328]; see also White and Tokita [311]
$m_1(t) = \left(1 + \dfrac{\varepsilon}{2}\right) \sum_i \dfrac{G_i}{\tau_{i\text{eff}}} e^{-t/\tau_i}$ $m_2(t) = \dfrac{-\varepsilon/2}{1 + \varepsilon/2} m_1(t)$ $\tau_{i\text{eff}}$ depends upon Π_d	Spriggs *et al.* [267] Bird and Carreau [28] White [299]
$m_1(t) = \left(1 + \dfrac{\varepsilon}{2}\right) \sum_i \dfrac{G_i}{\tau_{i\text{eff}}} e^{-t/\tau_{i\text{eff}}}$ $m_2(t) = \dfrac{-\varepsilon/2}{1 + \varepsilon/2} m_1(t)$ $\tau_{i\text{eff}}$ given by Eq. (38)	Bogue [33] Middleman [190] Bogue and White [34] Chen and Bogue [61] White [304] considers the case for $\tau_{i\text{eff}} = f(\prod_d)$
$m_1(t) = \left(1 + \dfrac{\varepsilon}{2}\right) \sum_i \dfrac{G_i}{\tau_{i\text{eff}}} e^{-\int \frac{dt}{f_i \tau_{i\text{eff}}}}$ f_i are functions of rate of deformation invariants $m_2(t) = \dfrac{-\varepsilon/2}{1 + \varepsilon/2} m_1(t)$	Carreau [55]

(see Fig. 4). Generally, the normal stresses are expressed in terms of the principal (or first) normal stress difference N_1 and the second normal stress difference N_2

$$N_1 = \sigma_{11} - \sigma_{22} = \Psi_1 \dot{\gamma}^2, \qquad N_2 = \sigma_{22} - \sigma_{33} = \Psi_2 \dot{\gamma}^2 \tag{42}$$

with normal stress coefficients Ψ_1 and Ψ_2.

Extensive experimental studies of the non-Newtonian viscosity function η exist for elastomers and compounds [65, 66, 81, 127, 196, 200, 208, 297, 305, 309,

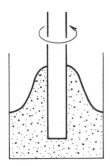

Fig. 4. Weissenberg (rod-climbing) normal stress effect.

317], solutions of elastomers [44, 138], and most other types of polymeric fluids, including molten plastics [20, 136, 159, 192, 193, 228], as well as polymer solutions and emulsions. At low shear rates in unfilled polymer fluids the viscosity function η is a constant. As the rate of shear increases, the viscosity function decreases. Typically the function $\eta(\dot{\gamma})$ has been represented by expressions such as

$$\eta = K\dot{\gamma}^{n-1} \quad \text{(or} \quad \sigma_{12} = K\dot{\gamma}^n) \tag{43a}$$

$$\eta = \frac{\eta_0}{1 + A\dot{\gamma}^{1-n}} \tag{43b}$$

Equation (43a) is known as the power law where n is the power law exponent and represents behavior in intermediate shear rate regions. η_0 is the zero shear viscosity and Eq. (43b) represents behavior in both low and intermediate shear rate regions. In the case of polymer solutions the function frequently becomes constant again at higher shear rates. Typical experimental data are shown in Fig. 5.

Much less rheological information is available for normal stresses. There is a large quantity of data available on elastomer solutions for both N_1 and N_2 [43, 44, 102, 138, 277] (see Fig. 5a). At low shear rates Ψ_1, which is positive, is constant but decreases with increasing rate of shear. Ψ_2 is negative and about 10% of Ψ_1. Data for polymer melts are in general agreement with these responses [20, 31, 159, 192, 216a, 253]. Oda *et al.* [216a] have recently been able to correlate together normal stress N_1 data with the shear stress σ_{12} for polystyrene melts. The behavior is independent of temperature. Polymers of similar molecular weight distributions exhibit almost identical correlations.

It may be shown rigorously for the most general types of viscoelastic fluid that at low shear rates (second-order fluid asymptote) [64, 298]:

$$\lim_{\dot{\gamma}\to 0} \eta(\dot{\gamma}) = \eta_0 = \int_0^\infty G(s)\,ds = \int_0^\infty H(\tau)\,d\tau \tag{44a}$$

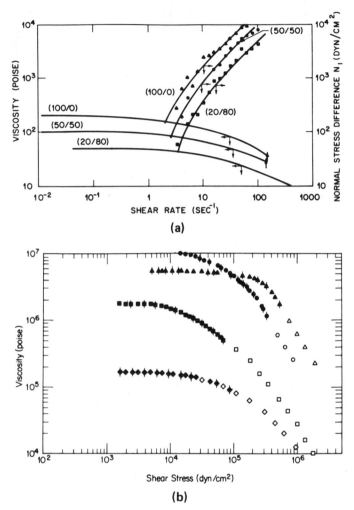

Fig. 5. (a) Viscosity and principal normal stress difference N_1 versus shear rate for a 40/60 SBR in a 10% solution in decalin–decane with various solvent ratios. (From Kotaka and White [138].) (b) Viscosity–shear stress data for polymer melts: ●○● (PS), polystyrene; ◆◇◆ (LDPE), low-density polyethylene; ■□■ (HDPE), high-density polyethylene; and ▲△▲ (PMMA), poly-methyl methacrylate used in this study. The solid points are for the cone–plate Weissenberg rheogoniometer data; the points containing superposed vertical lines are for the parallel disk geometry. Open points are Instron capillary rheometer data. (From Lee and White [159].)

$$\lim_{\dot\gamma \to 0} \Psi_1(\dot\gamma) = 2 \int_0^\infty sG(s)\, ds = 2 \int_0^\infty \tau H(\tau)\, d\tau \tag{44b}$$

$\eta(\dot\gamma)$ and $\Psi_1(\dot\gamma)$ are predicted to be constant at low shear rates.

In terms of the constitutive equation Eq. (26) the three rheological properties η, Ψ_1, and Ψ_2 are

$$\eta = \int_0^\infty z[m_1(z) + m_2(z)]\,dz \tag{45a}$$

$$\Psi_1 = \int_0^\infty z^2[m_1(z) + m_2(z)]\,dz \tag{45b}$$

$$\Psi_2 = \int_0^\infty z^2 m_2(z)\,dz \tag{45c}$$

and if we represent $m_1(z)$ and $m_2(z)$ through Eqs. (37) we obtain

$$\eta = \sum_i G_i \tau_{i\text{eff}} \tag{46a}$$

$$\Psi_1 = 2\sum_i G_i \tau_{i\text{eff}}^2 \tag{46b}$$

$$\Psi_2 = \varepsilon \sum_i G_i \tau_{i\text{eff}}^2 \tag{46c}$$

In the limit of small shear rates $\tau_{i\text{eff}}$ becomes equal to τ_i. Note that Ψ_1 may be expressed as

$$\Psi_1 = 2\bar{\tau}\eta = 2J_e\eta^2 \tag{47a}$$

where

$$\bar{\tau} = \frac{\sum_i G_i \tau_{i\text{eff}}^2}{\sum_i G_i \tau_{i\text{eff}}} \tag{47b}$$

$$J_{\text{eapp}} = \frac{\sum_i G_i \tau_{i\text{eff}}^2}{(\sum_i G_i \tau_{i\text{eff}})^2} \tag{47c}$$

is a characteristic relaxation time representing the ratio of the second to the first moment of the relaxation kernel functions and J_{eapp} is the apparent steady state compliance. For small $\dot{\gamma}$, J_{eapp} becomes the linear viscoelastic J_e (Oda et al. [216a]). The N_1–σ_{12} correlation reflects the temperature independence of J_{eapp}.

The variation of shear stress σ_{12} with time at the start-up of flow is a well-known rheological experiment in the rubber industry. Studies were reported by Dillon and Cooper [74] during the 1930s and by Taylor et al. [278] in the succeeding decade. At low shear rates σ_{12} monotonically increases to a steady state. At higher shear rates it overshoots the steady state (see Fig. 6). This is nicely shown in the experiments of Nakajima and Collins [209] on butadiene–styrene copolymer. Bogue [33] was the first to show that this stress overshoot is primarily a nonlinear viscoelastic effect. Middleman [190], Chen

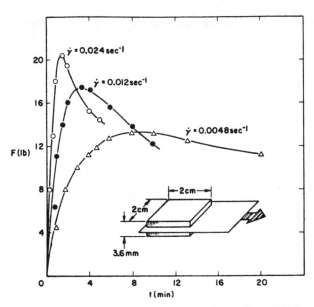

Fig. 6. Transient shear force buildup and stress overshoot. (From Middleman [190].)

and Bogue [61], Carreau [55], Furuta *et al.* [98], and Goldstein [103] have tried to use stress overshoot as a critical test for distinguishing the applicability of different nonlinear viscoelastic constitutive equations. Middleman, Chen and Bogue, and Furuta *et al.* find Eqs. (36)–(38) to be a useful form. If $v_1 = 0$ for times less than zero and $v_1 = \dot{\gamma}x_2$ for times greater than zero, this constitutive equation predicts

$$\sigma_{12} = \sum \{G_i \tau_{i\text{eff}}[1 - e^{-t/\tau_{i\text{eff}}} + (t/\tau_{i\text{eff}})^2 b\dot{\gamma}\tau_{i\text{eff}} Ei(t/\tau_i)e^{-b\dot{\gamma}t}]\}\dot{\gamma}$$

$$Ei(x) = \int_x^\infty (e^{-x}/x)\,dx \tag{48}$$

Equation (48) is compared to experimental data in Fig. 6. It should be clear that for large times, Eq. (48) approaches Eq. (46a). For small $\dot{\gamma}$, $\tau_{i\text{eff}}$ approaches τ_i and Eq. (48) becomes

$$\sigma_{12} = [\sum G_i \tau_i(1 - e^{-t/\tau_i})]\dot{\gamma} \tag{49}$$

which represents a monotonically increasing stress that reaches an asymptotic value $(\sum G_i \tau_i)\dot{\gamma}$. For larger values of $\dot{\gamma}$, Eq. (48) predicts overshoot of the stress.

2. Elongational Flows (*Uniaxial and Biaxial Extension*)

A second class of important deformations in polymer processing operations are the elongational flows, for which

$$v_1 = a_1 x_1, \qquad v_2 = a_2 x_2, \qquad v_3 = a_3 x_3, \qquad \mathbf{d} = \begin{vmatrix} a_1 & 0 & 0 \\ 0 & a_2 & 0 \\ 0 & 0 & a_3 \end{vmatrix} \qquad (50)$$

with $a_1 + a_2 + a_3 = 0$. Special cases of interest are uniaxial extension, for which

$$a_2 = a_3 = -\tfrac{1}{2}a_1, \quad a_1 > 0 \qquad (51a)$$

biaxial extension

$$a_1 = a_2 > 0, \qquad a_3 = -2a_1 \qquad (51b)$$

and planar extension (pure shear)

$$a_2 = -a_1, \qquad a_3 = 0 \qquad (51c)$$

Let us consider uniaxial extension in greater detail. This is one of the most common deformation fields encountered in polymer processing (e.g., in melt spinning) and in experimental measurements (tensile tester). Experimental studies for elastomers and compounds are given by Vinogradov et al. [289a], Stevenson [269a], and Cotton and Thiele [67a]. A growing literature also exists on experimental studies of molten plastics [20a, 62a, 128a, 182].

For long duration flows with constant a_1, the stress σ_{11} is given by

$$\sigma_{11} = -p + P_{11} = P_{11} - P_{22} = \chi \frac{dv_1}{dx_1} \qquad (52)$$

where χ is a function of a_1. According to Eqs. (36) and (37) we have

$$\chi = \left(1 + \frac{\varepsilon}{2}\right) \sum_i \frac{3G_i \tau_{i\text{eff}}}{(1 - 2a_1\tau_{i\text{eff}})(1 + a_1\tau_{i\text{eff}})}$$

$$- \frac{\varepsilon}{2} \sum \frac{3G_i \tau_{i\text{eff}}}{(1 + 2a_1\tau_{i\text{eff}})(1 - a_1\tau_{i\text{eff}})} \qquad (53)$$

For low deformation rates this leads to

$$\chi \rightarrow 3 \sum G_i \tau_i = 3\eta_0 \qquad (54)$$

i.e., the tensile viscosity is three times the zero shear viscosity. For large deformation rates, it may not be possible to attain a steady state.

We may compute the time variation of the tensile stress using the constitutive equation, Eq. (36). For the special case of a Lodge fluid we obtain (cf.

Denn and Marrucci [71] and Chang and Lodge [59])

$$\sigma_{11}(t) = \sum_i G_i \left[\frac{3a_1\tau_i}{(1 - 2a_1\tau_i)(1 + a_1\tau_i)} - \frac{2a_1\tau_i e^{2a_1t}e^{-t/\tau_i}}{1 - 2a\tau} - \frac{a\tau_i e^{at}e^{-t/\tau_i}}{1 + a_1\tau_i} \right] \quad (55a)$$

The predicted variation of stress with time is sketched in Fig. 6.

When a_1 is large

$$\sigma_{11}(t) = (\sum_i G_i)e^{2a_1t} \quad (55b)$$

and the stress increases in an unbounded manner. There is a critical deformation rate a_{crit}

$$a_{crit} = 1/2\tau_m \quad (56)$$

where τ_m is the maximum relaxation time.

For the more general fluid of Eqs. (36)–(38), we obtain in place of Eq. (56)

$$a_{crit} = \frac{1}{2\tau_{meff}} = \frac{1}{2\tau_m}\left[1 + b\tau_m \frac{\sqrt{3}}{2} a_{crit} \right] \quad (57)$$

If b is larger than $2/\sqrt{3}$, then there will be no critical value at a_1, above which there is unbounded stress growth.

An equivalent development for biaxial extension is possible and an expression similar to Eq. (53) may be derived. Details are given by the author [304] elsewhere. For small deformation rates and large times t this leads to a biaxial extensional viscosity which is six times the shear viscosity. We again find that above a critical deformation rate, which is specified by Eq. (56), the stress increases in an unbounded manner. The problem of biaxial deformation stress analysis in polymer processing is discussed by Alfrey [5].

3. Sinusoidal Oscillations

A widely used class of deformations in rheological instruments comprises small sinusoidal deformations of the type

$$\gamma = \gamma_0 \sin \omega t \quad (58)$$

These are variously applied in shear or in extension and may be applied to other geometric conditions, such as biaxial stretching. The interpretation of these experiments is generally carried out in terms of the theory of linear viscoelasticity and a dynamic storage modulus $G'(\omega)$ and a dynamic viscosity $\eta'(\omega)$. The stress is out of phase with the input strain by an angle δ and we may write

$$\sigma = G^*\gamma_0 \sin(\omega t + \delta) \quad (59a)$$

$$= G'(\omega)\gamma_0 \sin \omega t + G''(\omega)\gamma_0 \cos \omega t \quad (59b)$$

$$= G'(\omega)\gamma(t) + \eta'(\omega) \, d\gamma/dt \quad (59c)$$

where $G''(\omega)$ is $\omega\eta'(\omega)$. The quantities $G'(\omega)$ and $\eta'(\omega)$ may be expressed in terms of the relaxation modulus $G(t)$ and the relaxation spectrum $H(\tau)$ as

$$G'(\omega) = \omega \int_0^\infty G(s) \sin \omega s \, ds = \omega^2 \int_0^\infty \frac{\tau H(\tau)}{1 + \omega^2\tau^2} \, d\tau \tag{60a}$$

$$\eta'(\omega) = \int_0^\infty G(s) \cos \omega s \, ds = \int_0^\infty \frac{H(\tau)}{1 + \omega^2\tau^2} \, d\tau \tag{60b}$$

The $G'(\omega)$ and $\eta'(\omega)$ functions may be computed if $G(t)$ and/or $H(\tau)$ are known. Alternatively, if one has $G'(\omega)$ and $\eta'(\omega)$, the integrals in Eqs. (60a,b) may be inverted to give $G(t)$ and $H(\tau)$. For more details see Ferry [84].

The dynamic properties specified by Eqs. (60a,b) are related to the normal stresses and steady shear viscosity discussed earlier. From Eqs. (44a,b) and (60a,b) it follows that [64]

$$\lim_{\omega \to 0} \left[G'(\omega)/\omega^2 \right] = (\tfrac{1}{2}) \lim_{\dot\gamma \to 0} \Psi_1 \tag{61a}$$

$$\lim_{\omega \to 0} \eta'(\omega) = \lim_{\dot\gamma \to 0} \eta(\dot\gamma) \tag{61b}$$

Equation (61b) is a well-known result shown experimentally by numerous authors through the years. The validity of Eq. (61a) has been shown by Osaki *et al.* [222].

Some of the studies of the influence of finite amplitude sinusoidal oscillations on viscoelastic fluids have appeared in the literature and the data have been interpreted in terms of nonlinear single integral constitutive equations similar to Eq. (36) [165, 324]. Another series of related studies is the superposition of sinusoidal oscillations on steady shearing flows [38, 165, 275]. Generally studies of these flows are used to test constitutive equations and determine appropriate forms of $m_1(z)$ and $m_2(z)$.

E. Viscous Heating

Polymer fluids possess high viscosities and low thermal conductivities. Large amounts of power are required to induce polymers to flow and this power is dissipated as heat, which cannot be readily conducted away. Therefore the flow of polymers is often highly nonisothermal and entails large temperature gradients. Any quantitative study of flow under such conditions requires knowledge of both the variation of rheological properties with temperature and the temperature profiles.

To calculate temperature profiles we must apply the first law of thermodynamics, which relates the interaction of mechanical work, heat, and total energy changes in a system through the expression

$$dE = dQ + dW \tag{62a}$$

where E represents the sum of internal, kinetic, and potential energy, dQ is the heat absorbed, and dW is the work done on the system. It may be shown that this global expression may be converted by means of the divergence theorem to the equation [286]

$$\rho \, D\varepsilon/Dt = -\nabla \cdot \mathbf{q} + \boldsymbol{\sigma} : \nabla \mathbf{v} \tag{62b}$$

where ε is internal energy and $\mathbf{q}$ the heat flux vector. Accepting Fourier's law of heat conduction, incompressibility, and that the internal energy depends primarily upon temperature, we obtain

$$\rho c \, DT/Dt = k \, \nabla^2 T + \boldsymbol{\sigma} : \nabla v \tag{62c}$$

which is the generally applied form of the energy equation. It is implicit, in moving from Eq. (62b) to Eq. (62c), that the recoverable free energy of viscoelasticity is entropic in nature.

The importance of viscous dissipation heating may be shown by considering the rise of temperature under adiabatic conditions in a shearing flow. Specifically, Eq. (62) reduces to

$$\rho c \, \partial T/\partial t = \sigma \dot{\gamma} = \eta \dot{\gamma}^2 \tag{63a}$$

and

$$\Delta T = \int_0^t (\eta/\rho c)\dot{\gamma}^2 \, dt \tag{63b}$$

Let us consider a shear rate of 100 $\sec^{-1}$. For a typical rubber compound, we have a viscosity of 50,000 P at this shear rate. If we can neglect heat conduction and the influence of temperature on viscosity, this leads to a temperature rise of about 12°C $\sec^{-1}$.

More sophisticated analyses of viscous heating have appeared in the literature. Here the force balance and energy balance equations are solved simultaneously with consideration of temperature-dependent viscosities and heat conduction. Bird et al. [30] discuss the formulation of this type of problem. Brinkman's [41] early analysis of heat buildup in the flow of a Newtonian fluid through a capillary is significant, as is Bird's [27] extension of this analysis to non-Newtonian polymer melts. There have been numerous studies of more complex viscous heating problems in recent years [227, 274, 326].

There are other heating and cooling effects which may interact with viscous heating. We will simply mention them here. There is the Gough–Kelvin effect [88, 131, 285], which is the heating one finds in rubber bands when they are stretched and is an intrinsic part of entropy elasticity. This phenomenon is usually associated with the entropy elasticity of vulcanized rubber, but was actually discovered by Gough (in 1805) [105] more than 30 years before the invention of vulcanization. The effect has been studied quantitatively by Joule [131] and other researchers [285] and found to be small.

A second energetic effect is cooling due to volumetric expansion of polymer melts during flow from high-pressure to low-pressure regions. The phenomenon is well known in the flow of gases, but can also be significant in highly viscous polymer melts, where large pressure gradients are required to induce flow. The significance of this effect appears to have first been proposed by Toor [283], though his views have not been universally accepted.

III. Rheological Measurements

A. PARALLEL PLATE SIMPLE SHEAR TESTER

Perhaps the conceptually simplest type of rheometer can be constructed by sandwiching a material to be tested between two parallel plates that are separated by a distance H, and moving one plate parallel to the second at a fixed velocity V. The shear rate $\dot{\gamma}$ is V/H. For normal liquids this is not practical but for elastomers and compounds it is very much so. Apparatuses of this type have been designed and used by Zakharenko et al. [325], Middleman [190], and Goldstein [103]. The simplest method of design is to insert two slabs of rubber between three parallel plates, as shown in Fig. 7, and to install this fixture in a tensile testing machine. Tensile force may be simply related to shear stress. Obviously such an instrument can yield viscosity as a function of shear

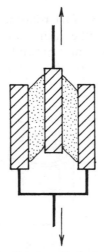

Fig. 7. Parallel plate simple shear tester.

rate, though in practice the shear rates achieved are rather low. Of great interest also is the application by Middleman and Goldstein to the study of stress transients at the start-up of flow (see Fig. 5), and the use of such results to interpret viscoelastic characteristics through expressions similar to Eqs. (48) and (49). Similar studies have recently been carried out by Furuta *et al.* [98] in our laboratories.

B. Coaxial Cylinder Rheometer

The next most sophisticated type of rheometer is one in which the parallel plates are curved to form coaxial cylinders (see Fig. 8). Mooney [196] used such an instrument to obtain the first quantitative rheological data on raw rubber. The instrument has been widely used for low-viscosity fluids such as emulsions and suspensions, but beyond the work of Mooney cited earlier and later studies by Philippoff and Gaskins [228] and Cogswell [63] it has not been widely used for bulk polymers.

The shear stress at the surfaces of a coaxial cylinder viscometer may be directly obtained from torque measurements. Clearly if θ is the 1 direction and r the 2 direction, the torque M at radius r is

$$M = 2\pi r L \cdot r\sigma_{12}(r) \qquad \text{and} \qquad \sigma_{12}(r) = M/2\pi r^2 L \qquad (64)$$

where L is the length of the cylinder and end effects have been neglected.

Calculation of the shear rate is straightforward if the thickness of the gap H between the inner cylinder, generally called the *bob*, and the outer cylinder, called the *cup*, is small. Here we may write

$$\dot{\gamma} = R\Omega/H \qquad (65)$$

where Ω is the angular velocity of the rotating member.

If the gap size is large, then the shear stress will vary significantly across the gap because of Eq. (64) and we must carefully determine $\dot{\gamma}$ at the surface of the bob or cup. Calculations based on Eq. (64) show that if H/R is of the order of

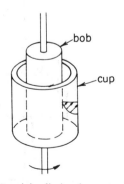

Fig. 8. Coaxial cylinder rheometer.

0.05, there will be a 10% variation in $\sigma_{12}(r)$ and if it is 0.1, there will be a 23% variation. Thus, in general we must be careful in our evaluation of $\dot\gamma$. A procedure for determining $\dot\gamma$ at the cup or bob has been developed by Mooney [196] and generalized by Krieger and co-workers [148–150] and Pawlowski [226]. Taking $\dot\gamma$ as $(-r\,d\omega/dr)$ we may write

$$\Omega_c - \Omega_b = \int_{\Omega_b}^{\Omega_c} d\omega = -\int_{\kappa R}^{R} \dot\gamma\, d\ln r = \frac{1}{2}\int_{\sigma_{12}(\kappa R)}^{\sigma_{12}(R)} \dot\gamma\, d\ln\sigma_{12} \tag{66}$$

For a rotating bob in any infinite sea, one may differentiate Eq. (66) to obtain (using Leibnitz's rule [318])

$$\dot\gamma(\kappa R) = \left[\frac{d\ln\Omega_b}{d\ln\sigma_{12}}(\kappa R)\right]\Omega_b \tag{67}$$

For the case of a stationary bob this may be differentiated with respect to the shear stress or torque M at the bob to give

$$\dot\gamma(\kappa R) = \dot\gamma(R) - 2\left[\frac{d\ln\Omega_c}{d\ln M}\right]\Omega_c \tag{68a}$$

The problem now is to relate $\dot\gamma(R)$ to $\dot\gamma(\kappa R)$. Krieger and Elrod [149] show that

$$\dot\gamma(\kappa R) = \frac{\Omega}{\ln 1/\kappa}\left[1 + \left(\ln\frac{1}{\kappa}\right)\frac{\partial\ln\Omega_c}{\partial\ln M} + \frac{(\ln 1/\kappa)^2}{3\Omega_c}\frac{\partial^2\Omega_c}{\partial\ln M^2} - \frac{(\ln 1/\kappa)}{45\Omega}\frac{\partial^4\Omega_c}{\partial\ln M^4} + \cdots\right]$$
$$\tag{68b}$$

More recently Krieger [148] has shown that power law fluid representations of the viscosity shear rate relationship are a good approximation and may be usefully directly applied to obtain viscosity shear rate data. For instance, for the foregoing case Krieger recommends

$$\dot\gamma(\kappa R) = (2/n)/(1 - \kappa^{2/n})\Omega_c \tag{69}$$

An interesting critique on measurement of viscosities in this instrument is given by Savins et al. [254].

Another rheological measurement of interest which can be made in this geometry is of the transient shear stresses at the beginning of flow. Mooney [196] used such an instrument to measure elastic recovery by removing the torque from the driven cylinder after the steady state was achieved.

C. CONE–PLATE VISCOMETER

The cone–plate geometry is one of the most widely used for characterizing the rheological properties of polymer fluids. The geometry was apparently first used and interpreted by Mooney and Ewart [204] to eliminate end effects in a coaxial cylinder viscometer. Piper and Scott [232] developed a biconical rotor for a viscometer used on rubber. However, its wide usage is due to Weissenberg [96, 296], who showed how the geometry could be used to measure normal stresses and commercialized an instrument to carry out such measurements [247]. The development of Weissenberg's "rheogoniometer" has recently been reviewed by Roberts [247] and Russell [251], two of his early colleagues. In recent years a new instrument of improved design called the Mechanical Spectrometer has been widely used for normal stress measurements [168, 169]. Numerous measurements of both viscosity and normal stresses have been carried out with the cone–plate instrument for both polymer solutions [44, 102, 138, 267] and melts [20, 136, 159] but apparently not for elastomers. The polymer solution studies involve polyisobutylene and butadiene–styrene copolymer solutions.

The great advantage of this instrument is that for small cone angles the shear rate is constant throughout the gap between the cone and the plate, and is given by (Fig. 9)

$$\dot{\gamma} = \frac{\Omega}{h(r)} = \frac{\Omega}{r \tan \Psi} = \frac{\Omega}{\tan \Psi} \tag{70}$$

where $h(r)$ is the vertical distance from the plate to the cone and Ψ is the angle between the cone and plate.

The shear stress may be computed from the torque M through the expression

$$M = \int_0^R 2\pi r^2 \sigma_{12} \, dr = \frac{2\pi R^3}{3} \sigma_{12}, \qquad \frac{3M}{2\pi R^3} = \sigma_{12} \tag{71}$$

The first normal stress difference N_1 may be determined from the thrust force F pushing apart the cone and the plate. This has the form

$$F = -\int_0^R 2\pi r \sigma_{\theta\theta} \, dr \tag{72a}$$

where $\sigma_{\theta\theta}$ is related to the other components of the stress tensor through the

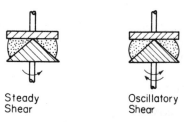

Steady
Shear

Oscillatory
Shear

Fig. 9. Cone–plate rheometer.

radial component of the Cauchy equation of motion (in spherical components):

$$\frac{\partial \sigma_{rr}}{\partial r} + \frac{2\sigma_{rr} - \sigma_{\theta\theta} - \sigma_{\phi\phi}}{r} = 0 \qquad (72b)$$

Noting that $\partial\sigma_{\theta\theta}/\partial r$ is the same as $\partial\sigma_{rr}/\partial r$, integrating, then substituting into Eq. (72a), we obtain

$$F = (\pi R^2/2)(\sigma_{\phi\phi} - \sigma_{\theta\theta}), \qquad N_1 = 2F/\pi R^2 \qquad (73)$$

where we have taken $\sigma_{rr}(R)$ to be zero and noted that ϕ is the 1 direction and θ the 2 direction, or direction of shear.

As shown by Markovitz and Williamson [172], it is possible to measure normal stresses using the radial pressure distribution across the surface of one platen. However, early measurements are apparently in error because of pressure hole errors [43]. More recently these have been eliminated by Christiansen and his colleagues [62, 191], who have combined pressure distribution and total force measurements to give N_1 and N_2 independently.

It has been found by experimenters [20, 136, 138, 159] that it is not possible to operate the cone–plate instrument at high shear rates because the polymer fluid balls up and exits from the gap. Normal stresses seem also to create vortexlike secondary flows in the gap between a cone and plate. These have been illustrated by Giesekus [101].

D. PARALLEL DISK AND MOONEY VISCOMETERS

A closely related geometry is torsional flow between parallel disks. The first instrument of this type applied to polymer fluids would be the "disk in cavity" Mooney viscometer [195, 278] (Fig. 10b). A simple parallel disk instrument (Fig. 10a) was introduced and used by Russell [250, 251] to measure shear stresses and to give the first normal stress measurements for viscoelastic fluids. Later applications of instruments based on this geometry have been dis-

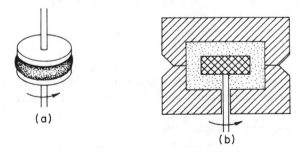

Fig. 10. (a) Parallel disk and (b) Mooney viscometers.

cussed for polymer solutions [102, 109, 139] and melts [31, 159, 253]. Macosko and Weissert [169] present normal stresses obtained on a butadiene–styrene copolymer in this geometry.

The shear rate between the two disks in this instrument is a function of radial position according to the expression

$$\dot{\gamma} = r\Omega/h \tag{74}$$

where h is the distance between the disks and Ω is the rotation rate of the moving disk.

The torque M in this geometry is given by

$$M = \int_0^R 2\pi r^2 \sigma_{12} \, dr \tag{75}$$

where σ_{12} varies with radius r. As has been shown by several researchers, Eq. (74) may be substituted into Eq. (75) and the integral differentiated with respect to the shear rate $\dot{\gamma}(R)$ at the outer perimeter of the disk to yield

$$\sigma_{12}(R) = \frac{S + 3}{4} \frac{2M}{\pi R^3} \tag{76a}$$

where

$$S = d \log M / d \log \Omega \tag{76b}$$

For a power law fluid $S \sim n$.

The theory of normal stress measurements in this geometry dates to Russell [250, 251] but the generally accepted final form is due to Kotaka *et al.* [139]

$$N_1 - N_2 = \frac{2F'}{\pi R^2} \left[1 + \frac{1}{2} \frac{d \log F'}{d \log \dot{\gamma}(R)} \right] \tag{77}$$

where F' is the force pushing apart the disks. Equation (77) may be used in conjunction with Eq. (73) to obtain both N_1 and N_2. Such experiments have been carried out on polymer melts by Lee and White [159] and on solutions by Ginn and Metzner [102].

The use of the geometry of a disk rotating in a cavity as a method of measurement of viscosity of rubber is due to Mooney [195, 201, 278]. Later discussions of the instrument have been given by Mooney and Wolstenhome [205], White and Tokita [299, 309], Wolstenhome [317], and Nakajima and Collins [209]. The geometry of the Mooney viscometer is summarized in Fig. 10.

To a first approximation, the Mooney viscometer would seem to represent the sum of torsional flow along the surfaces of the disk and Couette flow along the periphery of the disk. The torque M on the disk may be expressed as

$$M = 2\left[\underbrace{\int_0^R 2\pi r^2 \sigma_{\theta z}\, dr}_{\text{surface}}\right] + \underbrace{2\pi R^2 H(\sigma_{\theta r})_R}_{\text{periphery}} \tag{78}$$

where H is the thickness of the disk. Equation (78) may be solved for the torsional shear stress at the outer radius of the disk and an expression analogous to Eq. (76a) obtained:

$$\sigma_{\theta z}(R) = \frac{S+3}{4}\frac{M}{\pi R^3}F \tag{79}$$

where the function F is less than unity and has the value

$$F = \left[1 + \frac{(n+3)}{2}\frac{H}{R}\left(\frac{\dot{\gamma}_w}{\dot{\gamma}_R}\right)^n\right]^{-1} \tag{80}$$

Here n is the power law exponent, $\dot{\gamma}_R$ is the flow shear rate at the outer radius of the disc, and $\dot{\gamma}_w$ is the Couette flow shear rate. To a first approximation the ratio $\dot{\gamma}_w/\dot{\gamma}_R$ is h/Δ. For better approximations see [201, 309].

In a traditional ML-4 Mooney viscometer experiment $\dot{\gamma}(R) - R\Omega/h$ is about 1.5 sec^{-1}.

Mooney [201] has developed a theory of determining slippage at the disk surface in this geometry. He has carried out experimental studies of torque variations induced by changing the smoothness and type of metal in the disk.

E. DYNAMIC MEASUREMENTS AND ECCENTRIC ROTOR RHEOMETERS

Our discussions of instruments in Subsections A–D emphasized steady shear flow. It should be apparent that small-oscillation dynamic measurements, i.e., determination of $G'(\omega)$ and $\eta'(\omega)$, are possible in each of these geometries if the moving member is oscillated sinusoidally with a small amplitude rather than being rotated at a constant velocity. Such experiments between coaxial cylinders, cone and plate, and parallel plates are common. (See Fig. 10b.) The expressions for shearing stresses developed may be directly applied to infinitesimal oscillating strains.

There is another approach to oscillatory measurements of viscoelastic properties. One may have a steady flow in which the moving fluid is periodically subjected to higher deformation rates. This basic concept is due to Bryce Maxwell. Maxwell and Chartoff [179] have pointed out that it is possible to obtain such dynamic measurements in steady rotational flows by using eccentric parallel disks rotating at the same angular velocity (see Fig. 11). Kepes has

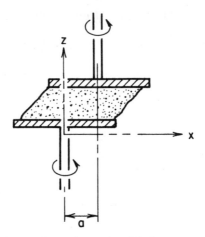

Fig. 11. Eccentric rotating disk rheometer.

shown that this is also possible using a rotating sphere in a rotating cavity with different axes (balance rheometer) and Abbott *et al.* [2, 3] have suggested eccentric cylinders or spheres. Broadbent and Walters [42] have presented experimental studies for a number of these instruments. Theoretical studies of these geometries have been given by Blyler and Kurtz [32], Bird and Harris [29], Yamamoto [321], and Abbott *et al.* [1–3].

The development of the mechanical spectrometer [168, 169] has made eccentric disk measurements of linear viscoelastic properties straightforward, and this method is being used in several laboratories. Macosko and Weissert [169] discuss such a study for rubber compounds. For small displacements of axes *a* (relative to the disk separation *h*), the dynamic modulus G' and dynamic viscosity η' are given by

$$G'(\Omega) = \frac{F_x h}{\pi R^2 a} \tag{81a}$$

$$\eta'(\Omega) = \frac{F_y h}{\pi R^2 a} \frac{1}{\Omega} \tag{81b}$$

where F_x is the force in the direction of the displacement of the axes, F_y the force in the perpendicular direction, and Ω the rotation speed.

F. Capillary Rheometer

Capillary rheometer (see Fig. 12) is one of the oldest and most widely used experimental methods for measuring the viscosity of fluids. It was used extensively on almost all classes of complex fluids by the founders of modern rheology

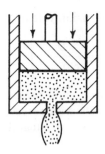

Fig. 12. Capillary rheometer.

during the 1920s and 1930s. Its application to rubber dates to the work of Marzetti [175] and Dillon [73, 75] and it has been widely used in recent years [65, 66, 81, 127, 305].

The basic idea of the instrument is to relate the pressure loss for extrusion through a small-diameter tube of diameter D and length L to the shear stress at the capillary wall and the extrusion rate to a wall shear rate. The total pressure drop p_T through a die is a sum of a pressure loss within the die Δp_e at the ends (i.e., at the entrance and the exit):

$$p_T = \Delta p_e + \Delta p \tag{82}$$

We may relate Δp to the wall shear stress $(\sigma_{12})_w$ by a simple force balance

$$\pi D L (\sigma_{12})_w = (\pi D^2/4)\,\Delta p \tag{83}$$

which allows us to write

$$(\sigma_{12})_w = D\,\Delta p/4L \tag{84a}$$

$$p_T = \Delta p_e + 4(\sigma_{12})_w L/D \tag{84b}$$

By using a series of dies with varying L/D, one may determine $(\sigma_{12})_w$ from the slope of a graph of p_T versus L/D, which is called a Bagley plot [17]. The wall shear rate must be kept constant, but this may be ensured by using dies of constant diameter and maintaining Q.

Implicit in Eq. (84) is the idea that the diameter of the reservoir preceding the die is much greater than the die. If this is the case, reservoir pressure losses need to be considered. This is discussed by Metzger and Knox [184].

The capillary wall shear rate may be obtained in a manner devised by Weissenberg [83, 189, 240] (see also Mooney [194]). Noting that Eqs. (83) and (84a) apply to a telescoping flow at each radius, we see that the shear stress varies linearly with the radius

$$\sigma_{12}(r) = (\sigma_{12})_w r/R \tag{85}$$

This allows us to rewrite the extrusion rate as

$$Q = \int_0^R 2\pi r u \, dr = \frac{D^3}{8(\sigma_{12})_w} \int_0^{(\sigma_{12})} z^2 \left(-\frac{du}{dr} \right) dz \qquad (86)$$

where we have integrated by parts and presumed the material to adhere to the wall. Differentiation of $Q/\pi D^3$ with respect to σ_w and application of the Leibnitz rule for the differentiation of integrals [318] allows us to solve for the capillary wall shear rate:

$$\dot{\gamma}_w = \left(-\frac{du}{dr} \right)_w = \left(\frac{3n' + 1}{4n} \right) \frac{32Q}{\pi D^3} \qquad (87a)$$

with

$$n' = \frac{d \ln(\sigma_{12})_w}{d \ln(32Q/\pi D^3)} \qquad (87b)$$

Application of Eqs. (84) and (87) allows the evaluation of the viscosity function $\eta(\dot{\gamma})$. Mooney [194] has modified Eqs. (86) and (87) to allow for slip and has shown how slip at the capillary wall may be determined.

Through the years, various analyses have been suggested for obtaining normal stresses from capillary measurements. These methods have been based variously on jet expansion-extrudate swell [183, 185, 276, 307a], exit pressure loss [69, 113, 118], and entrance pressure loss [18, 20, 155, 229]. None of these measurements, at least for bulk polymers, has received broad acceptance. It would seem at present that the exit pressure loss is probably the most useful method and is likely to be eventually utilized for melts, whereas jet expansion will probably be used for solutions.

G. TENSILE TESTING—ELONGATIONAL FLOW

In recent years there has been extensive interest in elongational flow experiments carried out at constant elongation rate a_1 or stress χ. These studies determine the elongation viscosity. Such experimental studies were initiated by Ballman [20a], Cogswell [62a], Meissner [182], and Vinogradov and his co-workers [289a]. Studies have mainly been reported on molten plastics [20a, 62a, 128a, 182]. However experimental studies for elastomers are described by Vinogradov et al. [289a,b] and by Stevenson [269a]. Cotton and Thiele [67a] have investigated the behavior of rubber–black compounds while Lobe and White [161a] have investigated carbon black filled polystyrene melts.

In these experiments, one achieves a constant elongation rate a_1 by (i) noting that

$$a_1 = \frac{1}{L}\frac{dL}{dt} \rightarrow L(t) = L(0)e^{a_1 t} \tag{88a}$$

and having one clamp move away from the other at an exponential rate [20a, 269a, 289a] or (ii) having a filament taken up on a rotating roll or otherwise exited from the test at constant velocity [128a, 182a] so that

$$a_1 = V/L \tag{88b}$$

Measurements may be made at constant stress using a special cam system; a rotating roll removes the filament [62a]. Experiments of this type are usually carried out horizontally in silicone oil constant temperature baths.

Tensile experiments on elastomers are frequently carried out in constant clamp velocity experiments, as in an Instron tensile tester. In such experiments

$$a_1 = V/(L_0 + Vt) \tag{88c}$$

so that a_1 decreases with time. Such results are difficult to interpret.

Other types of elongational deformation experiments are described in the literature. Stress relaxation measurements for elastomers are frequently carried out in this manner. Such experiments yield a Young's relaxation modulus which is three times the shear modulus.

IV. Molecular Structure and Rheological Properties of Uncompounded Elastomers

A. CRYSTALLINE TRANSITIONS AND STRESS-INDUCED CRYSTALLIZATION

One of the most important features of polymer structure is the uniformity of microstructure in the polymer backbone. Very regular backbone structures in the sense of identical character of all or most structural units will lead to the occurrence of crystallinity. The identity of structural units implies not only chemical character and ordering of atoms but identity of geometric and stereoregular isomeric character. Thus not all polybutadienes (nor indeed all 1,4 addition or 1,2 addition polybutadienes crystallize. 1,4-Polybutadienes must possess near complete cis-1,4 or trans-1,4 microstructures and 1,2-polybutadienes must be isotactic or syndiotactic in order to crystallize. Similar rules hold for the polydienes, such as polyisoprene and presumably polychloroprene. Only isotactic and syndiotactic vinyl polymers crystallize, but vinylidene polymers with identical substituents, such as polyvinylidene chloride, generally crystallize.†

A most important feature of the crystallization transition for elastomer rheology and processing is stress-induced crystallization. It is well known that many elastomers, including natural rubber (cis-1,4-polyisoprene) [6, 77, 133, 135], trans-1,4-polyisoprene [99], trans-1,4-polychloroprene [54, 98, 151], and polyisobutylene [97], can exhibit crystallization when stretched in a tensile testing unit. There are two related physical phenomena which act to cause strain-induced crystallization. First, the equilibrium melting temperature rises due to the decreased entropy of the amorphous polymer induced by the stretch. Consider the thermodynamic expression

$$T = (H_a - H_c)/(S_a - S_c) = \Delta H_f/\Delta S_f \tag{89}$$

where the subscripts a and c denote amorphous and crystalline. When an amorphous polymer is stretched, its entropy ΔS_a decreases, resulting in an overall decrease in the entropy of fusion ΔS_f with relatively little change in the amorphous enthalpy H_a of heat of fusion ΔH_f. This increases the crystallization temperature T_m. Theories of this effect have been developed by Flory [86] and later researchers [151, 323]. A more striking and important aspect is the increased rate of crystallization caused by strains or—probably better—stresses. Experimental studies [6, 70, 267a] on a variety of polymers show that the effect may involve several orders of magnitude. A careful study of the mechanism of stress-induced crystallization in elastomers has been given by Andrews [6], who has traced it to increased rates of nucleation.

The practical implication of stress-induced crystallization is the formation of a new phase with greatly increased modulus and memory. The crystallized phase exhibits elastic–plastic rather than viscoelastic behavior. Further crystallization occurs first in the region of highest stresses. Thus if a deforming elastomer tensile specimen has a defect, such as a tear, stress-induced crystallization will occur at the tip, preventing its propagation. Stress-induced crystallization acts both to increase modulus and resistance to deformation, and

† Minor variations in the microstructure lead to reductions in the crystalline melting temperature in the same manner that impurities in a low molecular weight liquid reduce it. As shown by Flory [87], the melting point with random variations in microstructure is depressed according to $1/T_m = (1/T_m°) - (R/\Delta H_u) \ln(1 - X)$ where $T_m°$ is the melting temperature of the pure polymer, X is the mole fraction of the impurity, and ΔH_u is the heat of fusion per mole of structural unit. An experimental study showing the validity of the foregoing equation for high-trans-1,4 polybutadienes has been carried out by Mandelkern et al. [171] and by Berger and Buckley [23], and similar effects have been described in cis-1,4-polybutadienes [19, 210]. It must be remembered that this analysis and interpretation is limited to random distributions of unlike structural units. If the impurities occur in blocks, there will be similar long pure blocks of the homopolymer which could crystallize by themselves as a pure substance. Increased blocking increases T_m beyond the predictions of Flory's equation. This effect has been observed in the so-called alfin polybutadienes (prepared by a catalyst made from sodium compounds of an alcohol and an olefin) [76, 308].

to stabilize the system, preventing the propagation of defects. Elastomers showing stress-induced crystallizations exhibit enormous elongations to break in tensile tests.

One of the more interesting experimental studies of stress-induced crystallization in elastomers is the Folt et al. [91] of extrusion of high-cis polyisoprenes in which crystallizations occur in the capillary entrance region. Similar observations have been made on polyethylene melts [266].

B. GLASS TRANSITIONS

Elastomers manifest a second type of transition which appears not to be thermodynamic in character. At the so-called glass transition, elastomers exhibit rapidly increasing modulus and viscosity through several orders of magnitude within a small temperature range. There are, however, no discontinuous changes in volume, entropy, or enthalpy, although there seem to be such changes in their rates of change with temperature. It is believed that the glass transition temperature marks a loss in molecular mobility. In essence, at this temperature the flexible macromolecules of an elastomer transform into a glass. It may be seen that in crystallizing polymers, T_g is always lower than the melting temperature T_m. Characteristic of elastomers is that T_g is below room temperature and indeed usually below $-50°C$.

The importance of T_g in rheology is that it plays the role of a characteristic or normalizing temperature similar to the critical temperature in gases. Basically, the greater the value of $T - T_g$, the lower the value of the modulus. This is readily seen by comparing "soft" 1,4-polybutadienes, which have T_g's of about $-100°C$, and "hard" high-styrene butadiene–styrene copolymers, where T_g is $-50°C$ or more. This idea has been in part quantitatively formulated, as indicated in the next section.

C. TIME–TEMPERATURE SHIFT AND WLF EQUATION

During the 1940s experimental studies of linear viscoelastic properties of elastomers by Andrews and Tobolsky [7, 8] showed that there was a simple relationship between temperature and time in the relaxation modulus. Data obtained at one temperature could simply be shifted along the time axis to coincide with data obtained at a second temperature T_s; i.e.,

$$G(t, T) = G(t/a_T, T_s) \qquad (90a)$$

where a_T is a temperature-dependent shift factor. Frequently one sees instead expressions of the form

$$G(t, T) = (\rho T/\rho_s T_s)G(t/a_T, T_s) \qquad (90b)$$

Williams, Landel, and Ferry [315] have shown that the shift factors a_T for most amorphous polymers are interrelated. They find that a_T may be expressed as

$$\log a_T = -8.86(T - T_s)/[101.6 + (T - T_s)] \tag{91}$$

where T_s is a characteristic temperature. A reasonable but less good fit of a_T data may be obtained by an alternative expression which replaces T_s with T_g:

$$\log a_T = -17.44(T - T_g)/[51.6 + (T - T_g)] \tag{92}$$

Since the time–temperature superposition shift may be used for the relaxation modulus, it follows that similar shifting must be possible with other linear viscoelastic properties, since these are simply integrals of $G(t)$. Thus it may be shown that

$$\eta_0(T) = \int_0^\infty G(s, T)\, ds = a_T \frac{\rho T}{\rho_s T_s} \int_0^\infty G\left(\frac{s}{a_T}, T_s\right) d\frac{s}{a_T}$$

$$= a_T \frac{\rho T}{\rho_s T_s} \eta_0(T_s) \tag{93a}$$

$$G'(\omega, T) = \frac{\rho T}{\rho_s T_s} G'(a_T\omega, T_s) \tag{93b}$$

$$\eta'(\omega, T) = a_T \frac{\rho T}{\rho_s T_s} \eta'(a_T\omega, T_s) \tag{93c}$$

The time–temperature shift in the relaxation modulus implies a frequency–temperature shift in the dynamic properties.

The extension of such shifting procedures to nonlinear rheological properties is not clear-cut. It depends upon the form of the constitutive equation. Bernstein et al. [24] and later White and Tokita [311] discussed specific forms of constitutive equations where such shifting is possible. The main premise of these theories is to use constitutive equations of the type given by Eq. (36), only specifying

$$m_j(s, T) = (\rho T/\rho_s T_s)m_j(s/a_T, T_s) \tag{94}$$

for $j = 1, 2$. The nonlinearities are included through strain invariants. If one presumes simple shear flow, one then has

$$m_j(s, \dot{\gamma}s, T) = (\rho T/\rho_s T_s)m_j(s/a_T, \dot{\gamma}s/a_T, T_s) \tag{95}$$

From Eq. (44a) it follows that

$$\eta(\dot{\gamma}, T) = a_T(\rho T/\rho_s T_s)\eta(a_T\dot{\gamma}, T_s) \tag{96}$$

This demands that plots of η for $\dot{\gamma}$ may be shifted at a 45° angle to obtain data for different temperatures. Such a procedure has indeed been proposed and used by Bueche [47].

D. COPOLYMER MICROSTRUCTURE

The distribution of structural units along the length of a polymer chain plays a major role in the thermodynamic and mechanical behavior of the bulk polymer. If the distribution of units is random, the copolymer behaves very much like an atactic homopolymer (e.g., like normal polystyrene) and exhibits glassy, rubbery, and molten states but no crystalline character. Generally, the value of T_g for a random copolymer specifies the modulus as a function of temperature. For butadiene–styrene copolymers with allowance for variations in butadiene microstructure, the T_g–composition behavior has been given by Bahary et al. [19].

The situation is very different for block copolymers, which are becoming increasingly important as thermoelastic elastomers [174]. If the copolymers have long segments of identical units in the polymer backbone, they will tend to form separate phases. If the copolymer has an ABA or more complex backbone structure and the A segment (or a segment occurring twice in the backbone) undergoes a glass or crystalline transition, the material solidifies. The existence of domain structures in such solids has been shown by electron microscopy [125, 206]. It is believed that such domain structures continue to exist in melts and concentrated solutions. A number of studies of melt and solution rheological properties of block copolymers have appeared [125, 138, 145, 147, 225]. Kotaka and White [138] have determined both normal stresses and viscosity in solutions of block copolymers. It is clear that the existence of domain structures in solutions and melts leads to abnormally high viscosities and elasticities.

E. BULK POLYMER MOLECULAR MACROSTRUCTURE

1. Viscosity

It had been realized from the time of the acceptance of the macromolecular hypothesis that melt rheological properties are strongly dependent upon molecular weight. The first investigations of viscosity as a function of molecular weight are those of Flory [85, 258]; since then, work has been done by Fox and Flory [93], by Fox and other co-workers [92, 94], and more recently by several authors [52, 143, 176, 177, 238, 239]. For linear polymers, it was found that at low shear rates the viscosity varies with about the first power of the molecular weight until a critical molecular weight at which there was a transition to an approximately 3.4-power dependence, i.e.,

$$\eta_0 \sim M \quad (M < M_c), \qquad \eta_0 \sim M^{3.4} \quad (M < M_c) \tag{97}$$

This value of M_c is referred to as the entanglement molecular weight and the phenomenon as the entanglement transition. The value of M_c varies from polymer to polymer but seems generally to correspond to the same number of atoms in the backbone. Thus M_c in vinyl polymers rises with the mass of pendant side groups, being about 4000 for polyethylene and 38,000 for polystyrene.

At low molecular weights, polymers are Newtonian fluids. However, at molecular weights in the range of M_c, non-Newtonian viscosity is observed with viscosities decreasing with increasing shear rate. As the molecular weight increases, the non-Newtonian viscosity behavior becomes more pronounced.

The viscous properties of polymer melts depend upon the distribution of molecular weights as well as the absolute molecular weight. For reference, we define the number average M_n; weight average M_w; z average M_z; $(z + 1)$ average M_{z+1}; and viscosity average molecular weight M_η

$$M_n = \frac{\sum N_i M_i}{\sum N_i}, \qquad M_w = \frac{\sum N_i M_i^2}{\sum N_i M_i}$$

$$M_z = \frac{\sum N_i M_i}{\sum N_i M_i^2}, \qquad M_{z+1} = \frac{\sum N_i M_i^4}{\sum N_i M_i^3} \tag{98}$$

$$M_\eta = \left(\frac{\sum N_i M^{a+1}}{\sum N_i M_i}\right)^{1/a}$$

where N_i is the number of molecules of molecular weight M_i and a is the Mark–Houwink exponent in the intrinsic viscosity–molecular weight relationship.

In the studies of Fox and co-workers [93, 94], weight average molecular weights were used to characterize the viscosity–molecular weight relationship; i.e.,

$$\eta_0 = k M_w^{3.4} \qquad (M > M_c) \tag{99}$$

However, in studies of polyethylene where molecular weight distributions are especially broad, Busse and Longworth [52] concluded that the viscosity average molecular weight gave a better representation on the experimental data.

The foregoing discussion is limited to the dependence of the low shear rate viscosity upon molecular weight. As we have noted, polymer melts generally exhibit non-Newtonian viscosity behavior. There have been numerous attempts to relate the non-Newtonian viscosity to molecular weight and molecular weight distribution. The influence of molecular weight distribution on the viscosity–shear rate behavior of elastomers has received considerable study. Qualitatively, as shown by Weissert and Johnson [297], narrow molecular weight distribution elastomers are more Newtonian at low shear rates but high shear rate behavior seems similar.

The early study of the influence of branching on the rheological properties of polymer melts was Shaefgen and Flory's study [258] of the influence of branching on the viscosity of condensation polymers. Branching was found to decrease the magnitude of the low shear rate viscosity. This view remained common until the work of Kraus and Gruver [143], who found, for branched polybutadiene with equivalent trichain and tetrachain branches from a single branching point, that the viscosity is lower than that of linear polymers with the same molecular weight only in lower molecular weight ranges. Beyond a critical molecular weight, the branched polymers increase rapidly in viscosity with molecular weight and surpass the linear polymer viscosity. A similar result was found by Folt [89]. The author's experience [308] in studying polybutadienes generally agrees with the results of Kraus and Gruver. More recent studies of Masuda et al. [177] on branched polystyrenes generally show, however, lower viscosities than linear polystyrenes. It is probable that one must go to much higher molecular weights in polystyrene to achieve the crossovers.

Numerous theories of the viscous and viscoelastic properties of polymer melts have appeared. The theories of Yamamoto [319, 320], Lodge [163], Graessley [106], and Carreau [55], among others, for bulk polymer melts may be mentioned. The polymer melt is considered in the theories of these authors in terms of a network of entangled chains in which the rheological response is determined by the entanglement density and the rate of formation and breakdown of entanglements. Although none of these theories is particularly rigorous, they do generally predict a constant low shear rate viscosity that increases with molecular weight approximately as in Eqs. (97) and (99). At higher shear rates the viscosity is predicted to decrease. Viscosity–shear rate–molecular weight master curves have been predicted by Bueche and Harding [48] and Graessley [106].

2. Viscoelastic Behavior

There have been numerous studies of the influence of molecular weight and its distribution upon viscoelastic properties. The earliest studies of this type were by Tobolsky and his co-workers, notably Andrews [7, 8, 279, 280], and resulted in the representation of the spectrum of relaxation times $H(\tau)$ for a narrow distribution polymer by a wedge box distribution, i.e.,

$$H(\tau) = A/\sqrt{\tau} \ \ (\tau < \tau_1), \qquad H(\tau) = H_0 \ \ (\tau_2 < \tau < \tau_m) \qquad (100)$$

The values of A and H_0 are nearly independent of molecular weight. For narrow molecular weight distribution samples it was found that

$$\tau_m \propto M^{3.4} \qquad (101)$$

similar to the Fox–Flory viscosity relationship. The similarity of dependence of τ_m and η_0 on molecular weight is generally to be expected, though, as

$$\eta_0 = \int_0^\infty G(s)\, ds = \int_0^\infty H(\tau)\, d\tau \sim H_0 \tau_m \tag{102}$$

Tobolsky's results have generally been confirmed by later researchers.

However it must be remembered that Tobolsky did not use very narrow distribution samples in his studies.

In more recent years investigations of the influence of molecular weight and its distribution on linear viscoelastic properties of polymers have been made by Masuda and his co-workers [177, 220] and by Vinogradov [287a]. The latter author's experiments have been primarily on polybutadienes. Studies by Masuda et al. [177, 220] using very narrow distribution polymers seem to exhibit a peak in their $H(\tau)$–τ data near the end of the box, i.e., at $\tau = \tau_m$. Bogue (personal communication, 1974) has pointed out to the author that it is possible, with the data treatment techniques used, that there may be no box at all, but only a spike, for narrow distribution homopolymers.

Studies of the influence of molecular weight distribution on $H(\tau)$ were pioneered by Ninomiya [211], who proposed a linear blending law of the form

$$H_b(\tau) = \sum w_i H_i(\tau/\lambda_i) \tag{103}$$

where w_i is the weight fraction and the λ_i are molecular-weight-dependent coefficients. This has been found insufficient and was replaced by Masuda et al. [177] and later by Bogue et al. [35] with quadratic blending laws containing cross terms. According to Bogue et al., Eq. (103) is valid for low molecular weight components, but for high molecular weights one has (for two components)

$$H_b(\text{large } \tau) = (1 - f_1)w_1 H_1 + f_1 w_1^2 H_{11} + 2f_1^{1/2} f_2^{1/2} w_1 w_2 H_{12}$$
$$+ f_2 w_2^2 H_{22} + (1 - f_2)w_2 H_2 \tag{104}$$

where f_1 and f_2 are adjustable parameters reflecting participation in the network which increases with molecular weight and the H_{ij} are cross-spectra. We refer the reader to the original paper for more details.

The influence of small amounts of high molecular weight polymers on viscoelastic properties is quite striking. However, the significance is not immediately realizable from mathematical representatives such as those discussed above. Elastic memory is greatly increased by small amounts of high molecular weight polymer. A striking study of this has been made by Leaderman et al. [158], who found the elastic shear recovery per unit stress (i.e., J_e, the steady-state shear compliance) in blended polyisobutylenes to vary with $(M_z M_{z+1}/M_w)$ to the second power.

Oda et al. [216a] find the principal normal stress difference N_1 to depend on shear stress in a temperature independent manner. From Eq. (47c) this functionality is determined by J_{eapp}. Oda et al. find the value of N_1 at fixed

σ_{12} to increase greatly with broadening molecular weight distribution. This is consistent with Leaderman et al.'s results.

F. POLYMER SOLUTIONS

The dependence of polymer solution viscosity upon concentration and molecular weight has received considerable study, beginning with the researches of Ferry and his students [130]. At low concentrations the viscosity varies linearly with concentration and with a power between 0.5 and 0.8 of molecular weight. With increasing concentration the low shear rate viscosity rises more rapidly and fifth-power dependence upon concentration (see Fig. 13) and a 3.4-power dependence upon molecular weight is obtained; i.e.,

$$\eta_0 \sim M_w^{3.4} c^5 \tag{105}$$

This result has been confirmed by several researchers using different polymer systems [128, 176].

The dependence of normal stress coefficient Ψ_1 in polymer solutions upon concentration and molecular weight has been analyzed by Tanner [277] and by Ide and White [128] (see Fig. 13). The concentration dependence is much higher than in Eq. (105), being of the order of 7.5–10.

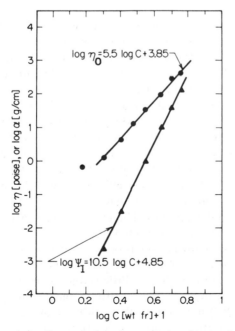

Fig. 13. Influence of concentration on the low shear rate viscosity η_0 and principal normal stress difference coefficient Ψ_1 in polystyrene–styrene solutions. (From [128], with permission of John Wiley and Sons.)

The foregoing results are restricted to flexible polymer chains. Rigid macromolecules such as poly-p-benzamide or poly(γ-benzyl glutamate) exhibit more complex behavior. Liquid crystal structures form at moderate concentrations and viscosity–concentration plots possess maxima (122, 154, 224].

G. MECHANOCHEMICAL DEGRADATION

Mastication softening of natural rubber was discovered around 1820 by Thomas Hancock [119], who also showed that at any concentration level, solutions of masticated rubber would possess a lower viscosity than the initial polymer. It was not, however, until the acceptance of the macromolecular hypothesis and the work of Cotton [67] and Busse [50, 51] in the 1930s that the importance of oxygen (as well as the mechanism involved) in the mastication process came to light. The results of these researchers were subsequently verified by Pike and Watson [231, 292]. Essentially it is thought that the natural rubber degrades to form free radicals under the action of stresses and that the radicals are stabilized through reaction with oxygen:

$$R—R \xrightarrow{\text{stress}} R\cdot + R\cdot$$
$$R\cdot + O_2 \longrightarrow RO_2$$

One of the major points made, notably by Busse, was that the temperature has a peculiar influence on breakdown. At low temperatures, degradation is induced by applied stresses and if the temperature is increased while the kinematics of mastication are unchanged, the rate of polymer degradation will decrease. If the temperature continues to be raised, the rate of degradation will pass through a minimum and begin to increase again. This is due to the interaction of the simultaneous decrease of viscosity and increase in rate of oxidative attack with temperature. While most investigations of stress-induced degradation have been carried out on natural rubber, studies on other elastomers (synthetic cis-1,4-polyisoprene, cis-1,4-polybutadiene, butadiene–styrene copolymer) have appeared [39, 89, 120, 126, 146]. Folt [89] has studied the influence of mastication-induced degradation on the rheological properties of elastomers, especially polybutadienes and polyisoprenes. The greatest breakdown rates are found in cis-1,4-polyisoprene, a fact attributed to (1) stress-induced crystallization occurring during the deformation process, which results in especially high stresses, and (2) resonance stabilization of the radicals produced.

Generally it is found that there is little reduction in molecular weight when mastication is carried out in nitrogen. Busse and Cunningham [51] and Pike and Watson [231, 292] have found that addition of free radical acceptors to natural rubber masticated in nitrogen leads to molecular weight reduction. Pike and Watson suggest that the free radical acceptors play a role similar to oxygen. Quantitative verification of this has been obtained by Ayrey et al. [16]

by relating spectral and radiochemical determination of end groups to molecular weight reduction.

The influence of mastication on molecular weight distribution has been studied by Angier et al. [12] and later investigators [120, 126, 146]. They have consistently found that high molecular weight species are preferentially degraded and narrower molecular weight distribution materials are formed.

Angier and Watson [9, 11] found that graft copolymers could be prepared by masticating elastomer blends in the absence of oxygen. The systems studied included natural rubber/polychloroprene, natural rubber/butadiene–styrene copolymer, and polychloroprene/butadiene–styrene copolymer. The mechanism hypothesized was

$$A\!-\!A' \xrightarrow{\text{stress}} A\cdot + A', \qquad B\!-\!B' \xrightarrow{\text{stress}} B\cdot + B\cdot'$$

$$A\cdot + B\cdot \longrightarrow A\!-\!B$$

It has been shown by Angier et al. [10, 13] that if an elastomer is swollen with a vinyl monomer (styrene, chlorostyrene, acrylic acid, methyl acrylate, methacrylic acid, methyl methacrylate, vinyl pyridine, methyl vinyl ketone, etc.), mastication in the absence of oxygen can lead to the formation of block copolymers. This would seem to be through the mechanism

$$R\!-\!R' \xrightarrow{\text{stress}} R\cdot + \dot{R}'$$

$$R\cdot + M \longrightarrow RM\cdot$$

$$RM\cdot + M \longrightarrow RMM\cdot$$

$$RMM\cdot + M \longrightarrow RM_{n+1}M\cdot$$

$$RM_{n+1}M\cdot + R''M_{m+1}M\cdot \longrightarrow RM_{n+m+2}R''$$

Angier et al. [14, 58] have shown that this process may also be applied to glassy vinyl plastics if they are processed in a rubbery state.

Most of the references in this section are to the studies of mechanochemical degradation of elastomers by the British (now Malaysian) Rubber Producers Research Association. There have, however, been valuable studies of this phenomenon by other research groups, of which we may mention Baraboim and his school [21] in the Soviet Union and Porter and Johnson [56, 237] in the United States. Casale, Porter, and Johnson [56] have given a review of mechanochemical degradation with application to a greater range of polymer systems (including solutions and glasses) than discussed here.

V. Flow of Blends and Compounds

A. Two-Phase Polymer Melt Flow

Generally elastomers are processed in the form of compounds which are blends of more than one polymer melt, contain significant quantities of fillers such as carbon black, and are oil extended. No discussion of elastomer rheology can be complete without discussion of the influence of these additives. We first turn to the problem of two-phase polymer melt flow.

The key problem in two-phase polymer melt flow is to characterize the distribution of the phases. There have been a number of studies of stratified two-phase flow of polymer melts in recent years [112, 115, 159, 193, 256, 257, 265, 306, 313]. These studies are the result of the commercial development of co-extrusion of plastic sheet and film [15, 256, 257], bicomponent fibers [46, 124], and sandwich molding [223], and are not directly related to the rubber industry. Although coextrusion of tire treads and sidewalls seems to have been long practiced by several rubber companies, there has never been any open discussion of it, nor have there been any published scientific studies of the process. The most striking aspect of the stratified two-phase studies is that during flow through a duct the high-viscosity melts seem to push into the low-viscosity melts and the low-viscosity melts creep around the periphery of the cross section, resulting in encapsulation [112, 115, 159, 193, 265]. Experimental studies have shown that other possible causes of this behavior, such as interfacial tensions and normal stresses, do not play a major role.

The experimental studies referred to in the preceding paragraph represent an extreme case. Of more interest is the disperse flow of polymer blends. There have been few studies of the formation of elastomer blends and the resulting dispersion. The most thorough study is that of Walters and Keyte [291]. White and Tokita [308, 312] relate the continuous and discontinuous phase to elongation to break and phase ratio, respectively. Lee and White [160] have investigated the variation in phase distribution across the cross section of a series of polystyrene–polyethylene blends formed by mixing of pellets and mill mastication. It was found that there was a trend for low-viscosity melt to migrate out to the walls and that this trend was a decreasing function of the degree of dispersion. A problem of major interest is the treatment of disperse polymer blends as continuous media, so that the definition of viscosity functions has a meaning independent of geometry. Some work in this area has been carried out [160], but much more is needed.

Most studies of disperse two-phase flow deal with the problem of the variation of apparent composite properties with composition. Plochocki [235], Han and Yu [116, 117], and Lee and White [160] have studied viscosity as a function of composition in molten plastic blends. Folt and Smith [90] have investigated this effect in elastomer blends. The results of these experiments are complex. The viscosity–composition function often exhibits minima and sometimes maxima.

Han and Yu [117] have investigated the viscoelastic properties of polymer melt blends as a function of composition using exit pressure and extrudate swell measurements. The response is complex and maxima and minima can exist as composition is varied.

B. INFLUENCE OF FILLERS AND OILS

The addition of fillers to polymer melts has a major influence on the rheological properties. The nature of the interaction, as well as the influence of the fillers, is complex. The major filler used with elastomers is carbon black and it has long been known that rubber is adsorbed from solutions onto carbon blacks and activated carbon [25, 142, 252]. In addition, Watson [293] has shown that when rubber–carbon black compounds are masticated, the black surface acts as a free radical acceptor. A concise and critical review on carbon black reinforcement has been written by Kraus [141], who has also edited a treatise on the subject [140].

Generally, filler-reinforced polymer melts exhibit an increase in viscosity, an apparent decrease in elastic memory, and the development of thixotropy, which is termed the *Mullins effect* [207]. We shall discuss these in turn. The viscous behavior of filled polymer melts has received considerable study but our understanding of it is still far from complete. On one side there is a highly developed hydrodynamic theory dating to the work of Einstein [82], Jeffreys [129], and Guth *et al.* [111]. More recent work is reviewed by Barthes–Biesel and Acrivos [22]. These theories allow the representation of the viscosity of a dilute suspension in terms of a series involving a volume fraction

$$\eta = \eta(0)[1 + A\phi + B\phi^2 + \cdots] \qquad (106)$$

where the coefficients depend upon particle shape and orientation. For spheres, A is 2.5 and B has been reported to be in the range of 10–14. Theories of concentration suspensions have been developed by Simha [260] and by Frankel and Acrivos [95].

Experimental studies of the influence of fillers on the viscosity show extraordinary behavior that does not obviously follow from any hydrodynamic theory. First, for small particles ($d_p < 1000$ Å) the viscosity is strongly dependent upon particle size. The smaller the size of the particle, the larger the viscosity at any volume loading. This phenomenon is quite striking in carbon blacks. Thus, using the classical carbon black designations [65, 263, 304],

$$\eta(\text{SAF}) > \eta(\text{ISAF}) > \eta(\text{HAF}) > \eta(\text{FEF}) > \eta \begin{pmatrix} \text{Guth} \\ \text{Gold} \\ \text{Simha} \end{pmatrix}$$

(For SAF black $\bar{d}_p \sim 170$ Å; for ISAF, ~ 200 Å; for FEF, ~ 560 Å.) Equation (106) generally underestimates the viscosity of rubber–black compounds. Second, filled polymers exhibit unusual viscosity–shear rate dependence. At low volume loadings, the behavior is similar to that of the pure polymer melt,

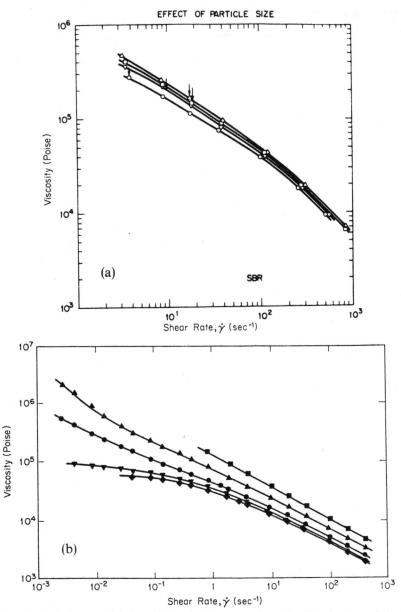

Fig. 14. (a) Influence of carbon black particle size on the viscosity–shear rate behavior of SBR compounds at constant loading and structure: $\bigcirc$, no carbon black (milled); $\triangle$, 20 ISAF-LS (23 mμm); $\square$, 20 FEF-LS (47 mμm); $\triangle$, 20 SRF-LM (80 mμm); $\downarrow$, extrudate distortion (From White and Crowder [305]). (b) Viscosity–shear rate concentration behavior for TiO$_2$–polyethylene: $\blacklozenge$, 0; $\blacktriangledown$, 4.3; $\bullet$, 13; $\blacktriangle$, 22.1; $\blacksquare$, 36.0%. (From Minagawa and White [192].)

but at moderate and high volume loadings, yield values are observed. This result has been documented for various filler–polymer systems [60, 192, 221, 289, 325]. Thus the viscosity is very large at low shear rates and decreases rapidly with increasing shear rate. This behavior is summarized in Fig. 14b for TiO_2–polyethylene. It may be seen that a result of this response is that the filler has a much greater influence on viscosity at low shear rates than at high shear rates.

The reason for the enhanced contribution to the viscosity of the very small carbon black particles has been the subject of considerable discussion through the years. One view which has gained attention and some acceptance is that, as argued by Medalia [181a,b] and Kraus [141a], carbon black consists of complex primary aggregates of fused small particles [181a] and the effective volume of the black includes not only the black volume itself but that of the polymer occluded onto the internal void volume associated with the primary aggregate. Medalia and Kraus have correlated the influence of carbon black on diverse mechanical properties. There may also be immobilized regions of thickness Δ about black particles and the volume of particles of radius R is $4\pi(R + \Delta)^3/3$. For a given volume fraction of black, the effective volume fraction would increase with decreasing particle size. This view has recently been argued by Pliskin and Tokita [234], who use it to analyze an experimental study of rubber–black compounds.

Another striking effect is the influence of fillers on the viscoelastic and extrusion properties of polymer melts. It has long been realized by rubber technologists that fillers, especially carbon black, "deaden" rubber and other polymer melts and cause improvements in extrusion characteristics, such as decreased extrudate distortion and extrudate swell. The author found in unpublished experiments many years ago that elastic recovery in rubber compounds measured in a modified Mooney viscometer decreased with black loading [308]. Only in recent years, however, has the filler effect on viscoelastic properties really been discussed in the open literature [114, 192, 289, 305]. Important extensive studies are given by Vinogradov et al. [289] and by Nakajima [209a]. Normal stress measurements indicate that the effective relaxation time τ as determined by Eq. (47) decreases with filler loading [114, 192]. Addition of carbon black reduces extrudate swell and retards the development of extrudate distortion.

The influence of viscous aromatic, naphthanic, and other oils on rheological behavior of rubber and rubber compounds has been considered by Kraus and Gruver [144], Hopper [127], Collins and Oetzel [66], and Derringer [72]. The addition of oil is found to decrease viscosity. Kraus and Gruver present arguments leading to a quantitative correlation of the form

$$\eta = \phi^{3.4} F(\dot{\gamma}\phi^{1.4}) \tag{107}$$

where ϕ is the volume fraction polymer.

VI. Rheological Applications to Processing

A. GENERAL

There are many ways in which elastomer rheology may be applied to processing operations. This is because processing is in large part simply flow and forming of compounds. The most obvious use is to determine the sources of problems when difficulties arise in factory production. Frequently the breakdown of regular production can be traced to some cause such as high viscosity levels or elastic memory of compounds. This is, of course, how the early rubber rheologists were first employed and to a good extent what they do to this day. A second application of rheology is to the design of processes and equipment. This function of rheologists has become highly developed in the plastics industry, especially in the area of screw extruder design [274]. Its beginnings can be seen in the fiber industry with the publication of papers analyzing melt spinline dynamics. Origins can perhaps also be noted in recent rubber industry publications [36, 264].

In this section we will discuss some of these areas. We begin with consideration of dimensional analysis of the force balance and energy equations describing flow in processing equipment and their application to scale-up. We will then turn to some specific applications.

B. DIMENSIONAL ANALYSIS AND SCALE-UP

One of the most useful ideas for the industrial polymer rheologist is dimensional analysis. This is a traditional method of engineering analysis in which the conservation laws (force-momentum, energy, etc.) are placed in dimensionless form and the dimensionless groups that arise are interpreted. There were applications of this approach to the development of model basins for ships and to the flow of liquids in pipes in the nineteenth century [243, 248]. Indeed, the general philosophy of dimensional analysis in engineering was developed by Reynolds [243] for the latter application. These ideas were extended in the early years of the present century by Nusselt [216]. The dimensionless correlations developed by these authors can be used as a guide in the design of pipelines, heat exchangers, and packed columns. There are numerous texts describing this subject matter which are an integral part of university engineering education [30, 255].

Unfortunately, the Reynolds–Nusselt dimensional studies are not directly applicable to polymer melt and rubber processing for two important reasons. First, they are based upon Newtonian fluid behavior, and second, they do not include viscous dissipation heating.

The earliest considerations of the dimensional analysis of isothermal viscoelastic fluids were by Karl Weissenberg [123, 294, 295] during the period from 1928 to 1948. Weissenberg suggested that the recoverable elastic strain represents fluid elasticity in the same way that the Reynolds number represents inertia. A large recoverable strain represents high viscoelastic forces. Reiner [242] later suggested that the basic dimensionless group correlating viscoelastic phenomena was a "Deborah number" which may be represented as a ratio of the characteristic relaxation time τ_{ch} of the deforming material to the duration t of the deformation. A large value of τ_{ch} or a small value of t will, according to Reiner, accentuate viscoelastic response. A more quantitative formulation of the intuitive ideas of Weissenberg and Reiner resulted after the modern theory of nonlinear viscoelasticity had been developed. White [298], Slattery [262], White and Tokita [299, 310], and Metzner et al. [186, 187] have published analyses that have brought these ideas together into a formal scheme.

We begin with Cauchy's law of motion, Eq. (10), and a general constitutive equation such as Eq. (36), which yields

$$\rho\left[\frac{\partial \mathbf{v}}{\partial t} + (\mathbf{v}\cdot\nabla)\mathbf{v}\right] = -\nabla p + \nabla \cdot \int_0^\infty [m_1(z)\mathbf{c}^{-1} - m_2(z)\mathbf{c}]\, dz + \rho\mathbf{g} \quad (108)$$

where $m_1(z)$ and $m_2(z)$ are given by an expression similar to Eq. (37). A characteristic velocity U and length L are introduced into this equation and the entire expression is put in dimensionless form. These dimensionless groups are

$$\frac{LU\rho}{G_m\tau_m}, \quad \frac{U^2}{gL}, \quad \tau_m\frac{U}{L}, \quad \frac{\tau_m}{t}, \quad \frac{\tau_i}{\tau_m}, \quad \frac{G_i}{G_m}, \quad \varepsilon, \quad b$$

Here $LU\rho/G_m\tau_m$ is a Reynolds number signifying the ratio of inertial to viscous forces; U^2/gL is a Froude number representing a ratio of inertial to gravitational forces; $\tau_m U/L$ is known as the Weissenberg number and τ_m/t as the Deborah number, both representing the ratio of viscoelastic to viscous forces. The other groups are viscoelastic ratio numbers representing the detailed character of the functional.

Generally, inertial forces are negligible in polymer melts and elastomers and gravitational forces need not be considered. This leaves the Weissenberg, Deborah, and viscoelastic ratio numbers. If we consider an application to a processing operation in which the material flows through some equipment, it must be recognized that t is a residence time and

$$\tau_m/t = [\tau_m/(L_{11}/V)] = \tau_m U/L_{11} \quad (109)$$

and the Deborah number is equivalent to a Weissenberg number with the characteristic length in the direction of flow. Our dimensionless groups thus reduce to

$$\tau_m U/L, \quad \tau_i/\tau_m, \quad G_i/G_m, \quad \varepsilon, \quad b$$

The results of the preceding paragraph have some important implications.

First, the characteristic velocity and length arise in only one group and this occurs in the form of a ratio V/L. This means that for any particular visco-elastic fluid there will be dynamic similarity between large and small geo-metrically scaled *isothermal* systems only if velocities are scaled with lengths. Furthermore, if any instabilities or changes in regime occur within such a system, they will occur at a critical value of a dimensionless group of the form $\tau V/L$ where

$$\tau = \tau_{\mathrm{m}} f [\tau_i/\tau_{\mathrm{m}}, G_i/G_{\mathrm{m}}, \varepsilon, b] \tag{110}$$

Here the function f may be quite complex, so that τ may differ significantly from material to material (with differing viscoelastic ratio numbers) even if they possess the same τ_{m}.

The discussion of the foregoing paragraphs is limited to isothermal systems, with which, however, we are generally not concerned. We must also consider the energy equation Eq. (62), which we may write as

$$\rho c \left[\frac{\partial T}{\partial t} + (\mathbf{v} \cdot \nabla) T \right] = k \nabla^2 T + \boldsymbol{\sigma} : \nabla \mathbf{v} \tag{111}$$

Introducing characteristic velocity V, length L, and temperature difference θ, we may show that the dimensionless groups arising in this equation are

$$LV\rho c/k, \qquad \eta V^2/k\theta, \qquad G_{\mathrm{m}}\tau_{\mathrm{m}}/\eta, \qquad \text{viscoelastic ratio numbers}$$

Here $LV\rho c/k$ is known as the Peclet number [30, 27] and signifies the ratio of convective heat flux to heat conduction, while $\eta V^2/k\theta$ is the Brinkman number, representing the ratio of viscous heating to heat conduction. The ratio of the Brinkman number to the Peclet number has been called the Hersey number [299].

The Brinkman number is of great importance in the interpretation of polymer processing since it represents the ability of the system to conduct away heat produced by viscous dissipation. It is useful to rewrite it in the form

$$\frac{\eta V^2}{k\theta} = L^2 \left(\frac{\eta}{k\theta} \right) \left(\frac{V}{L} \right)^2 \tag{112}$$

If we compare small and large systems that have been scaled at constant V/L, as we would expect for dynamic similarity, the Brinkman number rises as L^2. The larger the system, the greater the tendency for viscous heating to override heat conduction and for the temperature to rise. Another way of putting this is that a large system will tend to behave adiabatically. Thus the isothermal dimensional analysis and interpretation is most suitable to systems with small dimensions. For large systems we have to consider viscous dissipation as well.

The origin of the effect described in the preceding paragraph may be viewed in another way. Viscous dissipation heating is proportional to the volume of the system L^3, whereas heat transfer is proportional to the surface area, which varies as L^2. As one goes to larger systems, surface area per unit volume decreases, making heat transfer more difficult and the system more adiabatic.

A final point of importance, which has already been implied but which should be stated more explicitly, is the inability to scale at both constant Weissenberg number and constant Brinkman number, i.e., at constant

$$\tau V/L, \qquad \eta V^2/k\theta$$

The only way one may scale at constant Brinkman number is to maintain the characteristic velocity constant as L increases, which would decrease the Weissenberg number.

C. COLD FLOW

The first problem encountered in rubber processing occurs not in the processing equipment per se but in the storage or warehouse area, where a bottom bale of rubber flows out over the floor due to the weight of bales on top of it (see Fig. 15). This is closely related to a classical fluid mechanics problem solved by Stefan [269] using the methods of what we now call lubrication theory [255]. The problem is also identical to the analysis of a Williams compression plastometer. Basically, the bale being squeezed flows out slowly as a lubricant. Since (1) the deformation rates are small and (2) the time scale is large, i.e., the time required for sufficient flow $t_{1/4}$ (the time required for the height of the bottom bale to be reduced 75% in thickness) is much greater than τ_m, we may consider the bale of rubber to behave as a Newtonian fluid. It follows that if we consider the initial bale as a disk with radius R and thickness h, then (compare Furuta *et al.* [98])

$$t_{1/4} = \frac{45}{8} \eta_0 \frac{R^2}{Fh^2} \tag{113}$$

where F is the force of the upper bales. The only rheological property arising is η_0, the zero shear viscosity. Increasing η_0 increases $t_{1/4}$ and decreases cold flow. This might be done by increasing weight average molecular weight [Eq. (99)] or, following Kraus and Gruver [143], by increasing long chain branching. Both procedures seem to have been followed.

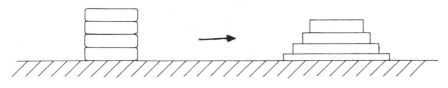

Fig. 15. Cold flow.

D. Mill Behavior and Mixing

One of the basic problems of the industrial rubber rheologist is the determination of the processability of gum elastomers which are to be compounded with filler in an internal mixer or mill. Two major processing difficulties may arise. First, if the elastomer has a very high elastic memory, it will respond as a solid and resist the moving blade or rotor in an internal mixer or the entrance into the nip between the rolls of a mill. In such a case filler will not be incorporated into the elastomer. A second problem which arises is that the stresses in the region between the rotor and the wall in the internal mixer or in the nip between the mill rolls may tear apart the elastomer into crumbs. Experimental studies of the mill problem have been done by Bulgin [49], Tokita and White [282, 312], Wagner [290], and Shih [259]. Useful discussions by White [299, 300] and by Tokita and Pliskin [281] are also available. Tokita and White have represented mill behavior in terms of four regions, or better, regimes, which are summarized in Fig. 16. In regime 1 the highly elastic rubber will not move into the mill nip without difficulty. In regime 2, a good elastic band is obtained on one of the rolls. In regime 3, the elastomer crumbles and "bagging" (or crumbs) results. In regime 4 a melt fluid coats one of the rolls. Some elastomers seem to exhibit two or three of these regimes, whereas others exhibit all four regimes. The two problems mentioned at the outset relate to regimes 1 and 3.

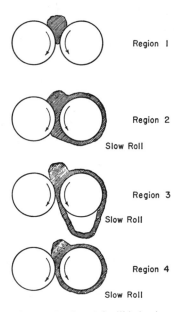

Fig. 16. Regimes of mill behavior.

The regime 1–2 mill problem is addressed by White [299, 300], who suggests that it may be simplified to creep under the action of applied stress. The rotating rolls apply frictional stresses to the rubber in the bank of the mill and gradually pull it down to the nip. The time t required for the material to enter the nip determines whether the material is rated as in regime 1 (large t) or regime 2 (small t). Since the deformation rates for regime 1 are small, but t is probably less than τ_m, we may proceed by using linear viscoelasticity. The time t required for a shear strain γ to be applied to a fluid with relaxation time τ_m and compliance function $J(t)$ is

$$J(t)\sigma = \sum_{i=1}^{m} J_i(1 - e^{-t/\tau_i})\sigma \tag{114}$$

Generally J_i plays a role similar to H_0 in the wedge box distribution and would be independent of molecular weight and approximately of temperature. In any series of polymers with fixed G, τ_m will determine t. Large τ_m will lead to regime 1 and difficult processing.

The problem of regime 3 and bagging on mills has been related to small elongations to break in tensile tests [281, 312]. The rubber seems to fracture and crumble in the nip. Small elongations to break correlate with narrowness of molecular weight distributions, and not surprisingly it is the anionic alkyllithium and other narrow distribution elastomers which exhibit this problem. It may be largely eliminated by broadening the molecular weight distribution. The broader distribution elastomers do not seem to exhibit a regime 3. The regime 4–3 transition obtained by decreasing temperature was observed by Tokita and White to occur at a critical value of τ_m and was suggested to be related to a hydrodynamic instability. Later, the same authors [312] reconsidered the problem and suggested that failure is due to fracture resulting from high elastic energies, which also correlates with τ_m. The latter view would seem the correct one.

We now turn to internal mixers where the rubber is masticated by rotating blades. As pointed out by Tokita and Pliskin [281], the rubber in an internal mixer will also be in states equivalent to those of regimes 1–4. If the rubber is in regime 1 or 3, good mixing will not be possible. The best mixing will be in regime 2, where high shear stresses exist, but good mixing in regime 4 is also possible. This prediction is borne out by mill mixing experiments cited by White and Tokita [312].

If the material deforms smoothly as described by regime 4 (and perhaps 2), the flow within an internal mixer may probably be treated in the manner described by Bolen and Colwell [36a] (see also McKelvey [167]). The melt is sheared between the blade and the wall in the manner described by lubrication theory. Viscous heating caused by the shearing action results in temperature rises. The goodness of mixing will be determined by the intensity of the shear stress history.

The discussion in Section II.B about the impossibility of scaling when viscous heating is important finds direct application to internal mixers. Attempts to scale at constant V/L or $R\Omega/R = \Omega$, where R is the radius and Ω the angular velocity of the rotor, result in the strikingly different responses of large and small mixers. The difference is due to increasing Brinkman number $\eta V^2/k\theta$ [see Eq. (119)] with scale factor. The large mixer behaves as if adiabatic and becomes hot, while the small mixer is nearly isothermal. Thus higher shear stresses (proportional to $\eta\Omega$) occur in the small mixer because the temperature is lower. Better mixing results. Hence recipes that mix well in the laboratory often prove a disaster on a factory scale.

The internal mixer and the mill have often traditionally been used in conjunction as indicated in Fig. 17. Here the rubber and fillers are compounded in the internal mixer and then dumped for further mixing on a mill. The response of the rubber on the mill is in part determined by the extent of mixing in the internal mixer. This in turn depends on whether the rubber responded as if in regimes 1, 2, 3, or 4 while in the internal mixer.

E. Extrusion

1. Flow in Screw

The screw extruder was originally developed by the rubber industry [134] and continues to be its most important forming apparatus (see Fig. 18). Generally rubber is extruded on a single-screw machine in which a screw rotates in a tightly fitted heated barrel. The rubber is fed into a hopper and is conveyed along the length of the barrel by the rotation of the screw. The rubber is transported along a conduit formed by the screw flights and the inner surface of the barrel. As the rubber is conveyed, it builds up pressure and, reaching the end of the screw, allows itself to be forced through a die.

As compared with the extrusion of plastics, the process of extrusion of rubber compounds is reasonably simple: strips of rubber are added to the

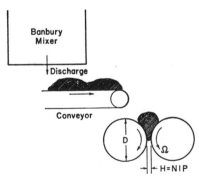

Fig. 17. Discharge from a large Banbury mixer onto a conveyor belt leading to a mill.

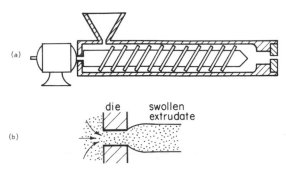

Fig. 18. (a) Screw extruder. (b) Die swell.

hopper and the behavior of the compound along the screw changes little except that the material generally increases in temperature and becomes less viscous and more fluid as it moves through the screw. In the extrusion of plastics, there are usually three distinct regions within a screw. Pellets or powder are added to the hopper and the section of the screw near the hopper is called the *solid-conveying region*. However, the heated barrel soon begins to melt the pellets and this results in a second region, a *melting region*. In glassy polymers, such as polystyrene, the material goes through its glass transition in this region. After the pellets have completely melted there is a third region, generally referred to as the *metering region*, where the extruder acts as a screw pump. With rubber, the entire screw presumably behaves as a metering region.

A detailed theory relating the rheological and thermal properties of polymers to their screw extrusion characteristics has been developed. As early as 1922, Rowell and Finlayson [249] worked out the theory of screw pumps for viscous liquids. Pigott [230] and Carley *et al.* [53] applied this theory to the analysis of screw extruders. Generally these studies make a parallel plate approximation. Booey and Squires [268] considered the influence of channel curvature on Newtonian fluids. Middleman [152, 188] derived a theory of a screw pump with a power law viscosity function using the parallel rectangular channel approximation. The influence of curvature on power law fluids has been analyzed by Tadmor and Klein [274]. Nonisothermal flow in the metering region for power law fluids with temperature-dependent rheological properties was studied by Griffith [110], Zamodits and Pearson [326], and Martin [173a]. The melting region in a screw extruder was studied experimentally by Maddock [70] and by Marshall and Klein [173] and was modeled by Tadmor and Klein [272, 274]. Models of the solid-conveying region have been developed by Darnell and Mol [68] and by Tadmor *et al.* [132]. Integrated models of all three regions of the screw extruder have been developed and put in the form of digital computer programs by Klein *et al.* [137, 274] and later authors.

In a rubber extruder, the material presumably responds mainly as if it were in a metering region. Here we will concisely summarize the analytical formulation and application of rheology in this region. Following most authors, we neglect the influence of curvature and stretch out the channel. Taking x as the direction of flow and y as the direction perpendicular to the screw surface, we have the flow field

$$\mathbf{v} = u(y, z)\mathbf{i} + v(y, z)\mathbf{j} + w(y, z)\mathbf{k} \tag{115}$$

with

$$u > v \sim w$$

The Cauchy equations of motion [Eq. (10)] reduce to

$$0 = -\frac{\partial p}{\partial x} + \frac{\partial \sigma_{xy}}{\partial y} + \frac{\partial \sigma_{xz}}{\partial z} \tag{116a}$$

$$0 = -\frac{\partial p}{\partial z} + \frac{\partial \sigma_{yz}}{\partial y} \tag{116b}$$

$$0 = -\frac{\partial p}{\partial y} + \frac{\partial \sigma_{yz}}{\partial z} \tag{116c}$$

where

$$\sigma_{xy} = \eta \frac{\partial u}{\partial y}, \qquad \sigma_{xz} = \eta \frac{\partial u}{\partial z}, \qquad \sigma_{yz} = \eta \left(\frac{\partial v}{\partial z} + \frac{\partial w}{\partial y} \right) \tag{117}$$

and the shear viscosity depends upon the second invariant of the rate of deformation Π_d:

$$\Pi_\mathrm{d} = \left(\frac{\partial u}{\partial y} \right)^2 + \left(\frac{\partial u}{\partial z} \right)^2 + \left(\frac{\partial v}{\partial z} + \frac{\partial u}{\partial y} \right)^2 \tag{118}$$

The pressure gradients are presumed constant. The screw is traditionally considered stationary and the barrel to move, so that we have boundary conditions

$$u(0, z) = v(0, z) = w(0, z) = 0$$

$$u\left(y, \pm \frac{W}{2} \right) = v\left(y, \pm \frac{W}{2} \right) = w\left(y, \pm \frac{W}{2} \right) = 0 \tag{119}$$

$$u(H, z) = \pi D N \cos \theta, \qquad v(H, z) = -\pi D N \sin \theta$$

where W is the distance between the flights, H the screw depth, θ the screw helix angle, N the screw rotation rate, and D the diameter. We must solve these equations coupled with an energy balance. If we neglect the convective terms and heat conduction only in the y direction is considered, this gives

$$0 = k \frac{\partial^2 T}{\partial y^2} + \eta \Pi_\mathrm{d} \tag{120}$$

where η depends upon temperature. The boundary conditions are

$$\frac{\partial T}{\partial y}(0) = 0, \qquad T(H) = T_w \tag{121}$$

Martin [173a] has considered the importance of including the convective term in the energy balance.

Typically we presume the viscosity to exhibit non-Newtonian behavior in the form of a power law and to depend upon temperature through an activation energy.

$$\eta = K[\Pi_d]^{(n-1)/2} = K_0 e^{E/RT}[\Pi_d]^{(n-1)/2} \tag{122}$$

These equations may be solved for the velocity field $u(y, z)$ and the temperature field $T(x, y, z)$. The velocity field may be integrated through the channel to give the extrusion rate Q. The results may be expressed in dimensionless form as

$$\frac{Q}{WH\pi DN} = F\left[\frac{dp}{dx}\frac{H^{n+1}}{K_0(\pi DN)^n}, \underbrace{\frac{K_0(\pi DN)^{n+1}H^2}{k(T_w - T_0)}}_{\text{Brinkman number}}, \theta, \frac{H}{W}, \frac{E}{RT_0}, n\right] \tag{123}$$

As the size of extruders increases, the Brinkman number (proportional to D^{n+1}) plays an increasingly important role.

In the simplest model of the process, which dates back to Rowell and Finlayson [249] and Carley et al. [53], the fluid is taken as Newtonian and the viscosity as a constant. H/W and E/RT_0 are zero and $n = 1$. This yields a parabolic velocity field with superposed drag and pressure flow

$$u = \pi DN \cos \theta \left(\frac{y}{H}\right) - \frac{1}{2\eta}\frac{dp}{dx}(y^2 - Hy) \tag{124}$$

and dimensionless extrusion rate

$$\frac{Q}{WH\pi DN} = \frac{\cos \theta}{2} - \frac{1}{6}\left(\frac{dp}{dx}\frac{H^2}{\eta\pi DN}\right) \tag{125}$$

Computations based on modelling similar to Eq. (123) are given by Griffith [110] and Zamodits and Pearson [326].

Few experimental data on the flow of rubber compounds in extruder screws exist in the open literature. We may cite, though, the work of Vila [287], Pigott [230], and Zamodits and Pearson [326].

2. Flow in Die

The extrusion of polymer melts through dies involves three specific problems. First, the polymer must be pushed from a reservoir into the die. Second, it must be extruded and shaped by the die. Third, it emerges from the die and exhibits recovery and sometimes becomes distorted. Each of these is a distinct and complex rheological problem.

Let us first turn to the entrance flow problem. Although there have been numerous studies of flow patterns in polymer solutions and molten plastics at a die entrance [302], only the work of Vinogradov et al. [288] may be cited for elastomers. This involves very narrow molecular weight distribution uncompounded polybutadienes. In molten plastics such as low density polyethylene and polystyrene large vortices are observed in die entry regions [302, 305a]. These are not observed by Vinogradov et al. in the polybutadienes considered.

As originally noted by Mooney [198, 203], total pressure losses in a die are in large part due to ends losses and we must write

$$P_T = \Delta p_e + \Delta p \tag{126}$$

where Δp_e is the sum of entrance and exit pressure losses. A more extensive study of Δp_e for rubber compounds is given by White and Crowder [305]. Generally Δp_e is about 10–30 times the wall shear stress in a capillary die. From Eq. (84) we may write

$$p_T = 4(\sigma_{12})_w \left(\frac{L}{D} + \frac{N}{4}\right) \tag{127}$$

where N is a number from 10 to 30. Thus Δp_e adds an effective L/D ratio of 2.5–7.5 to the die. The large contributions to Δp_e at the entry have been associated with the above-mentioned vortex flows [19a].

The pressure loss in a capillary die for isothermal power law viscosity behavior is [260]

$$\Delta p = 4K \left(\frac{32Q}{\pi D^3}\right)^n \left(\frac{3n+1}{4n}\right)^n \frac{L}{D} \tag{128}$$

Skelland [261] summarized analytical results for other cross sections. For a slit die

$$\Delta p = 2K \left(\frac{6Q}{WH^2}\right)^n \left(\frac{2n+1}{3n}\right)^n \frac{L}{H} \tag{129}$$

where the slit dimensions are $W \times H$ with $W \gg H$.

Extrudate swell is a most difficult rheological problem to analyze. The phenomenon seems associated with elastic recovery. The few studies of extrudate swell in rubber generally note the inability to extrude an uncompounded rubber without severe extrudate distortion [20, 305]. Addition of carbon black improves extrudate character [127, 289, 305] but decreases swell. Extrudate swell generally increases with extrusion rate and decreases with increasing L/D ratio. Experimental studies have recently been done by Pliskin [233] and White and Crowder [305]. There are several theoretical studies of extrudate swell for viscoelastic fluids in the literature [183, 185, 276]. The most pertinent for molten polymers would seem to be that of Tanner [276], which is based on a nonlinear integral constitutive equation similar to Eq. (36) $[m_2(z) = 0]$ and presumes unconstrained recovery from Poiseuille flow with $L/D \to \infty$. Basically it is found that the extrudate swell B is

$$B = \frac{d}{D} = C + \left[1 + \frac{1}{2}\left(\frac{N_1}{2\sigma_{12}}\right)^2\right]^{1/6} = C + [1 + \tfrac{1}{2}(\tau\dot{\gamma}_w)^2]^{1/6} \qquad (130)$$

where C is a constant of order 0.12 which accounts for the swell of a Newtonian fluid. There has been no comparison of theory with experiment for any theories of extrudate swell for rubber. A critical comparison with unfilled molten plastics is given by White and Roman [307a].

The problem for filled polymer melts including rubber compounds is far more complex because no constitutive equations have been established for these systems. Extrudate swell is generally greatly decreased and as shown by Minagawa and White [192], Eq. (130) tends to greatly overestimate swell.

The problem of extrudate distortion and its mechanisms in polymer solutions and molten plastics has received considerable attention in the literature [18, 20, 284, 302]. However, there have been few investigations for rubber [18, 305] and these seem only to suggest that the characteristics and mechanism exhibit both similarities and differences.

ACKNOWLEDGMENT

The writing of this chapter was supported in part by the National Science Foundation under Grants GK 43921 and ENG 76-19815.

REFERENCES

1. T. N. G. Abbott and K. Walters, *J. Fluid Mech* **40**, 203 (1970); **43**, 257 (1970).
2. T. N. G. Abbott and K. Walters, *J. Fluid Mech.* **48**, 257 (1970).
3. T. N. G. Abbott, G. W. Bowen, and K. Walters, *J. Phys. D.* **4**, 190 (1971).
4. T. Alfrey, "Mechanical Behavior of High Polymers." Wiley (Interscience), New York, 1948.
5. T. Alfrey, *SPE Trans.* **6**, 68 (1965).
6. E. H. Andrews, *Proc. R. Soc., Ser. A* **277**, 562 (1964).
7. R. D. Andrews and A. V. Tobolsky, *J. Polym. Sci.* **7**, 221 (1952).
8. R. D. Andrews, N. Hofman-Bang, and A. V. Tobolsky, *J. Polym. Sci.* **3**, 669 (1948).
9. D. J. Angier and W. F. Watson, *J. Polym. Sci.* **18**, 129 (1955).
10. D. J. Angier and W. F. Watson, *J. Polym. Sci.* **20**, 235 (1956).
11. D. J. Angier and W. F. Watson, *Trans. Inst. Rubber Ind.* **33**, 22 (1957).
12. D. J. Angier, W. T. Chambers, and W. F. Watson, *J. Polym. Sci.* **25**, 129 (1957).
13. D. J. Angier, E. D. Farlie, and W. F. Watson, *Trans. Inst. Rubber Ind.* **34**, 8 (1958).
14. D. J. Angier, R. J. Ceresa, and W. F. Watson, *J. Polym. Sci.* **34**, 699 (1958).

15. Anonymous, *Plast. World* **31**(1), 48 (1973).
16. G. Ayrey, C. G. Moore, and W. F. Watson, *J. Polym. Sci.* **19**, 1 (1956).
17. E. B. Bagley, *J. Appl. Phys.* **28**, 624 (1957).
18. E. B. Bagley, *Trans. Soc. Rheol.* **5**, 355 (1961).
19. W. S. Bahary, D. I. Sapper, and J. H. Lane, *Rubber Chem. Technol.* **40**, 1529 (1967).
19a. T. F. Ballenger and J. L. White, *J. Appl. Polym. Sci.* **15**, 1949 (1971).
20. T. F. Ballenger, I. J. Chen, J. W. Crowder, G. E. Hagler, D. C. Bogue, and J. L. White, *Trans. Soc. Rheol.* **15**, 195 (1971).
20a. R. L. Ballman, *Rheol. Acta* **4**, 137 (1965).
21. N. K. Baraboim, "Mechanochemistry of Polymers." MacLaren, London, 1964. (Transl. from the Russ.)
22. D. Barthes-Biesel and A. Acrivos, *Int. J. Multiphase Flow* **1**, 1 (1973).
23. M. Berger and D. J. Buckley, *J. Polym. Sci. Part A* **1**, 2945 (1963).
24. B. Bernstein, E. A. Kearsley, and L. J. Zapas, *Trans. Soc. Rheol.* **7**, 391 (1963).
25. B. Bernstein, E. A. Kearsley, and L. J. Zapas, *J. Res. Natl. Bur. Stand., Sect. B* **68**, 103 (1964).
26. J. S. Binford and A. M. Gessler, *J. Phys. Chem.* **63**, 1376 (1959).
27. R. B. Bird, *SPE J.* Sept., 35 (1955).
28. R. B. Bird and P. J. Carreau, *Chem. Eng. Sci.* **23**, 427 (1968).
29. R. B. Bird and E. K. Harris, *AIChE J.* **14**, 758 (1968).
30. R. B. Bird, W. E. Stewart, and E. W. Lightfoot, "Transport Phenomena." Wiley, New York, 1960.
31. L. L. Blyler, *Trans. Soc. Rheol.* **13**, 39 (1969).
32. L. L. Blyler and S. J. Kurtz, *J. Appl. Polym. Sci.* **11**, 127 (1967).
33. D. C. Bogue, *IEC Fundam.* **5**, 253 (1966).
34. D. C. Bogue and J. L. White, Engineering Analysis of Non-Newtonian Fluids, NATO, Brussels, Belgium. No. 144 (1970).
35. D. C. Bogue, T. Masuda, Y. Einaga, and S. Onogi, *Polym. J.* **1**, 563 (1970).
36. J. J. Boguslawski, *Rubber Chem. Technol.* **45**, 1421 (1972).
36a. W. R. Bolen and R. E. Colwell, *SPE Antec Tech. Papers* **4**, 1004 (1958).
37. L. Boltzmann, *Sitzungsber. Kaiserl. Akad. Wiss. Wien* **70**, 275 (1874).
38. H. C. Booij, *Rheol. Acta* **5**, 215 (1966).
39. C. Booth, *Polymer* **4**, 471 (1963).
40. H. H. Bowerman, E. A. Collins, and W. Nakajima, *Rubber Chem. Technol.* **47**, 307, 318 (1973).
41. H. C. Brinkman, *Appl. Sci. Res., Sect. A* **2**, 120 (1951).
42. J. M. Broadbent and K. Walters, *J. Phys. D.* **4**, 1863 (1971).
43. J. M. Broadbent, A. Kaye, A. S. Lodge, and D. G. Vale, *Nature (London)* **217**, 55 (1968).
44. J. G. Brodnyan, F. H. Gaskins, and W. Philippoff, *Trans. Soc. Rheol.* **1**, 109 (1957).
45. R. A. Brooman, Engl. Patent 10.582 (1845).
46. A. H. Bruner, N. N. Roe, and P. Byrne, *J. Elastoplast.* **5**, 201 (1973).
47. F. Bueche, *J. Appl. Phys.* **30**, 1114 (1959).
48. F. Bueche and S. W. Harding, *J. Polym. Sci.* **32**, 177 (1958).
49. D. R. Bulgin, *Rubber Plast. Age* **42**, 715 (1961).
50. W. F. Busse, *Ind. Eng. Chem.* **24**, 140 (1932).
51. W. F. Busse and R. N. Cunningham, *Proc. Rubber Technol. Conf., London* p. 288 (1938).
52. W. F. Busse and R. Longworth, *Trans. Soc. Rheol.* **6**, 179 (1962).
53. J. F. Carley, R. S. Mallouk, and J. M. McKelvey, *Ind. Eng. Chem.* **51**, 974 (1953).
54. W. H. Carothers, I. Williams, A. M. Collins, and J. E. Kirby, *J. Am. Chem. Soc.* **53**, 4203 (1931).
55. P. Carreau, *Trans. Soc. Rheol.* **16**, 99 (1972).
56. A. Casale, R. S. Porter, and J. F. Johnson, *Rubber Chem. Technol.* **44**, 534 (1971).

57. A. L. Cauchy, *Exam. Math.* **2**, 42 (1827); *Oeurues* **7**(2), 60 (1889); see also *Exam. Math.* **2**, 108 (1827); *Oeurues* **7**(2), 141 (1889).
58. R. J. Ceresa, *Polymer* **1**, 477 (1960).
59. H. Chang and A. S. Lodge, *Rheol. Acta* **11**, 127 (1972).
60. F. M. Chapman and T. S. Lee, *SPE J.* **26**(1), 37 (1970).
61. I. J. Chen and D. C. Bogue, *Trans. Soc. Rheol.* **16**, 59 (1972).
62. E. B. Christiansen and R. W. Leppard, *Trans. Soc. Rheol.* **18**, 65 (1974).
62a. F. N. Cogswell, *Plast. Polym.* **36**, 109 (1968); *Rheol. Acta* **8**, 187 (1969).
63. F. N. Cogswell, *Plast. Polym.* **41**, 39 (1973).
64. B. D. Coleman and H. Markovitz, *J. Appl. Phys.* **35**, 1 (1964).
65. E. A. Collins and J. T. Oetzel, *Rubber Age* (*N.Y.*) **102**, 64 (1970).
66. E. A. Collins and J. T. Oetzel, *Rubber Age* (*N.Y.*) **103**, Feb. (1971).
67. F. H. Cotton, *Trans. Inst. Rubber Ind.* **5**, 487 (1931).
67a. G. R. Cotton and J. L. Thiele, Paper presented at ACS Rubber Division Meeting, Montreal, May, 1978.
68. W. H. Darnell and E. A. J. Mol, *SPE J.* **12**(4), 20 (1956).
69. J. M. Davies, J. F. Hutton, and K. Walters, *J. Phys. D.* **6**, 2259 (1973).
70. J. R. Dees and J. E. Spruiell, *J. Appl. Polym. Sci.* **18**, 1053 (1974).
71. M. M. Denn and G. Marrucci, *AIChE J.* **17**, 101 (1971).
72. G. C. Derringer, *Rubber Chem. Technol.* **47**, 825 (1974).
73. J. H. Dillon, *Physics* (*N.Y.*) **7**, 73 (1936).
74. J. H. Dillon and L. V. Cooper, *Rubber Age* (*N.Y.*), 1306 (1937).
75. J. H. Dillon and N. Johnston, *Physics* (*N.Y.*) **4**, 225 (1933).
76. J. D. D'Ianni, F. J. Naples, and J. E. Field, *Ind. Eng. Chem.* **42**, 95 (1950).
77. D. J. Dunning and P. J. Pennells, *Rubber Chem. Technol.* **40**, 1381 (1967).
78. S. Eccher, *Ind. Eng. Chem.* **43**, 479 (1951).
79. S. Eccher, *Proc. Meet. Ital. Soc. Rheol.*, *1st* **2**, 79 (1971).
80. S. Eccher and A. Valentinotti, *Ind. Eng. Chem.* **50**, 829 (1958).
81. S. Einhorn and S. B. Turetzky, *J. Appl. Polym. Sci.* **8**, 1257 (1964).
82. A. Einstein, *Ann Phys.* (*Leipzig*) **19**, 289 (1906).
83. R. Eisenschitz, B. Rabinowitsch, and K. Weissenberg, *Mitt. Dtsch. Materialpruefungsanst.*, *Sonderh.* **9**, 91 (1929).
84. J. D. Ferry, "Viscoelastic Properties of Polymer," 2nd Ed. Wiley, New York, 1969.
85. P. J. Flory, *J. Am. Chem. Soc.* **62**, 1057 (1940).
86. P. J. Flory, *J. Chem. Phys.* **15**, 397 (1947).
87. P. J. Flory, *J. Chem. Phys.* **17**, 223 (1949).
88. P. J. Flory, "Principles of Polymer Chemistry." Cornell Univ. Press, Ithaca, New York, 1953.
89. V. L. Folt, *Rubber Chem. Technol.* **42**, 1294 (1969).
90. V. L. Folt and R. W. Smith, *Rubber Chem. Technol.* **46**, 1193 (1973).
91. V. L. Folt, R. W. Smith, and C. L. Wilkes, *Rubber Chem. Technol.* **44**, 1, 12, 26 (1971).
92. T. G. Fox and V. R. Allen, *J. Chem. Phys.* **41**, 344 (1964).
93. T. G. Fox and P. J. Flory, *J. Am. Chem. Soc.* **70**, 2384 (1949).
94. T. G. Fox and S. Loschaek, *J. Appl. Phys.* **26**, 1082 (1955); T. G. Fox, S. Loschaek, and S. Gratch, *in* "Rheology: Theory and Applications" (R. R. Eirich, ed.), Vol. 1, pp.   . Academic Press, New York, 1956.
95. N. A. Frankel and A. Acrivos, *Chem. Eng. Sci.* **22**, 847 (1967).
96. S. M. Freeman and K. Weissenberg, *Nature* (*London*) **161**, 324 (1948).
97. C. S. Fuller, C. J. Frosch, and N. R. Pape, *J. Am. Chem. Soc.* **62**, 1905 (1940).
98. I. Furuta, V. M. Lobe, and J. L. White, *J. Non-Newt. Fluid. Mech.* **1**, 207 (1976).
99. A. N. Gent, *J. Polym. Sci., Part A* **3**, 1387 (1965).
100. J. W. Gibbs and E. B. Wilson, "Vector Analysis." Scribner, New York, 1901.

101. H. Giesekus, *Proc. Int. Rheol. Congr.*, *4th* **1**, 249 (1965).
102. R. F. Ginn and A. B. Metzner, *Proc. Int. Rheol. Congr.*, *4th* **2**, 583 (1965); *Trans. Soc. Rheol.* **13**, 429 (1969).
103. C. Goldstein, *Trans. Soc. Rheol.* **18**, 357 (1974).
104. C. Goodyear, "Gum Elastic and its Varieties with a Detailed Account Of Its Applications and Uses and the Discovery of Vulcanization." New Haven, Connecticut, 1855.
105. J. Gough, *Proc. Lit. Philos. Soc., Manchester* (2) **1**, 288 (1805).
106. W. W. Graessley, *J. Chem. Phys.* **43**, 2696 (1965).
107. A. E. Green and R. S. Rivlin, *Arch. Ration. Mech. Anal.* **1**, 1 (1957).
108. M. S. Green and A. V. Tobolsky, *J. Chem. Phys.* **14**, 180 (1946).
109. H. W. Greensmith and R. S. Rivlin, *Philos. Trans. R. Soc. London, Ser. A* **245**, 399 (1953).
110. R. M. Griffith, *IEC Fundam.* **1**, 180 (1962).
111. E. Guth and R. Simha, *Kolloid-Z.* **74**, 266 (1936).
112. C. D. Han, *J. Appl. Polym. Sci.* **17**, 1289 (1973).
113. C. D. Han, *Trans. Soc. Rheol.* **18**, 163 (1974).
114. C. D. Han, *J. Appl. Polym. Sci.* **18**, 821 (1974).
115. C. D. Han, "Rheology in Polymer Processing." Academic Press, New York (1976).
116. C. D. Han and T. C. Yu, *J. Appl. Polym. Sci.* **15**, 1163 (1971).
117. C. D. Han and T. C. Yu, *Polym. Eng. Sci.* **12**, 81 (1972).
118. C. D. Han, M. E. Charles, and W. Philippoff, *Trans. Soc. Rheol.* **14**, 393 (1970).
119. T. Hancock, "Personal Narrative of the Origin and Progress of the Caoutchoucor India Rubber Manufacture in England." Longman, Brown, Green, Longmans & Roberts, London, 1857.
120. D. J. Harmon, *J. Franklin Inst.* **209**, 519 (1955).
121. H. Hencky, *Ann. Phys. (Leipzig)* **2**, 617 (1929).
122. J. Hermans, *J. Colloid Sci.* **17**, 638 (1962).
123. R. O. Herzog and K. Weissenberg, *Kolloid-Z.* **46**, 277 (1928).
124. E. M. Hicks, E. A. Tippetts, J. V. Hewett, and R. H. Brand, *in* "Man Made Fibers" (H. Mark, S. Atlas, and E. Cernia, eds.), Vol. 1, p. 375. Wiley, New York, 1967.
125. G. Holden, E. T. Bishop, and N. R. Legge, *J. Polym. Sci., Part C* **26**, 37 (1969).
126. T. Homma, N. Tagata, and H. Hibino, *Nippon Gomu Kyokaishi* **41**, 242 (1968).
127. J. R. Hopper, *Rubber Chem. Technol.* **40**, 463 (1967).
128. Y. Ide and J. L. White, *J. Appl. Polym. Sci.* **18**, 2997 (1974).
128a. Y. Ide and J. L. White, *J. Appl. Polym. Sci.* **22**, 1061 (1978).
129. G. B. Jeffreys, *Proc. R. Soc., Ser. A* **102**, 161 (1922).
130. M. F. Johnson, W. W. Evans, I. Jordan, and J. D. Ferry, *J. Colloid Sci.* **7**, 498 (1952).
131. J. P. Joule, *Philos. Trans. R. Soc. London, Ser. A* **149**, 91 (1859).
132. L. Kacir and Z. Tadmor, *Polym. Eng. Sci.* **12**, 387 (1972); L. Kacir, Z. Tadmor, and E. Broyer, *Polym. Eng. Sci.* **12**, 378 (1972).
133. J. R. Katz, *Naturwissenschaften* **13**, 410 (1925).
134. M. Kaufman, *Plast. Polym.* **37**, 243 (1969).
135. H. G. Kim and L. Mandelkern, *J. Polym. Sci., Part A-2* **6**, 181 (1968).
136. R. G. King, *Rheol. Acta* **5**, 35 (1966).
137. I. Klein and D. I. Marshall, *Polym. Eng. Sci.* **6**, 198 (1966).
138. T. Kotaka and J. L. White, *Trans. Soc. Rheol.* **17**, 587 (1973).
139. T. Kotaka, M. Kurata, and M. Tamura, *J. Appl. Phys.* **30**, 1705 (1969).
140. G. Kraus, ed., "Reinforcement of Elastomers." Reinhold, New York, 1965.
141. G. Kraus, *Rubber Chem. Technol.* **28**, 1070 (1965).
141a. G. Kraus, *Polym. Lett.* **8**, 601 (1970); *Rubber Chem. Technol.* **49**, 199 (1971).
142. G. Kraus and J. Dugone, *Ind. Eng. Chem.* **47**, 1809 (1955).
143. G. Kraus and J. T. Gruver, *J. Polym. Sci., Part A* **3**, 105 (1965).

144. G. Kraus and J. T. Gruver, *Trans. Soc. Rheol.* **9**(2), 17 (1965).
145. G. Kraus and J. T. Gruver, *J. Appl. Polym. Sci.* **11**, 1581 (1967).
146. G. Kraus and K. W. Rollman, *J. Appl. Polym. Sci.* **8**, 2585 (1964).
147. G. Kraus, F. E. Naylor, and K. W. Rollman, *J. Polym. Sci., Part A-2* **9**, 1839 (1971).
148. I. M. Krieger, *Trans. Soc. Rheol.* **12**, 5 (1968).
149. I. M. Krieger and H. Elrod, *J. Appl. Phys.* **24**, 134 (1953).
150. I. M. Krieger and S. H. Maron, *J. Appl. Phys.* **23**, 147 (1952).
151. W. R. Krigbaum and R. J. Roe, *J. Polym. Sci.* **2**, 4391 (1964).
152. F. W. Kroesser and S. Middleman, *Polym. Eng. Sci.* **5**, 230 (1965).
153. W. Kuhn, O. Kunzle, and A. Preissman, *Helv. Chim. Acta* **30**, 307, 464, 839 (1947).
154. S. L. Kwolek, U.S. Patent 3,671,542 (1972).
155. H. L. LaNieve and D. C. Bogue, *J. Appl. Polym. Sci.* **12**, 353 (1968).
156. H. Leaderman, *Ind. Eng. Chem.* **35**, 374 (1943).
157. H. Leaderman, "Elastic and Creep Properties of Filamentous Materials and Other High Polymers." Textile Found., Washington, D.C., 1973.
158. H. Leaderman, R. G. Smith, and L. C. Williams, *J. Polym. Sci.* **36**, 233 (1959).
159. B. L. Lee and J. L. White, *Trans. Soc. Rheol.* **18**, 467 (1974).
160. B. L. Lee and J. L. White, *Trans. Soc. Rheol.* **19**, 481 (1975).
161. R. S. Lenk, "Plastics Rheology." Wiley (Interscience), New York, 1968.
161a. V. M. Lobe and J. L. White, unpublished research (1975–1978).
162. A. S. Lodge, *Proc. Int. Rheol. Congr., 2nd* p. 229 (1954).
163. A. S. Lodge, *Trans. Faraday Soc.* **53**, 120 (1956).
164. A. S. Lodge, "Elastic Liquids." Academic Press, New York, 1964.
165. I. F. MacDonald, B. D. Marsh, and E. Ashare, *Chem. Eng. Sci.* **24**, 1615 (1969); also *Trans. Soc. Rheol.* **17**, 535 (1973); **18**, 299, 313 (1974).
166. C. Macintosh, Engl. Patent 4,804 (1823).
167. J. M. McKelvey, "Polymer Processing." Wiley, New York, 1963.
168. C. W. Macosko and J. M. Starita, *SPE J.* **27**(Nov.), 38 (1971).
169. C. W. Macosko and F. C. Weissert, *in* "Rubber and Related Products, New Methods for Testing and Analyzing" (W. H. King and T. H. Rogers, eds.), p. 127. Amer. Soc. Test. Mater., Philadelphia, Pennsylvania, 1974.
170. B. Maddock, *SPE Antec Tech. Pap.* **15**, 383 (1959).
171. L. Mandelkern, M. Tryon, and F. A. Quinn, *J. Polym. Sci.* **19**, 77 (1956).
172. H. Markovitz and R. B. Williamson, *Trans. Soc. Rheol.* **1**, 25 (1957).
173. D. I. Marshall, I. Klein, and R. H. Uhl. *SPE J.* **21**, 1192 (1965).
173a. B. Martin, *Plast. Polym.* **38**, 113 (1970).
174. R. Martino, *Mod. Plast.* **51**(12), 50 (1974).
175. B. Marzetti, *Rubber Age (N.Y.)* **15**, 454 (1923).
176. T. Masuda, Ph.D. Thesis, Kyoto Univ., Kyoto, 1973.
177. T. Masuda, K. Kitagawa, T. Inoue, and S. Onogi, *Macromolecules* **3**, 116 (1970).
178. T. Masuda, Y. Ohta, and S. Onogi, *Macromolecules* **4**, 763 (1971); T. Masuda, Y. Ohta, S. Onogi, and Y. Nakagawa, *Polym. J.* **3**, 92 (1972).
179. B. Maxwell and R. Chartoff, *Trans. Soc. Rheol.* **9**(1), 41 (1965).
180. J. C. Maxwell, *Philos. Trans. R. Soc. London* **157**, 249 (1866); see also "Scientific Papers," Vol. 2, p. 26. Dover, New York, 1890.
181. J. C. Maxwell, Constitution of Bodies *in* "Encyclopedia Britannica," 1878; see also "Scientific Papers," Vol. 2, p. 616. Dover, New York, 1890.
181a. A. Medalia, *J. Colloid. Interf. Sci.* **24**, 393 (1967).
181b. A. Medalia, *J. Colloid. Interf. Sci.* **32**, 115 (1970).
182. J. Meissner, *Trans. Soc. Rheol.* **16**, 405 (1972).
183. R. A. Mendelson, F. L. Finger, and E. B. Bagley, *J. Polym. Sci., Part C* **35**, 177 (1971).

184. A. P. Metzger and J. Knox, *Trans. Soc. Rheol.* **9**(1), 13 (1965).
185. A. B. Metzner, W. T. Houghton, R. A. Sailor, and J. L. White, *Trans. Soc. Rheol.* **5**, 133 (1961).
186. A. B. Metzner, J. L. White, and M. M. Denn, *Chem. Eng. Prog.* **62**(12), 81 (1966).
187. A. B. Metzner, J. L. White, and M. M. Denn, *AIChE J.* **12**, 863 (1966).
188. S. Middleman, *Trans. Soc. Rheol.* **9**(1), 83 (1965).
189. S. Middleman, "The Flow of High Polymers." Wiley, New York, 1968.
190. S. Middleman, *Trans. Soc. Rheol.* **13**, 123 (1969).
191. M. J. Miller and E. B. Christiansen, *AIChE J.* **18**, 600 (1972).
192. N. Minagawa and J. L. White, *Appl. Polym. Sci.* **20**, 501 (1976).
193. N. Minagawa and J. L. White, *Polym. Eng. Sci.* **15**, 825 (1975).
194. M. Mooney, *J. Rheol.* **2**, 210 (1931).
195. M. Mooney, *Ind. Eng. Chem., Anal. Ed.* **6**, 147 (1934).
196. M. Mooney, *Physics* (*N.Y.*) **7**, 413 (1936).
197. M. Mooney, *Ind. Eng. Chem., Anal. Ed.* **17**, 514 (1945).
198. M. Mooney, *J. Colloid Sci.* **2**, 69 (1947).
199. M. Mooney, *J. Colloid Sci.* **6**, 96 (1951).
200. M. Mooney, *in* "Rheology: Theory and Applications" (F. R. Eirich, ed.), Vol. 2. Academic Press, New York, 1958.
201. M. Mooney, *Proc. Int. Rubber Conf., Washington, D.C.* p. 368 (1959).
202. M. Mooney, *Rubber Chem. Technol.* **35**(5), XXVII (1962).
203. M. Mooney and S. A. Black, *J. Colloid Sci.* **7**, 204 (1952).
204. M. Mooney and R. H. Ewart, *Physics* (*N.Y.*) **5**, 350 (1934).
205. M. Mooney and W. E. Wolstenhome, *J. Appl. Phys.* **25**, 1098 (1954).
206. M. Morton, J. E. McGrath, and P. Juliano, *J. Polym. Sci., Part C* **25**, 99 (1969).
207. L. Mullins, *J. Phys. Chem.* **54**, 239 (1950).
208. N. Nakajima and E. A. Collins, *Polym. Eng. Sci.* **14**, 137 (1974).
209. N. Nakajima and E. A. Collins, *Rubber Chem. Technol.* **47**, 333 (1974).
209a. N. Nakajima, H. H. Bowerman, and E. A. Collins, *J. Appl. Polym. Sci.* **21**, 3063 (1977).
210. G. Natta, G. Crespi, G. Guzzetta, S. Leghisa, and F. Sabioni, *Rubber Plast. Age* **42**, 402 (1961).
211. K. Ninomiya, *J. Colloid Sci.* **14**, 49 (1959).
212. K. Ninomiya and I. Furuta, *Rubber Chem. Technol.* **44**, 1395 (1971).
213. K. Ninomiya and G. Yasuda, *Rubber Chem. Technol.* **40**, 493 (1967).
214. W. Noll, *J. Ration. Mech. Anal.* **4**, 3 (1955).
215. W. Noll, *Arch. Ration. Mech. Anal.* **2**, 197 (1958).
216. H. Nusselt, *Verh. Dtsch. Ing.* **53**, 1750 (1909).
216a. K. Oda, J. L. White, and E. S. Clark, *Polym. Eng. Sci.* **18**, 25 (1978).
217. J. G. Oldroyd, *Proc. R. Soc., Ser. A* **200**, 523 (1950).
218. J. G. Oldroyd, *Q. J. Mech. Appl. Math.* **4**, 271 (1951).
219. J. G. Oldroyd, *Proc. R. Soc., Ser. A* **245**, 278 (1958).
220. S. Onogi, T. Masuda, and K. Kitagawa, *Macromolecules* **3**, 109 (1970).
221. S. Onogi, T. Matsumoto, and Y. Warashina, *Trans. Soc. Rheol.* **17**, 175 (1973).
222. K. Osaki, M. Tamura, T. Kotaka, and M. Kurata, *J. Phys. Chem.* **69**, 3642 (1965).
223. D. E. Oxley and D. J. H. Sandiford, *Plast. Polym.* **39**, 288 (1971).
224. S. P. Papkov, V. G. Kulichin, V. D. Kalmykoua, and A. Y. Malkin, *J. Polym. Sci., Polym. Phys.* **12**, 1753 (1974).
225. D. R. Paul, J. E. St. Laurence, and J. H. Troell, *Polym. Eng. Sci.* **10**, 70 (1970).
226. J. Pawlowski, *Kolloid-Z.* **130**, 129 (1953).
227. J. R. A. Pearson, "Mechanical Principles of Polymer Melt Processing." Pergamon, Oxford, 1966.

228. W. Philippoff and F. H. Gaskins, *J. Polym. Sci.* **21,** 205 (1956).
229. W. Philippoff and F. H. Gaskins, *Trans. Soc. Rheol.* **2,** 263 (1958).
230. W. T. Pigott, *Trans. ASME* **73,** 947 (1951).
231. M. Pike and W. F. Watson, *J. Polym. Sci.* **9,** 229 (1952).
232. G. H. Piper and J. R. Scott, *J. Sci. Instrum.* **22,** 206 (1945).
233. I. Pliskin, *Rubber Chem. Technol.* **46,** 1218 (1973).
234. I. Pliskin and N. Tokita, *J. Appl. Polym. Sci.* **16,** 472 (1972).
235. A. Plochocki, *J. Appl. Polym. Sci.* **16,** 987 (1972).
236. W. F. O. Pollett, *Br. J. Appl. Phys.* **6,** 199 (1955).
237. R. S. Porter and J. F. Johnson, *J. Phys. Chem.* **63,** 202 (1959); *J. Appl. Phys.* **35,** 15, 3149 (1964).
238. R. S. Porter and J. F. Johnson, *Proc. Int. Rheol. Congr., 4th* **2,** 467, 479 (1965).
239. R. S. Porter and J. F. Johnson, *Chem. Rev.* **66,** 1 (1966).
240. B. Rabinowltch, *Z. Phys. Chem., Abt. A* **145,** 1 (1929).
241. M. Reiner, "Deformation Strain and Flow," 2nd Ed. Lewis, London, 1960. 1st Ed., 1948.
242. M. Reiner, *Phys. Today* **17**(Jan.), 62 (1964).
243. O. Reynolds, *Philos. Trans. R. Soc. London, Ser. A* **174,** 935 (1883).
244. R. S. Rivlin, *J. Appl. Phys.* **18,** 444 (1947).
245. R. S. Rivlin, *J. Ration. Mech. Anal.* **5,** 179 (1956).
246. R. S. Rivlin and J. L. Ericksen, *J. Ration. Mech. Anal.* **4,** 323 (1955).
247. J. E. Roberts, *in* "Karl Weissenberg 80th Birthday Celebration Essay" (J. Garris, ed.), p. East Afr. Lit. Bur. Nairobi, Kenya, 1973.
248. H. Rouse, "History of Hydraulics." Dover, New York, 1957.
249. H. S. Rowell and D. Finlayson, *Engineering* **114,** 606 (1922); **126,** 249, 385 (1928).
250. R. J. Russell, Ph.D Thesis, Univ. of London, London, 1946.
251. R. J. Russell, *in* "Karl Weissenberg 80th Birthday Celebration Essays" (J. Harris, ed.), p. East Afr. Lit. Bur., Nairobi, Kenya, 1973.
252. G. Sadakne and J. L. White, *J. Appl. Polym. Sci.* **17,** 453 (1973).
253. K. Sakamoto, N. Ishida, and Y. Fukusawa, *J. Polym. Sci., Part A-2* **6,** 1999 (1968).
254. J. G. Savins, G. C. Wallick, and W. R. Foster, *Soc. Pet. Eng. J.* **3**(1), 14 (1963).
255. H. Schlichting, "Boundary Layer Theory," 2nd Ed. McGraw-Hill, New York, 1960.
256. W. J. Schrenk and T. Alfrey, *Polym. Eng. Sci.* **9,** 939 (1969).
257. W. Schrenk and T. Alfrey, *SPE J.* **29**(June), 38 (1973); **29**(July), 43 (1973).
258. J. R. Shaefgen and P. J. Flory, *J. Am. Chem. Soc.* **70,** 2709 (1948).
259. C. K. Shih, *Trans. Soc. Rheol.* **15,** 759 (1971).
260. R. Simha, *J. Appl. Phys.* **23,** 1020 (1952).
261. A. H. P. Skelland, "Non-Newtonian Flow and Heat Transfer." Wiley, New York, 1967.
262. J. C. Slattery, *AIChE J.* **11,** 831 (1965).
263. P. P. A. Smit, *Rheol. Acta* **8,** 277 (1969).
264. D. H. Smith and R. L. Christy, *Rubber Chem. Technol.* **45,** 1434 (1972).
265. J. H. Southern and R. L. Ballman, *Appl. Polym. Symp.* **20,** 175 (1973).
266. J. H. Southern and R. S. Porter, *J. Macromol. Sci., Phys.* **4,** 541 (1970); *J. Appl. Polym. Sci.* **14,** 2305 (1970).
267. T. W. Spriggs, J. D. Huppler, and R. B. Bird, *Trans. Soc. Rheol.* **10**(1), 191 (1966).
267a. J. E. Spruiell and J. L. White, *Polym. Eng. Sci.* **15,** 660 (1975).
268. P. H. Squires, *SPE Trans.* **3,** 7 (1964).
269. J. Stefan, *Sitzungsber. Kaiserl. Akad. Wiss. Wien* **69,** 713 (1874).
269a. J. Stevenson, *AIChE J.* **18,** 540 (1972).
270. S. C. Stillwagon, *India Rubber World* (Oct. 1), 50 (1939).
271. G. G. Stokes, *Trans. Cambridge Philos. Soc.* **8,** 286 (1845).
272. Z. Tadmor, *Polym. Eng. Sci.* **6,** 185 (1966).

273. Z. Tadmor, *Polym. Eng. Sci.* **6**, 203 (1966).
274. Z. Tadmor and I. Klein, "Engineering Principles of Plasticating Extrusion." Van Nostrand-Reinhold, New York, 1970.
275. R. I. Tanner, *Trans. Soc. Rheol.* **12**, 155 (1968).
276. R. I. Tanner, *J. Polym. Sci., Part A-2* **8**, 2067 (1970).
277. R. I. Tanner, *Trans. Soc. Rheol.* **17**, 365 (1973).
278. R. H. Taylor, J. H. Fielding, and M. Mooney, *Symp. Rubber Test., ASTM* p. 36 (1947).
279. A. V. Tobolsky, *J. Am. Chem. Soc.* **74**, 3786 (1952).
280. A. V. Tobolsky, "Properties and Structure of Polymers." Wiley, New York, 1960.
281. N. Tokita and I. Pliskin, *Rubber Chem. Technol.* **46**, 1166 (1973).
282. N. Tokita and J. L. White, *J. Appl. Polym. Sci.* **10**, 1011 (1966).
283. H. Toor, *Ind. Eng. Chem.* **48**, 922 (1956); *Trans. Soc. Rheol.* **1**, 177 (1957).
284. J. Tordella, *J. Appl. Phys.* **27**, 454 (1956); *Trans. Soc. Rheol.* **1**, 203 (1957); *Rheol. Acta* **1**, 216 (1958); *Trans. Soc. Rheol.* **7**, 215 (1963).
285. L. R. G. Treloar, "Physics of Rubber Elasticity," 2nd Ed. Oxford Univ. Press (Clarendon), London and New York, 1958).
286. C. Truesdell and R. A. Toupin, The Classical Field Theories, *in* "Handbuch der Physik," Vol. III/1. Springer-Verlag, Berlin and New York, 1960.
287. G. R. Vila, *Ind. Eng. Chem.* **36**, 113 (1944).
287a. G. V. Vinogradov, *Pure Appl. Chem.* **39**, 115 (1974).
288. G. V. Vinogradov, N. L. Insarova, B. B. Boiko, and E. K. Borisenkora, *Polym. Eng. Sci.* **12**, 323 (1972).
289. G. V. Vinogradov, A. Ya. Malkin, E. P. Plotnikova, O. Y. Sabasi, and N. E. Nikolayava, *Int. J. Polym. Mater.* **2**, 1 (1972).
289a. G. V. Vinogradov, B. V. Radushkevich, and V. P. Fikham, *J. Polym. Sci., Part A-2* **8**, 1 (1970).
289b. G. V. Vinogradov, A. Ya. Malkin, V. V. Volosevitch, V. P. Shatalov, and V. P. Yudin, *J. Polym. Sci. Polym. Phys. Ed.* **13**, 1721 (1975).
290. M. Wagner, *J. Appl. Polym. Sci.* **19**, 757 (1970).
291. M. H. Walters and D. N. Keyte, *Rubber Chem. Technol.* **38**, 62 (1965).
292. W. F. Watson, *Trans. Inst. Rubber Ind.* **29**, 32 (1954).
293. W. F. Watson, *Ind. Eng. Chem.* **47**, 1281 (1956); *Proc. Rubber Technol. Conf., 3rd, London* p. 553 (1954).
294. K. Weissenberg, *Nature (London)* **159**, 310 (1947).
295. K. Weissenberg, *Proc. Int. Rheol. Congr., 1st* 1. 29 (1949).
296. K. Weissenberg, *Proc. Int. Rheol. Congr., 1st* 2. 114 (1949).
297. F. C. Weissert and B. L. Johnson, *Rubber Chem. Technol.* **40**, 590 (1967).
298. J. L. White, *J. Appl. Polym. Sci.* **8**, 2339 (1964).
299. J. L. White, *Rubber Chem. Technol.* **42**, 257 (1969).
300. J. L. White, *Proc. Int. Rheol. Congr., 5th* **3**, 17 (1970).
301. J. L. White, *Rubber Chem. Technol.* **44**(5), G70 (1971).
302. J. L. White, *Appl. Polym. Symp.* **20**, 155 (1973).
303. J. L. White, *J. Inst. Rubber Ind.* **8**(4), 148 (1974).
304. J. L. White, *Rheol. Acta* **14**, 600 (1975).
304a. J. L. White, *Rubber Chem. Technol.* **50**, 163 (1977).
305. J. L. White and J. W. Crowder, *J. Appl. Polym. Sci.* **18**, 1013 (1974).
305a. J. L. White and A. Kondo, *J. Non-Newt. Fluid Mech.* **3**, 41 (1977).
306. J. L. White and B. L. Lee, *Polym. Eng. Sci.* **15**, 481 (1975).
307. J. L. White and A. B. Metzner, *J. Appl. Polym. Sci.* **8**, 1367 (1963).
307a. J. L. White and J. F. Roman, *J. Appl. Polym. Sci.* **20**, 1005 (1976); **21**, 869 (1977).
308. J. L. White and N. Tokita, unpublished research (1964–1966).

309. J. L. White and N. Tokita, *J. Appl. Polym. Sci.* **9,** 1929 (1965).
310. J. L. White and N. Tokita, *J. Appl. Polym. Sci.* **11,** 321 (1967).
311. J. L. White and N. Tokita, *J. Phys. Soc. Jpn.* **22,** 719 (1967); **24,** 436 (1968).
312. J. L. White and N. Tokita, *J. Appl. Polym. Sci.* **12,** 1589 (1968).
313. J. L. White, R. C. Ufford, K. C. Dharod, and R. L. Price, *J. Appl. Polym. Sci.* **16,** 1313 (1972).
314. E. Wiechert, *Ann. Phys. Chem.* **90,** 335 (1893).
315. M. L. Williams, R. F. Landel, and J. D. Ferry, *J. Am. Chem. Soc.* **77,** 3701 (1955).
316. R. F. Wolf, *India Rubber World* (Aug. 1), 39 (1936).
317. W. E. Wolstenholme, *Rubber Chem. Technol.* **38,** 769 (1965).
318. C. R. Wylie, "Advanced Engineering Mathematics." McGraw-Hill, New York, 1951.
319. M. Yamamoto, *J. Phys. Soc. Jpn.* **11,** 413 (1956).
320. M. Yamamoto, *J. Phys. Soc. Jpn.* **12,** 1148 (1957).
321. M. Yamamoto, *Jpn. J. Appl. Phys.* **8,** 1252 (1969).
322. M. Yamamoto, *Trans. Soc. Rheol.* **15,** 331, 783 (1971).
323. M. Yamamoto and J. L. White, *J. Polym. Sci., Part A-2* **9,** 1390 (1971).
324. H.-C. Yeh and L. V. McIntire, *Trans. Soc. Rheol.* **16,** 711 (1972).
325. N. V. Zakharenko, F. S. Tolstukhina, and G. M. Bartenev, *Rubber Chem. Technol.* **35,** 326 (1962).
326. H. Zamodits and J. R. A. Pearson, *Trans. Soc. Rheol.* **13,** 357 (1969).
327. L. J. Zapas, *J. Res. Natl. Bur. Stand., Sect. A* **70,** 525 (1966).
328. L. J. Zapas and T. Craft, *J. Res. Natl. Bur. Stand., Sect. A* **69,** 541 (1965).
329. S. Zaremba, *Bull. Acad. Cracov.* p. 594 (1903).

Chapter 7

Vulcanization

A. Y. CORAN

MONSANTO INDUSTRIAL CHEMICALS COMPANY
RUBBER CHEMICALS RESEARCH LABORATORIES
AKRON, OHIO

Copyright © 1978 by Academic Press, Inc.
All rights of reproduction in any form reserved.
ISBN 0-12-234360-3

I. General Considerations

A. DEFINITION OF VULCANIZATION

Vulcanization is a process by which elastomeric materials are generally prepared; it consists of the formation of a molecular network by a chemical tying together of independent chain molecules. The resulting rubbers retract forcibly to their approximately original shape after large mechanically imposed deformations. Vulcanization thus is an intermolecular reaction which increases the retractive force and reduces the amount of permanent deformation remaining after removal of the deforming force, i.e., increases elasticity while it decreases plasticity.

According to the theory of rubber elasticity [1], the retractive force resisting a deformation is proportional to the number of network-supporting polymer chains per unit volume of elastomer. A supporting chain is a segment of polymer backbone between network junctures. An increase in the number of junctures gives an increase in the number of supporting chains. In an unvulcanized high polymer, above its melting point, only molecular chain entanglements constitute junctures and their number per molecule increases with molecular weight. Vulcanization usually produces network junctures by the insertion of chemical cross-links between polymer chains. These cross-links may be chains of sulfur atoms, single sulfur atoms, carbon–carbon bonds, polyvalent organic radicals, or polyvalent metal ions.

B. THE EFFECTS OF VULCANIZATION ON VULCANIZATE PROPERTIES

Major effects of vulcanization [2–7] are illustrated by the idealization in Fig. 1. It should be noted that the static modulus increases with vulcanization

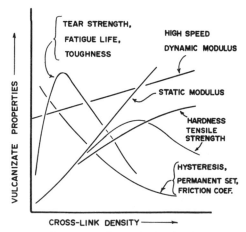

Fig. 1. The effects of vulcanization.

to a greater extent than the dynamic modulus [2]. The dynamic modulus is a composite of viscous and elastic responses, whereas the static modulus is a measure of the elastic component alone. Vulcanization, then, causes a shift from viscous or plastic behavior to elasticity.

Tear strength, fatigue life, and toughness are related to the *energy* to break. These properties increase with small amounts of cross-links but are reduced with increasing cross-link formation. Hysteresis diminishes with increasing cross-link formation and is a measure of the deformation energy that is not stored or borne by the network chains, but instead is converted to heat. Properties related to energy to break, then, increase with increases in the number of network chains and hysteresis; but since hysteresis decreases as more network chains are developed, the energy-to-break-related properties peak at some intermediate cross-link density.

It should be noted that the properties of Fig. 1 are not functions of cross-link density only. They are also affected by the type of cross-link, the nature of the polymer, the type and amount of filler, etc.

Reversion is a term applied to the loss of network structures by nonoxidative thermal aging. It is usually associated with isoprene rubbers vulcanized by sulfur. It can be the result of an overly long vulcanization or of hot aging of thick sections. It is most severe at temperatures above 150°C in vulcanizates containing a large number of polysulfidic cross-links. It is thought that cross-links break and reform into cyclic structures. In isoprene rubbers little or no chain scission occurs with reversion [7, 9, 10], which occurs probably by an ionic mechanism and is especially troublesome in the presence of basic amine compounds.

C. PARAMETERS OF THE VULCANIZATION PROCESS

Critical parameters related to the process of vulcanization are the time elapsed before it starts, the rate at which it occurs, and the extent. There must be sufficient delay (called scorch time or scorch resistance) before the outset of vulcanization to permit mixing, forming, and molding. Afterward, the formation of cross-links should be rapid and its extent controlled. This is illustrated by Fig. 2.

Scorch resistance is usually measured by the time at a given temperature required for the onset of cross-link formation as measured by an abrupt increase in viscosity. The Mooney viscometer is mostly used. It consists of a rotating disk in a cavity which contains the test rubber. The temperature is selected to be a typical, or average, processing temperature. The extent and rate of vulcanization are measured by determination of mechanical properties after various cure periods.

In recent years, specialized rheometers (or cure meters) have been used. These devices are capable of measuring the extent of cure of a single sample

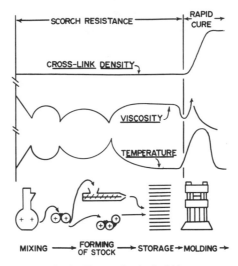

Fig. 2. Thermorheological history.

at a given temperature on a continuous basis. The most popular of these instruments are of the oscillating disk variety [11]. An instrument of this type is illustrated schematically in Fig. 3. The test sample of rubber is cured automatically in a cavity, while a disk embedded in the sample rotates in an oscillatory fashion through a small arc. The frequency of oscillation can be varied, as can the size of the arc. The resistance to oscillation is measured and recorded as a function of time on a so-called rheometer chart like the one shown in Fig. 4. More recently, cure simulators have been introduced [12]. They are modifications of the oscillating disk rheometer which allow the cure temperature to be programmed. The cure temperature–time profile of an industrial

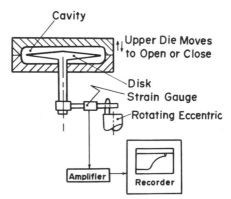

Fig. 3. Schematic diagram of cure meter.

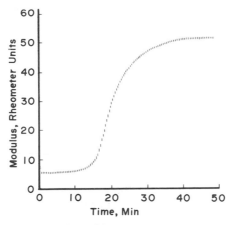

Fig. 4. Rheometer chart.

mold can be imposed on the curing cavity of the rheometer. The test sample can then be cured in the rheometer under the same conditions as those encountered in the manufacturing plant. The temperature–time profile is displayed along with the extent of cure.

Using the newer techniques together with Mooney scorch measurements, a comprehensive picture of the curing characteristics of a stock can be obtained [13]. Further, samples of stocks are being pulled from every part of the plant and very rapidly tested in rheometers. Thus, the use of rheometers is no longer limited to the laboratory, but has been extended to the routine assessment of individual operations in the plant.

II. Sulfur Vulcanization

Sulfur was the first agent used to vulcanize the first commercial elastomer, natural rubber (NR) [14]. Vulcanization was accomplished by mixing 8 parts of sulfur per 100 parts (phr) of rubber, and required 5 hr at 140°C. The addition of zinc oxide reduced the time to 3 hr. The use of accelerators in concentrations as low as 0.5 phr has since reduced the time to as little as 2–5 min. As a result, elastomer vulcanization by sulfur with no accelerator is no longer of commercial significance.

Our knowledge of the chemistry of unaccelerated vulcanization is somewhat confused. Many slow reactions occur during slow unaided vulcanization. Some investigators have felt that the mechanisms involve free radicals (Scheme 1) [15–18]; others have postulated mechanisms involving ions [19] (Scheme 2).

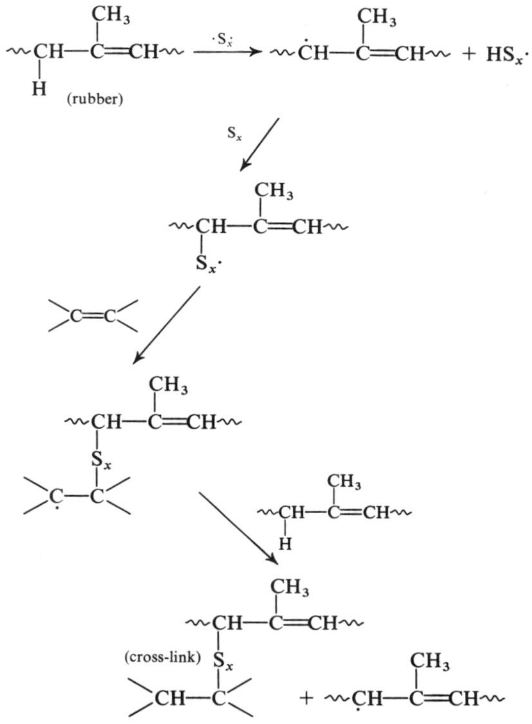

Scheme 1.

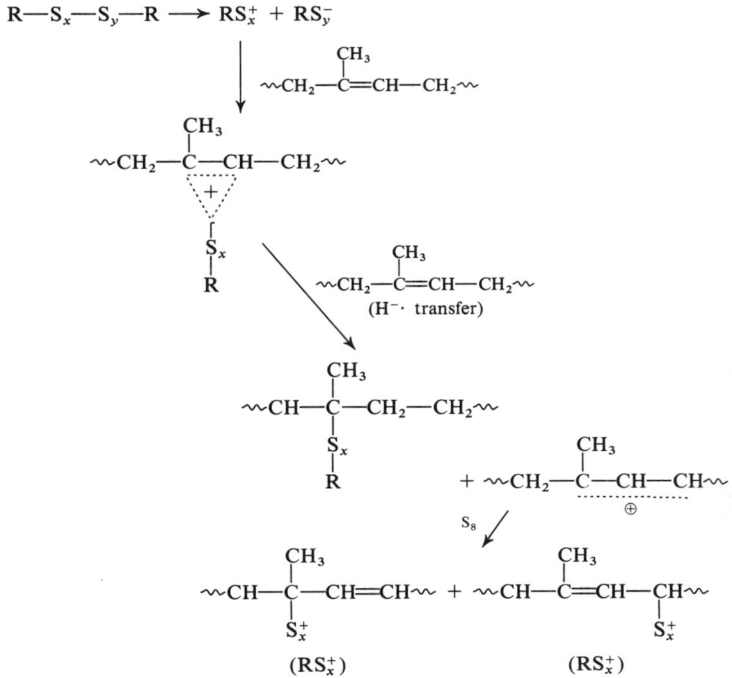

Scheme 2.

The postulation of the the intermediates

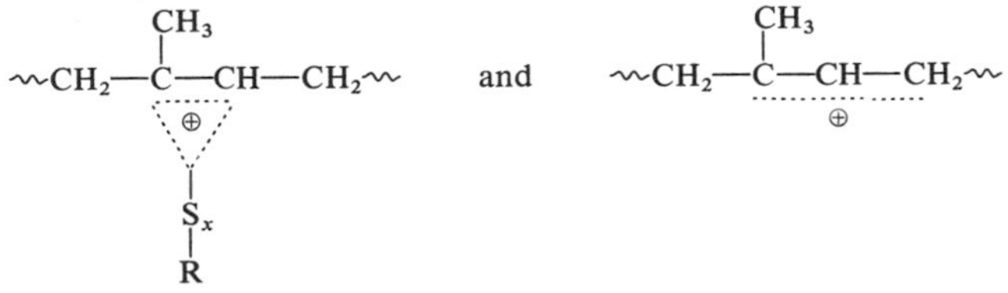

$$
\text{~CH}_2\text{—}\underset{\overset{\displaystyle\oplus}{|}}{\underset{\displaystyle \underset{R}{\overset{|}{S_x}}}{\overset{CH_3}{C}}}\text{—CH—CH}_2\text{~} \qquad \text{and} \qquad \text{~CH}_2\text{—}\underset{\displaystyle \oplus}{\overset{CH_3}{C}}\text{—CH—CH}_2\text{~}
$$

was necessary to explain the fact that model compound reactions gave both unsaturated products with sulfur atoms connected to secondary and tertiary carbon atoms.

III. Accelerated Sulfur Vulcanization

A. Types of Elastomers and Accelerators Used in Accelerated Sulfur Vulcanization

Accelerated sulfur vulcanization is suitable for the following types of elastomers:

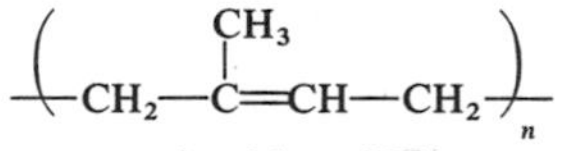

$$
\left(\text{CH}_2\text{—}\overset{CH_3}{\underset{}{C}}\text{=CH—CH}_2\right)_n \qquad\qquad \left(\text{CH}_2\text{—CH=CH—CH}_2\right)_n
$$

natural rubber (NR) or   polybutadiene (BR)
synthetic isoprene rubber (IR)

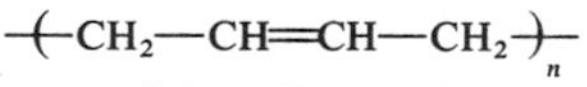

$$
\left(\text{CH}_2\text{—CH=CH—CH}_2\right)_n\left(\text{CH}_2\text{—CH}\right)_{0.3n}
$$

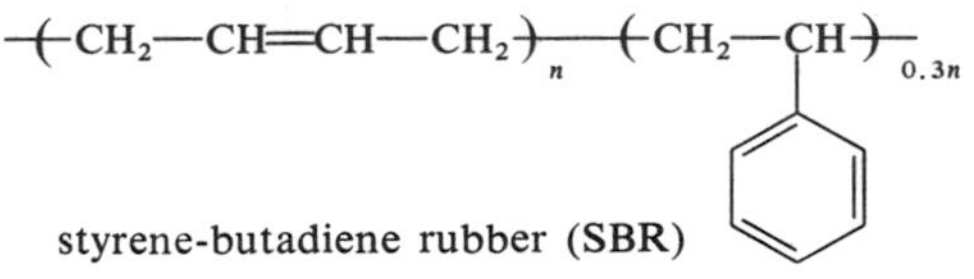

styrene-butadiene rubber (SBR)

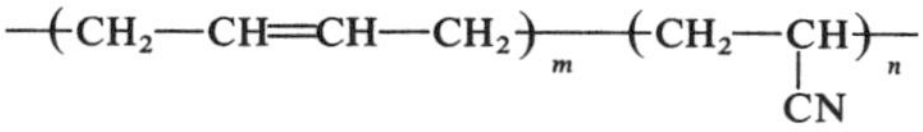

$$
\left(\text{CH}_2\text{—CH=CH—CH}_2\right)_m\left(\text{CH}_2\text{—}\underset{CN}{\overset{}{CH}}\right)_n
$$

nitrile rubber (NBR)

$$
\left(\underset{CH_3}{\overset{CH_3}{\underset{}{C}}}\text{—CH}_2\right)_n\left(\text{CH}_2\text{—CH=CH—CH}_2\right)_{0.1n}
$$

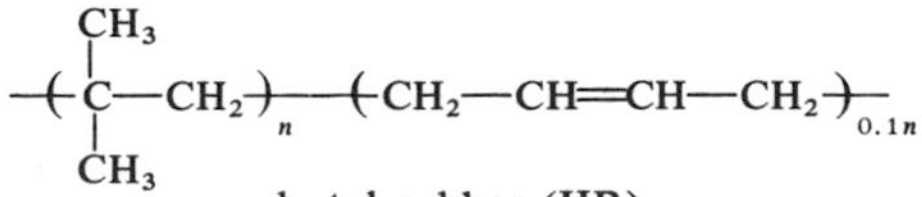

butyl rubber (IIR)

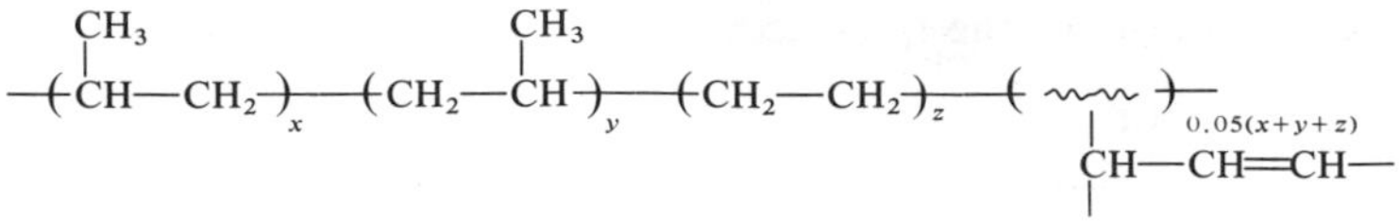

$$-(\underset{\overset{|}{CH_3}}{CH}-CH_2)_x-(CH_2-\underset{\overset{|}{CH_3}}{CH})_y-(CH_2-CH_2)_z-(\sim\!\!\sim)_{0.05(x+y+z)}$$
$$\underset{\overset{|}{CH-CH=CH-}}{}$$

ethylene propylene diene monomer rubber (EPDM)

The reactive portions of all of these elastomers can be simply represented by $>CH-CH=CH-$.

TABLE I[a]

Zinc oxide	2–10
Fatty acid	1–4
Sulfur	0.5–4
Accelerator	0.5–2

[a] Concentration in parts per 100 parts of elastomer.

A typical recipe for the vulcanization system is shown in Table I. Zinc oxide and fatty acid constitute the activator system, where zinc ion is made soluble by salt formation between the acid and oxide. Accelerators are illustrated by the following examples.

Thiazole types

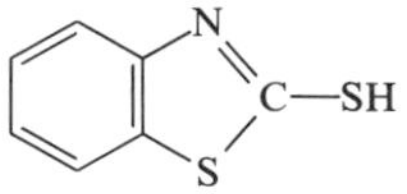

2-mercaptobenzothiazole (MBT)

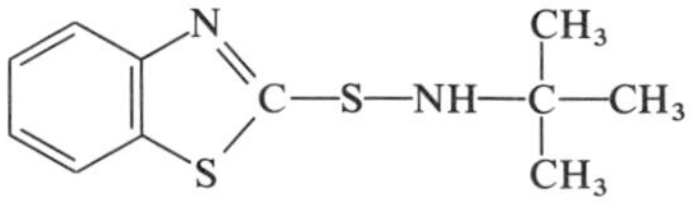

N-ε-butylbenzothiazole-2-sulfenamide (BBS)

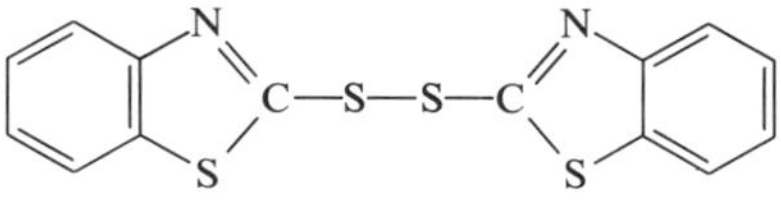

2,2′-dithiobisbenzothiazole (MBTS)

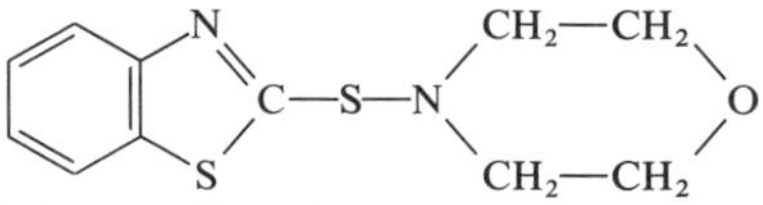

2-(morpholinothio)benzothiazole (MTB)

Dithiocarbamate types

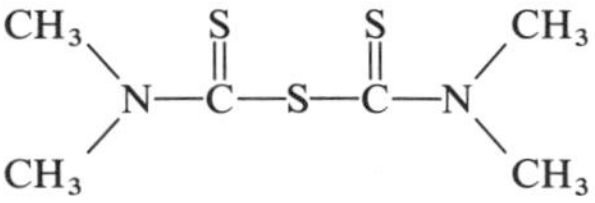

tetramethylthiuram monosulfide (TMTM)

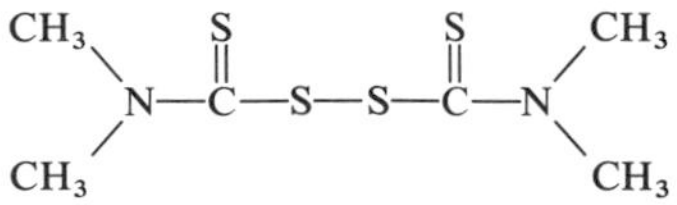

tetramethylthiuram disulfide (TMTD)

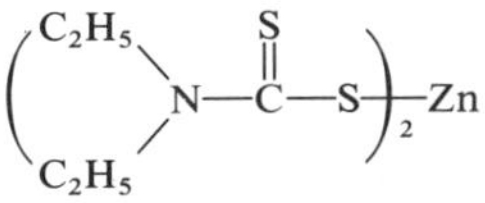

zinc diethyldithiocarbamate (ZEDC)

Amine types

diphenylguanidine (DPG)   di-*o*-tolylguanidine (DOTG)

In many cases mixtures of accelerators are used. Typically, a thiazole type is used with smaller amounts of a dithiocarbamate or an amine type. In such mixtures the two types activate one another, causing faster than average rates of vulcanization. If two accelerators of the same class are used, average characteristics will be obtained.

Different types of accelerators give vulcanization characteristics which differ in resistance to scorch (premature vulcanization) and the rate of vulcanization after cross-link formation starts. These differences are illustrated by Fig. 5, which is an idealized map of accelerator system characteristics. It should be noted that within groups or types, differences can be achieved by the selection of specific accelerators. For example, in the group of sulfenamide accelerators the scorch resistance of MTB is somewhat greater than that of BBS, but faster rates of cross-link formation can be achieved with BBS than with MTB.

Included in Fig. 5 is the effectiveness of *N*-(cyclohexylthio)phthalimide, known as PVI, as an inhibitor of premature vulcanization. This retarder is frequently used to control scorch with little or no effect on the rate of cross-

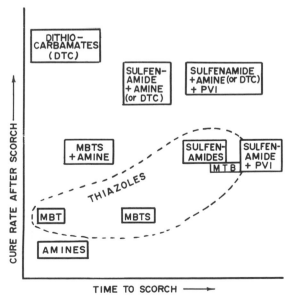

Fig. 5. Schematic acceleration map for diene rubbers.

link formation. It differs from the acidic retarders, such as salicylic acid, which also reduce the rate of cross-link formation, and from N-nitrosodiphenylamine, which is less active and is almost ineffective in the presence of phenylenediamine antidegradants [20].

B. THE CHEMISTRY OF ACCELERATED SULFUR VULCANIZATION

Accelerated sulfur vulcanization is thought to proceed by the following [21–23] steps:

(a) The accelerator reacts with sulfur to give monomeric polysulfides of the type Ac—S_x—Ac, where Ac is an organic radical derived from the accelerator. Certain initiating species may be necessary to start the reaction, which then appears to be autocatalytic.

(b) The polysulfides can interact with rubber to give polymeric polysulfides of the type rubber—S_x—Ac. During this reaction, the formation of mercaptobenzothiazole (MBT) was observed when an accelerator derived from MBT had been used. When MBT itself is used, it first disappears, then reappears during the formation of rubber polysulfides [21] (see also [23a]).

(c) The rubber polysulfides then react, either directly or through a reactive intermediate, to give cross-links or rubber polysulfides of the type rubber—S_x—rubber. If a sulfenamide accelerator is used, the reaction can be represented as in Scheme 3.

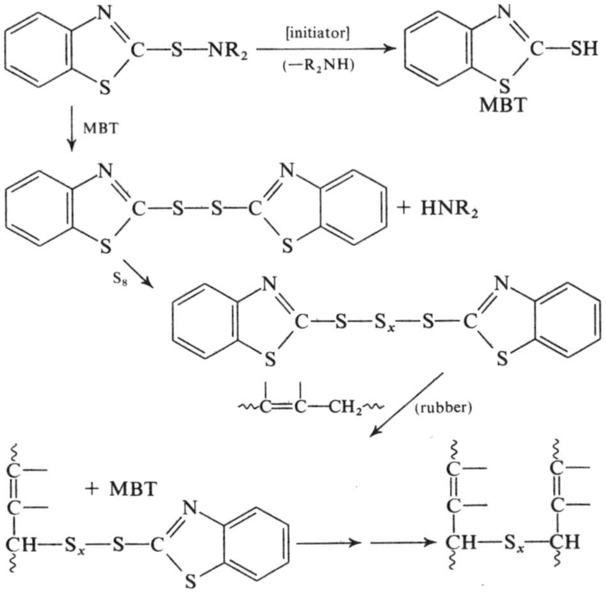

Scheme 3.

It is interesting to contrast accelerated with unaccelerated sulfur vulcanization. An obvious difference is the large amount of sulfur required to obtain a given state of cure without accelerator. Equally obvious is the increased rate of cross-link formation in the presence of an accelerator. But more subtle differences exist. Work with model olefins suggests that in the case of accelerated sulfur vulcanization, the sulfur attacks the rubber hydrocarbon almost exclusively at the allylic positions [24]. If no accelerator is used, most of the substitution is on other carbon atoms [25]. These observations suggest that the effectiveness of accelerators lies in the mobilization of the allylic positions. A mechanism for this could be written as follows [26] (see also [26a]):

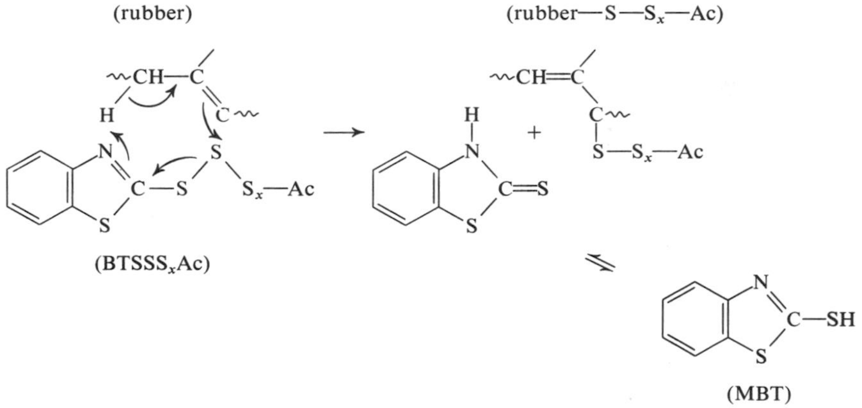

(It should be noted that an ionic mechanism similar to that for unaccelerated vulcanization could also be written [10]; however, some evidence exists against ionic mechanisms for cross-link formation [27].)

Cross-link formation could occur by the sequences of reactions given by Schemes 4–6.

Scheme 4 is the same reaction as the previously mentioned mechanism for the formation of polymeric polysulfides [26, 26a]. It requires that MBT be

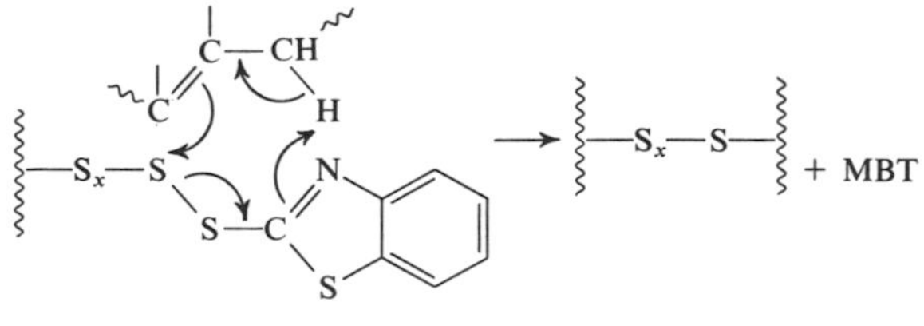

Scheme 4.

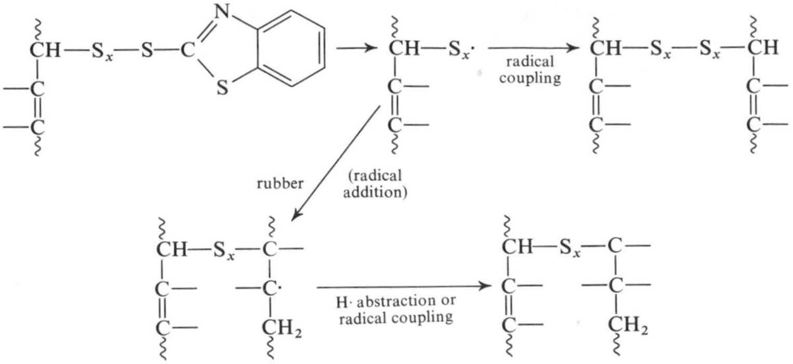

Scheme 5.

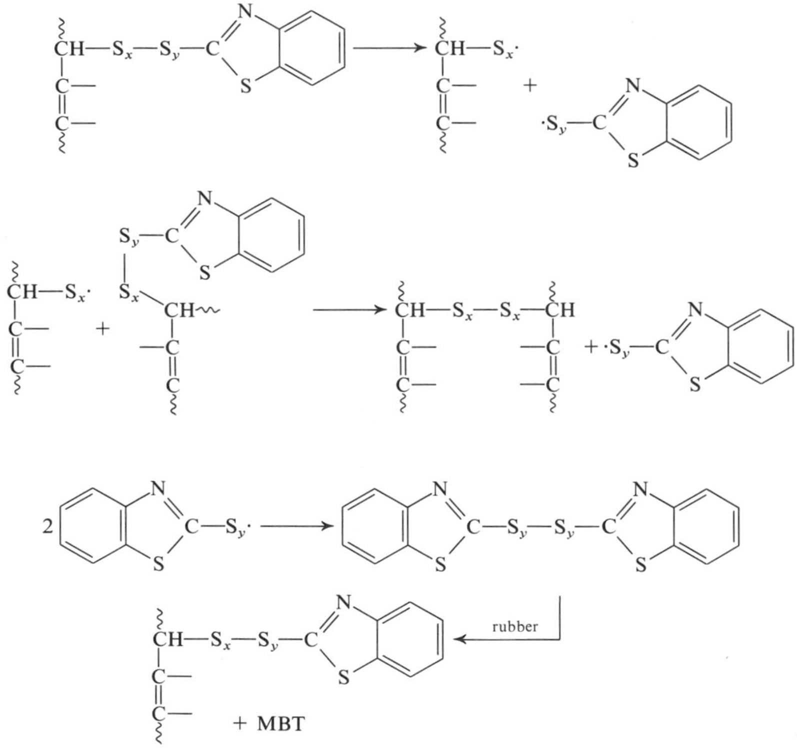

Scheme 6.

liberated simultaneously with cross-link formation. This indeed happens in the case of morpholinothiobenzothiazole-accelerated vulcanization of NR in the absence of Zn^{++} ions [21, 23a]. However, in the case of dithiobisbenzo-thiazole-accelerated vulcanization in the presence of Zn^{++} ion, the liberation of MBT or of its salts stops with the onset of rapid cross-link formation [22].

At normal vulcanization temperatures, there may be some doubt whether the conformational requirements for the approach between two polymeric species can be met. However, it has been demonstrated that vulcanization could occur by such a mechanism at higher temperatures [26]. Further, Scheme 4 would give exclusively allylic attachments, in agreement with the studies of accelerated sulfuration of model olefins [24].

Scheme 5 is similar to the free radical mechanism proposed for vulcanizations with unaccelerated sulfur. If the radical addition suggested by this scheme were to predominate, the attachments due to cross-link formation would not be in agreement with the model olefin studies, which support allylic attachments. A backbone double bond would have to shift. The radical coupling between two polymeric polythiyl radicals, on the other hand, would give allylic links, provided that the connections of the polymeric polysulfides, rubber—S_x—Ac, were allylic.

Scheme 6 is also a free radical mechanism, but the cross-link is formed by direct exchange between a polymeric polythiyl radical, rubber—S_x, and a polymeric polysulfide, rubber—S_x—Ac. This reaction was investigated with model polysulfides. It was found to be fairly rapid and the kinetic chain was long [28]. In addition, heating methyl tetrasulfide gives methyl polythiyl radicals [29].

If cross-link formation occurs by a free radical mechanism (Scheme 5 or Scheme 6), delayed action could be the result of a quenching action by monomeric polythiyl radicals [30]. If the polymeric radicals are rapidly quenched by an exchange reaction before they are able to form cross-links, formation of the latter will be impeded until substantial depletion of the monomeric polysulfides. This is illustrated by Scheme 7. According to this theory for delayed action, if even polysulfides such as bisalkylpolysulfides are mixed with uncured rubber stocks, more delay would result [31].

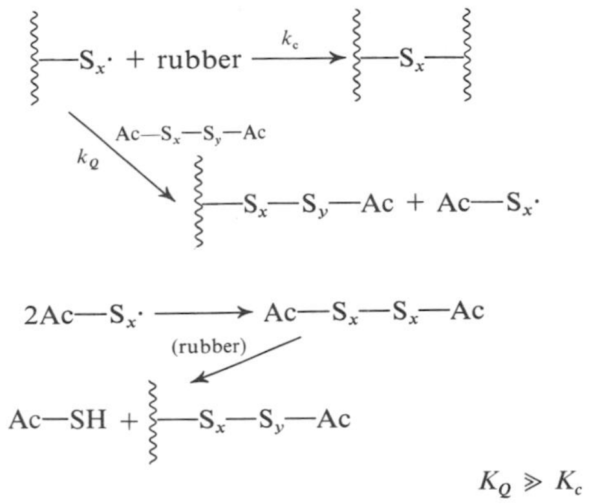

Scheme 7.

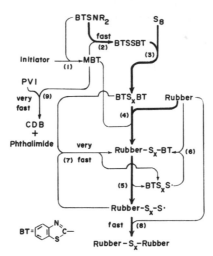

Fig. 6. Scheme for delayed action vulcanization.

In the case of benzothiazole sulfenamide acceleration, the delayed action is largely based on the disappearance of the accelerator, which becomes depleted in an autocatalytic fashion with the formation of MBT [32] (see also [32a]). The rate of depletion is about proportional to the amount of MBT present. If MBT could be taken out of the system as fast as it forms, substantial increases in processing safety would result. Such is the case when cyclohexylthiophthalimide is present. This compound, referred to as PVI (for prevulcanization inhibitor), reacts rapidly with MBT to form cyclohexyldithiobenzothiazole (CDB) and phthalimide. If the depletion of accelerator, and hence the formation of intermediates of the type $Ac-S_x-Ac$, are delayed sufficiently, delayed action can result even without the quenching action evoked in the former theory. In other words, all that is necessary to describe delayed action is an autocatalytic disappearance of a starting material.

The various mechanisms for delayed action could, of course, coexist. This situation is illustrated by Fig. 6. There is also the possibility that a compound such as PVI could react to give polysulfides which would quench radicals of the type rubber—S_x—S. Indeed, the analytical results given by Leib *et al.* [32; see also 32a] indicate that only about one quarter of the PVI is recovered as CBD. The other three quarters could very well form more highly sulfurated polysulfides capable of quenching. Also it is interesting that PVI can inhibit unaccelerated sulfur vulcanization. The PVI concept, then, is not limited to the thiazole-type acceleration. The complexity of PVI-induced delayed action has been further illustrated in a recent report [33].

C. THE ROLE OF ZINC ION IN THIAZOLE ACCELERATION

Studying the kinetics of the vulcanization of natural rubber accelerated by thiazolesulfenamide by varying concentrations in the vulcanization system, it was found that increases in the concentration of fatty acid, and hence increases in the concentration of soluble Zn^{++}, cause an increased overall rate in the early formation of rubber—S_x—Ac [27]. Further, a decrease in the *rate* but an *increase* in the extent of cross-link formation were noted. Associated with the same studies was the finding that Zn^{++} forms complexes with accelerators and accelerator polysulfides. As a result, the increase in the rate of the early reactions was explained by the reaction

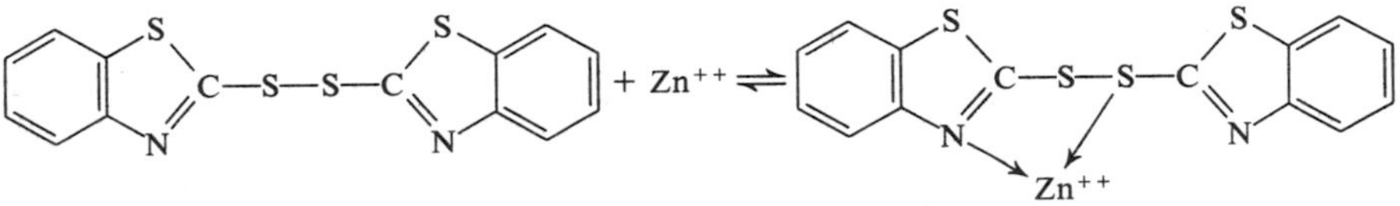

where the chelated form of the accelerator was postulated to be more reactive than the free accelerator in the early reactions:

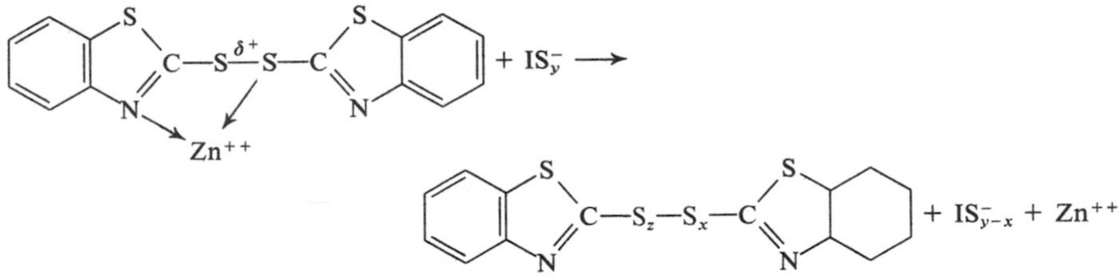

Here, IS^-_y is an ionized form of sulfur which is assumed to be rapidly formed in a reaction between sulfur and any of a number of initiating species. In the case of dithiocarbamate acceleration, it has been proposed that the presence of zinc ion can increase the rate of sulfurization through the formation of species of the type shown [7], where L is a ligand such as an amine molecule.

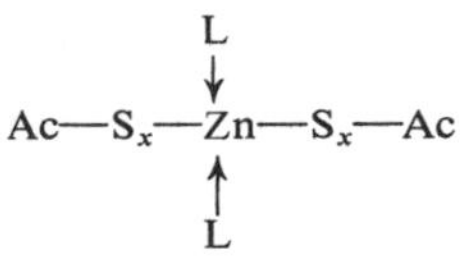

$$\begin{array}{c} L \\ \downarrow \\ Ac—S_x—Zn—S_x—Ac \\ \uparrow \\ L \end{array}$$

The decreased specific rate and increased extent of cross-link formation have been explained by the following proposition [27].

rubber—S_xS_y—S—C (benzothiazole)   + Zn^{++} ⇌ rubber—S_x—S_y—S—C (benzothiazole) Zn^{++}

slow ↓   very slow ↓

rubber—S_xS_y· + ·S—C (benzothiazole)   rubber—S_x· + ·S_y—S—C (benzothiazole)

[rubber] fast ↓   [rubber] fast   [rubber] fast ↓   [rubber] fast

cross-link   cross-link
rubber—S_x—S_y—rubber   rubber—S_x—rubber

rubber—S_y—S—C (benzothiazole)

rubber—S—C (benzothiazole)

not a cross-link precursor   a new cross-link precursor

Here zinc chelation changes the position of the S—S bond most likely to break. Since now a stronger bond must break, the rate is slower. Further, the monomeric radical is more highly sulfurated.

D. FORMULATING TO ACHIEVE SPECIFIED VULCANIZATION PARAMETERS

The following serves to illustrate the application of some of the principles discussed earlier. The object is to be able to obtain a desired degree of process safety (scorch resistance) and at the same time obtain a desired rate of cure in the press. This can be accomplished to some degree by accelerator selection. Another method is to change the concentration of a given accelerator. By either method, however, changes in cure rate and process safety occur at the same time. A change in accelerator selection may improve cure rate but give an unacceptable scorch problem. It has been reported that in the case of natural rubber gum stocks accelerated by sulfenamides, an increase in accelerator concentration gives an increased rate of cross-link formation. However, this can cause either an increase or a decrease in delay time [27, 34]. In NR or SBR black stocks, similar effects are observed, though to a lesser extent. This might be traceable to an overshadowing effect of carbon black. Indeed, carbon black can greatly activate the early vulcanization reactions [35]. At any rate, it has been difficult to formulate a specified rate of cross-link formation and, at the same time, a specified amount of delay.

The use of PVI, however, enables the independent adjustment of cure rate and processing safety [32, 32a, 36]. Figure 7 illustrates this. The recipe of the stock used is given in Table II. The specific rate of cross-link formation is

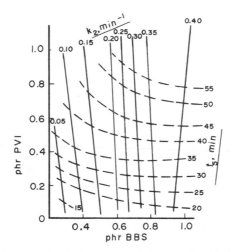

Fig. 7. The effect of vulcanization system design on vulcanization parameters.

TABLE II

NR	70.0
Polybutadiene	30.0
LS HAF carbon black	60.0
Stearic acid	1.0
Zinc oxide	5.0
Antidegradant	1.0
Oil	10.0
PVI	Variable
Sulfur	Variable
Sulfenamide accelerator	Variable

described by a first-order rate constant k_2 obtained from isothermal rheometer charts. The cure was at 149°C. Processing safety is given by t_5, the time for a five-unit increase in Mooney viscosity at 135°C. In Fig. 7, the accelerator is N-t-butylbenzothiazolesulfenamide (BBS). Similar, though not identical, results were obtained with morpholinothiobenzothiazole (MTB) and N,N-dicyclohexylbenzothiazolesulfenamide (DCBS). It can be concluded from the fact that the rate and delay contours are almost perpendicular to each other, and run almost parallel and perpendicular to the compositional or formulation axes, that rates and delay time can be varied rather independently.

In the study underlying Fig. 7, the product of sulfur concentration times accelerator concentration was kept constant at 1.5 (phr)^2 to obtain a relatively constant modulus. With the exceptions of the few stocks containing very small amounts of accelerator but large amounts of sulfur, a 300% modulus of about 1000 psi $\pm$ 15% was obtained. (Stress at 300% strain was 1000 $\pm$ 150 psi.)

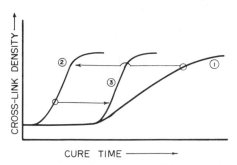

Fig. 8. The use of activating accelerator/retarder combinations: ①, sulfenamide accelerator; ② dithiocarbamate or amine added to ①; ③, PVI added to ②.

In stocks containing only synthetic rubbers such as SBR, the effects of compounding variables on cure rate and Mooney scorch may not be as pronounced. However, in stocks containing even a relatively small amount of natural rubber, the effects resemble most closely those observed in predominantly natural rubber stocks. Other accelerators, such as MBT, bis-2,2-dithiobenzothiazole (MBTS), and thiuram types, also respond to PVI, as do mixtures such as MBTS/DPG.

Another approach to the problem is to replace a minor portion of primary accelerator with an equivalent amount of activating secondary accelerators such as DPG or a thiuram type [37] which will produce the same modulus. This will increase the cure rate. The addition of PVI will increase the delay time. With such combinations, it is possible to increase scorch time at given processing temperatures without changing the time needed for curing. Indeed, the total time for a desired cure can be decreased. This is illustrated by Fig. 8.

E. The Effect on Adhesion to Brass-Plated Steel

The adhesion of vulcanized rubber to brass-plated steel wire is generally considered to be the result of interactions between copper and the vulcanization system [38–42]. Indeed, changes in the vulcanization system can induce great changes in the degree of adhesion, especially after aging.

The stocks described in the preceding subsection have been used in adhesion studies. Brass-plated (70% Cu, 30% Zn) steel tire cords were embedded in rubber strips. The strips were vulcanized to the optimum cure (selected by the oscillating disk rheometer). The blocks containing the embedded tire cords were heated to 100°C. The force to pull the cords from the hot rubber was measured shortly after heating and after 24 hr of 100°C aging. The "pullout," or adhesion force, for the unaged specimens was generally high and did not strongly reflect changes in the curing system. The effects on aged adhesion, however, were rather dramatic. This is illustrated by Fig. 9, which gives adhesion force contours (pounds per $\frac{3}{8}$-in. embedment) for specimens contain-

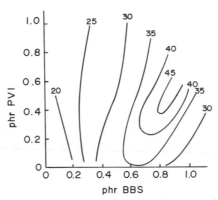

Fig. 9. The effect of changes in the concentration of accelerator and PVI on aged adhesion to brass-plated wire.

ing *t*-butylbenzothiazole sulfenamide. Similar contours were also obtained for compositions containing morpholinothiobenzothiazole and *N,N*-dicyclohexylbenzothiazole sulfenamide.

In the case of each accelerator, optimal or maximal pullout values after aging were achieved at different concentrations of curing ingredients. However, all aged adhesion data could be expressed as a function of the specific rate constant of cross-link formation k_2 (at 149°C) and the amount of scorch delay (t_5 at 135°C). This is seen in Fig. 10. The contours of Fig. 10 show that a narrow range of rate and delay correspond to an area of maximum aged adhesion. The dotted contour is for 45 lb pullout of unaged specimens. One notices that wide ranges of formulations give good initial adhesion, but even the mild aging condition used here gives rise to a critical situation with respect to compounding.

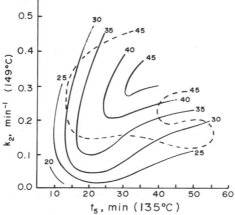

Fig. 10. The effect of changes of vulcanization parameters on adhesion to brass-plated wire. Numbers are pounds to pullout through $\frac{3}{8}$ in. at 100°C; ———, aged 24 hr at 100°C; ----, unaged.

By varying the vulcanization kinetics through compounding variations, one may arrive at a composition that will give good aged adhesion to brass. However, rather small changes in plant conditions could cause significant losses. The balance between the reactivities of the metallic surface, of the vulcanization system, and probably of the rubber itself must be maintained. Small changes in heat history, in concentrations of compounding ingredients, and in the state of the metallic surface could add up to real problems.

It should be noted that the rates of reactions associated with vulcanization are not the only parameters affecting adhesion. The type of rubber is important. Natural rubber gives very good results, whereas other rubbers are difficult to make adhere to brass-plated tire cords. The type of carbon black may also be critical, and certain accelerators, such as dithiocarbamates, should be avoided even in very low concentrations.

F. THE EFFECT ON VULCANIZATE PROPERTIES

It has long been recognized that increases in sulfur and accelerator concentrations give rise to higher cross-link densities and, hence, to higher moduli. However, as the ratio of the concentration of a sulfenamide accelerator to the concentration of sulfur increases, so does the proportion of monosulfidic cross-links in natural rubber stocks [43]. This is illustrated by Fig. 11. Higher accelerator concentrations also lead to an abundance of pendent groups of the type rubber—S_x—Ac. Higher concentrations of sulfur (in respect to accelerator concentrations) give both more polysulfide cross-links and more sulfur combined with the rubber chains in cyclic modifications. Other changes in vulcanizate properties have been attributed to the changes in network structure and chain modifications [5, 44–47]. Types of cross-links and chain modification are given in Fig. 12.

The effects of changes in sulfur and accelerator concentrations on vulcanizate properties have been studied [48] using the recipe given in Table III.

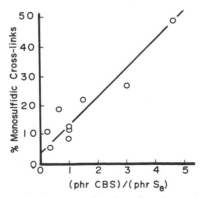

Fig. 11. Cross-link type as a function of accelerator/sulfur ratio.

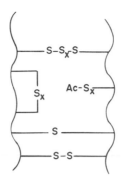

Fig. 12. Cross-link types and chain modifications.

TABLE III

NR	100
N330 carbon black	50
N-2-propyl-N'-phenyl-p-phenylenediamine (IPPD)	2
Zinc oxide	5
Stearic acid	3
Plasticizer	3
Sulfur	Variable
N-cyclohexylbenzothiazolesulfenamide (CBS)	Variable

The effects on modulus, thermal aging (ASTM D865-62), and DeMattia flex fatigue life (ASTM D813-59) are given in Fig. 13. The effect of compounding on modulus is indicated by the approximately diagonal contours of negative slope. They are approximately parallel and suggest modulus increases with an increase in either sulfur or accelerator concentration. The concentrations are given on logarithmic coordinates. The average slope of the modulus contours is about -0.63. This indicates that a constant modulus will be obtained if the product of (phr sulfur) $\times$ (phr accelerator)$^{0.63}$ is held constant.

The contours that reflect retention of ultimate elongation indicate that thermal aging resistance (at 100°C in air for 2 days) depends only on the concentration of sulfur. Higher concentrations of sulfur give poor aging characteristics, associated with the number of points of chain sulfuration. It has been said that chain modification can direct or activate chain scission [49]. Another view is that sulfur interferes with the antidegradant (in this case IPPD) [50]. It is interesting that these contours are parallel to the accelerator concentration axis and perpendicular to the sulfur concentration axis. This indicates a more or less pure dependence on sulfur concentration per se, and no dependence at all on the ratio of sulfur concentration to accelerator concentration.

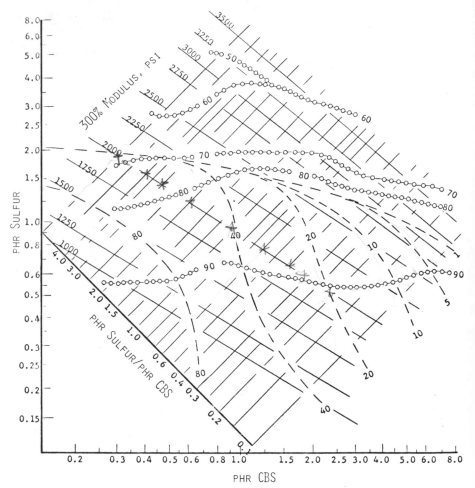

Fig. 13. Vulcanizate properties: ----, De Mattia flex fatigue life (kc × 10⁻¹ to 0.5-in. crack); ○, percent retention of elongation after 2 days at 100°C.

The contours for DeMattia flex fatigue life are more complex. The test is run so that the specimens are about equally strained. There is some question as to whether we should consider fatigue life data at equal strain or at equal strain energy [51–53]. The answer probably depends on the intended use of the rubber stock. In some cases, where strain is restricted by fabric reinforcement, fatigue life should be considered at equal strain. However, for other applications, where the strain is not limited, strain energy would be more relevant. At any rate, the contours as presented here can be interpreted in terms of either constant strain or constant strain energy per cycle. All points on the chart can be compared at an approximately equal strain per cycle.

However, if we derive values of flex life by interpolating between the DeMattia contours but only along a constant modulus contour, we can extract values corresponding to an approximately equal strain energy per cycle. If we choose the higher modulus contours, we are considering higher strain energies. The reason for using the words "approximately constant" is that a crack is continuously growing in the specimen during the test. Also, since the DeMattia test is a bending test, the amount of strain depends greatly on the thickness of the specimen. The exact thickness of the specimen is difficult to control.

If we consider the group of flex life contours as a whole, or at approximately constant strain energy per cycle, the following may be concluded: Too high a concentration of either sulfur or accelerator gives poor flex life. However, by the selection of the proper ratio of sulfur to accelerator concentration, higher modulus vulcanizates can be obtained with a minimum loss of flex life. The flex life at approximately equal strain energy per cycle can be illustrated by extracting values along the 2000 psi modulus contour line. We can then construct Table IV. It is apparent that at equal strain energy per cycle corresponding to a modulus of 2000 psi, the maximum flex life can be obtained with a sulfur-to-accelerator concentration ratio of 2.5. Other optimum ratios for various moduli can be obtained from the contours. This is illustrated by Table V. These optimum ratios, of course, should not be considered as universal. Different base recipes, different types of antidegradants, different types of filler, and different types of fatigue tests are certain to change the position of the optima.

The loss of fatigue life at low levels of sulfur but high levels of accelerator has been attributed to a high concentration of accelerator-terminated appending groups [9] and a high concentration of monosulfidic cross-links. Monosulfidic cross-links are not able to exchange or rearrange to relieve the highly localized stresses associated with the growth of a flaw. However, polysulfidic

TABLE IV

(phr Sulfur)	Flex life
(phr Sulfenamide)	(kc × 10^{-1} to 0.5 in. crack)
6	40
5	50
4	53
3	55
2	55
1	40
0.6	35
0.4	27
0.3	25
0.2	19

TABLE V

Modulus (psi)	Optimum sulfur accelerator ratio	Maximum flex life (kc $\times 10^{-1}$ to 0.5 in. crack)
1000	3.5	80
1500	3.0	80
2000	2.5	55
2250	1.0	30
2500	0.45	12
2750	0.27	7

cross-links are able to rearrange under stress [7, 54, 55]. Hence, where moderately high concentrations of sulfur (in respect to accelerator) are used, flex life improves. However, even at constant modulus (or approximately constant strain energy), when even higher concentrations of sulfur are used, flex life decreases. It is possible that this is due to extensive cyclic chain modifications.

The properties of cured stocks containing low sulfur but high accelerator concentrations are similar to those reported for the so-called EV (for efficient vulcanization) and semi-EV vulcanizates. These stocks do, in fact, give EV and semi-EV type vulcanizates. However, when large amounts of sulfenamide accelerator are used, flexibility in formulation for specific curing parameters is reduced. To avoid this problem and others, rather than use high accelerator levels, a conventional sulfur level can be replaced or partially replaced by a sulfur donor compound. Examples of these donors are tetramethylthiuram disulfide (TMTD) and 4,4'-dithiodimorpholine (DTDM). This type of vul-

canization system design was reported by McCall [56]. He found that by judiciously balancing the levels of accelerator, sulfur, and donor compound (4,4'-dithiodimorpholine (DTDM)), he could obtain good vulcanization characteristics, good thermal stability, good flex life, and superior retention of flex life.

G. SELECTED ACCELERATED SULFUR SYSTEM RECIPES

Examples of starting point recipes are given in Table VI. The recipes of Table VI are not intended as ultimate solutions to compounding problems. Variations will undoubtedly be necessary to meet particular requirements.

TABLE VI

RECIPES FOR ACCELERATED SULFUR OR "CONVENTIONAL" VULCANIZATION SYSTEMS[a]

	NR	SBR		Nitrile		Butyl	EPDM
Sulfur	2.5	1.8	1.5	0.5	0.25	2.0	1.5
DTDM	—	—	—	—	1.0	—	—
Zinc oxide	5.0	5.0	5.0	3.0	2.0	3.0	5.0
Stearic acid	2.0	2.0	2.0	0.5	0.5	2.0	1.0
BBS	0.6	0.9	—	—	—	—	—
MBTS	—	—	1.2	2.0	—	0.5	—
MBT	—	—	—	—	—	—	0.5
DPG	—	—	0.4	—	—	—	—
TMTD	—	—	—	1.0	1.0	1.0	1.5
Typical vulcanization conditions[b]							
T (°C)	148	153	153	140	140	153	160
t (min)	25	30	20	60	60	20	20

[a] Concentrations are in parts per 100 parts of rubber (phr).

[b] Conditions change depending on other aspects of the compositions (filler, antidegradant, etc.).

It should be further noted that other compounding ingredients such as fillers, antidegradants and plasticizers alter curing characteristics.

IV. Vulcanization by Phenolic Resins, Quinone Derivatives, or Maleimides

A. VULCANIZATION BY PHENOLIC RESINS

Diene rubbers can be cross-linked by compounds of the general structure

where X is —OH, halogen atom, or

and R is usually an alkyl radical. These materials are usually of a resinous nature. Elastomers appropriate for this type of vulcanization are generally the same as for accelerated sulfur systems. The vulcanization scheme, adapted from Van der Meer [57; Thelamon, 58; see also 58a], is believed to be that in Scheme 8.

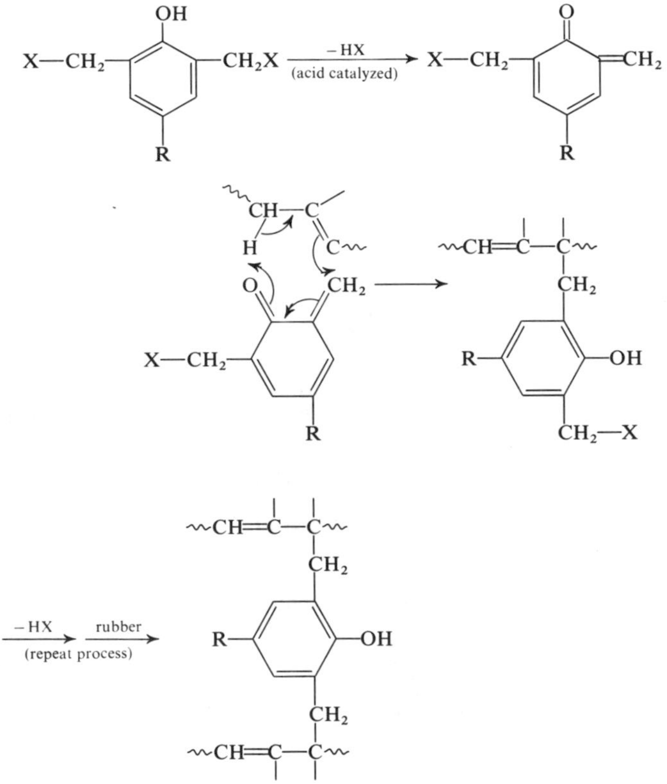

Scheme 8.

Resin curing systems are used extensively with butyl rubber for high-temperature applications. Resin cures are slower than accelerated sulfur cures and higher temperatures are required, but they can be activated by zinc oxide and halogen atoms in position X. An example of recipes for butyl rubber are given in Table VII where SP-1055 is the phenolic curative.

The chemistry of vulcanization by resins is no doubt similar to that obtained when a resorcinol–formaldehyde–rubber latex adhesive system is used for bonding rubber to tire cords during vulcanization. Representative of this is Scheme 9. Thus the resin can be chemically bound to the adhesive latex rubber.

TABLE VII

RECIPES FOR VULCANIZATION BY RESINS, QUINONE DERIVATIVES, AND MALEIMIDES[a]

	Butyl		SBR			
	1	2	1	2	Nitrile	EPR
ZnO	5	5	—	—	—	—
PbO$_2$	—	2	—	—	—	—
Stearic acid	1	—	—	—	—	—
Resin SP-1055	12	—	—	—	—	—
Benzoquinone dioxime	—	2	—	—	—	—
m-Phenylenebismaleimide	—	—	0.85	0.85	3	3
MBTS	—	4	2	—	—	—
Dicumyl peroxide	—	—	—	0.3	0.3	1.6
Typical vulcanization conditions[b]						
T (°C)	182	182	153	153	153	160
t (min)	80	80	25	25	30	15

[a] Concentrations are in parts per 100 parts of rubber (phr).
[b] Conditions change depending on other aspects of the compositions.

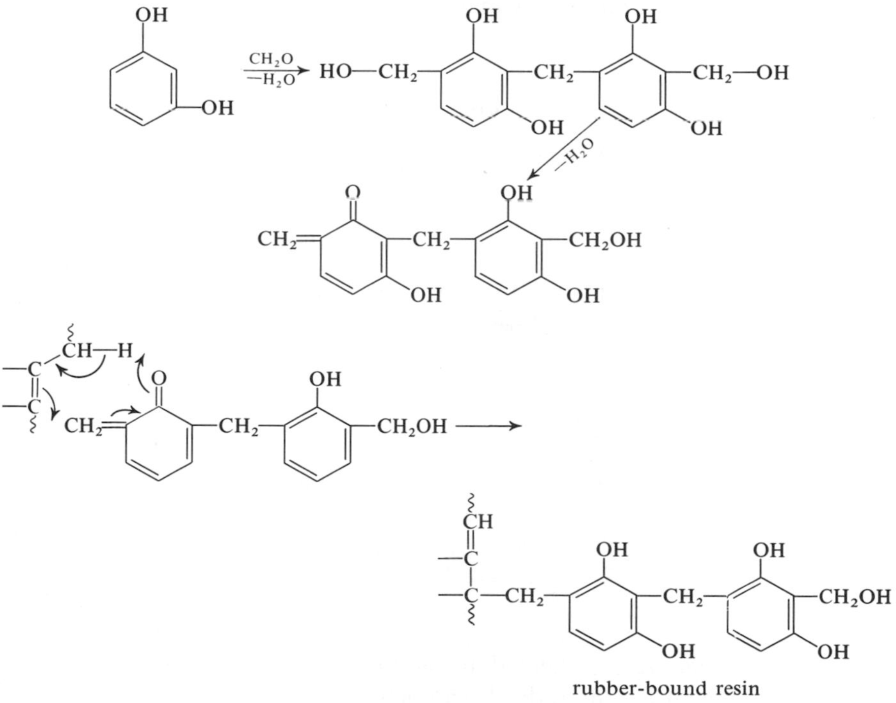

rubber-bound resin

Scheme 9.

The resin also adheres to the tire cord. In addition, low molecular weight resin molecules can diffuse into the carcass rubber, causing stiffening in the immediate region of the tire cord by additional curing.

B. VULCANIZATION BY QUINONE DERIVATIVES

Benzoquinone and many of its derivatives can vulcanize diene rubbers. In most cases, oxidizing agents are required. Only benzoquinone dioxime is now of technical importance. It is useful with the same diene rubbers that are curable by accelerated sulfur, and is sometimes used with sulfur vulcanization.

The mechanism is thought to involve the formation of nitroso groups by oxidation [59]. Indeed, p-dinitrosobenzene will vulcanize diene rubbers rapidly.

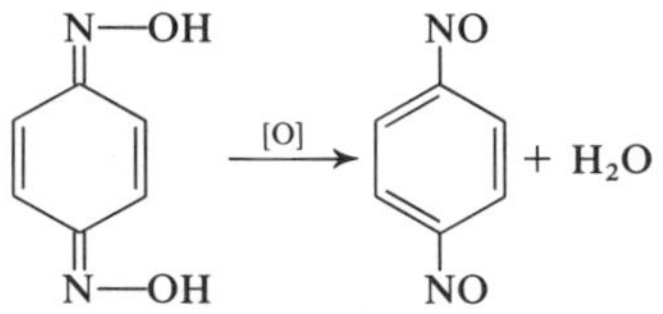

Early workers felt that the nitroso group reacted with rubber to give nitrones [60–63]. For example, with natural rubber, structures of the type

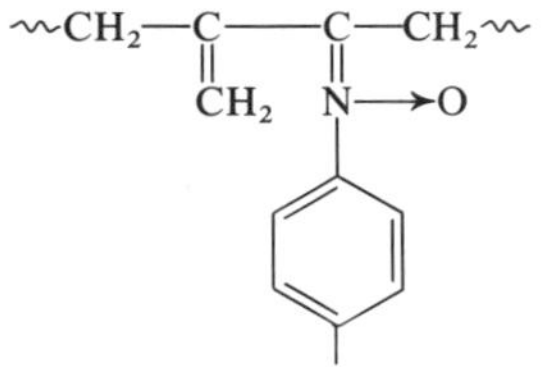

shown here would be formed. More recently, Sullivan has postulated that the nitroso group reacts with rubber as follows [64]:

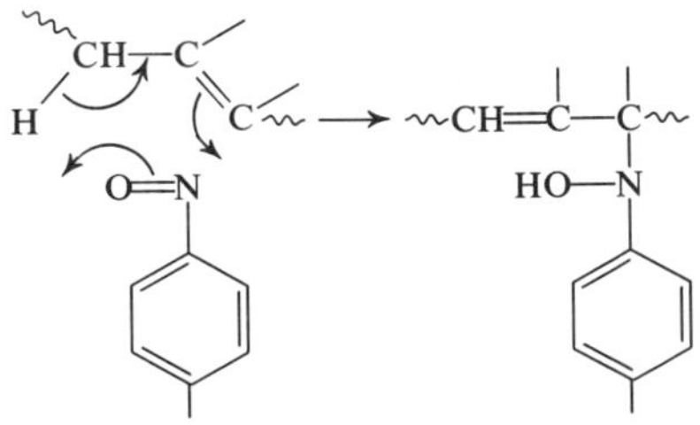

Stable free radicals were indicated by electron spin resonance spectra. A structural assignment for the radicals can be represented as

$$\text{\textasciitilde\textasciitilde CH=C-C\textasciitilde\textasciitilde}$$

$$\text{N—O·}$$

This could conceivably be oxidized to a nitrone. However, there is recent evidence that ultimately the attachment to rubber occurs through an amine linkage which is formed by an unknown mechanism [65]. When using the

$$\text{\textasciitilde\textasciitilde CH=C-C\textasciitilde\textasciitilde} \xrightarrow{?} \text{\textasciitilde\textasciitilde CH=C-C\textasciitilde\textasciitilde}$$

$$\text{N—OH} \qquad \text{NH}$$

dioxime as a vulcanizing agent, an oxidizing agent is used. This is typically lead peroxide. The addition of zinc oxide improves properties. MBTS can be used to increase cross-linking efficiency and to improve scorch resistance. A sample recipe is given in Table VII.

C. VULCANIZATION BY MALEIMIDE DERIVATIVES

Derivatives of maleimide are cross-linking agents for diene rubbers. The most efficient ones contain more than one maleimide moiety per molecule. Probably the best known of these is *m*-phenylenebismaleimide.

A catalytic free radical source such as dicumyl peroxide is usually required to initiate the reaction. A mechanism, adapted from Kovacic *et al.*, is given by Scheme 10 [66].

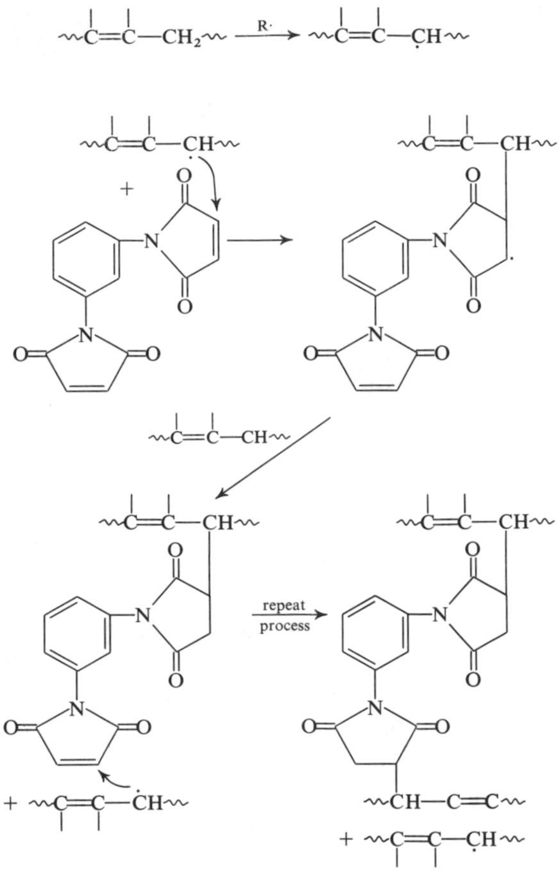

Scheme 10.

Although a peroxide is frequently used with a maleimide vulcanizing agent, other agents, such as MBTS, can also catalyze the reaction. At temperatures well above the usual vulcanization temperatures, the maleimides will react with rubber without catalyst. This could occur as shown.

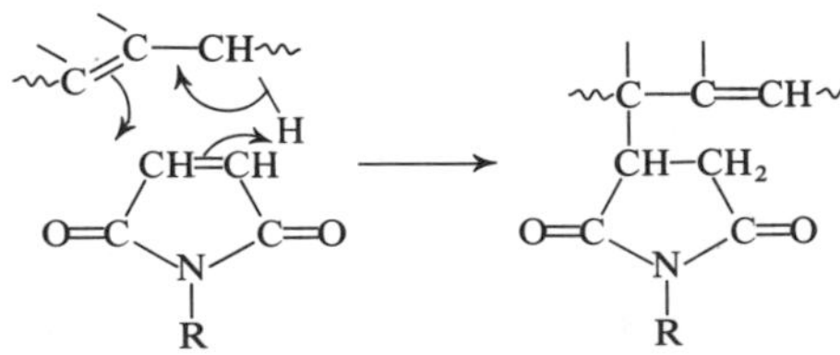

This is another example of what has variously been called a pseudo-Diels–Alder, ene, or "no-mechanism" reaction [67, 68].

D. Chemical Similarities between Vulcanization by Accelerated
   Sulfur, Phenolic Resins, Quinone Derivatives, and Maleimides

The attack of the vulcanization system upon rubber molecules can be visualized to proceed in each of the foregoing systems by similar routes.

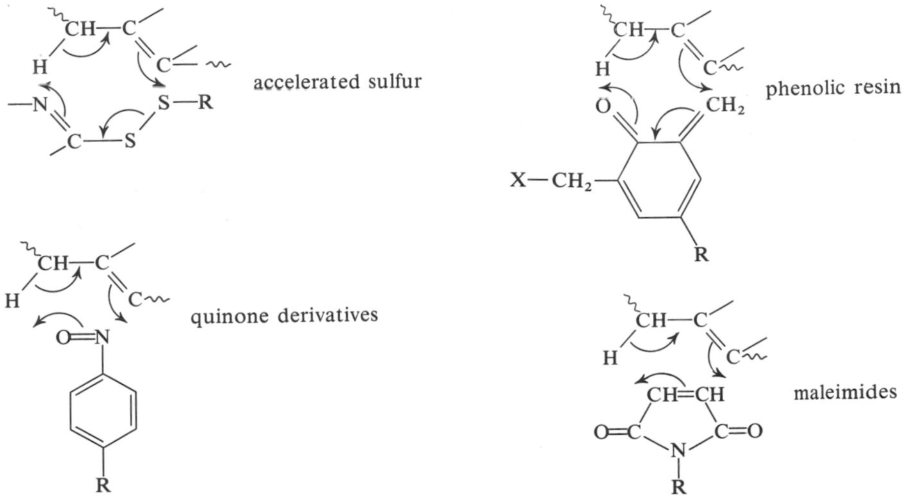

Scheme 11.

Chemical structural requirements for these types of vulcanization can then be simply stated. The elastomer molecules must contain allylic hydrogen atoms (see Scheme 5). The attacking species from the vulcanization system must contain sites for proton acceptance and electron acceptance in proper steric relationship to permit the following rearrangement, where A is the proton acceptor site and B is the electron acceptor site.

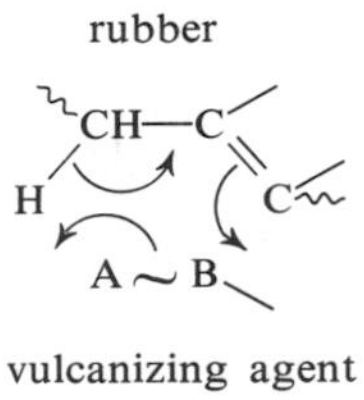

Double bonds *per se*, without allylic hydrogens in the elastomer molecules, will not permit the types of vulcanization in Scheme 11. Even highly activated double bonds do not suffice. For example, a polyglycol fumarate ester is not readily cross-linked by accelerated sulfur systems, though it is very efficiently cross-linked by an organic peroxide [69] (see also [69a]).

E. SELECTED RECIPES FOR VULCANIZATION BY RESINS, QUINOIDS,
    OR MALEIMIDES

Vulcanizates based on cross-linking by phenolic resins, quinone derivatives, or derivatives of maleimide are particularly useful in cases where thermal
stability is required. In addition, some of these systems are important in
rubber adhesion technology. Examples of such recipes are given by Table VII.

V. Vulcanization by Metal Oxides

Chlorobutadiene rubbers (or chloroprene rubbers) are generally vulcanized
by the action of metal oxides. The structure of chlorobutadiene rubber (CR)
can be represented by

$$\left(CH_2-\underset{\underset{}{\overset{\overset{Cl}{|}}{C}}=CH-CH_2\right)_n \left(CH_2-\underset{\underset{\underset{CH_2}{\parallel}}{\overset{\overset{Cl}{|}}{\underset{CH}{|}}}}{C}\right)_{0.015n}$$

The primary cross-linking agent is usually zinc oxide, which is used along with
magnesium oxide. (Lead oxides are sometimes used when low water absorption
is required.) The reaction is thought to involve the vinyl group of the elastomer,

$$-CH_2-\underset{\underset{\underset{CH_2}{\parallel}}{\overset{\overset{Cl}{|}}{\underset{CH}{|}}}}{CH}-$$

which is the result of 1,2 polymerization. Two conflicting routes to cross-link
formation have been proposed. One requires the incorporation of zinc atoms
into the cross-link [70]; the other leads to ether cross-links [71]. The first is
summarized by the overall reaction:

$$2CH_2=CH-\underset{\underset{}{\overset{\overset{CH_2}{|}}{C}}}{C}-Cl + ZnO + MgO \longrightarrow CH_2=CH-\underset{}{\overset{\overset{CH_2}{|}}{C}}-O-Zn-O-\underset{}{\overset{\overset{CH_2}{|}}{C}}-CH=CH_2 + MgCl_2$$

The second can be represented by

$$
\begin{array}{ccc}
\underset{\displaystyle \underset{\displaystyle \underset{\displaystyle CH_2}{\overset{\|}{CH}}}{\overset{|}{C}}}{-CH_2-C-}
& \rightleftharpoons &
\underset{\displaystyle \underset{\displaystyle \underset{\displaystyle Cl}{\overset{|}{CH_2}}}{\overset{\|}{CH}}}{-CH_2-C-}
\xrightarrow{\;ZnO\;}
\underset{\displaystyle \underset{\displaystyle \underset{\displaystyle OZnCl}{\overset{|}{CH_2}}}{\overset{\|}{CH}}}{-CH_2-C-}
\end{array}
$$

$$
\underset{\displaystyle \underset{\displaystyle \underset{\displaystyle OZnCl}{\overset{|}{CH_2}}}{\overset{\|}{CH}}}{-CH_2-C-} \; + \;
\underset{\displaystyle \underset{\displaystyle \underset{\displaystyle -CH_2-C-}{\overset{|}{CH}}}{\overset{|}{CH_2}}}{Cl} \longrightarrow
-CH_2-C- \; + \; ZnCl_2
$$

$$
ZnCl_2 + MgO \longrightarrow ZnO + MgCl_2
$$

It should be noted that either magnesium oxide or zinc oxide alone will vulcanize chloroprene rubbers; however, the combination is better than either one alone. Zinc oxide by itself tends to be too scorchy, whereas magnesium oxide alone is inefficient.

TABLE VIII

METAL OXIDE SYSTEMS FOR CHLOROPRENE RUBBER[a]

ZnO	5	5	5
MgO	4	—	4
Calcium stearate	—	5.5	—
Stearic acid	—	—	1
TMTM	—	—	1
DOTG	—	—	1
ETU	0.5	0.5	—
Sulfur	—	—	1
Typical vulcanization conditions[b]			
T (°C)	153	153	153
t (min)	15	15	15

[a] Concentrations in phr.
[b] Conditions change depending on other aspects of the compositions.

The accelerators used in accelerated sulfur vulcanization are generally not applicable to metal oxide vulcanization of chloroprene rubbers. Some of them are too fast, some too slow, and others act as retarders. It is interesting that a mixed accelerated sulfur (TMTM + DOTG + S)/metal oxide vulcanization can be used to obtain relatively high states of cure [20]. This may be desirable for high resilience or for good dimensional stability.

The accelerator most widely used today with metal oxide cures is ethylene-thiourea (ETU) or 2-mercaptoimidazoline. A mechanism for ETU acceleration has been proposed by Pariser [72] (Scheme 12).

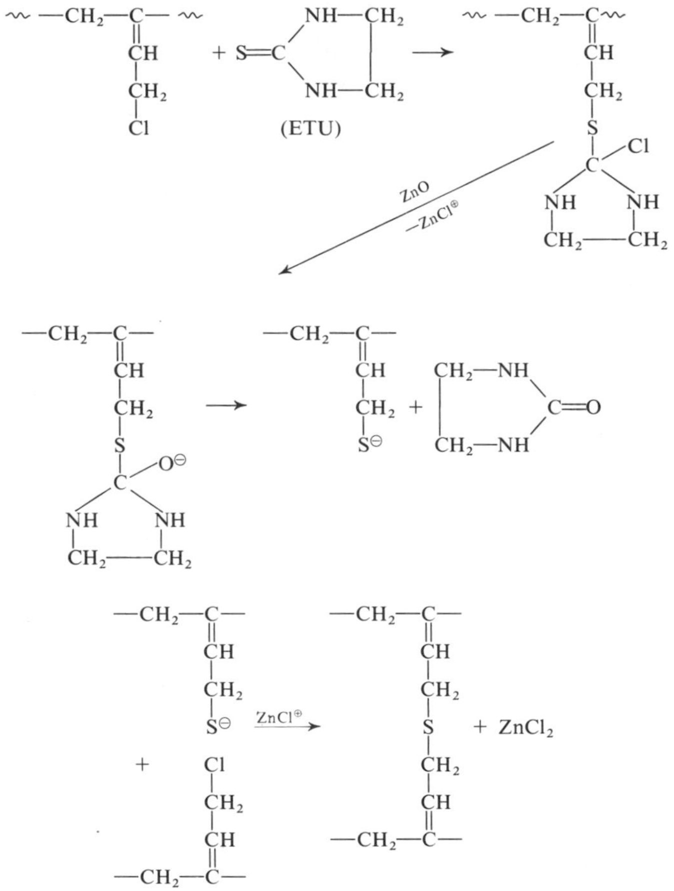

Scheme 12.

Examples of recipes for metal oxide vulcanization are given in Table VIII. It should be noted that in one case calcium stearate was used instead of magnesium oxide to obtain better aging characteristics [73]. Neoprene W is the grade of chloroprene rubber intended for the recipes.

VI. Vulcanization by Peroxides

A. THE SCOPE OF PEROXIDE VULCANIZATION

Most types of elastomers can be vulcanized by the action of organic peroxides. Those that have been considered for use with elastomers are diacyl peroxides, dialkyl peroxides, and peresters. Of these, dialkyl peroxides and *t*-butyl perbenzoate give efficient cross-link formation; di-*t*-butyl peroxide and dicumyl peroxide give good vulcanizates in compounds which contain reinforcing carbon blacks, but di-*t*-butyl peroxide is a volatile liquid and inconvenient to use, so that dicumyl peroxide remains as the generally useful peroxide for vulcanizing elastomers. (Other nonvolatile alkyl peroxides are preferred when the odor left from dicumyl peroxide cure is objectionable.) In the course of the reaction, the peroxides decompose to give free radicals in a reasonable length of time at a reasonable vulcanization temperature. Dicumyl peroxide meets both of these requirements [74]. It should be noted that acidic compounding ingredients (fatty acids, certain carbon blacks, and acidic silicas) can catalyze a nonradical-generating, wasteful decomposition of the peroxides. Other compounding ingredients such as antidegradants can reduce the cross-linking efficiency by quenching or altering the free radicals before they can attack the polymer. Compensation for such effects may be necessary.

Peroxides are of importance mainly because of their ability to cross-link elastomers that contain no sites for attack by other types of vulcanizing agents. They are particularly useful for ethylene propylene rubbers, certain millable urethanes, and silicone rubbers. They are not generally applicable to polyisobutylene elastomers (butyl rubber) because of a tendency to produce too many polymer chain scissions [75–77].

Elastomers derived from isoprene and butadiene are readily cross-linked by peroxides. However, many of the properties are inferior to those of vulcanizates cured by accelerated sulfur. Nevertheless, peroxide vulcanizates of these diene rubbers may be desirable in applications where creep resistance (especially at elevated temperatures) is required.

B. PEROXIDE VULCANIZATION OF UNSATURATED HYDROCARBON ELASTOMERS

The first step in peroxide-induced vulcanization is the decomposition of the peroxide to give free radicals

$$Peroxide \longrightarrow 2R^{\cdot}$$

where $R^{\cdot}$ is an alkoxyl, an alkyl, or an acyloxyl radical, depending on the type of peroxide used. For example, dibenzoyl peroxide gives benzoyloxyl radicals, but dicumyl peroxide gives cumyloxyl and methyl radicals [78].

In the case of elastomers derived from butadiene or isoprene, the next step is either the abstraction of a hydrocarbon atom in the allylic position from the elastomer molecule or the addition of the peroxide-derived radical to a double bond [79–81].

$$R\cdot + \text{〜CH}_2-\overset{|}{\text{C}}=\overset{|}{\text{C}}\text{〜} \longrightarrow \text{〜}\overset{\cdot}{\text{C}}\text{H}-\overset{|}{\text{C}}=\overset{|}{\text{C}}\text{〜} + RH$$

$$R\cdot + \text{〜CH}_2-\overset{|}{\text{C}}=\overset{|}{\text{C}}\text{〜} \longrightarrow \text{〜CH}_2-\overset{\overset{\displaystyle R}{|}}{\text{C}}-\overset{|}{\overset{\cdot}{\text{C}}}\text{〜}$$

In the case of isoprene rubber the abstraction route predominates over radical addition. Two polymeric free radicals then unite to give a cross-link.

$$2\text{〜}\overset{\cdot}{\text{C}}\text{H}-\overset{|}{\text{C}}=\overset{|}{\text{C}}\text{〜} \longrightarrow \begin{array}{c}\text{〜CH}-\overset{|}{\text{C}}=\overset{|}{\text{C}}\text{〜}\\ |\\ \text{〜CH}-\overset{|}{\text{C}}=\overset{|}{\text{C}}\text{〜}\end{array}$$

Alternately, cross-links could form by a chain reaction which involves the addition of polymeric free radicals to double bonds [74, 82, 83]. Here, a

$$\begin{array}{c}\text{〜}\overset{\cdot}{\text{C}}\text{H}-\overset{|}{\text{C}}=\overset{|}{\text{C}}\text{〜}\\ +\\ \text{〜CH}_2-\overset{|}{\text{C}}=\overset{|}{\text{C}}\text{〜}\end{array} \longrightarrow \begin{array}{c}\text{〜CH}-\overset{|}{\text{C}}=\overset{|}{\text{C}}\text{〜}\\ |\\ \text{〜C}-\overset{\cdot}{\text{C}}-\text{CH}_2\text{〜}\end{array}$$

$$\begin{array}{c}\text{〜CH}-\overset{|}{\text{C}}=\overset{|}{\text{C}}\text{〜}\\ |\\ \text{〜C}-\overset{\cdot}{\text{C}}-\text{CH}_2\text{〜}\\ +\\ \text{〜CH}_2-\overset{|}{\text{C}}=\overset{|}{\text{C}}\text{〜}\end{array} \longrightarrow \begin{array}{c}\text{〜CH}-\overset{|}{\text{C}}=\overset{|}{\text{C}}\text{〜}\\ |\\ \text{〜C}-\text{CH}-\text{CH}_2\text{〜}\\ +\\ \text{〜}\overset{\cdot}{\text{C}}\text{H}-\overset{|}{\text{C}}=\overset{|}{\text{C}}\text{〜}\end{array}$$

cross-link has formed without the loss of the free radical, so that the process can be repeated until termination by radical coupling occurs. Coupling can be between two polymeric radicals to form a cross-link or by unproductive processes; e.g., a polymeric radical can unite with a radical derived from the peroxide, thus wasting two radicals. If a polymeric radical decomposes to give a vinyl group and a new polymeric radical, the net result is a scission, or cut, in the chain. When dialkyl peroxides are used, few monomeric radicals are lost by coupling with polymeric radicals, and if the elastomer is properly chosen, the scission reaction is not excessive either [74]. For dicumyl peroxide in isoprene rubber, the cross-linking efficiency is about 1. One "mole" of cross-links is produced per mole of peroxide, so that cross-linking is mainly by the coupling of two polymeric radicals [84–86]. (One mole of peroxide gives two moles of monomeric radicals, which give rise to one "mole" of cross-links.)

If BR or SBR is used, the efficiency can be much greater than 1, especially if all antioxidant materials are removed. This indicates a chain reaction for the cross-link formation, and can be explained at least in part by steric considerations. In butadiene-based rubbers, the double bonds are quite accessible. Radical addition to the double bonds gives highly reactive radicals which are likely to add to other polymer double bonds. Such a chain of additions is more likely in butadiene rubber than in the presence of hindering methyl groups in isoprene rubbers. One might expect that nitrile rubber would also be vulcanized with efficiencies greater than 1. However, though the double bonds in nitrile rubber are highly accessible, the presence of nitrile groups impedes the cross-link reaction [82].

C. PEROXIDE VULCANIZATION OF SATURATED HYDROCARBON ELASTOMERS

Peroxides can also cross-link saturated hydrocarbon polymers, though the efficiency is reduced by branching. For example, polyethylene is cross-linked by dicumyl peroxide at an efficiency of about 1.0, saturated ethylene propylene rubber at an efficiency of about 0.4, while butyl rubber cannot be cured at all.

The reactions are similar to those for butadiene and isoprene rubbers.

$$peroxide \longrightarrow 2R\cdot$$

$$R\cdot + -CH_2-CH_2- \longrightarrow RH + -CH_2-\overset{\cdot}{C}H-$$

$$2 -CH_2-\overset{\cdot}{C}H- \longrightarrow \begin{matrix} -CH_2-CH- \\ | \\ -CH_2-CH- \end{matrix}$$

However, the reactions [74, 75] become different for branched polymers.

$$2 \cdot CH_2- \longrightarrow -CH_2-CH_2-$$

Here, though the peroxide has reacted and has been depleted, no cross-links have been formed between polymer chains, and the average molecular weight of the polymer has even been reduced. It is now of great interest that sulfur, or the so-called coagents [74, 87, 88], can be used to suppress scission reactions. Many of the coagents are cross-linking agents for diene rubbers. Their mechanism of action may be visualized as in Scheme 13, where Q is sulfur or a coagent. If the ratio of reaction rates k_c/k_s is sufficiently high, scission reactions will be suppressed. Examples of successful coagents are m-phenylenebismaleimide, benzoquinonedioxime, polybutadiene (containing a large amount of 1,2-butadiene units),

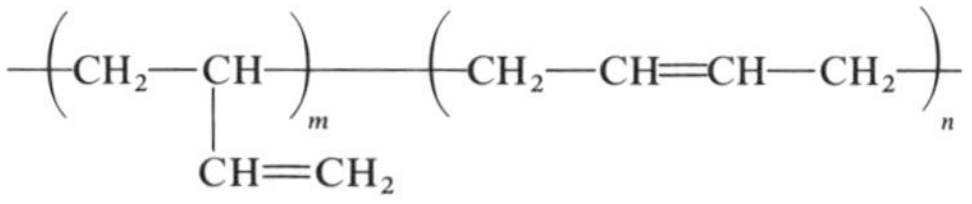

$$-\left(CH_2-\underset{\underset{CH=CH_2}{|}}{CH}\right)_m-\left(CH_2-CH=CH-CH_2\right)_n-$$

triallyl cyanurate, diallyl phthalate, and ethylene dimethacrylate. They all have strong radical acceptor sites.

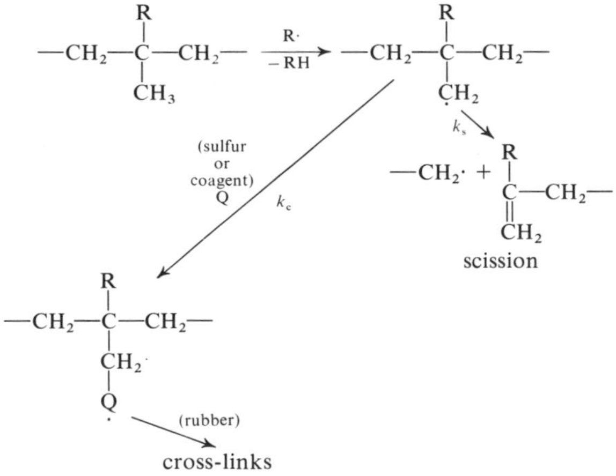

Scheme 13.

D. Peroxide Vulcanization of Silicone Rubbers

The structure of silicone rubbers can be represented by

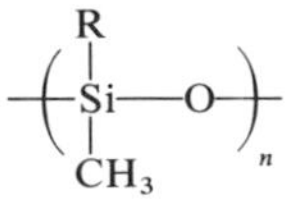

$$-\left(\underset{\underset{CH_3}{|}}{\overset{\overset{R}{|}}{Si}}-O\right)_n-$$

where R can be methyl, phenyl, vinyl, trifluoropropyl, or 2-cyanoethyl [74]. Silicone rubbers containing vinyl methyl siloxane groups can be cured by dialkyl peroxides such as dicumyl peroxide. Others, such as the dimethyl siloxane types, require diacyl peroxides such as bis-(2,4-dichlorobenzoyl) peroxide.

In the case of the saturated siloxanes, the mechanism is hydrogen atom abstraction by a peroxide-derived radical, followed by polymeric radical coupling to give cross-links. Inefficiencies or nonproductive use of peroxide result from the coupling of peroxide-derived radicals with polymeric radicals or from the decomposition of the primary radicals [89]. The incorporation of unsaturated sites into the polymer in the form of vinyl groups provides further opportunities for cross-linking. Both the initial attack upon the polymer chains and cross-linking can now proceed by radical addition reactions as well as by abstraction. Thus the process becomes more efficient [90–94].

Vulcanization is usually done in two steps: A preliminary vulcanization in a mold followed by a high-temperature (180°C) postcure in air. The latter is applied for two reasons: to remove acidic materials, which are detrimental to vulcanization and may catalyze hydrolytic decomposition of the vulcanizate [95]; and because the high temperature in air causes formation of additional cross-links [96] of the following type:

$$\begin{array}{ccc} & \underset{|}{CH_3} & \underset{|}{CH_3} \\ \sim\!\!\!-Si\!-\!O\!-\!Si\!-\!O\!\sim \\ & \underset{|}{CH_3} & \underset{|}{O} \\ & \underset{|}{CH_3} & \underset{|}{} \\ \sim\!\!\!Si\!-\!O\!-\!Si\!-\!O\!\sim \\ & \underset{|}{CH_3} & \underset{|}{CH_3} \end{array}$$

E. PEROXIDE VULCANIZATION OF URETHANE ELASTOMERS

Urethane elastomers suitable for peroxide vulcanization are typically prepared from poly(mixed ethylene and propylene) adipate and 4,4′-methylenediphenylisocyanate [97].

$$O\!\!\left(\!R\!-\!O\!-\!\overset{O}{\overset{\|}{C}}\!-\!C_4H_8\!-\!\overset{O}{\overset{\|}{C}}\!-\!O\!\right)_{\!x}\!\!R\!-\!OH$$

$$+\; O\!\!=\!\!C\!\!=\!\!N\!-\!\!\langle\!\!\bigcirc\!\!\rangle\!-\!CH_2\!-\!\!\langle\!\!\bigcirc\!\!\rangle\!-\!N\!\!=\!\!C\!\!=\!\!O \longrightarrow$$

$$\left[\overset{O}{\overset{\|}{C}}\!-\!NH\!-\!\!\langle\!\!\bigcirc\!\!\rangle\!-\!CH_2\!-\!\!\langle\!\!\bigcirc\!\!\rangle\!-\!NH\!-\!\overset{O}{\overset{\|}{C}}\!-\!O\!\!\left(\!R\!-\!O\!-\!\overset{O}{\overset{\|}{C}}\!-\!C_4H_8\!-\!\overset{O}{\overset{\|}{C}}\!-\!O\!\right)_{\!x}\!\!R\!-\!O\right]_{\!y}$$

Peroxide can cure by abstracting hydrogen atoms from arylated methylene groups [98], but hydrogen atoms may also be abstracted from α-methylene groups of the adipate moieties [99]. Vulcanization efficiencies, though usually adequate, can be increased by incorporating urea structures into the polymer chain [91, 97].

F. Sample Recipes for Peroxide Vulcanization

Examples of starting point recipes for peroxide vulcanization are given in Table IX. It is apparent that peroxide vulcanization is most versatile. This is especially true if one uses coagents or small amounts of sulfur. Outstanding characteristics of peroxide vulcanizates are low permanent set and thermal stability of the network. Butyl rubber is possibly the only type that cannot be cured well by peroxides [74].

TABLE IX

Sample Recipes for Peroxide Vulcanization[a]

	NR	SBR	EPR		Silicone	Millable urethane
			1	2		
Dicumyl peroxide	1.0	1.5	2.7	2.7	—	1.6
Bis(2,4-dichloro benzoyl) peroxide	—	—	—		1.0	—
Buton 150 (high 1,2-PB)	—	—	—	5	—	—
Sulfur	—	0.3	0.32		—	—
Typical vulcanization conditions[b]						
T (°C)	150	150	160	160	115, 250[c]	153
t (min)	45	45	30	30	15, 1440[c]	45

[a] Concentrations in phr.
[b] Conditions change depending on other aspects of the compositions.
[c] Temperature or time of postcure in air.

VII. Cross-Linking by Chain Extension Reactions

Chain extension is the coupling of low molecular weight polymer chains by nearly quantitative end group reactions [97]. It is represented as

$$X\text{\textasciitilde\textasciitilde}X + Y\text{\textasciitilde\textasciitilde}Y$$
$$\downarrow$$
$$-X\text{\textasciitilde\textasciitilde}X-Y\text{\textasciitilde\textasciitilde}Y-X\text{\textasciitilde\textasciitilde}X-Y\text{\textasciitilde\textasciitilde}Y-$$

where X and Y are functional groups that react with one another to form links between chain ends. Alternately, one of the two components can be a small difunctional molecule.

$$X\text{\textasciitilde\textasciitilde}X + Y-Y$$
$$\downarrow$$
$$-X\text{\textasciitilde\textasciitilde}X-Y-Y-X\text{\textasciitilde\textasciitilde}X-Y-Y-X\text{\textasciitilde\textasciitilde}X-$$

Such reactions can be either addition or condensation. Examples are:

X groups	Y groups	Type of polymer
—COOH	—OH	Polyesters
—COOH	—NH$_2$	Polyamides
—N=C=O	—OH	Polyurethane
—N=C=O	—NH$_2$	Polyurea
$-\overset{\mid}{\underset{\mid}{Si}}-OR$	$-\overset{\mid}{\underset{\mid}{Si}}-OH$	Silicones

If now some of the difunctional groups are replaced by trifunctional molecules, cross-linking will occur in addition to chain extension. It is, though, necessary that the number of X groups be about equal to the number of Y groups; otherwise, a very incomplete network would be formed. The scheme for this type of cross-linking can be written:

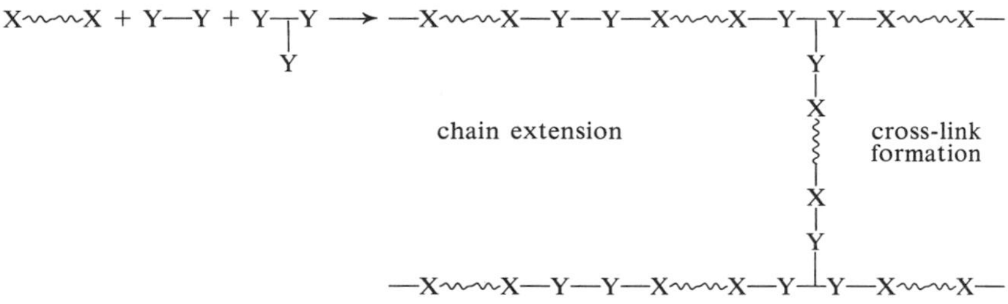

X〰〰X + Y—Y + Y⊤Y → —X〰〰X—Y—Y—X〰〰X—Y⊤Y—X〰〰X—
           Y   Y

chain extension   X   cross-link
                                X   formation

—X〰〰X—Y—Y—X〰〰X—Y⊥Y—X〰〰X—

In liquid, or "cast," urethane elastomers, triols (such as trimethylolpropane) or trifunctional hydroxy-terminated polyethers (such as polypropylene oxide modified with glycerine) are used [97].

Additional opportunities for cross-link formation exist in urethane elastomer technology because after an isocyanate reacts to give a urethane linkage, the hydrogen of the urethane linkage itself can further react with additional diisocyanate to give cross-links by allophanate formation:

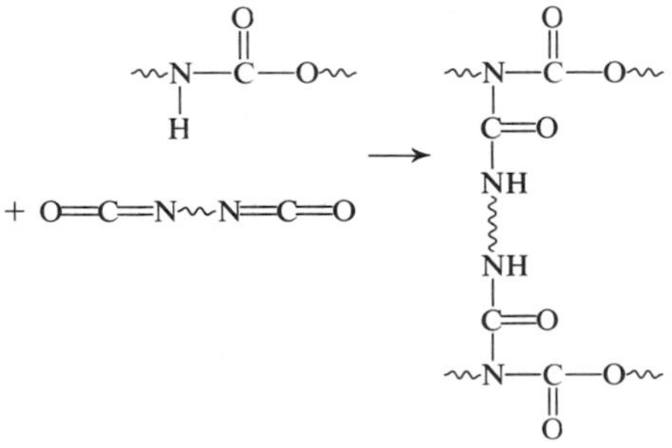

Similarly, urea or amide linkages in linear polymer chains can react with diisocyanate to give biuret or acyl urea cross-links.

$$\text{biuret} \qquad \text{acyl urea}$$

An example of cross-linking reactions by chain extension is the use of the so-called liquid rubbers, which are frequently hydroxy-terminated low molecular weight (of order 3000) polybutadienes or SBR [100] and generally contain trihydroxylated molecules. Typical hydroxyl contents are 2.2–2.6 OH groups per molecule. If such a material is mixed with an equivalent amount of a diisocyanate such as toluene diisocyanate (TDI) and a small amount of catalyst such as dibutyltin dilaurate, chain extension and cross-link formation will occur rapidly at 100°C. Such cross-linked compositions, however, are generally lacking in ultimate properties unless the liquid rubber is first reinforced by thoroughly mixing in a reinforcing carbon black. Precautions such as complete removal of water and vacuum degassing are essential to prevent foaming, since isocyanates react rapidly with water to give amine and carbon dioxide. This is especially true in the presence of the catalyst.

The ultimate properties of cross-linked liquid rubber may also be improved by the use of low molecular weight glycols or diamines. The liquid rubber is first mixed with an excess of the diisocyanate and aged for a day or two.

$$\text{HO}\text{\textasciitilde\textasciitilde}\text{OH} + \text{HO}\text{\textasciitilde\textasciitilde}\text{OH} + \text{TDI (excess)}$$
$$\underset{\text{OH}}{|}$$
$$\downarrow$$
$$\text{OCN}\text{\textasciitilde\textasciitilde}\text{NCO} + \text{OCN}\text{\textasciitilde\textasciitilde}\text{NCO} + \text{TDI (unused excess)}$$
$$\underset{\text{NCO}}{|}$$

This mixture, or "prepolymer," is then mixed with enough glycol or diamine to react with the catalyst and all of the —NCO groups. If a glycol is used, the network structure formed contains rubber polymer segments interspersed between glycol urethane polymer segments,

$$\left(\text{G—O—}\overset{\overset{\text{O}}{\|}}{\text{C}}\text{—NH—}\underset{}{\overset{\text{CH}_3}{\bigcirc}}\text{—NH—}\overset{\overset{\text{O}}{\|}}{\text{C}}\text{—O}\right)_n$$

where G is an alkylene group derived from the glycol.

The aromatic urethane, or "hard," segments have a reinforcing function, since they tend to agglomerate into hard domains, which act as well-bonded reinforcing filler particles. It should be noted that if a diamine is used, the hard repeating unit would consist of polyurea sections. At any rate the effect is called the block polymer effect. It is important in considering block polymers.

VIII. Thermoplastic Elastomers

Thermoplastic elastomers are not vulcanized by the usual chemical means. The network of supporting chains is developed during the agglomeration of "hard" segments or blocks within the polymer chains. These segments form hard domains which remain connected to soft or rubbery blocks. The polymer molecules thus are composed of two types of segments, referred to as "hard" and "soft." Hard segments have melting points (if crystallizable) or glass transition temperatures (if not crystallizable) that are well above use temperatures. Soft segments are not crystallizable or have very low melting points and have glass transition temperatures well below use temperatures. The hard domains then act as both cross-links and reinforcing filler particles [101].

The great advantage of these block elastomers is that they can be processed as thermoplastics. They are heated above the melting point (T_m) and/or glass transition temperatures (T_g) of the hard segment to permit flow for forming and molding. Upon cooling, crystalline or glass domains develop which act as junctures in a molecular network of supporting chains.

The sequence or arrangement of types of segments or blocks in the polymer molecule is important. If the hard segment is designated A (—) and the soft segment B (~) possible arrangements are:

A—B

A—B—A

B—A—B

—(A—B)$_n$—

A—A—A—A—A—
 | | | |
 B   B   B   B

B—B—B—B—B—
 | | | |
 A   A   A   A

For good elastomeric properties, terminal soft segments and soft segment branches should be avoided, since they become nonsupporting chains or loose ends after the hard segments are agglomerated into network junctures. The formation of hard domains and the resulting network might be expressed by the schematic representation in Fig. 14. Many examples can be found in the literature [102–104].

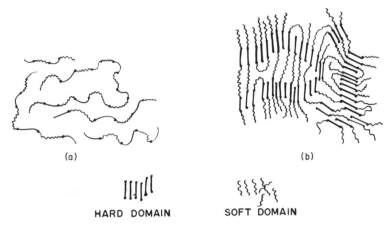

(a)   (b)

HARD DOMAIN   **SOFT DOMAIN**

Fig. 14. Formation of hard domains (————): (a) molten; (b) solidified.

The earliest cases of thermoplastic block elastomers were, as stated, de-rived from diisocyanates. Hydroxy-terminated polyalkylene ethers, or adipate polyesters treated first with excess diisocyanate, then with low molecular weight glycol, result in a polyether or polyester–polyglycol urethane block copolymer wherein the polyglycol urethane sections are the hard segments. More recently, ionic polymerization techniques have permitted the synthesis of polystyrene–poly(1,4-butadiene) A—B—A thermoplastic elastomers [102, 105–107]. They have the disadvantage of a failing dimensional stability at temperatures over about 60°C, because the polystyrene segments agglomerate to noncrystalline glassy islands of a T_g of only about 100°C.

Another, the more recent type, is the thermoplastic elastomer composed of aromatic polyester–polyalkylene ether $-(A—B)_{\overline{n}}$. Unlike the A—B—A styrene–butadiene type, they have good dimensional stability at elevated temperatures, since their crystallizable hard segments melt between 158°C and about 201°C, depending on their composition [108].

Generally, as the proportion of hard segment increases, elastomer stiffness increases. In the case of thermoplastic elastomers based on crystalline hard blocks, increases in hard-segment chain length lead also to higher melting points.

The advantages of the thermoplastic elastomers consist, as stated, of their ability to be processed as thermoplastics and of the fact that no additional processing is required for chemical vulcanization. Disadvantages relate to dimensional instability at elevated temperatures and lack of recovery from large deformations. It should be noted, however, that if hard segments of high enough melting points (or glass transition temperatures) are used, the dimen-sional stability at elevated temperatures is greatly improved and some limits on recovery from very large deformations remain as the only disadvantage.

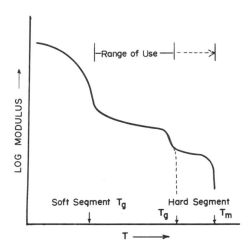

Fig. 15. The effect of segment transition temperatures on block polymer modulus.

The effect of segment transition temperatures on the modulus of block polymers is illustrated by Fig. 15 [102]. The effect on the temperature range of useful properties is apparent. Below the T_g of the soft segment, the polymers are not elastomeric; they are quite stiff and may even be brittle. Above the highest-temperature transition of the hard segments, they have the characteristics of viscous liquids; but in the middle range of hard block islands embedded in a rubbery soft block matrix, they are highly useful elastomers.

REFERENCES

1. P. J. Flory, "Principles of Polymer Chemistry." Cornell Univ. Press, Ithaca, New York, 1953.
2. E. V. Kuvshinskiĭ and E. A. Sidorovich, *Rubber Chem. Technol.* **32,** 662 (1959).
3. F. Bueche, *J. Polym. Sci.* **24,** 189 (1957).
4. F. Bueche, *J. Polym. Sci.* **33,** 259 (1958).
5. M. L. Studebaker, *Rubber Chem. Technol.* **39,** 1359 (1966).
6. W. L. Cox and C. R. Parks, *Rubber Chem. Technol.* **39,** 785 (1966).
7. L. Bateman, ed., "The Chemistry and Physics of Rubber-Like Substances." MacLaren, London, 1963.
8. K. W. Scott, O. Lorenz, and C. R. Parks, *J. Appl. Polym. Sci.* **8,** 2909 (1964).
9. D. S. Campbell, *J. Appl. Polym. Sci.* **14,** 1409 (1970).
10. C. G. Moore and M. Porter, *J. Appl. Polym. Sci.* **11,** 2227 (1967).
11. G. E. Decker, R. W. Wise, and D. Guerry, *Rubber Chem. Technol.* **36,** 451 (1963); see also "Annual Book of ASTM Standards," Part 37, Method D-2084-75 (1975).
12. W. R. Deason and R. W. Wise, *Rubber Chem. Technol.* **42,** 1481 (1969).
13. D. G. Lloyd, *J. Rubber Res. Inst. Malaya* **22,** 399 (1969).
14. C. Goodyear, U.S. Patent 3633 (1844).
15. E. H. Farmer and F. W. Shipley, *J. Polym. Sci.* **1,** 293 (1946).
16. E. H. Farmer, *J. Chem. Soc.* p. 1519 (1947).

17. E. H. Farmer, *J. Soc. Chem. Ind.* **66**, 86 (1947).
18. G. F. Bloomfield and R. F. Naylor, *Int. Congr. Pure Appl. Chem.*, *11th, London* **2**, 7 (1947).
19. L. Bateman, C. G. Moore, and M. Porter, *J. Chem. Soc.* p. 2866 (1958).
20. J. Dillhoefer, personal communication (1970).
21. R. H. Campbell and R. W. Wise, *Rubber Chem. Technol.* **37**, 635 (1964).
22. R. H. Campbell and R. W. Wise, *Rubber Chem. Technol.* **37**, 650 (1964).
23. P. L. Hu and W. Scheele, *Kautsch. Gummi* **15**, 440 (1962).
23a. M. M. Coleman, J. R. Skelton, and J. L. Koenig, *Rubber Chem. Technol.* **46**, 938 (1973).
24. T. D. Skinner, *Rubber Chem. Technol.* **45**, 182 (1972).
25. R. T. Armstrong, J. R. Little, and K. W. Doak, *Rubber Chem. Technol.* **17**, 788 (1944).
26. C. D. Trivette, personal communication (1963).
26a. J. R. Wolfe, Jr., T. L. Pugh, and A. S. Killian, *Rubber Chem. Technol.* **41**, 1329 (1968).
27. A. Y. Coran, *Rubber Chem. Technol.* **38**, 1 (1965).
28. C. D. Trivette and A. Y. Coran, *J. Org. Chem.* **31**, 100 (1966).
29. I. Kende, T. L. Pickering, and A. V. Tobolsky, *J. Am. Chem. Soc.* **87**, 5582 (1965).
30. A. Y. Coran, *Rubber Chem. Technol.* **37**, 689 (1964).
31. C. D. Trivette and A. Y. Coran, U.S. Patent 3,354,131 (1967).
32. R. T. Leib, A. B. Sullivan, and C. D. Trivette, *Rubber Chem. Technol.* **43**, 1188 (1970).
32a. C. D. Trivette, Jr., E. Morita, and O. W. Maendes, *Rubber Chem. Technol.* **50**, 570 (1977).
33. P. N. Son, *Rubber Chem. Technol.* **46**, 999 (1973).
34. E. Morita, J. J. D'Amico, and E. J. Young, *Rubber Chem. Technol.* **41**, 721 (1968).
35. G. R. Cotten, B. B. Boonstra, D. Rivin, and F. R. Williams, *Dtsch. Kautsch. Ges., West Berlin, 1968; Kaut. Gummi Kunstst.* **22**, 477 (1969).
36. S. J. Stewart, R. I. Leib, and J. E. Kerwood, *Rubber Age (N.Y.)* **102**(10), 56 (1970).
37. R. R. Hickernell, unpublished work (1970).
38. S. Buchanan and W. D. Rae, *Trans. Inst. Rubber Ind.* **21**, 323 (1945).
39. W. A. Gurney, *Trans. Inst. Rubber Ind.* **21**, 31 (1945).
40. W. A. Gurney, *Rubber Chem. Technol.* **19**, 199 (1946).
41. A. E. Hicks and F. Lyon, *Adhes. Age* **12**(May), 21 (1969).
42. A. Maeseele and E. Debruyne, *Rubber Chem. Technol.* **42**, 613 (1969).
43. A. Y. Coran, *Rubber Chem. Technol.* **37**, 673 (1964).
44. C. G. Moore, L. Mullins, and P. McL. Swift, *J. Appl. Polym. Sci.* **5**, 293 (1961).
45. C. G. Moore and B. R. Trego, *J. Appl. Polym. Sci.* **5**, 299 (1961).
46. T. D. Skinner and A. A. Watson, *Rubber Chem. Technol.* **42**, 404 (1969).
47. R. M. Russell, T. D. Skinner, and A. A. Watson, *Rubber Chem. Technol.* **42**, 418 (1969).
48. A. Y. Coran, *Jubilee Conf. Inst. Rubber Ind., Leamington, Engl., 1971*.
49. C. R. Parks and O. Lorenz, *Ind. Eng. Chem., Prod. Res. Dev.* **2**(4), 279 (1963).
50. C. L. M. Bell and J. I. Cuneen, *J. Appl. Polym. Sci.* **11**, 2201 (1967).
51. A. N. Gent, P. B. Lindley, and A. G. Thomas, *J. Appl. Polym. Sci.* **8**, 455 (1964).
52. G. J. Lake and P. B. Lindley, *Rubber J.* **146**(10), 24 (1964).
53. G. J. Lake and P. B. Lindley, *Rubber J.* **146**(11), 30 (1964).
54. W. Cooper, *J. Polym. Sci.* **28**, 195 (1958).
55. B. A. Dogadkin, Z. N. Tarasova, and I. I. Goldberg, *Proc. Rubber Technol. Conf., 4th, London, 1962* p. 65 (1963).
56. E. B. McCall, *J. Rubber Res. Inst. Malaya* **22**, 354 (1969).
57. S. Van der Meer, *Rev. Gen. Caoutch. Plast.* **20**, 230 (1943).
58. C. Thelamon, *Rubber Chem. Technol.* **36**, 268 (1963).
58a. A. Giller, *Kaut. Gummi Kunstst.* **19**, 188 (1966).
59. J. Rehner and P. J. Flory, *Ind. Eng. Chem.* **38**, 500 (1946).
60. G. Bruni and E. Geiger, *Rubber Age (N.Y.)* **22**(4), 187 (1927).

61. R. Pummerer and W. Grundel, *Rubber Chem. Technol.* **2**, 373 (1929).
62. J. Rehner and P. J. Flory, *Rubber Chem. Technol.* **19**, 900 (1946).
63. R. F. Martell and D. E. Smith, *Rubber Chem. Technol.* **35**, 141 (1962).
64. A. B. Sullivan, *J. Org. Chem.* **31**, 2811 (1966).
65. C. S. L. Baker, D. Barnard, and M. Porter, *Rubber Chem. Technol.* **43**, 501 (1970).
66. P. Kovacic and P. W. Hein, *Rubber Chem. Technol.* **35**, 528 (1962).
67. H. M. R. Hoffmann, *Angew. Chem., Int. Ed. Engl.* **8**, 556 (1969).
68. R. K. Hill and M. Rabinovitz, *J. Am. Chem. Soc.* **86**, 965 (1964).
69. A. Y. Coran, unpublished work (1971).
69a. F. P. Baldwin, P. Borzel, C. A. Cohen, H. S. Makowoski, and J. F. van Castle, *Rubber Chem. Technol.* **43**, 522 (1970).
70. W. Hofmann, *in* "Kautschuk-Handbuch" (S. Boström, ed.), Vol. 4, p. 324. Verlag Berlin. Union, Stuttgart, 1961.
71. A. G. Stevenson, *in* "Vulcanization of Elastomers" (G. Alliger and I. J. Sjothun, eds.), p. 271. Rheinhold, New York, 1964.
72. R. Pariser, *Kunststoffe* **50**, 623 (1960).
73. R. O. Becker, *Rubber Chem. Technol.* **37**, 76 (1964).
74. L. D. Loan, *Rubber Chem. Technol.* **40**, 149 (1967).
75. R. Rado and D. Simunkova, *Vysokmol. Soedin.* **3**, 1277 (1961).
76. D. K. Thomas, *Trans. Faraday Soc.* **57**, 511 (1961).
77. L. D. Loan, *J. Polym. Sci., Part A* **2**, 2127 (1964).
78. J. Scanlan and D. K. Thomas, *J. Polym. Sci., Part A* **1**, 1015 (1963).
79. E. H. Farmer and S. E. Michael, *J. Chem. Soc.* p. 513 (1942).
80. E. H. Farmer and C. G. Moore, *J. Chem. Soc.* p. 131 (1951).
81. E. H. Farmer and C. G. Moore, *J. Chem. Soc.* p. 142 (1951).
82. L. D. Loan, *J. Appl. Polym. Sci.* **7**, 2259 (1963).
83. B. M. E. Van der Hoff, *Ind. Eng. Chem., Prod. Res. Dev.* **2**, 273 (1963).
84. D. K. Thomas, *J. Appl. Polym. Sci.* **6**, 613 (1962).
85. C. R. Parks and O. Lorenz, *J. Polym. Sci.* **50**, 287 (1961).
86. K. W. Scott, *J. Polym. Sci.* **58**, 517 (1962).
87. A. E. Robinson, J. V. Marra, and L. O. Amberg, *Ind. Eng Chem., Prod. Res. Dev.* **1**, 78 (1962).
88. L. P. Lenas, *Rubber Chem. Technol.* **37**, 229 (1964).
89. A. A. Miller, *J. Polym. Sci.* **42**, 441 (1960).
90. F. M. Lewis, *Rubber Chem. Technol.* **35**, 1222 (1962).
91. P. G. Bork and C. W. Roush, *in* "Vulcanization of Elastomers" (G. Alliger and I. J. Sjothun, eds.), p. 366. Reinhold, New York, 1964.
92. M. L. Dunham, D. L. Bailey, and R. Y. Mixer, *Ind. Eng. Chem.* **49**, 1373 (1957).
93. T. R. Harper, A. D. Chipman, and G. M. Konkle, *Rubber World* **137**, 711 (1958).
94. D. K. Thomas, *Polymer* **7**, 243 (1966).
95. C. W. Roush, J. Kosmider, and R. L. Baufer, *Rubber Age* (*N.Y.*) **94**, 744 (1964).
96. W. Hofmann, "Vulcanization and Vulcanizing Agents," p. 242. Palmerton Publ. Co., New York, 1967.
97. J. H. Saunders and K. C. Frisch, "Polyurethanes Chemistry and Technology, Part II: Technology." Wiley (Interscience), New York, 1964, and references therein.
98. O. Bayer and E. Mueller, *Angew. Chem.* **72**, 934 (1960).
99. S. V. Urs, *Ind. Eng. Chem., Prod. Res. Dev.* **1**, 199 (1962).
100. P. W. Ryan, *J. Elastoplast.* **3**, 57 (1971).
101. M. Morton, J. E. McGrath, and P. C. Juliano, *J. Polym. Sci., Part C* **26**, 99 (1969).
102. D. C. Allport and W. H. Janes, eds., "Block Copolymers." Wiley (Halsted Press), New York, 1973.

103. G. Holden, E. T. Bishop, and N. R. Legge, *J. Polym. Sci., Part C* **26**, 37 (1969).
104. E. T. Bishop and S. Davidson, *J. Polym. Sci., Part C* **26**, 59 (1967).
105. H. L. Hsieh, *Plast. Age* **40**, 394 (1965).
106. A. V. Snider, *Rubber World* **152**(4), 90 (1965).
107. R. D. Deonin, *SPE J.* **23**(1), 45 (1967).
108. M. Brown and W. K. Witsiepe, *Rubber Age* (*N. Y.*) **104**(Mar.), 35 (1972).

Chapter **8**

Reinforcement of Elastomers by Particulate Fillers

GERARD KRAUS

PHILLIPS PETROLEUM COMPANY
BARTLESVILLE, OKLAHOMA

I. Introduction

It is not at all uncommon for particulate fillers such as carbon black to increase the strength of vulcanized rubbers more than tenfold. Thus it is hardly surprising that relatively few applications of elastomers utilize the polymer in the unfilled state.

The degree of reinforcement provided by a filler depends on a number of variables, the most important of which is the development of a large polymer–filler interface. Such can only be furnished by particles of colloidal dimensions. Spherical particles 1 μm in diameter have a specific surface area of 6 m^2/cm^3. This constitutes roughly the lower limit of significant reinforcement. The upper limit of useful specific surface area is of the order of 300–400 m^2/cm^3, and is set by considerations of dispersibility, processability of the unvulcanized mix, and serious loss of rubbery characteristics of the composite.

The most reinforcing fillers known are carbon blacks and silicas. Silicates, clays, whiting (calcium carbonate), and other "mineral fillers" are used exten-

339

Copyright © 1978 by Academic Press, Inc.
All rights of reproduction in any form reserved.
ISBN 0-12-234360-3

sively where a high degree of reinforcement is not essential. These are discussed in more detail in Chapter 9. Since all reinforcement phenomena can be amply demonstrated with carbon blacks, we shall direct our attention primarily to this class of materials, with occasional reference to other fillers wherever characteristic differences exist.

II. Carbon Black*

A few introductory remarks on carbon black will serve to make clear some of the observations discussed later. Carbon blacks are prepared by incomplete combustion of hydrocarbons or by thermal cracking.

In the now obsolete and virtually extinct channel process, natural gas is burned in small burners with a sooty diffusion flame and the carbon collected on a cool surface—a section of channel iron. Relatively free access of the hot product to air leads to oxidation of the surface, one of the unique features of this process. The surface oxygen complexes give *channel black* acidic character, manifesting itself in high adsorption capacity for bases and low pH of aqueous slurries. Surface acidity has a repressing effect on sulfur vulcanization.

At present almost all rubber-reinforcing carbon blacks are manufactured by oil furnace processes. A fuel, either gas or oil, is burned in an excess of air, producing a turbulent mass of hot gases into which the conversion oil is injected. Reaction to finely divided carbon is complete within milliseconds and the hot smoke stream is quenched with water. *Furnace blacks* have low oxygen contents (usually less than 1%) and their surface is neutral or alkaline.

In the thermal process, oil or, more frequently, natural gas is cracked in absence of oxygen at a hot refractory surface. The process produces unoxidized blacks of relatively small specific surface area ($6-15$ m^2/g) (*thermal blacks*).

The American Society of Testing Materials has established a carbon black classification system based essentially on particle size and degree of surface oxidation (the factor that primarily affects the curing time of rubber–carbon black compounds) with the intention of replacing an earlier letter-type nomenclature. The N series comprises "normal-curing" furnace and thermal blacks, the S series "slow-curing" channel blacks and deliberately oxidized furnace blacks. A three-digit suffix identifies particle size as shown in the accompanying tabulation. Different ASTM numbers inside each category identify blacks by other characteristics, mainly "structure" (see Section III). For example, the designation N-220 immediately identifies a carbon black as normal curing with particle size in the 20–25 size range, but additional specifications are needed to distinguish it from, say, N-234.

The ASTM classification is not a precise one, because mean particle sizes are almost impossible to measure accurately. Consequently there is frequently

---

* See also Burgess *et al.* [1] and Kraus [2, 2a].

considerable overlap, particularly between the 200 and 300 series, each of which covers a range of only 5 nm.

Range of ASTM numbers	Nominal number average particle diameter (nm)	Old classification
900–999	201–500	MT
800–899	101–200	FT
700–799	61–100	SRF
600–699	49–60	GPF, HMF
500–599	40–48	FEF
400–499	31–39	FF
300–399	26–30	HAF, EPC
200–299	20–25	ISAF
100–199	11–19	SAF
000–099	1–10	—

III. Filler Morphology and Its Characterization*

Figure 1 shows an electron micrograph of a typical reinforcing oil furnace carbon black. The material consists of irregular chainlike, branched aggregates of nodular subunits. The latter are the "particles" of the ASTM classification. This terminology is unfortunate, since it implies their existence as independent entities. In fact, the nodular subunits are firmly fused together, as is clearly seen in the high-resolution electron micrograph in Fig. 2. Thus, the smallest discrete entity existing in rubber is always the aggregate. The extent of inter-nodular fusion is commonly termed "structure." "High structure" implies a large number of nodules per aggregate, but it is impractical to define the degree of structurization in this manner. A practical scale of structure can be constructed in terms of the results of vehicle demand tests (discussed later in this section) and is widely used in the industry. The gross morphology of reinforcing silicas is similar to that of carbon blacks, except that fusion tends to be more extensive, with some aggregates resembling coiled rods. Clays and other mineral fillers exhibit essentially no structure in the sense considered here.

Figure 2 reveals clear signs of internal order. Each nodule is a paracrystalline domain of concentrically arranged quasi-graphitic layer planes, the spacing of which (~ 3.5 Å) slightly exceeds that of graphite. Subtle differences in para-crystalline structure exist between carbon blacks, but they are of little consequence to their behavior. Only when crystallinity is enhanced by extreme heat treatment to effect graphitization are significant changes in reinforcing ability observed.

* See also Kraus [3], Hess et al. [4], Medalia and Heckman [5], and Hess et al. [6].

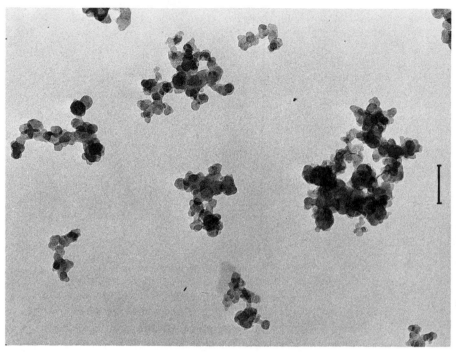

Fig. 1. Transmission electron micrograph of typical reinforcing oil furnace carbon black. (The bar represents 100 nm.)

It would be logical to attempt to relate physical properties of reinforced rubber directly to the sizes and shapes of the primary aggregates. The great variety of these, however, makes this a very difficult task, necessitating characterization of carbon black morphology by computer-automated image analysis. This technique, which is the subject of much current research, yields an enormous amount of geometrical detail [6]. Fortunately, an understanding of the essential features of the reinforcement process does not require so detailed a description.

It is adequate for most purposes to describe filler morphology by two parameters—one related to a mean particle size, and one to structure. The traditional particle size count of electron micrographs, in which each nodule is counted as a discrete sphere, is unrealistic and has justly lost favor in recent years. A far superior technique is the direct determination of the specific surface area. This is most conveniently accomplished by adsorption methods [7].

Structure may be expressed quantitatively by *vehicle demand*. When a moderately viscous vehicle (e.g., dibutylphthalate, DBP) is added gradually to dry carbon black in an internal mixer fitted with a device measuring the torque on the mixer blades, no appreciable torque is developed until the inter-

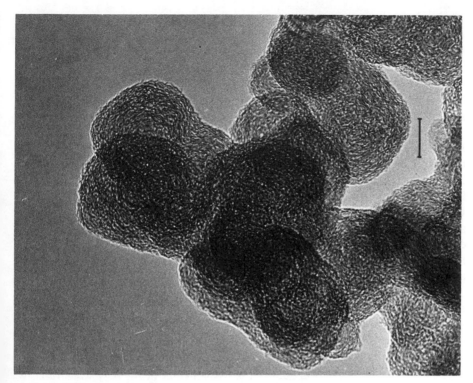

Fig. 2. High-resolution phase-contrast electron micrograph of same carbon black as in Fig. 1. (The bar represents 10 nm.)

stices between carbon black particles are filled with vehicle. The latter point is marked by an almost instantaneous rise in torque and is taken as the end point of the "titration." The quantity of vehicle added up to the end point is a measure of void volume of the filler. It is related to structure in the following manner. It has been shown that the mass density of carbon black aggregates decreases radially outward from the center of mass, a characteristic of random flocs [8]. In other words, the packing of the nodules inside an aggregate becomes more open with increasing distance from the center. Consequently, a large aggregate containing many nodules (high structure) contains a smaller quantity of solid matter per volume of a circumscribed sphere pervaded than a small one, made up of a few nodules (low structure). The vehicle demand, expressed as volume per weight of pigment, comprises this internal void volume, augmented by additional void volume resulting from the manner in which the aggregates pack at the end point. If the latter does not vary too much from one carbon black to another, the vehicle demand gives a good estimate of structure. It has been found that subjecting the black to high pressure before measuring its vehicle demand tends to equalize the between-aggregate packing density [9].

The standard industry test for structure uses dibutylphthalate (DBP) as the vehicle, either with or without precompression of the black.

Of many possible adsorption methods for determination of specific surface area, three deserve special mention [7]. Iodine adsorption is widely used in quality control of carbon black. It is subject to occasionally large systematic errors arising from differences in surface chemical characteristics of the black and so provides only a rough estimate of the true specific surface area. Nitrogen adsorption is far superior and provides an excellent measure of total specific surface. However, in some carbon blacks a significant fraction of the total surface is contained in micropores, only a few angstroms in size, which are not accessible to rubber. This pore surface can contribute up to 20 m^2/g to the nitrogen surface area of N-100 and N-200 series blacks. Fortunately, this error can be eliminated by use of larger adsorbate molecules. Particularly suited for this third method are large surfactant molecules which are adsorbed on carbon black from aqueous solution, giving rise to Langmuir-type isotherms, from which the monolayer coverage is easily determined. Hexadecyltrimethylammonium bromide (CTAB) is a good example of this [7].

Another method frequently used in carbon black quality control is determining the tinting strength [10]. In tint tests, fixed small amounts of carbon black are mixed with a large constant amount of white pigment in the presence of a vehicle to prepare gray pastes the optical reflectances of which are measured. The fraction of incident radiation remitted to the detector depends on the extent of absorption and scattering by the pigments in the paste. The design of the experiment assures that practically all the absorption is due to carbon black, while almost all the scattering is due to the white pigment. The latter is substantially constant. The optical absorbance of a pigment increases with decreasing particle size, so that smaller particles give rise to darker pastes or higher tinting strength. What makes the tint method difficult to interpret quantitatively are the effects of structure. The absorbing units are not the carbon black nodules, but the primary structure aggregates. Tinting strength increases with both increasing specific surface area and decreasing structure, both changes tending to produce smaller aggregates. However, tinting strength can vary significantly even at constant specific surface area and vehicle demand, since aggregate sizes and shapes are widely distributed and the method senses a complex average of both.

IV. Physical and Chemical Interactions at the Filler Surface*

Whether or not reinforcement requires the formation of primary valence bonds, anchoring polymer chains to the filler surface, has been the subject of much controversy. The origin of this controversy can be traced to differing

* See also Kraus [2, 3].

views of what constitutes "true" reinforcement. The experimental evidence is quite clear:

(1) chemical bonding at the filler–rubber interface is *not* a necessary condition for reinforcement:

(2) chemical bonding leads to the unique combination of mechanical properties associated with carbon black, the best overall reinforcing agent known.

Let us examine the evidence in more detail. If the existence of chemical bonding can be established for a filler–polymer system with a high degree of probability, and if means for "deactivating" the surface can be found which do not fundamentally alter the filler morphology, then a comparison of vulcanizate physical properties permits the effects of bonding to be established. Conversely, chemical filler–polymer linkages can be introduced into a system devoid of (or at least deficient in) such linkages.

In the example of carbon black the evidence overwhelmingly points to the existence of chemisorptive linkages [3]. First, carbon black surfaces contain functional groups capable of reacting with polymer molecules to form grafts during processing and vulcanization [11]. Second, numerous reactions of hydrocarbon polymers with carbon black have been demonstrated. In fact, several possible mechanisms exist by which grafts may be formed. For example, carbon blacks chemisorb olefins at vulcanization temperatures [12]. The degree of chemisorption is increased in the presence of sulfur. Shear-generated polymeric free radicals have been shown to graft to the carbon black surface during mixing [13]. Carbon blacks inhibit free radical reactions through quinonic surface groups; these reactions also result in formation of polymer grafts [14, 15]. Hydrogen can be exchanged between rubber and bound hydrogen situated at the edges of the large aromatic ring molecules that make up the layer planes of carbon black [16].

When carbon black is heated to 2700°C or higher in an inert atmosphere, all surface functional groups are removed [11, 17]. Oxygenated species are driven off first, being evolved as CO and CO_2, followed by loss of combined hydrogen. The change is accompanied by formation of a more regular graphitic crystal structure and by the disappearance of active sites for physical adsorption. Thus, the surface of graphitized carbon black is not only chemically unreactive, but also energetically homogeneous with regard to van der Waals interactions.

Table I shows results of one of several reported studies with graphitized carbon blacks [18], all of which are in excellent agreement. Although the gross morphology (surface area and structure) changes only little upon graphitization, the amount of bound rubber (rubber unextractable from the unvulcanized mix) drops dramatically, as does the stress at 300% extension (σ_{300}) and the resistance to abrasion. However, the tensile strength (σ_b) remains remarkably high, particularly when compared to the value for an unfilled SBR vulcanizate

TABLE I

EFFECTS OF GRAPHITIZATION OF CARBON BLACK ON REINFORCEMENT[a]

	Carbon black A		Carbon black B	
	Original	Graphitized	Original	Graphitized
Specific surface area[b] (m^2/g)	116	86	108	88
Vehicle demand[c] (cm^3/g)	1.72	1.78	1.33	1.54
Bound rubber (%)	34.4	5.6	30.6	5.8
σ_{300} (MPa)	14.4	3.5	10.3	2.9
σ_b (MPa)	26.2	23.4	27.6	22.7
ε_b (%)	450	730	630	750
Hysteresis[d]	0.204	0.297	0.238	0.315
Relative abrasion resistance[e]	100	34	100	47

[a] SBR 1500.
[b] By N$_2$ adsorption.
[c] Linseed oil.
[d] Method not specified in Brennan et al. [18].
[e] Laboratory test.

(~ 1.4 MPa), and the elongation at break (ε_b) increases. In related studies, tire treads containing graphitized black were found to wear some 30–35% faster than treads containing the precursor black [19, 20]. Clearly, although graphitization leads to deterioration of some features of reinforcement,* the graphitized material is by no means nonreinforcing.

Colloidal silica is a good reinforcing agent, imparting high tensile strength and excellent resistance to tearing [21]. However, in comparison to carbon black the material is deficient in modulus (σ_{300}) and resistance to abrasion. It can be made to approximate the characteristics of carbon black by use of bridging agents capable of tying the filler surface into the network by primary valence bonds. Table II shows an example in which bis(3-ethoxysilylpropyl)-tetrasulfide was added to a silica-reinforced rubber compound. As expected, the changes are in exactly the opposite direction from those accompanying the graphitization of carbon black. Deactivation has also been demonstrated with silica. Esterification of fumed silica renders it unreactive toward the polymer in amine-cured bromine-terminated poly(butadiene). The result is stress–strain behavior strikingly resembling that of graphitized carbon black [22].

Experiments with polymeric fillers have also shown that covalent bonding is not a necessary condition for the development of many strengthening effects by dispersed particles of higher modulus [23, 24].

* That is, it affects those that depend on intact carbon–rubber interfacial bonding.

TABLE II

EFFECT OF A BRIDGING AGENT ON
SILICA REINFORCEMENT OF SBR 1502[a]

	Without additive	With additive[b]
σ_{300} (MPa)	4.1	8.3
σ_b (MPa)	17.4	21.7
ε_b (%)	630	520
Abrasion loss[c] (mm^3)	126	89

[a] Data quoted from [22a]. Conventional sulfur recipe with 50 phr (parts per 100 parts of rubber) of Ultrasil VN3 silica.

[b] Bis(3-ethoxysilylpropyl)tetrasulfide, 2 phr (active material).

[c] Laboratory test.

Not infrequently *bound rubber* [25] is used as a criterion of polymer–filler interaction, or filler "activity." It is determined by solvent extraction of un-vulcanized mixes of fillers and rubbers. Good solvents are used. If the filler loading is high enough, a coherent gel is left behind containing all the filler plus the bound rubber. At smaller loadings the filler–gel disperses and has to be centrifuged to remove it from the free rubber solution, but this complication is usually avoided. Bound rubber, in this simplest form, is the result of several possible phenomena—physical adsorption, different forms of chemi-sorption, physical entrapment of free molecules, and cross-linking of free molecules to the filler–gel complex. The amount of bound rubber observed is also very dependent on experimental conditions—the chemical constitution of the rubber, its molecular weight and distribution, mixing technique, tem-perature, time, and others. Thus, a bound rubber value has no meaning per se, except in relative comparisons of fillers under closely controlled conditions. Even here, proper interpretation of results requires knowledge of a great deal of additional information [26].

Bound rubber is a strong function of the specific surface area of carbon blacks, as expected, and shows significant dependence on the degree of structure development, increasing with both variables. The latter appears to be a con-sequence of polymer penetrating the internal void space of the structure aggre-gates; there, as a result of high local carbon black concentration, molecules are adsorbed more efficiently than on the exterior surfaces. Heat treatment of rubber–black mixes raises the bound rubber value and makes it increasingly dependent on the specific surface area, thereby deemphasizing the structure dependence.

It is not difficult to see why bound rubber values tend to correlate with carbon black performance, since higher structure and larger surface areas simul-taneously lead to higher bound rubber and increased reinforcing action (see Section X). However, the fact that bound rubber is so heavily dominated by morphological features makes it a poor criterion of *specific* activity, unless differences in the latter are large and morphological differences are small (as in the example of Table I).

Several authors [27, 28] have described more elaborate techniques for estimating "bonded rubber," in which attempts are made to isolate the contribution of the most energetic adsorption mechanisms to the general bound rubber phenomenon.

With polymers of broad molecular weight distribution (the usual case) adsorption of the larger molecules at the filler surface is statistically favored [29].

V. The State of the Filler in the Rubber Mix*

It is almost axiomatic that the full effect of a reinforcing filler will be realized only when its surface is thoroughly "wetted" by the polymer in which it is completely dispersed. This requires penetration of the polymer into the occluded void space of the structure aggregates and random spatial arrangement of these aggregates in the polymer matrix.

When carbon black is mixed into rubber in conventional equipment, the first step is the penetration of rubber into the void space. This occurs before the black is randomly dispersed. If at this stage considerable rubber–black interaction occurs, subsequent dispersion is rendered more difficult since bound rubber cements many primary aggregates together. For this reason low-structure, high-surface-area blacks are difficult to disperse; their small void space and dense packing leads to high local black concentration and their large surfaces provide ample opportunity for early interaction with polymer. At the same time such blacks are quite rapidly incorporated; it is only their subsequent dispersion into individual structure aggregates that is rendered difficult. High-structure blacks are more slowly incorporated, but more easily attain eventually a satisfactory degree of dispersion.

Opinions differ as to the degree of breakdown, or attrition of structure, during high-shear mixing of blacks with rubber. The question is not an easy one to answer because it is exceedingly difficult to determine when a carbon black is dispersed completely into its primary structure aggregates. This applies not only to dispersion in rubber, but also to any preparation of carbon black dispersions for the purpose of evaluating primary aggregate sizes (e.g., by electron microscopy). Secondary agglomeration (flocculation) is an ever present obstacle to quantitative interpretations. At present it appears that primary structure breakdown during compounding is a real, but relatively small, effect.

Even in perfectly random dispersions of fillers in rubber there exist many interaggregate physical contacts. Consider a carbon black of 1.00 cm³/g DBP absorption (vehicle demand) at a volume loading of 20% in rubber. Taking

---

* See also Boonstra and Medalia [30].

1.85 g/cm^3 as the density of carbon black, the void space associated with a unit volume of black is 1.85 cm^3, and the volume of the rubber mix occupied or pervaded by the carbon black structure aggregates is 57%. It is easy to see that many interaggregate contacts will exist. In fact, in this example, their number is so great that a continuous network of carbon is formed, leading to significant electrical conductivity of the composite [31]. We shall have occasion to comment further on the secondary agglomeration network in connection with the mechanical behavior of reinforced rubbers at small strains (Section VIII).

VI. The State of the Rubber in the Filled Composite*

The adsorption of polymer segments on the filler surface must lead to some loss of mobility of the chains. Since rubbery behavior of a polymer is predicated on the ability of its molecules to engage in long-range convolutions, any loss of mobility must affect the physical properties of the rubber. The question of the degree of immobilization is, therefore, of fundamental importance. Studies of proton magnetic resonance in carbon-black-reinforced rubbers have shown that there is, indeed, loss in segmental mobility, but *severe* restriction of motion is confined to a layer about 5 Å thick at the surface, i.e., a distance of a few carbon bond lengths along the polymer chain [32]. In a vulcanizate containing 50 phr (parts per hundred or rubber) of a black with 80 m^2/g specific surface area, this represents roughly 2% of the total rubber. Obviously, mobility must increase gradually outward from the filler surface so that a thicker layer, estimated at 30 Å, finds itself under the influence of the surface [3]. These conclusions are consistent with other observations. Thus the glass transition temperature of carbon-black-filled rubbers is raised typically only by 2–3°C; substantial immobilization of a large portion of the polymer would require a much larger change [33]. Somewhat larger changes have been reported for polymers with polar groups along the chain, reinforced with inorganic fillers [3].

It would be totally incorrect to identify the immobilized rubber with bound rubber. In a filler–rubber mix, chains of the bound rubber extend far into the polymer matrix where they freely intermix with unadsorbed rubber molecules. Upon vulcanization they become part of the network, indistinguishable from (originally) free rubber.

A most useful concept in the reinforcement of elastomers by highly structurized fillers is *occluded rubber* [34, 35]. This is the rubber that finds itself in the internal void space of the structure aggregates. It assumes a special character because it is partially shielded from deformation when the rubber is strained. Again, occluded rubber is not synonymous with bound or immobilized rubber, though obviously some portion of the occluded rubber is both.

* See also Kraus [3].

In studying reinforcement effects it is often desirable to know whether the rubber matrix has material properties analogous to an otherwise identical vulcanizate in which the filler has simply been omitted. Almost invariably it has not, because fillers are not generally inert in vulcanization, accelerating or retarding the cure, and affecting the ultimate yield of cross-links. Since fillers raise the viscosity of a mix, they can also accentuate chain scission during processing.

Silicas and carbon blacks containing substantial amounts (roughly $> 1\%$) of combined oxygen repress or retard sulfur vulcanization. Modern oil furnace carbon blacks usually contain less, unless they are deliberately oxidized. Some vulcanization accelerators require carbon black for activation and, when this is true, a corresponding gum vulcanizate would be severely undercured. On the other hand, in other vulcanization systems the effects of furnace blacks on cross-link yield are small. Notable examples are dicumyl peroxide [36] and some sulfur–sulfenamide recipes [37].

VII. Properties of Unvulcanized Rubber–Filler Mixes*

The properties of unvulcanized compounded rubbers containing fillers are of great interest in connection with processing operations. In particular, one is interested in the viscosity and elastic response of these compounds.

The viscosity of a suspension of spherical particles in a medium of viscosity η_0 is given by the equation [38, 39]

$$\eta = \eta_0(1 + 2.5c + 14.1c^2) \tag{1}$$

where c is the volume fraction of filler. At very small c, the third term in the parenthesis vanishes and the equation reduces to the familiar Einstein viscosity law. Note that particle size does not enter the equation. Micron-size glass beads and medium thermal black (N-990), an almost structureless carbon black of ~ 7 m^2/g surface area, obey Eq. (1) rather well [26]. High-surface-area fillers with significant structure never do. Their behavior is highly complex and has so far defied all attempts at theoretical treatment. It is not difficult to find reasons for this. First, structure aggregates of carbon black and silica are not spheres. Moreover, the aggregates are not "free draining"; the medium cannot move freely through their internal void space and rubber occluded therein augments the effective filler concentration [34, 35]. Further, secondary agglomeration of the (structure) aggregates causes highly non-Newtonian and thixotropic behavior. Finally, adsorption of polymer molecules at the filler surface causes a lowering of the molecular weight of the free rubber as large molecules are adsorbed preferentially [29]; at the same time, loops of the attached polymer molecules interact with the free rubber through entanglement

---

* See also Kraus [2, 3].

coupling. Thus, the contribution the medium makes to the overall viscosity is no longer given by η_0 of the pure rubber. In spite of these and other complications, clear-cut trends exist between filler concentration, morphology, and rheological behavior.

In general, structure leads to higher viscosity and lower melt elasticity. This behavior can be rationalized in terms of occluded rubber. Even in the most ideal cases, the viscosity must be increased over and above that given by Eq. (1), since some portion of the occluded rubber must be effectively included in c. The amount of occluded rubber increases with structure level, as does the vehicle demand. Consequently, in a fixed formulation, good correlation is obtained between viscosity and DBP absorption, irrespective of specific surface area, as shown by data on a random selection of carbon blacks (all of the ASTM N series) differing widely in morphology (Fig. 3a).

The filled rubber mix is a composite of a continuous phase exhibiting considerable long-range elasticity and a rigid filler. If occluded rubber is, indeed, shielded from deformation, then the fraction of the total mass capable of undergoing large reversible deformation must be decreased by occlusion of part of the rubber [34]. There results a decrease of elastic recovery of the composite

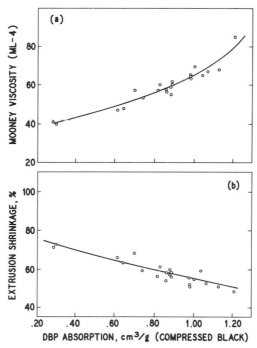

Fig. 3. (a) Mooney viscosity and (b) extrusion shrinkage as a function of carbon black structure. SBR 1500 rubber with 50 phr black. Carbon blacks range in specific surface area (CTAB) from 14 to 164 m²/g.

as a whole, which manifests itself in smaller mill and extrusion shrinkage and lower die swell (Fig. 3b). Extrusion shrinkage is also affected by specific surface area and cognizance has been given to this variable by applying a correction to empirical structure–shrinkage correlations [9]. Extrusion shrinkage data for various black loadings have been treated successfully by assigning a fraction of the occluded rubber to the filler concentration [34].

With rare exceptions, the viscosity of high molecular weight rubbers is shear rate dependent (non-Newtonian). The presence of a filler always increases the shear dependence; i.e., the increase in viscosity caused by the filler is greatest at low shear rates. Indeed, the viscosity at vanishingly small shear rates may become effectively infinite. The phenomenon is the result of secondary filler agglomeration, leading to the establishment of a weak filler network held together by van der Waals forces; this network is progressively disrupted by increasing shearing stresses.

VIII. Mechanical Behavior of Filled Rubber Vulcanizates at Small and Moderate Strains*

Reinforced rubbers are viscoelastic materials; their elastic moduli are time-dependent quantities. Let us consider first the *equilibrium* Young's modulus, i.e., the tensile modulus at infinitesimal strain (ε) measured at vanishingly small strain rates ($\dot{\varepsilon}$). It has been shown [40] that for an elastic medium filled with rigid spherical particles, Eq. (1) will apply with, viscosities replaced by moduli; i.e.

$$E = E_0(1 + 2.5c + 14.1c^2); \qquad \varepsilon, \dot{\varepsilon} \to 0 \qquad (2)$$

As in the case of viscosity, glass beads or medium thermal black conform to the theory [41], small-particle and structured fillers do not. Nonspherical shape, occluded rubber, immobilization of polymer segments at the interface, and secondary agglomeration all combine to increase the modulus over and above the prediction of Eq. (2). Modification of the properties of the matrix by the filler (cure effects) may increase or decrease the modulus.

Much insight into the physical situation is provided by dynamic viscoelastic measurements. Consider an experiment in which a filled vulcanizate is subjected at constant frequency to a periodic sinusoidal shear strain, the strain amplitude γ_0 being increased stepwise. The dynamic shear modulus is a complex quantity, with a component G' in phase with the strain (the storage modulus) and a component G'' 90° out of phase with the strain (the loss modulus). The tangent of the phase angle δ between stress and strain is equal to G''/G'. Results of such an experiment [42] are shown in Fig. 4. Note that the effect of the carbon black on the dynamic storage modulus is large and constant up to a double

* See also Kraus [2, 3].

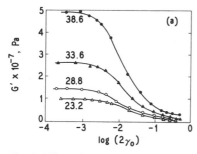
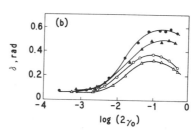

Fig. 4. Effect of strain amplitude (γ_0) on (a) dynamic storage modulus (G') and (b) phase angle (δ) of butyl rubber vulcanizates reinforced with N-330 black at various loadings (in volume percent). (After Kraus [2a], data from Payne [42].)

strain amplitude of about 0.001. In this region the secondary agglomeration network remains intact and the large modulus is caused almost entirely by the elastic response of this network. In fact, pastes prepared from carbon black and nonpolymeric vehicles exhibit moduli of the same order of magnitude [2]. As the amplitude of deformation is increased, disruption of the agglomeration network causes G' to fall until a new plateau value is approached (not strictly attained in the data of Fig. 4a). At this point the filler network may be regarded as destroyed; the change, however, is reversible and the original modulus may be recovered by resting the sample in an unstrained condition. The second plateau value of the modulus has been explained satisfactorily by Eq. (2), in which the actual filler concentration is replaced by an effective concentration augmented by occluded rubber [43]. This means that the dynamic modulus at moderate strain amplitudes is a function of primary filler structure. On the other hand, the modulus of the secondary agglomeration network must increase with the volume density of filler-to-filler contacts. This is a function of the state of subdivision of the filler and hence of its specific surface area. At very small amplitudes, therefore, the storage modulus increases roughly in proportion to the specific surface area of carbon black, with some additional dependence on structure.

The behavior of the phase angle δ is shown in Fig. 4b. The maximum energy stored during a deformation cycle is proportional to G', while the energy dissipated is proportional to G''. Since

$$\tan \delta = G''/G' \tag{3}$$

the phase angle depends directly on the ratio of viscous to elastic response of the material. At very small strains $\tan \delta$ is small and constant; the agglomeration network (whether or not fully developed) is essentially an elastic structure. As the strain amplitude is increased, both G'' and $\tan \delta$ pass through maximum values. The high energy dissipation in the region of the maxima is the result of frictional forces active in the agglomeration–deagglomeration of the filler.

This mechanism decreases in importance at high strain amplitudes, where the agglomeration network is broken down. However, at all strain amplitudes reinforcing fillers contribute to energy dissipation. This fundamental fact is closely related to the very mechanism of the reinforcing action; see Section X and Chapter 10.*

Strain amplitudes of the order of 0.1–0.5 are common in many applications of rubber, e.g., the flexing of a running automobile tire. Even smaller amplitudes are involved in vibration damping. Thus, the results of Fig. 4 are of more than academic interest, as are the trends of the viscoelastic functions with varying filler morphology. The latter are substantially the same in various modes of deformation (shear, extension, compression) and are illustrated in Figs. 5 and 6, using compression data for the vulcanizates of Fig. 3. In this example an oscillation of 3% amplitude was superimposed on a constant compressive strain of 18.75%. The relatively large background strain itself serves to diminish the effect of agglomeration. Note that the storage modulus E' is substantially a function of structure alone, whereas tan δ is a nearly structure-independent function of specific surface area. The latter observation means that storage and dissipation of energy increase with structure in the same proportion. It is also

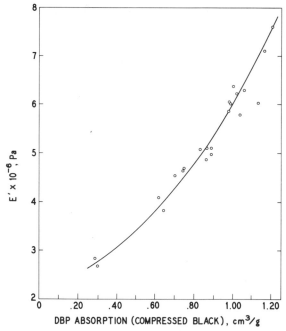

Fig. 5. Dependence of dynamic storage modulus on carbon black structure. SBR 1500 with 50 phr black. Data in compression [2a].

* In Chapter 10 also the effect of frequency will be discussed. See also Eirich and Smith [52].

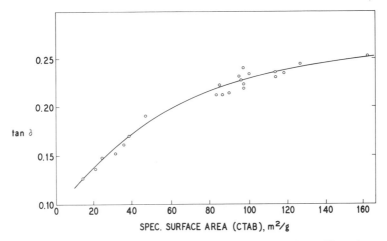

Fig. 6. Loss tangent of vulcanizates of Fig. 5 as function of carbon black specific surface area [2a].

clear that the loss modulus (E' tan δ) must be a function of both specific surface area and structure; the energy dissipated per cycle increases with both. If, however, a cyclic *stress* of constant amplitude is imposed upon the specimen, the energy dissipation becomes porportional to the loss *compliance* D'', which is given by

$$D'' = E''/(E'^2 + E''^2) = \tan \delta/E'(1 + \tan^2 \delta) \qquad (4)$$

Because E' increases with structure, while tan δ remains constant, the effect of higher structure is to lower the energy dissipation. The role of structure is thus dependent on the type of mechanical excitation.

IX. Large Deformation Behavior*

The most commonly performed mechanical test on rubber vulcanizates is the uniaxial extension of a strip, at constant rate of strain, to its breaking point. Figure 7 shows, schematically, the result of such an experiment with an amorphous rubber incapable of crystallization under strain. Curve A is for a gum vulcanizate, B for a vulcanizate reinforced with a high-surface-area structure carbon black, and C for a vulcanizate reinforced with the same black *after graphitization*. Following standard practice, the engineering stress σ is computed on the original undeformed cross section.

All three vulcanizates are viscoelastic materials. The stress, therefore, is a function not only of strain, but also of time; stress relaxation occurs throughout the entire experiment. If the strain rate of the test is increased, the stresses attained at any given extension will be greater than in experiments at smaller

* See also Kraus [2] and Mullins [44].

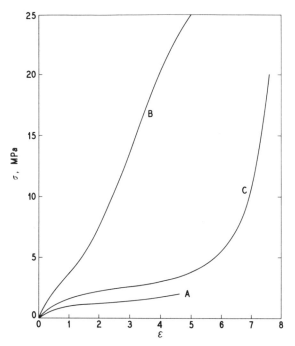

Fig. 7. Stress–strain curves (schematic) of vulcanizates from a noncrystallizing rubber: A, un-filled; B, reinforced with carbon black; C, reinforced with graphitized carbon black.

strain rates. If at any extension short of the breaking strain the strain rate is reversed (from $\dot\varepsilon$ to $-\dot\varepsilon$), the descending branch of the stress–strain curve will fall below the ascending branch, tracing out a hysteresis loop. All these phenomena are common to filled and unfilled rubbers alike.

Consider now the behavior of sample B. The initial slope of the stress–strain curve is much greater than that of A, because of the filler contributions already discussed. Once the effects of secondary agglomeration are overcome, several mechanisms remain which will keep the stress rising faster than in the unfilled sample. The first is *strain amplification*. Since the rigid filler does not share in the deformation [45], it is clear that the inclusion of a rigid adhering filler in a soft matrix will cause the average local strain in the matrix to exceed the macroscopic strain. Thus the rubber in the filled vulcanizate finds itself more highly strained and responds with a higher stress. Strain amplification also increases the mean rate at which the matrix is strained, leading to a further increase in stress.

It has been proposed that the quantity in parentheses in Eq. (2) is, in fact, a strain amplification factor [41]. With highly structured fillers, the concentration c should include the contribution of occluded rubber, making the amplification effect a function of structure level.

Consideration of the filled vulcanizate on a molecular scale provides insight into additional stress-raising mechanisms. The equilibrium retractive force of a rubbery network is proportional to the network chain density v (see Chapter 4): $f \propto vRT$. Adsorption of polymer segments at several sites of a filler particle effectively introduces giant multifunctional cross-links into the system and increases the network chain density. While the simple statistical theory of rubber elasticity (Chapter 4) is not quantitatively applicable to filled systems, qualitatively any increase in network density must produce an increase in stress. Filler particles also cause inhomogeneities in the retractive force, contributed to the stress by network chains attached to them [45, 46]. Consider the situation depicted in Fig. 8. As the network is strained and the two particles (aggregates of carbon black) move apart, chain 1 will be more highly extended and will attain its limit of finite extensibility long before chain 2; it will contribute more to the retractive force of the network. Eventually the chain will break (either at the surface or internally), providing a mechanism of stress relaxation. The breaking of chains at macroscopic extensions short of breaking has been demonstrated by electron spin resonance measurements [47].

Returning to Fig. 7, we note that at $\varepsilon \cong 4$ the slope of stress–strain curve B begins to decrease. At such large strains there is some "de-wetting" from the filler surface. Not merely individual chains, but all chains in a given area of contact may be detached, opening a vacuole. De-wetting is another mechanism for stress relief. It is directly observable under the electron microscope by stretching microtome sections of filled rubber supported by a clear polymeric film [48]. Such studies have shown that de-wetting is delayed to higher and higher elongations as specific surface area and structure of carbon black increase. Defining an *adhesion index* as the stress at which 20% of the black had undergone de-wetting, Hess *et al.* [48] obtained estimates of de-wetting stresses of the order of 10 MPa for a furnace black of 30 m^2/g surface area, but only 1–2 MPa for the same black graphitized. It is now clear why the stress–strain curve of vulcanizate (C) in Fig. 7 lies so far below the fully reinforcing black (B): the cause is weak bonding (recall Table I). Even in the absence of outright

TO NETWORK

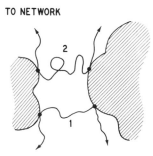

Fig. 8. Schematic of network chains attached to filler particles [2a].

de-wetting, molecular slippage [49] provides ample opportunity for stress relief. Furthermore, occluded rubber tends to be pulled out of the interstices of the aggregate, diminishing the strain amplification effect. Another consequence of these processes is the high elongation at break.

A phenomenon often discussed in connection with filler reinforcement is stress softening [50]. When a filled rubber is extended to a strain ε_1, returned to zero strain, and stretched again, the second stress–strain curve lies below the first one, but rejoins it at ε_1 (Fig. 9). This happens even if the vulcanizate is rested after the first extension for an extended period of time in an attempt to ensure full recovery. There has been considerable controversy about the causes of stress softening and it appears that more than one mechanism may contribute to the effect. One of these is simply incomplete elastic recovery [50]. This occurs also in unfilled rubbers, but is greatly increased in filled rubbers by strain amplification. Another mechanism is the progressive detachment, or breaking, of network chains attached to filler particles [46]. Referring to Figs. 8 and 9, if chain 1 breaks at strain ε_1, then in the second extension this chain is no longer supporting stress and the vulcanizate is softened, but only up to strain ε_1; at $\varepsilon > \varepsilon_1$ the chain is no longer intact in either the first or second extension. At some higher strain ε_2, chain 2 breaks. Stress softening after extension to ε_2 is greater than after extension to ε_1, and the second extension curve joins the original stress–strain curve at ε_2, etc. In this mechanism, some strong bonds are considered to be broken at all levels of strain, including small strains. However, chain slippage of attached polymer segments along the surface [49] would have a similar effect. Instead of a chain breaking or detaching from the surface, the point of attachment is visualized as slipping along chain and/or surface in such a way as to lengthen the distance between adsorbed segments. Were this to happen to chain 1 in Fig. 8, then in the second extension the chain would contribute a smaller retractive force to the stress, but again only up to

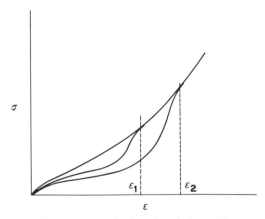

Fig. 9. The stress-softening effect (schematic).

the strain of the first extension. This is, of course, a form of incomplete recovery, but one which is specific to the presence of a filler. Finally, it is possible to account for the essential features of stress softening by a purely phenomenological analysis, using as a model the conversion of a hard phase to a soft one, the extent of the conversion increasing continuously with imposed stress [51].

A widely used practical criterion for the stiffness of rubber vulcanizates is the stress at 300% extension, the so-called 300% modulus (σ_{300}). With highly reinforcing carbon blacks (good rubber–filler bonding) this strain is in the region where secondary agglomeration effects have ceased to be important and before significant de-wetting occurs. In this region, occluded rubber and strain amplification exert a dominant effect on the modulus, which increases with the DBP absorption of the black but is relatively insensitive to specific surface area. There are, of course, additional factors which govern the value of σ_{300}. For example, if the filler influences the vulcanization reaction, it will affect the cross-link density and with it the modulus.

X. Failure Properties*

The basic physics of failure of rubbery materials under stress are discussed in Chapter 10. Essential to the development of high strength is the ability of the rubber to dissipate strain energy near the tip of growing cracks by viscoelastic processes. In unfilled single-phase amorphous vulcanizates incapable of crystallizing under strain, viscoelastic processes are effective only in the transition region between glassy and rubbery behavior. Thus, at elevated temperatures such materials are weak, but they become progressively stronger as the temperature is lowered or the strain rate is increased.

The addition of a filler introduces additional mechanisms by which strain energy is dissipated. Viscoelastic stress analysis of two-component systems shows that the inclusion of particles in a viscoelastic medium increases hysteresis (mechanical energy dissipation) even when the inclusions are perfectly elastic and when the viscoelastic properties of the medium are the same as in the unfilled state [54]. Any loss of segmental mobility in the polymer matrix resulting from interaction with the filler will further increase hysteresis. Motions of filler particles, chain slippage or breakage, and de-wetting at high strains also accentuate hysteretic behavior. In addition to providing for increased energy dissipation, dispersed particles finally serve to deflect or arrest growing cracks, thereby further delaying the onset of catastrophic failure.

A consequence of the hysteretic nature of filler reinforcement is a general lack of additivity to other strengthening processes. Thus, fillers do not materially increase the strength of amorphous rubbers near the glass transition, where the

---

* See also Kraus [3], Mullins [44], Eirich and Smith [52], and Thomas [53].

polymer is already highly hysteretic [3, 53]. They are most effective in the rubbery region, where the need for reinforcement is greatest.

The connection between hysteresis and strength is illustrated by a particularly simple empirical law [55]:

$$U_b \cong B(H_b \varepsilon_b)^{1/2} \tag{5}$$

where U_b is the energy input per unit volume to break a vulcanizate (area under the stress–strain curve), H_b the energy dissipated in the deformation prior to break (area of the hysteresis loop), ε_b the strain at break, and B a constant. The equation has been found to hold approximately for filled and unfilled rubbers alike, independent of temperature and strain rate. Its practical utility is, however, small because none of the quantities related can be predicted from first principles. Successful correlations between strength, hysteresis, and filler adhesion (the latter judged by dilatation caused by de-wetting) have also been demonstrated by using polymeric model fillers [56].

It was shown earlier that viscous losses in carbon-black-filled rubbers increase with the specific surface area of the black. Consequently the strength also increases, as illustrated by the uniaxial tensile strength data displayed in Fig. 10. Here σ_b, ε_b, and hysteresis (measured as temperature rise in a flex-

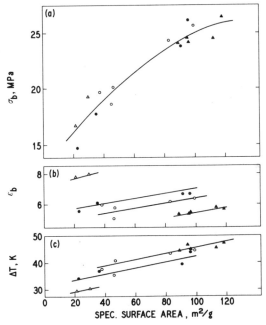

Fig. 10. (a) Tensile strength (σ_b), (b) breaking strain (ε_b), and (c) temperature rise (ΔT) as functions of carbon black surface area and structure. Black loading 50 phr in SBR 1500. **DBP** value (pressed): $\bigcirc$, 0.85; $\bullet$, 0.70; $\triangle$, 0.30; $\blacktriangle$, 1.00 cm³/g [2a].

ometer test) are plotted against specific surface area with structure as parameter. At this loading (50 phr), the effect of structure is to increase hysteresis but to decrease ε_b. These trends combine to make the tensile strength relatively insensitive to variations in structure, as suggested superficially by Eq. (5). The situation is, however, more complicated.

When carbon black loading is varied, hysteresis increases continuously with the black concentration, but ε_b and σ_b pass through maxima. Because of the effect of occluded rubber, increasing structure at fixed loading results in rising effective filler concentrations, and one would expect σ_b to be structure dependent, which is not obvious in the data of Fig. 10. The reason for this is that, at 50 phr, the effective filler concentrations are in a range bracketing the (rather flat) maximum in σ_b. At both low and high loadings, however, structure has a demonstrable effect on tensile strength (Table III). Because the effective loadings now fall on opposite sides of the maximum, σ_b increases with structure at 20 phr, but decreases at a loading of 80 phr.

The uniaxial tensile test is a type of tearing experiment and much of what has been said here concerning tensile strength applies to tearing as well.

Two important related phenomena in rubber are cut growth and fatigue. The former is the growth of a deliberately introduced cut upon repeated cyclic deformation of insufficient magnitude to cause immediate failure. The latter is cut growth originating from a natural flaw in the specimen, such as a surface scratch or an agglomerate of undispersed filler. The effects of carbon black properties on fatigue life (cycles to catastrophic failure) are illustrated in Table IV [57]. The data [57] show that fatigue life increases with structure level and decreases with specific surface area. High structure and small surface area are just the morphological features favoring good black dispersion (see Section V). Theory requires the fatigue life to be inversely related to the size of the initiating flaw; the agglomerate sizes measured in this study were consistent

TABLE III

CARBON BLACK STRUCTURE AND TENSILE STRENGTH[a]

DBP absorption[b]	Specific surface area[c]	20 phr		80 phr	
		σ_b	ε_b	σ_b	ε_b
(cm^3/g)	(m^2/g)	(MPa)		(MPa)	
0.46	117	6.0	5.7	26.3	6.0
0.67	101	10.1	6.1	24.2	4.7
0.90	98	12.0	6.2	21.6	3.3
1.00	110	13.8	6.4	—	—

[a] At 25°C, SBR 1500 test recipe similar to that of Fig. 10.
[b] On precompressed samples.
[c] By N_2 adsorption.

TABLE IV

FATIGUE TEST RESULTS ON SBR REINFORCED
WITH DIFFERENT CARBON BLACKS[a]

Carbon black	Specific surface area[b] (m^2/g)	DBP absorption (cm^3/g)	Mean fatigue life (Kc)
N-327	95	0.51	11
N-347	94	1.25	53
N-761	23	0.68	280
N-765	33	1.17	450

[a] Data of Dizon et al. [57].
[b] By N_2 adsorption.

with the observed fatigue behavior. However, cut growth and fatigue are also inversely related to strain energy density and to a cut growth "constant," which is a complex function of the chemical environment, oxygen and ozone strongly increasing the growth rate (see Chapter 10). Nevertheless, the observations of Table IV are qualitatively in agreement with the general experience on groove cracking of tire treads made from SBR or SBR–polybutadiene blends [58]. Structure has the opposite effect in natural rubber treads (truck tires).

Natural rubber and other strain-crystallizing elastomers generally display more complex behavior in their response to filler reinforcement [44, 53]. Whenever crystallinity develops as a result of deformation, the crystallites assume the role of reinforcing filler particles. For this reason, unfilled natural rubber vulcanizates may exhibit room temperature tensile strength so high that even the most reinforcing carbon blacks barely increase it. However, when strain-induced crystallization is absent or not fully developed, as in a very rapid extension at high temperature, fillers exert their normal reinforcing effects. A classic example of this is the abrasion or wear of automobile tires, where strain rates and temperatures encountered are such that crystallization cannot occur. As a result, carbon black reinforcement is just as important in natural rubber treads as it is in SBR or SBR–polybutadiene.

Because of its enormous technological importance, reinforcement against wear deserves special comment. It is well established that the relative resistance of various tread compounds to wear depends on the severity of the conditions causing wear, such as roughness of the abrading surface, slip, normal load, and temperature. Thus, the mean wear rate during the life of a tire or in a treadwear test is the result of a very complicated average of many conditions. For this reason it has become customary in wear testing of tires to rate tread compounds relative to a control or standard. The absolute wear rate of the latter is taken as a measure of the severity of the abrading conditions, and the relative wear rates as measures of individual abradability.

In a constant vulcanization recipe, wear rates resulting from use of different carbon blacks are closely correlated with specific surface area and structure, but the quantitative relationships change with severity. Figure 11 illustrates the trends with data interpolated from a computer analysis of tire tests on nearly 150 different furnace blacks [59]. At low severity, wear resistance is relatively insensitive to carbon black properties; in particular, structure has almost no effect (compare B with C and A with R). As conditions become more severe, the effect of specific surface area increases steadily (compare C with R and B with A). The influence of structure, however, manifests itself clearly only at high severity.

Because some of the results of Fig. 11 were obtained by interpolation, A, B, and C are hypothetical blacks chosen to facilitate illustration of the trends of treadwear with black properties. However, their assigned surface areas and structure levels are well within the range of blacks actually used in tires. Carbon blacks with less than 80 m^2/g surface area and 0.90 cm^3/g DBP (pressed) are rarely used in treads. Currently preferred for passenger treads are blacks of ~ 100 m^2/g surface and 1.00 cm^3/g (pressed) DBP. In truck treads, somewhat higher surface areas (~ 120 m^2/g) and lower structure (~ 0.95 cm^3/g) are favored. For a detailed treatment of the carbon black treadwear problem the reader is referred to Janzen and Kraus [59].

Specific surface area and structure may not be the only morphological features affecting wear. There is evidence that at constant surface area and

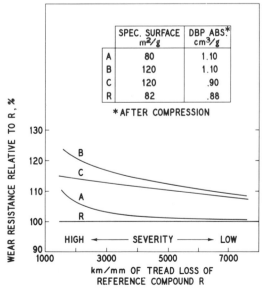

Fig. 11. Trends in wear resistance with carbon black specific surface area and structure for a typical SBR–polybutadiene tread formulation [2a].

vehicle demand, narrower nodule and aggregate size distributions *may* produce small improvements in wear resistance [60–62]. Surface oxidation in carbon blacks is undesirable and oxidized blacks are not used in highway treads. Colloidal silicas, even when compounded with bonding agents (Section IV), do not equal carbon black in wear-reinforcing capability [63] and so far have made only very minor inroads into tire applications.

REFERENCES

1. K. A. Burgess, F. Lyon, and W. S. Stoy, *in* "Encyclopedia of Polymer Science and Technology," Vol. 2 (H. F. Mark *et al.* eds.), pp. 820–836. Wiley (Interscience), New York, 1965.
2. G. Kraus, "Reinforcement of Elastomers." Wiley (Interscience), New York, 1965.
2a. G. Kraus, *Angew. Makromol. Chem.* **60/61,** 215 (1977).
3. G. Kraus, *Fortschr. Hochpolyn.-Forsch.* **8,** 155 (1971).
4. W. M. Hess, L. Ban, and G. C. McDonald, *Rubber Chem. Technol.* **42,** 1209 (1969).
5. A. I. Medalia and F. A. Heckman, *J. Colloid Interface Sci.* **36,** 173 (1971).
6. W. M. Hess, G. C. McDonald, and E. Urban, *Rubber Chem. Technol.* **46,** 204 (1973).
7. J. Janzen and G. Kraus, *Rubber Chem. Technol.* **44,** 1287 (1971).
8. A. I. Medalia, *J. Colloid Interface Sci.* **24,** 393 (1967).
9. R. E. Dollinger, R. H. Kallenberger, and M. L. Studebaker, *Rubber Chem. Technol.* **40,** 1311 (1967).
10. A. I. Medalia and L. W. Richards, *J. Colloid Interface Sci.* **40,** 233 (1972).
11. D. Rivin, *Rubber Chem. Technol.* **36,** 729 (1963); **44,** 307 (1971).
12. D. Rivin, J. Aron, and A. I. Medalia, *Rubber Chem. Technol.* **41,** 330 (1968).
13. W. F. Watson, *Ind. Eng. Chem.* **47,** 1281 (1955).
14. G. Kraus, J. T. Gruver, and K. W. Rollmann, *Proc. Conf. Carbon, 4th* pp. 291–300 (1960).
15. J. B. Donnet, G. Henrich, and G. Riess, *Rev. Gen. Caoutch. Plast.* **38,** 1803 (1961); **39,** 583 (1962); **41,** 519 (1964).
16. E. Papirer, A. Voet, and P. H. Given, *Rubber Chem. Technol.* **42,** 1200 (1969).
17. W. D. Schaeffer, W. R. Smith, and M. H. Polley, *Ind. Eng. Chem.* **45,** 1721 (1953).
18. J. J. Brennan, T. E. Jermyn, and B. B. Boonstra, *J. Appl. Polym. Sci.* **8,** 2687 (1964).
19. W. D. Schaeffer and W. R. Smith, *Ind. Eng. Chem.* **47,** 1286 (1955).
20. C. W. Sweitzer, K. A. Burgess, and F. Lyon, *Rubber Chem. Technol.* **34,** 709 (1961).
21. J. H. Bachmann, J. W. Sellers, M. P. Wagner, and R. F. Wolf, *Rubber Chem. Technol.* **32,** 1286 (1959).
22. D. C. Edwards and E. Fischer, *Kautsch. Gummi, Kunstst.* **26,** 46 (1973).
22a. Degussa Inc., brochure of Pigment Division, Frankfurt, Germany (1974).
23. M. Morton, J. C. Healy, and R. L. Denecour, *Proc. Int. Rubber Conf., 5th* p. 175 (1967).
24. M. Morton, *Adv. Chem. Ser.* **99,** 490 (1971).
25. P. B. Stickney and R. D. Falb, *Rubber Chem. Technol.* **37,** 1299 (1964).
26. G. Kraus, *Rubber Chem. Technol.* **38,** 1070 (1965).
27. A. K. Sircar and A. Voet, *Rubber Chem. Technol.* **43,** 973 (1970).
28. L. L. Ban, W. M. Hess, and L. A. Papazian, *Rubber Chem. Technol.* **47,** 858 (1974).
29. G. Kraus and J. T. Gruver, *Rubber Chem. Technol.* **41,** 1256 (1968).
30. B. B. Boonstra and A. I. Medalia, *Rubber Age (N.Y.)* **92,** 892 (1963); **93,** 82 (1963).
31. R. H. Norman, "Conductive Rubbers and Plastics." Appl. Sci. Publ., London, 1970.
32. S. Kaufmann, W. P. Slichter, and D. D. Davis, *J. Polym. Sci., Part A-2* **9,** 829 (1971).
33. G. Kraus and J. T. Gruver, *J. Polym. Sci., Part A-2* **8,** 571 (1970).
34. A. I. Medalia, *J. Colloid Interface Sci.* **32,** 115 (1970); *Rubber Chem. Technol.* **47,** 411 (1974).

35. G. Kraus, *J. Polym. Sci., Part B* **8,** 601 (1970); *Rubber Chem. Technol.* **44,** 199 (1971); *J. Appl. Polym. Sci.* **15,** 1679 (1971).
36. S. Wolff, *Kautsch. Gummi, Kunstst.* **23,** 7 (1970).
37. M. Porter, *Rubber Chem. Technol.* **40,** 866 (1967).
38. E. Guth, R. Simha, and O. Gold, *Kolloid-Z.* **74,** 266 (1936).
39. E. Guth and O. Gold, *Phys. Rev.* **53,** 322 (1938).
40. H. M. Smallwood, *J. Appl. Phys.* **15,** 758 (1944).
41. L. Mullins and N. R. Tobin, *J. Appl. Polym. Sci.* **9,** 2993 (1965).
42. A. R. Payne, *J. Appl. Polym. Sci.* **7,** 873 (1963).
43. A. I. Medalia, *Rubber Chem. Technol.* **46,** 877 (1973).
44. L. Mullins, *in* "The Chemistry and Physics of Rubberlike Substances" (L. Bateman, ed.), Ch. 11. Wiley, New York, 1963.
45. F. Bueche, *J. Appl. Polym. Sci.* **5,** 271 (1961).
46. F. Bueche, *J. Appl. Polym. Sci.* **4,** 107 (1960).
47. A. B. Sullivan and R. W. Wise, *Proc. Int. Rubber Conf., 5th* p. 235 (1967).
48. W. H. Hess, F. Lyon, and K. A. Burgess, *Kautsch. Gummi, Kunstst.* **20,** 135 (1967).
49. E. M. Dannenberg and J. J. Brennan, *Rubber Chem. Technol.* **39,** 597 (1966).
50. L. Mullins, *Rubber Chem. Technol.* **42,** 339 (1969).
51. A. N. Gent, *J. Appl. Polym. Sci.* **18,** 1397 (1974).
52. F. R. Eirich and T. L. Smith, *in* "Fracture" (H. Liebowitz, ed.), Vol. 7, Ch. 7. Academic Press, New York, 1972.
53. A. G. Thomas, *in* "The Chemistry and Physics of Rubberlike Substances" (L. Bateman, ed.), Ch. 10. Wiley, New York, 1963.
54. J. R. M. Radok and C. L. Tai, *J. Appl. Polym. Sci.* **6,** 518 (1962).
55. J. A. C. Harwood and A. R. Payne, *J. Appl. Polym. Sci.* **12,** 889 (1968).
56. M. Morton, R. J. Murphy, and T. C. Cheng, *Polym. Prepr., Am. Chem. Soc., Div. Polym. Chem.* **15**(1), 436 (1974).
57. E. S. Dizon, A. E. Hicks, and V. E. Chirico, *Rubber Chem. Technol.* **46,** 231 (1973).
58. M. L. Studebaker, *Rubber Chem. Technol.* **41,** 373 (1960).
59. J. Janzen and G. Kraus, *J. Elastomers Plast.* **6,** 142 (1974).
60. J. Janzen and G. Kraus, *Proc. Int. Rubber Conf., Brighton, Eng.* pp. G7-1-7 (1972).
61. A. I. Medalia, E. M. Dannenberg, F. A. Heckman, and G. R. Cotten, *Rubber Chem. Technol.* **46,** 1239 (1973).
62. C. J. Stacy, P. H. Johnson, and G. Kraus, *Rubber Chem. Technol.* **48,** 538 (1975).
63. S. Wolff, K. Burmester, and E. H. Tan, *Kautschuk Gummi Kunstst.* **29,** 691 (1976).

Chapter **9**

The Rubber Compound and Its Composition*

M. L. STUDEBAKER

PHILLIPS CHEMICAL COMPANY
STOW, OHIO

and

J. R. BEATTY

THE B. F. GOODRICH RESEARCH & DEVELOPMENT CENTER
BRECKSVILLE, OHIO

I. Introduction

Why and how compounding ingredients are selected and put together to yield a useful rubber† compound that will be satisfactory for the intended service are essential aspects of compounding. The major factors in compounding

* Permission to reproduce material from *Rubber Age* **107** (8) 20–35, (9) 39–45 (1975) is gratefully acknowledged.

† The terms "rubber" and "elastomer" are used interchangeably in this paper. The term "polymer" is more general but is frequently used to refer to the elastomer in the text.

Copyright © 1978 by Academic Press, Inc.
All rights of reproduction in any form reserved.
ISBN 0-12-234360-3

are *price, processing,* and *properties.* If the cost of the compound and the sales price of the finished article are not competitive, sales will not develop, and the compound will stay on the shelf even though it may have technical merit. Equally important, the compound must process well, i.e., must mix, extrude and/or calender, mold, etc., without undue difficulty from such common problems as pigment incorporation, scorch (premature vulcanization), lack of building tack, or excessive stickiness. Processing costs are an important element in production costs. In addition to meeting the foregoing criteria, the compound must also possess the physical properties required by the service in which the rubber article is to be used.

Compounding will probably always be part art and part science. The compounder must draw his conclusions from physical properties. The ability to select appropriate physical properties for a given service is his most important skill. We have selected a group of compounds, all of which have been used in the preparation of commercial articles, to exemplify the requirements which the compounder's product will have to satisfy. The data on these compounds will be used throughout this chapter to illustrate our discussion of various physical properties. To avoid infringing on proprietary rights, the names of antidegradants and curatives have been omitted, and oil and filler concentrations have been rounded off to the nearest 5 phr (parts per hundred parts of rubber).* However, it should be pointed out that rubber compounds which closely approximate these properties could be prepared satisfactorily with a cure system of a sulfenamide accelerator plus sulfur and a suitable antidegradant at appropriate levels. These compounds, although good starting points for formulating specific rubber articles, would have to be checked through for processing in the particular equipment available and adjusted if necessary. Lastly, the compounder might find it expedient to vary the ingredients slightly to arrive at the lowest compounding cost which would meet his specific requirements for physical properties.

The multitude of ingredients in rubber compounds can be classified into elastomers, cure systems, fillers, processing aids, and miscellaneous ingredients. These and their functions will be discussed briefly in the following pages.

II. Selection of Ingredients

A. POLYMER

The first and most important component of a rubber compound is the polymer. The lowest-priced polymer is normally selected into which the compounder can build the required properties to provide adequate (or acceptable)

---

* The properties of the raw stocks for this series of rubber compounds are presented in detail in [1a]; dynamic properties were presented graphically in [1b]. Concentrations have not been rounded off when all of the oil and/or black came from an oil-extended masterbatch. To simplify discussion, compounds based on polymer blends were not included.

service. The general price level and properties of elastomers are shown in Table I [1c]. The table is indicative—not absolute. One can compound outside the indicated values, but most commercial compounds will fall within the ranges given.

Another important factor in polymer selection is ease of mixing and the subsequent processability. Although the relative prices of natural rubber (NR) and (styrene–butadiene copolymers) SBR fluctuate, they are generally competitive; thus selection between these two types of elastomer is usually based on physical properties and current prices. However, there are some applications which require a property best provided by one specific polymer. For example, natural rubber is preferred in many applications requiring high tensile strength, high elongation, and certain types of fatigue resistance. Natural rubber provides superior fatigue resistance at comparatively high elongations, say, 100–200%. SBR generally provides excellent fatigue resistance at low elongations, 0–100%, and is superior to natural rubber when flexed under compression (Table II, [2]).

The next factors in polymer selection are the required tensile strength, elongation, and modulus or stiffness. Every manufactured article has specific stress–strain requirements. Table III lists 21 commercial rubber compounds arranged according to their tensile strength, which is a commonly used criterion of quality. Even so, the strength criterion must be used with judgment based on experience, which involves consideration of other properties as well, for instance, abrasion resistance. In general, the more heavily loaded compounds have lower tensile strength (cf. Fig. 1).

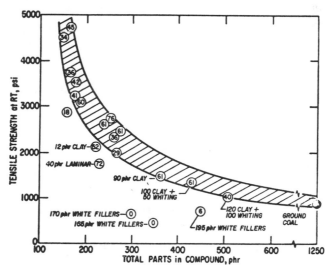

Fig. 1. Tensile strength versus total parts of the compounds for 21 commercial compounds. Numbers are square meters of carbon black surface per gram of rubber hydrocarbon. White filler loadings, when present, are indicated.

TABLE I

	Approx. price ($/lb)[b]	Specific gravity of base	Durometer hardness range	Tensile strength (psi at room temp.)	Elongation (% at room temp.)	Service temp. (°F)		Resistance[c]			
						Max.	Min.	Oil	Fuel	Water swell	Tear
Polyacrylate	1.50	1.10	40–100	1000–2200	100–400	300	0	VG	F	VP	F
Polyisobutylene	0.55	0.92	30–100	1000–3000	100–700	212	−65	VP	VP	VG	VG
Urethane	2.00	0.85	62–95	1000–8000	100–700	212	−65	E	VG	F	E
EPDM	0.60	0.85	30–100	1000–3000	100–300	300	−40	VP	VP	VG	G
Fluoroelastomers	12.0	1.4–1.95	60–90	1000–2400	100–350	450	−40	E	E	E	F
Chlorosulfonated polyethylene	0.85	1.10	50–95	1000–2800	100–500	250	−65	G	F	E	G
Natural rubber	0.45	0.93	20–100	1000–4000	100–700	180	−65	VP	VP	VG	VG
Polychloroprene	0.70	1.23	20–90	1000–4000	100–700	212	−65	G	F	G	G
Nitrile	0.80	1.00	30–100	1000–4000	100–600	250	−65	VG	G	VG	G
Polybutadiene	0.40	1.93	30–100	1000–3000	100–700	212	−80	VP	VP	VP	VG
Polyisoprene	0.50	0.94	20–100	1000–4000	100–750	180	−65	VP	VP	VG	VG
Polysulfide	1.50	1.34	20–80	500–1250	100–400	180	−65	E	E	VG	G
SBR	0.36	0.94	40–100	1000–3500	100–700	225	−65	VP	VP	G	G
Silicone	4.70	0.98	20–95	500–1500	50–800	450	−120	F	P	E	P
Epichlorohydrin	1.50	1.27	60–90	1000–2500	100–400	250	−50	G	G	G	G

[a] From *Rubber Age* **107,** 22, August 1975.
[b] Approximate prices, March, 1978.

Resistance[c]			Electrical properties	Adhesion to metal	Corrosion of metal	Principal advantage	Typical applications	Remarks
Adhesion	Ozone[d]	Weather						
G	E	VG	P	G	P	Oil and heat resistance to 350°F	High-temperature transmission seals	Low-temperature limit in transition because of polymer chemistry advances
G	E	VG	VG	G	G	Ozone resistance; good heat resistance	Insulation, diaphragms for washers, driers, vacuum cleaners	GP offering good heat, aging characteristics, low rebound
S	E	E	F	E	G	Oil resistance; excellent abrasion resistance	Grease seals, gears, insulation	Can be made in microcellular form
E	E	VG	VG	F	E	Lightest-weight polymer; ozone resistance	Drive belts, grommets for washer and drier drums	GP rubber offering good heat and weathering characteristics
G	E	E	E	VG	F	Maximum resistance to dry heat and oils	High-temperature oil-resistant seals (expensive)	Best for high-temperature and fuel and oil resistance applications
E	E	E	F	VG	F	Ozone resistance; chemical resistance	Chemical-resistant parts for driers, irons	Excellent aging or weathering with good color retention in sunlight
E	P	F	E	E	E	High resilience; good abrasion resistance	Shock mounts and drive belts; bushings for juicers	Historical choice for many applications
E	G	VG	G	E	F	Ozone and oil resistance; heat resistance to 212°F	Seals, grommets, bellows, and bushings for air conditioners	Good blend of oil, plus good weathering and ozone resistance
E	F	P	F	E	F	Oil resistance to 250°F	Seals, diaphragms for disposers, toothbrushes, washers	Automotive industry standard material for good oil resistance
E	P	F	G	E	G	Low-temperature and good abrasion resistance	Mounts, seals for juice dispensers, coffee makers	Equal to natural rubber in abrasion resistance; better low-temperature resistance
E	P	F	G	E	G	Synthetic natural rubber	Shock mounts and drive belts, bushings for juicers	Lower tensile strength than natural rubber; price more constant than natural rubber
P	G	E	VG	E	F	Excellent weather and oil resistance	Diaphragms, seals, gaskets; also gas regulators	Strong odor in many compounds
E	F	F	F	E	F	Good heat resistance to 212°F	Mounts, seals, for juice dispensers, coffee makers	GP polymer frequently used in place of natural rubber
P	E	E	E	E	G	Can be made white with no loss in tensile strength	Seals, gaskets for ranges, irons, television; also insulation	Good electrical properties; good dry heat resistance
F to G	G	G	F	E	P	Impermeability to air; good weather, low-temperature, and oil resistance	Gasket, seals for refrigerators; gas regulators; pump and valve parts	New elastomer being evaluated for a number of applications

[c] E, excellent; F, fair; G, good; P, poor; S, superior; VG, very good; VP, very poor.
[d] At room temperature, stressed.

TABLE II

EFFECT OF ELASTOMER TYPE AND STRAIN CYCLE ON FATIGUE LIFE[a]

| | | Life × 10^6 (cycles) | |
Compound	Elastomer	-75 to 0%	$+50$ to $+125\%$
7	OE-SBR	30	5.8
8	OE-SBR/BR Blend	13	11.3
9	EPT	30	23.5
2	SBR	22	0.2
1	NR	4	30.0

| | Compound formulations (parts by weight) | | | | |
	1	2	7	8	9
NR	100	—	—	—	—
SBR	—	100	—	—	—
OE-SBR	—	—	137.5	103	—
BR	—	—	—	25	—
EPT (Nordel 1070)	—	—	—	—	100
Zinc oxide	5	5	5	5	5
Stearic acid	3	—	1	1	1
EPC black	50	40	—	—	—
HAF black	—	—	75	75	—
SAF black	—	—	—	—	80
Petroleum oil	—	—	—	10	55
Antioxidant	1	—	1	1	—
MBTS	1	1.75	—	—	—
CBS	—	—	1	0.9	—
TMTM	—	—	—	—	1.5
MBT	—	—	—	—	0.75
Sulfur	3	2	2.5	2.5	1

[a] From Beatty [2].

Adequate loading of reinforcing carbon blacks gives high tensile strength for a given total loading. Adding unmodified clay, whiting, silica, etc. to a carbon black loaded stock reduces tensile strength. In compounds like those in Table III, blacks of large particle size do not provide sufficient carbon black surface per unit quantity of rubber to give adequate activation of cure and tensile reinforcement, especially when nonblack fillers are also present.

If increasing amounts of reinforcing pigments (in a practical way, this means carbon black) are added to a gum stock, the tensile strength usually increases to a maximum and then falls off at higher loadings (Figs. 2 and 3). Nonreinforcing pigments may or may not show a tensile strength–loading maximum, but the excessively high loadings commonly employed in low-cost rubber products invariably give markedly reduced tensile strength.

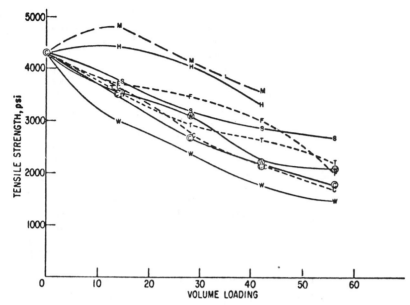

Fig. 2. Tensile strength relations for various fillers at comparable volume loadings in natural Rubber: M, MPC; H, HAF; F, FEF; S, SRF; T, MT; ⓦ, fine particle size $CaCO_3$; Ⓒ, hard clay; c, soft clay; w, medium particle size whiting. Data from Winspear [24].

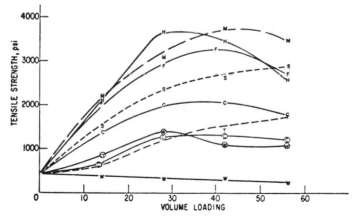

Fig. 3. Tensile strength relations for various fillers at comparable volume loadings in SBR: M, MPC; H, HAF; F, FEF; S, SRF; T, MT; ⓦ, fine particle size $CaCO_3$; Ⓒ, hard clay; C, soft clay; W, medium particle size whiting. Plot of data from Winspear [24].

The occurrence of a maximum in the tensile strength–loading curves for reinforcing blacks in natural rubber and in the neoprenes depends on the nature of the compound and the mixing conditions. Sometimes maxima are not observed where gums have high strength; Fig. 2 illustrates both cases. On the other hand, pronounced maxima are almost invariably present in the so-called

TABLE III

Compound no.	Compound for rubber article	Total parts in compound (polymer = 100)	Compound cost ($/lb)[b]	Specific gravity	Elastomer	Carbon black
188	Heavy-duty tread	163	0.1757	1.13	NR	45 N28
224	Automobile bushing	150	0.1706	1.11	NR	40 S31
189	Heat-resistant belt cover	164	0.1768	1.13	NR	45 N32
206	Grade 2 belting	175	0.1677	1.13	SBR 1606	50 N33
197	Heat- and abrasion-resistant belt cover	174	0.1592	1.12	SBR 1502	50 N33
227	Irrigation tube valve gasket	187	0.1610	1.11	SBR 1608 SBR 1603	35 N22 15 S30
194	Brake cup	158	0.1606	1.11	SBR 1503	40 N55
211	100-Level passenger tread	252	0.1362	1.15	SBR 6778	82.5 N33
208	Custom mix tread	239	0.1596	1.14	SBR 6779	75 N33
219	High-quality auto-motive mat	274	0.1364	1.19	SBR 1815	75 N3
190	Heat-resistant bèlt cover	161	0.1672	1.14	SBR 1500	45 N3 25 N7
191	Heavy-duty brake boot	220	0.1463	1.21	SBR 1801 SBR 1502	20 N3 35 N9
192	Bicycle tire	267	0.1091	1.14	SBR 1708	90 N6
217	Window gasket	229	0.1020	1.33	SBR 1815	30 N7 40 N9 75 N3
220	Automotive mat	361	0.1026	1.29	SBR 1815	75 N3
195	Garage door seal	430	0.0891	1.29	SBR 1805	75 N3
221	Brake pedal	510	0.0834	1.51	SBR 1821	15 N6 80 N5
228	Battery case	1250	0.0502	1.33	SBR 1712	—
229	Drop wire insulation	453	0.0741	1.53	SBR 1503	80 N
225	Proof goods	298	0.0947	1.55	SBR 1006	—
223	Shoe sole	348	0.1188	1.37	SBR 1506	—

[a] From *Rubber Age*, **107**, 24, August, 1975.
[b] Based on 1975 prices.
[c] In in square meters per gram of rubber hydrocarbon.

MMERCIAL COMPOUNDS ARRANGED ACCORDING TO TENSILE STRENGTH[a]

bon ck ace[c]	Nonblack filler (phr)	Total oil and resin (phr)	Tensile strength (psi)	300% Modulus (psi)	Elongation (%)	Shore A hardness
5	—	5	4700	2100	550	59
4	—	—	4490	1970	580	68
6	—	5	3730	1000	600	64
1	—	10	3530	1490	570	64
1	—	15	3210	1610	530	65
0		30	3070	950	640	57
8	—	—	2860	1360	530	66
6	—	62.5	2740	1330	570	64
1	—	50	2590	1130	610	59
1	25 Clay	60	2440	1380	490	59
6	—	5	2340	1330	460	62
2	12 Clay	17.5	2100	890	720	59
9	—	70	1980	1220	510	51
2	40 Laminar	80	1750	—	280	70
1	90 Clay	80	1480	1090	410	62
1	100 Clay 50 Whiting	75	1360	950	430	62
0	120 Clay 100 Whiting	80	1030	1030	300	72
—	900 Carbofil (ground coal)	175	920	—	10	95
6	95 Whitex No. 2 100 Atomite	65	700	570	470	82
—	75 Clay 75 Whiting 20 TiO₂	15	640	—	—	64
—	45 Zeolex 20 TiO₂ 45 Clay 45 Whiting	75	440	—	240	68

TABLE IV

Stress–Strain Values for Representative HAF (N330)-Loaded Compounds of Six Large-Volume Rubbers with Data for the Corresponding Gum Stocks[a]

	SBR 1500 (SBR)		Natural rubber (NR)		Hycar 1052 (NBR)		Neoprene WRT (CR)		Vistalon 6505 (EPDM)		Butyl 218 (IIR)	
	Gum	Black	Gum	Black	Gum	Black	Gum	Black	Gum	Black	Gum	Black
Optimum cure at 307°F (min)	25	20	20	20	25	20	80	60	25	25	60	60
Stress–strain data at 25°C												
100% Modulus (psi)	100	280	100	390	70	210	110	960	110	330	70	310
200% Modulus (psi)	140	830	170	1130	90	520	170	—	160	770	110	870
300% Modulus (psi)	180	1700	260	2120	110	1050	240	—	210	1350	160	1450
400% Modulus (psi)	—	2580	400	3120	130	1580	300	—	—	2090	260	2030
500% Modulus (psi)	—	3330	680	—	160	2180	460	—	—	—	—	—
600% Modulus (psi)	—	—	1480	—	—	2750	1130	—	—	—	—	—
Tensile strength (psi)	190	3430	2820	3960	200	2800	1290	2720	230	2240	320	2490
Elongation at break (%)	310	520	690	480	590	610	620	200	310	410	440	490
Formulations (parts by weight)												
Philprene 1500 (SBR)	100	100	—	—	—	—	—	—	—	—	—	—
No. 1 RSS (NR)	—	—	100	100	—	—	—	—	—	—	—	—
Hycar 1052 (NBR)	—	—	—	—	100	100	—	—	—	—	—	—
Neoprene WRT (CR)	—	—	—	—	—	—	100	100	—	—	—	—

Vistalon 6505 (EPDM)	—	—	—	—	—	—	—	—	100	100	—	—
Butyl 218 (IIR)	—	—	—	—	—	—	—	—	—	—	100	100
Philblack N330 (HAF)	—	50	—	50	—	50	—	50	—	50	—	50
Philrich 5 (HA oil)	10	10	5	5	—	—	—	—	—	—	—	—
Circo Light oil (NAPH oil)	—	—	—	—	—	—	10	10	—	—	—	—
Flexon 580 (NAPH oil)	—	—	—	—	—	—	—	—	25	25	—	—
Cumar P-25	—	—	—	—	10	10	—	—	—	—	—	—
Dibutyl sebecate	—	—	—	—	10	10	—	—	—	—	—	—
Neozone A	—	—	—	—	—	—	1	1	—	—	—	—
PBNA	—	—	—	—	1.5	1.5	—	—	—	—	—	—
Agerite Stalite S	2	2	2	2	—	—	—	—	—	—	—	—
Stearic acid	2	2	3	3	1	1	0.5	0.5	1	1	1	1
Magnesium oxide	—	—	—	—	—	—	4	4	—	—	—	—
Zinc oxide	4	4	5	5	5	5	5	5	5	5	3	3
NA-22	—	—	—	—	—	—	0.5	0.5	—	—	—	—
Methyl Tuads	—	—	—	—	—	—	—	—	1	1	1	1
Thionex	0.15	0.15	—	—	0.6	—	—	—	—	—	—	—
MBT	—	—	—	—	—	—	—	—	0.5	0.5	—	0.5
Santocure	1.1	1.1	0.6	0.6	—	—	—	—	—	—	—	—
Methyl Zimate	—	—	—	—	—	—	—	—	—	—	0.5	0.5
Sulfur	1.8	1.8	2.5	2.5	1.0	1.0	0.5	0.5	1.0	1.0	1.75	1.75

a From *Rubber Age*, **107,** 25, August 1975.

noncrystallizing rubbers, like the SBR's (Fig. 3). The tensile strength–loading maxima are generally not observed in rubbers of low unsaturation like butyl (IIR) and poly(ethene–propene–diene) (EPDM). The trade literature contains extensive data for constructing tensile strength–loading plots for various fillers and polymers. Soft, high tensile compounds are most readily compounded from crystallizing rubbers like natural rubber, neoprene, and the very weakly crystallizing butyl rubbers. Even the pure gum stocks of these rubbers have high tensile strength as contrasted with the so-called noncrystallizing rubbers; examples are SBR's, nitriles (NBR), and polyacrylates (ABR).

The data in Table IV illustrates typical stress–strain values in good, representative carbon black loaded compounds of some common rubbers. The data for the corresponding gum stocks are shown for comparison. An attempt was made to compound to provide roughly similar cure times, but the differing shapes of the Oscillating Disc Rheometer curves (Figs. 13 and 14) prevented this for butyl and neoprene.

Modulus is the resistance, per unit of cross section, to a deforming force. The so-called *static modulus* of the rubber compounders is obtained in stress–strain tests in which the samples are extended at a rate of 20 in./min.* The dynamic modulus is measured while the sample is oscillated about some given strain or stress, usually under some fixed superimposed load. *Hardness* is a modulus measured at very small deformations, commonly by the use of indenter devices. The static moduli at 300% extension, the dynamic moduli, and the hardness data of our 21 commercial samples are tabulated in Table V. For these compounds, the dynamic modulus was measured at comparatively small strains and thus correlates quite well with Shore hardness (see Fig. 4, in which we have plotted additional data on natural rubber and SBR compounds under the same test conditions to extend and amplify the curve).

The chief means of increasing modulus and hardness of a stock is through the use of fillers. Modulus and hardness depend so markedly on filler "structure" (or more specifically on carbon black structure)† and cross-link density that it is difficult to take data from random compounds like these in our "commercial compounds" group and draw generalizations. The most explicit method of illustrating modulus changes introduced through filler variations is through loading studies (cf. Figs. 5 and 6). The choice of filler(s) and loading

* The more nearly "static" modulus of the physicists is obtained by very slow stepwise loadings, in small deformation increments, and at slow rates.

† "Structure" is a loose operational term usually expressing the fact that carbon black particles are irregularly aggregated. The aggregation may be primary (i.e., permanent), resulting from agglomeration and fusing of particles in the gas stream in which the carbon black is formed; or it may be secondary agglomeration, which is virtually always present. This secondary agglomeration occurs after the carbon black is removed from the hot gases of the carbon black reactor and is readily altered during handling, pelleting, and mixing into rubber, and on deformation of the filled rubber articles. The secondary agglomerates are readily ruptured, but re-form immediately due to the action of attractive forces between carbon particles.

TABLE V

COMMERCIAL COMPOUNDS ARRANGED ACCORDING TO DYNAMIC MODULUS

Compound no.	Compound for rubber article	Hysteresis[a] (tan δ)	Dynamic modulus[a] $E' \times 10^{-7}$ (dyn/cm^2)	Shore A hardness	300% Modulus (psi)
228	Battery case	Too high to determine		95	—
229	Drop wire insulation	0.269	13.95	82	570
221	Brake pedal	0.354	9.83	72	1030
223	Shoe sole	0.145	8.84	68	[b]
189	Heat-resistant NR belt cover	0.247	7.46	64	1000
217	Window gasket	0.176	7.45	70	[c]
195	Garage door seal	0.270	7.33	62	950
197	Heat- and abrasion-resistant belt cover	0.176	7.15	65	1610
224	Automobile bushing	0.051	7.05	68	1970
206	Grade 2 belting	0.197	6.69	64	1490
220	Automobile mat	0.228	6.67	62	1090
194	Brake cup	0.109	6.59	66	1360
190	Heat-resistant belt cover	0.139	6.20	62	1330
211	100 Level passenger car tread	0.247	5.94	64	1330
208	Custom mix tread	0.263	5.93	59	1130
219	High-quality automobile mat	0.158	5.25	59	1380
188	Heavy-duty truck tread	0.112	4.78	59	2100
227	Irrigation tube valve gasket	0.176	4.77	57	950
191	Heavy-duty brake boot	0.134	4.44	59	890
192	Bicycle tire	0.062	3.28	51	1220
225	Proof goods	No test		64	—

[a] Determined on Roelig machine at 10% extension, 10 cps, and 70°C.
[b] Elongation 240%; 440 psi tensile strength.
[c] Elongation 280%; 1750 psi tensile strength.

will depend on the modulus required, a balance of other properties, and cost. Since rubbers, by definition, are highly elastic materials, superior rupture properties are not compatible with low elongation.

Another item of importance in polymer selection is abrasion resistance. Table VI lists the Pico abrasion [3] test results for our illustrative commercial compounds. For this group with widely differing compositions and properties,

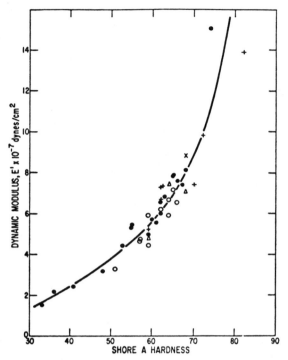

Fig. 4. Dynamic modulus versus Shore hardness for 21 commercial compounds: △, carbon-black-loaded natural rubber compounds; ○, SBR with carbon black only; +, SBR with black and white fillers; ×, SBR with white fillers only; ●, data from extensive study mentioned in Reference [1b] to extend the curve.

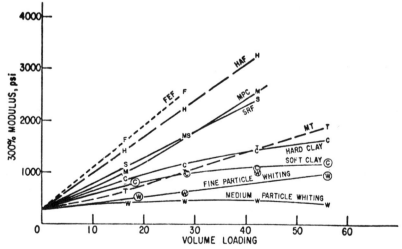

Fig. 5. 300% Modulus relations for various fillers, comparable volume loadings in NR. Plot of data from Winspear [24].

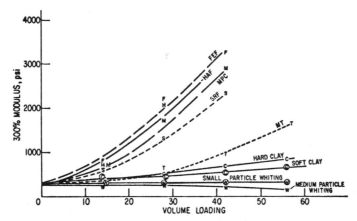

Fig. 6. 300% modulus relations for various fillers, comparable volume loadings in SBR 1500. Plot of data from Winspear [24].

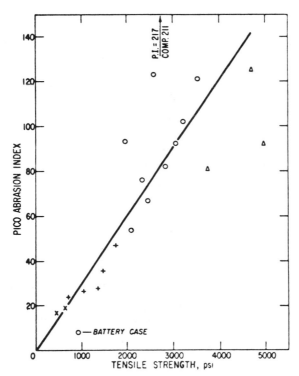

Fig. 7. Pico abrasion versus tensile strength for 21 commercial compounds. Same symbols as in Fig. 4.

TABLE VI

PICO ABRASION AND CUT GROWTH RESISTANCE OF COMMERCIAL COMPOUNDS[a]

Compound no.	Compound for rubber article	Total parts in compound (polymer = 100)	Elastomer	Carbon black (phr, type)	Nonblack filler	Total oil and resin	Tensile strength (psi)	Elongation at break (%)	Pico abrasion index	Cut growth resistance (hr to 5×)
188	Heavy-duty tread	163	NR	45 N285	—	5	4700	550	125	159
224	Automobile bushing	150	NR	40 S315	—	—	4490	580	92	34
189	Heat-resistant belt cover	164	NR	45 N326	—	5	3730	600	81	19
206	Grade 2 belting	175	SBR 1606	50 N330	—	10	3530	570	121	37
197	Heat- and abrasion-resistant belt cover	174	SBR 1502	50 N330	—	15	3210	530	102	22
227	Irrigation tube valve gasket	187	SBR 1608 SBR 1603	35 N220 15 S300	—	30	3070	640	92	31
194	Brake cup	158	SBR 1503	40 N550	—	—	2860	530	82	12
211	100 Level passenger tread	252	SBR 6778	82.5 N339	—	62.5	2740	570	217	23
208	Custom mix tread	239	SBR 6779	75 N330	—	50	2590	610	123	Average 28
219	High-quality auto-motive mat	274	SBR 1815	75 N330	25 Clay	60	2440	490	67	45
190	Heat-resistant belt cover	161	SBR 1500	45 N326	—	5	2340	460	76	Average 21

191	Heavy-duty brake boot	220	SBR 1801 SBR 1502	25 N774 20 N330 35 N990	12 Clay	17.5	2100	720	54	38
192	Bicycle tire	267	SBR 1708	90 N660	—	70	1980	510	93	24
217	Window gasket	229	SBR 1815	30 N770 40 N990 75 N330	40 Laminar	80	1750	280	47	17
220	Automotive mat	361	SBR 1815	75 N330	90 Clay	80	1480	410	36	Average 116
195	Garage door seal	430	SBR 1805	75 N330	100 Clay 50 Whiting	75	1360	430	28	45
221	Brake pedal	510	SBR 1321	15 N660 80 N550	120 Clay 100 Whiting	80	1030	300	27	16
228	Battery case	1250	SBR 1712	—	900 Carbofil (Ground coal)	175	920	10	8	Too brittle to test
229	Drop wire insulation	453	SBR 1503	80 N990	95 Whitex No. 2 100 Atomite	65	700	470	24	5
225	Proof goods	298	SBR 1006	—	75 Clay 75 Whiting 20 TiO_2	15	640	—	19	1.5
223	Shoe sole	348	SBR 1506	—	45 Zeolex 20 TiO_2 45 Clay 45 Whiting	75	440	240	17	1

[a] From *Rubber Age*, **107**, 26, August 1975.

the general relation between tensile strength and abrasion resistance holds (Fig. 7). Such data must be used with discretion, since many exceptions can be found. This is particularly true for NR/BR (polybutadiene) and SBR/BR blends, whose Pico abrasion resistance is decidedly higher than would be predicted from Fig. 7. The enhanced abrasion resistance in these cases is attributable to the presence of polybutadiene, whose beneficial effects are well known in tire compounding. Good tread compounds and high-quality belt covers, as examples, generally have markedly higher Pico abrasion resistance than other compounds for which cost was the decisive factor.

In many applications, such as tire treads and torsion springs, fatigue resistance is important. Cut growth is one aspect of fatigue resistance [2, 4]. Table VI lists cut growth data obtained on the belt fatigue tester [5]. Compounding for fatigue resistance is complex [2]. Compounds showing poor dispersion often show surprisingly slow cut growth. This is illustrated by the data from our commercial compounds (Fig. 8).

In SBR's especially, the flex life is markedly affected by filler dispersion and state of cure. Poor dispersion often results in improved cut growth properties because the propagation of the cut is interfered with by the agglomerates and results in a simulation of the knotty tear. Poor cut growth resistance is associated with low states of cure. As the state of cure increases, the cut growth

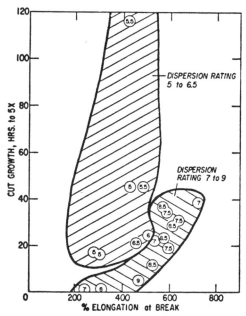

Fig. 8. Cut growth versus elongation for 21 commercial compounds. Numbers are dispersion ratings according to the Stumpe dispersion method [5a,b]. The Stumpe dispersion scale rates samples from 1 (poorest) to 10 (best).

resistance increases to a maximum and then falls off at very high states of cure. The latter observation is in line with the experience that a reasonably high elongation at break is important if a compound is to possess good cut growth resistance. This is a necessary but not sufficient requirement, since other factors are undoubtedly involved.

Oil resistance must be considered in polymer selection. Unfortunately, the oil-resistant polymers are comparatively high in price (Table I). Zelinski [6] illustrated this relation. In a series of oil-resistant polymers of a given type, low-temperature flexibility is generally inversely related to oil resistance.

Another important polymer property is tack. According to our definition, *tack* [7] is the tendency of two identical pieces of rubber to adhere to each other on contact. *Stickiness* is defined as the tendency of rubber to stick to unlike materials, such as mill rolls or metal parts. Both tack and stickiness can be measured directly [7]. The rubber compound must have sufficient tack for building operations, but must not have excessive stickiness, since this interferes with other processing steps. Too much tack can cause handling problems and trapped air.

The tack of the 21 illustrative commercial compounds (see footnote, page 368) generally decreases with increased total loading (Fig. 9). The more the polymer is diluted with filler and/or oil, the lower the tack. Some breakdown of the polymer is necessary for development of good tack. Hence, milling and mixing, which cause some molecular breakdown, will increase tack. Tack and stickiness often increase together, but the relationship has exceptions, such as the natural rubber stocks shown in Fig. 10.

Flammability is important in many applications. Reduction in flammability of polymers is normally achieved by incorporation of nonflammable or flame-retarding additives into the compound. Halogenated materials are among the principal types of additives used. When heated, these materials give off halogen

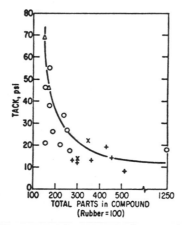

Fig. 9. Tack of 21 commercial compounds. Same symbols as in Fig. 4.

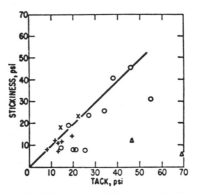

Fig. 10. Stickiness versus tack of 21 commercial compounds. Same symbols as in Fig. 4.

atoms, which act as free radical traps and interfere with oxidative reactions. These appear to work best when used in conjunction with antimony oxide. Phosphorus-containing additives as well as zinc salts and calcium borate are also used; these presumably act as nonvolatile Lewis acids which assist carbonization through dehydration and dehydrogenation. The neoprenes and fluoroelastomers contain sufficient halogen in the polymer itself to be quite flame resistant. In the polyurethane field, reductions in flammability have been achieved by using starting materials containing phosphorous or halogen. This type of approach has been extended in a limited way to other classes of polymers [8, 9].

In applications in which the rubber article is subjected to continuous stress, permanent set is an important property. Creep, stress relaxation, and set [10]* are manifestations of this same basic phenomenon, in that they involve permanent or irreversible flow in the rubber article when it is subjected to more or less prolonged stress. In a floor mat one can tolerate very high set, but this would be completely unsatisfactory in a torsion spring or motor mount. High stress relaxation is undesirable in gaskets.

* The ASTM definitions of these terms are:

   Creep, Drift, or Strain Relaxation—Creep, drift, or strain relaxation is that characteristic of an elastomeric vulcanizate of continuing to deform after the initial deformation under applied stress.
   Permanent Set—The amount by which an elastic material fails to return to its original form after a deformation. In the case of elongation, the difference between the length shortly after retraction and the original length, expressed as a percentage of the original length, is called the tension set, part of which is permanent and part subpermanent, which see. Permanent set is dependent on quality and type of rubber, degree and type of filler loading, state of vulcanization, and amount of deformation.
   Stress Relaxation—The spontaneous gradual lowering of stress imposed on a rubber article or specimen, due to structural rearrangements; also used to designate the decrease in stress resulting from plastic flow or deterioration such as attack by ozone.

(Reprinted by permission of the American Society for Testing and Materials, copyright, 1956.)

Hysteresis, the irreversible loss of mechanical energy by internal friction (see Chapter 5), is a basic property of every elastomer. When an elastomer or a rubber compound is deformed, only part of the energy of deformation is recovered when the deforming force is released; the remainder is lost as heat. It is the energy lost in repeated deformation that causes the temperature of a rubber article, like a tire, to rise. In many services, this property will govern the choice not only of the polymer but also the type and loading of the filler and other ingredients, the cure system, and the state of cure. All these factors are of particular importance, since the dynamic rather than the static properties are involved in most service applications. The hysteresis (tan δ) measurements on the 21 commercial compounds are listed in Table V.

Tear resistance is involved in sudden failures, fatigue, crack growth, cutting and chipping of tire treads, and many other performance-related rupture properties. In many low-priced articles and nonblack compounds such as crutch tips, rubber gloves, rubber footwear, bathing caps, and sleeves around automobile parts, tearing is a major factor which limits service life. Hardening due to aging, which reduces tear strength, is fatal to such items. Tear resistance requires good elongation at break, which is generally reduced by heat, overcure, and aging. Unfortunately, no generally useful tear test has been developed to date. See, however, Chapter 10. Thus, for some purposes hand tear testing by an experienced compounder may be best.

B. Polymer Blends

Blends of polymers (see Chapter 12) are the rule rather than the exception in today's compounds, since in most cases a wider range of properties or a better property balance can be obtained. For example, fatigue life and groove cracking may be affected markedly by small to medium percentages of one polymer in another. The addition of polybutadiene to SBR or natural rubber treads to improve the resistance to groove cracking is the outstanding example. Small amounts of SBR in a natural rubber stock also achieve this result. Further, marked improvements in extrusion, moldability, or mold flow can be achieved through polymer blending.

Blends of polymers may also be used to reduce cost with only a slight loss, and occasionally even a gain, in properties. Natural rubber is often added to other polymers to improve the building tack of polymers deficient in this property. Styrene–butadiene block copolymers are often blended with neoprene and other tough polymers to improve processing. When the block copolymer gets hot, the styrene blocks melt and the viscosity is drastically reduced. The polymer then acts as an effective plasticizer for the tough polymer. On cooling, the block structure and stiffness are regained. Many other examples of applied blending are cited in Chapter 12.

C. Curing System

The effect of the curing system on the properties of the resulting vulcanizate is very great. The factors involved in selecting cure systems are discussed in the following paragraphs. Many of the same criteria are involved as in the selection of the elastomer. Cost, though, is usually secondary compared to the importance of obtaining the desired rate and state of cure and the balance of vulcanizate properties. One can often make compromises on the type of rubber, but the cure system allows little choice if optimum properties are to be obtained. Factors involved in selection of cure systems are type of rubber; type of zinc oxide and amount of fatty acid; type of service (hot or cold); aging resistance required; cure rate; scorch control; ease of dispersion; and toxicity (tendency to cause dermatitis).

The type of rubber is important because different rubbers react differently to various cure systems, be they normally accelerated sulfur systems, semi-EV cures, or nonelemental sulfur cures. Efficient vulcanizing (EV) cures contain very high accelerator to sulfur ratios, making the sulfur rather efficient in the production of cross-links, which, in a properly designed EV cure system and at sufficiently long cures, are predominantly monosulfides. These, however, have their limitations, especially concerning flex life in natural rubber compounds. To overcome this, a compromise is often made by using cure systems which are intermediate between the EV and the conventional accelerator–sulfur systems; these are the so-called semi-EV cures, A high percentage of monosulfidic cross-links can also be obtained by using the so-called sulfurless cures based on TMTD (tetramethylthiuram disulfide), or Morfax [2-(morpholinodithio)benzothiazole], and prolonged cure times. TMTD stocks are notoriously scorchy, but Morfax provides all the scorch resistance normally needed. In fact, many Morfax cures usually require a "kicker," like TMTM or MBT, because they are rather slow curing (see Chapter 7). The amount and type of zinc oxide (coated or uncoated, particle size, French or American process, etc.) also affect the curing process.

The desired cure rate is another factor in the selection of a curative system. Cure rate, as measured with modern instruments like the oscillating disc rheometer (ODR) [11, 12], is the slope of the torque–time curve after it goes through the scorch period.* From this curve, one can estimate optimum cure

---

* An oscillating disc rheometer measures torque as a function of time at a given frequency and amplitude. The torque in the uncured stock is a measure of its viscosity and thus determines the same basic property that a Mooney viscometer does. It does so less precisely, but with the advantage that it does not apply a continuous torque, but a reciprocating one, and thus will not rupture the sample once cross-linking is complete. The point where the torque passes through a minimum and starts to increase is called the scorch point of the stock. As the torque increases, the stock is curing and, since the cured rubber has now become a solid body, the torque measurement here is that of a modulus. Reversion is indicated by a decrease following a maximum torque. A schematic diagram of this is shown in Fig. 15.

or the time for optimum cure. Many compounders consider the time to optimum cure to be that which gives a torque of 90% of the maximum torque. A comparison of relative cure rates of various accelerators is given in Chapter 7 on vulcanization.

Figure 11 illustrates how various rubbers with a given cure system have different scorch times, cure rates, cure times, and maximum torques when tested in the oscillating disc rheometer. Figure 12 shows the effect of 50 phr HAF (N330) carbon black on these same rubbers with the same cure system. The general promoting effect of carbon black on cure rate and cure time, as well as the increased moduli and torques, are illustrated. Figures 13 and 14

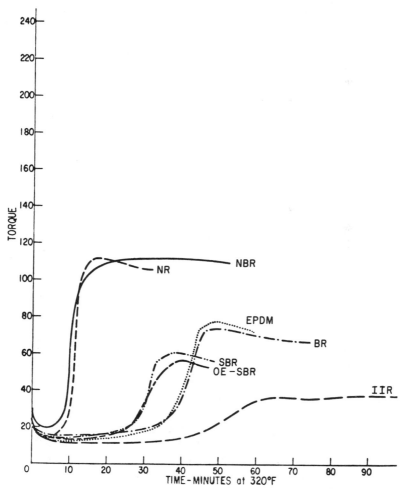

Fig. 11. Oscillating disc rheometer curves for various rubbers with identical cure system. Pure gum stocks. Recipe: rubber, 100; stearic acid, 2; zinc oxide, 5; Santocure MOR, 1; sulfur, 2.5.

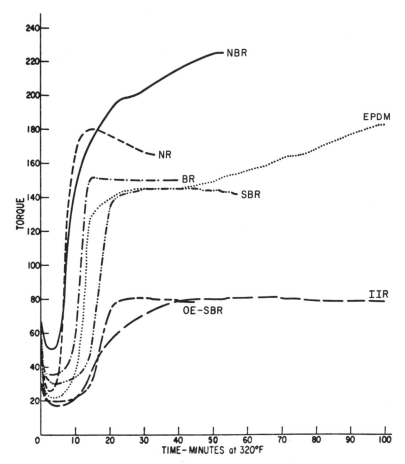

Fig. 12. Oscillating disc rheometer curves for various rubbers, identical cure system. 50 phr N330 carbon black. Recipe: rubber, 100; stearic acid, 2; zinc oxide, 5; N330 black, 50; Santocure MOR, 1; sulfur, 2.5.

show the results for gum and carbon black loaded vulcanizates obtained for various elastomers with cure systems selected for the polymer and designed to give roughly equal states of cure. The stress–strain data and formulations for these compounds are given in Table IV. The rheometer data are plotted in Figs. 13 and 14 as percentage of the maximum torque developed.

Fatty acids (usually stearic acid) are generally added to insure that adequate quantities of soluble zinc (as fatty acid salt) are present to develop good cures with low to moderate reversion. The stearic acid levels out differences between batches of polymer and ensures a uniform rate of cure. This practice was formerly of considerable importance when the older types of natural rubber were

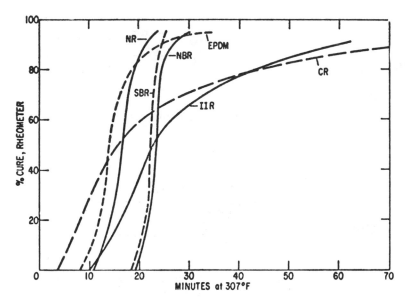

Fig. 13. Degree of cure attained for gum stocks of various polymers with so-called optimum cure systems.

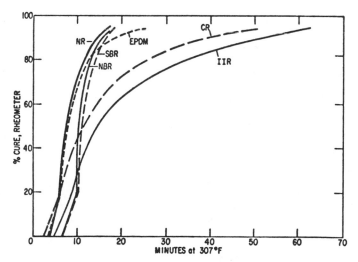

Fig. 14. Degree of cure attained for various polymers with 50 phr N330 with so-called optimum cure systems.

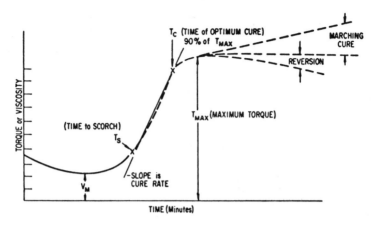

Fig. 15. Oscillating disk rheometer curve.

TABLE VII

A COMPARISON OF MOONEY SCORCH TIMES AT 250°F OF GUM AND
BLACK STOCKS WITH VARIOUS ACCELERATORS[a]

	Scorch time (min) for	
Accelerator (Trade name or abbreviation)	Pure NR gum	NR black stock[b]
1. N,N-Diisopropylbenzothiazole (DIBS)	63	21
2. 2,6-Dimethylmorpholinebenzothiazole-2-sulfenamide (Santocure 26)	60+	27
3. N-Oxydiethylene benzothiazole-2-sulfenamide (NOBS Special)	60+	24
4. Blend of 90% NOBS Special with 10% MBTS (NOBS #1)	60+	20
5. N-tert-butyl-2-benzothiazolesulfenamide (Santocure NS)	60+	23
6. 2-Benzothazyl-N,N-diethylthiocarbamyl sulfide (Ethylac)	26	15
7. N-Cyclohexyl-2-benzothiazolesulfenamide (Santocure)	60+	20
8. Benzothazyl disulfide (Altax)	73	12
9. 2-Mercaptobenzothiazole (Captax)	12	9
10. Tetramethylthiuram monosulfide (Monex)	25	12
11. Tetraethylthiuram disulfide (Ethyl Thiuram)	18	9
12. Zinc dibutyldithiocarbamate (Butyl Ziram)	5	4
13. Zinc dimethyldithiocarbamate (Methyl Ziram)	6	4
14. Butyraldehyde–aniline condensation product (Accelerator 808)	19	5
15. Diphenylguanidine (DPG)	16	10
16. Tetramethylthiuram disulfide (TMTD)	13	6

[a] Base compound: NR, 100; zinc oxide, 5; stearic acid, 3; sulfur, 3; accelerator, 1.
[b] Carbon-black-loaded with 50 phr HAF (N330).

the mainstay. Modern natural rubber and the synthetics can be obtained with much more uniform cure properties, but stearic acid is still commonly used for insurance.

The type of service dictates the required heat resistance of the compound and the degree of the retention of properties on aging. It is possible to make compounds much more resistant to hot environments through proper selection of cure systems. At equivalent cross-link densities, the cure system does not greatly affect the immediate low-temperature or high-temperature properties, such as the modulus; but it does affect the ability to withstand high-temperature service for a longer period of time [13].

One of the chief problems in processing is scorch control. The cure system must permit the stock to be processed in a normal manner without premature vulcanization (scorch). Table VII shows the effect of various accelerators on scorch [14]. The chemical name is given first, then the commonly used abbreviation or trade name. In SBR and BR these accelerators display essentially the same *relative* activity as in NR.

Carbon black promotes scorch, as illustrated dramatically in Table VII. The various types of carbon black can be placed in an order of increasing activity which will hold for most accelerated sulfur cures. At 50 phr loading, this order is shown in Table VIII. The high structure versions of these blacks are more scorchy in the factory than the regular structure blacks because they increase the viscosity of the batch more than the blacks of regular structure. Thus, while high structure blacks give improved dispersion, they also give rise to higher mixing and processing temperatures, and thus to earlier scorching.

Figure 16 illustrates a good mixing and processing stock on a mill. By contrast, Fig. 17 shows a poor processing stock which scorched on the mill.

TABLE VIII

GENERAL SCORCHINESS OF
SULFUR-CURED STOCKS
CONTAINING VARIOUS TYPES
OF CARBON BLACK

Type of black, in increasing scorch tendency
Thermal blacks
Channel blacks
SRF blacks
HMF blacks
FF blacks
FEF blacks
HAF blacks
ISAF blacks
SAF blacks

Fig. 16. A rubber batch that is milling well.

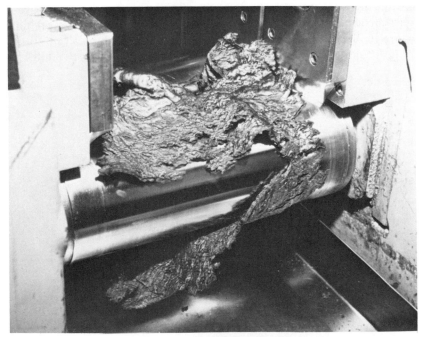

Fig. 17. Poor milling rubber batch.

Cure systems for butyl rubber and most EPDM's would be too fast for the more unsaturated rubbers like NR, BR, and SBR. Combinations of accelerators are commonly used. The accelerator present in the largest amount is called the primary accelerator. The other accelerator(s) are the secondary accelerator(s), or kicker(s). Thiurams and dithiocarbamates are commonly used as kickers for thiazoles or sulfenamides. MBT is a commonly used kicker for thiurams. DPG is used as a kicker for all three types of primary accelerators (thiurams, thiazoles, and sulfenamides). Kickers normally increase both rate of cure and tendency to scorch. One interesting secondary, Ethylac, can often be used to increase the rate of cure without drastic changes in scorchiness [15].

The use of so-called "triangular" acceleration in cure systems is increasing in NR, SBR, and particularly in EPDM. These triangular accelerator systems permit attaining a balance of acceptable scorch time, rate of cure, and desirable network maturity through the development of adequate monosulfidic cross-links.

The curing system is often selected partly on the basis of ease of dispersion. The suppliers of accelerators and other curatives have done much to improve the dispersibility of their products [16]. A homogeneous dispersion of the curing ingredients is necessary for good fatigue life and many other properties that are developed during vulcanization.

The final item in the selection of a curing system concerns toxicity. Some chemicals used in curing systems are somewhat toxic and may cause dermatitis in people working with them. This problem can be avoided by careful selection of the cure system, or by other measures taken to circumvent or prevent exposure.

D. CARBON BLACKS

Aside from the polymer and the cure system, the type and loading of carbon black is the most important factor by which the compounder can alter and control the properties of his compound. It should be reiterated that there is an activating effect of carbon black, in that the black actually assists in the curing of the rubber, reducing the needs for curative, cure time, and, thereby, also cost [17]. An important consideration for selecting a black is processing. This includes mixing, extruding, calendering, and other properties. Another is cost. One uses the lowest-cost black that gives the desired physical properties, such as abrasion resistance or modulus. In formulating very inexpensive compounds like floor matting, cents per square meter of carbon black surface is important rather than cents per pound.

Carbon blacks impart abrasion resistance to all rubbers, a fact that serves for one definition of reinforcement. Particle size is the major factor in determining abrasion resistance, since the extent of rubber–carbon interaction is determined by the latter's surface area, which increases inversely with the particle diameter. The smaller the particle, the greater the resistance to wear of a tread during moderate severity of service. This concept is illustrated by the data in Fig. 18a. At a severity of 95 miles/mil, the roadwear resistance of the tire treads increases with increasing specific surface area (i.e., area per gram). Carbon black structure is not important in this case, but at the very severe service of 15 miles/mil, carbon black structure (see p. 378) dominates the resistance to treadwear (Fig. 18b). These data were presented in a different manner in Studebaker [18]. High structure in carbon black leads to increased moduli, and within limits, any factor that increases moduli increases resistance to abrasion, particularly at high severities.

The increase in abrasion resistance through smaller particle size is not without its problems. Besides being more expensive, smaller-size blacks require

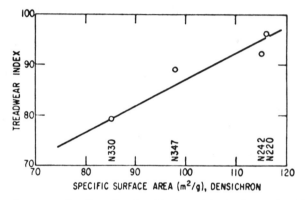

Fig. 18a. Treadwear as a function of carbon black particle size for moderately severe service (∼95 miles/mil).

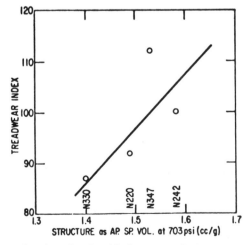

Fig. 18b. Treadwear as a function of carbon black structure for severe service (∼15 miles/mil).

more work on the rubber batch to achieve equal states of dispersion compared to coarser blacks. For tire treads, these disadvantages are balanced, up to a point, by better wear of the tread. For example, SAF blacks are difficult to disperse adequately. The problem can be handled by using carbon black masterbatches whereby the SAF blacks are incorporated into the polymer in the synthetic rubber plant. On subsequent mixing, good dispersions can be obtained. However, finer-particle blacks in all forms (free or masterbatch) generate more heat during mixing and are prone to produce carbon black scorch in high-speed high-pressure mixers and other processing equipment.

The effect of increased carbon black structure on processing and on rubber properties is rather extensive: mixing time is reduced; mixing temperature and peak power requirements are increased (power loads decrease at a faster rate for high structure blacks, which is due to an increased rate of breakdown); extrudate smoothness is increased; die swell is reduced; rate of extrusion is increased. Enhancement of treadwear due to structure increases with increasing severity of service; modulus and hardness are increased; hysteresis under severe test conditions is increased at constant deflection, but not at constant load; groove cracking of SBR or SBR/BR tire treads is reduced markedly. Many of these points are illustrated by data obtained in SBR 1500 with blacks of widely differing structure but of essentially constant surface area (Fig. 19). To a rubber

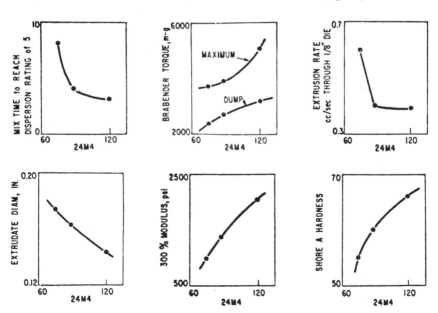

Fig. 19. Some effects of structure at essentially constant surface area. These carbon blacks have quite similar particle size, in the HAF range; they differ markedly in structure as measured by 24M4 DBP absorption in milliliters per 100 g. The values are, for N326, 69; N330, 88; and N358, 120.

compounder, structure in a carbon black is determined by the properties which it imparts to the raw rubber mix [19]. Specifically, it is best measured by determining the reduction in die swell or extrusion shrinkage of the filled rubber mix. Alternately, structure can be measured by 24M4 DBP (dibutyl phthalate) absorption on dry carbon black [20], or in the Brabender Plasti-Corder.

Particle size, or a particle-size-related specific surface area, can be measured with the electron microscope, by other optical methods, or by the adsorption of suitable large surfactant molecules [21].

There are several other properties which are especially influenced by furnace blacks. They will be discussed in Chapter 8. However, it is appropriate to mention that without the development of small-particle, high structure furnace blacks (high abrasion furnace, HAF, and fast extruding furnace, FEF) SBR might not be the general-purpose rubber that it is today. The original SBR, then called GR-S, was extremely tough, highly elastic ("nervy"), and difficult to process. The introduction of the high structure black, FEF, solved many GR-S processing problems. The first SBR tires were generally made using easy processing channel black (EPC) in the treads. They groove-cracked badly. The higher structure HAF greatly reduced this groove-cracking.

The development of quality in a rubber compound, e.g., judged by tensile strength, normally requires adequate carbon black surface per unit quantity of rubber.* This is illustrated in Fig. 1, where tensile strength is plotted against the total parts of material in the compound based on 100 parts of rubber. When adequate carbon black surface is present, the data fall into a narrow band. When the specific surface area is low, the tensile strength for a given quantity of polymer is also low.

Polymer chemists produce elastomers of widely differing molecular structure. These differences are usually quite definitely reflected in processing and raw stock properties of the unfilled elastomers. After incorporation of reinforcing fillers in suitable amounts, the differences between elastomers are greatly reduced. This is found to be true in cured and uncured stocks, and is referred to as the leveling action of carbon black.

E. PROCESSING AIDS AND EXTENDERS†

The factors involved in selecting processing aids and extenders may be listed as follows: compatibility, cost, efficiency, staining, and low-temperature properties. For the hydrocarbon rubbers manufactured in large volume, petroleum-based oils are commonly used as extenders and processing aids. Compatibility between the elastomer and the oils is a prime consideration. If

---

* Adequate tensile strength can be developed in crystallizing rubbers like natural rubber without carbon black. In the absence of reinforcing filler these gum stocks are quite soft.

† Plasticizing action is discussed in a variety of books on polymers. See Deanin [22a], Kurtz et al. [22b], and Barnhart [22c].

TABLE IX

COMPATIBILITIES OF SOLUBILITY PARAMETERS OF ELASTOMERS AND EXTENDING OILS

Elastomer	δ (cal/cc)$^{1/2}$	Highly compatible oil	δ (cal/cc)$^{1/2}$
Vistanex	$(7.5)^a$	Paraffinic oils	7.5
Butyl	(7.5)	Paraffinic oils	7.5
EPDM	(7.5)	Paraffinic oils	7.5
Natural rubber	8.15	Aromatic oils^b	8.9
SBR	8.54	Aromatic oils^b	8.9
cis-Polybutadiene	8.38	Aromatic oils^b	8.9
Neoprene	9.38	Highly aromatic oils or esters	8.9 or above
Nitrile	9.25	Highly aromatic oils or esters	8.9 or above

^{a} Values in parentheses are estimates.

^{b} Saturated portions of these oils should be mainly naphthenic rather than paraffinic.

the oil is not compatible with the polymer, it will bleed out and cause poor physical properties, a sticky surface, and poor adhesion in uses where it should adhere to materials like cord or metal. Compatibility is quite dependent on molecular characteristics which are manifested in solubility parameters* of the polymer and of the processing aid or extender. The compatibility is greatest when these solubility parameters of polymer and extenders are similar. Suggested processing or extending oils, based on compatibility, are listed in Table IX.

Available data on the solubility parameters for Vistanex (polyisobutylene) and butyl rubber are confusing, but from the properties imparted to these rubbers by various plasticizers, one must conclude that these rubbers behave as though their solubility parameters were close to those of paraffinic extender oils.

The interaction between various plasticizers and *polymers containing polar groups* will be strongly influenced by the polarity of the plasticizer. In these cases, hydrogen bonding will also have significant effects when molecular structure permits. This aspect of plasticizer action is of importance with such plastics as polyacrylonitrile, polystyrene, or the various polyacrylates. In these plastics, intermolecular forces are stronger than in elastomers; and plasticizers are important in moderating these forces and contributing flexibility at ambient temperatures. An excellent treatment by Deanin is recommended [22a]. In elastomers, such strong intermolecular bonding is not compatible with rubbery properties at temperatures where elastomers are commonly used. Monomers like styrene or acrylonitrile are generally copolymerized with butadiene to

* The solubility parameter δ is defined as the square root of the energy of evaporation per cc of material. The molecular cohesion is δ^2 multiplied by the molecular weight (or of the monomeric unit for polymers).

produce good rubbery properties over an acceptable range of ambient temperatures. The copolymers have molecular characteristics of an intermediate nature depending on monomer ratios. This has the effect of reducing the effects of hydrogen bonding and polarity associated with the plasticizer.

The question of cost enters when low-temperature elasticity can be obtained only by using the more expensive plasticizers, which are not petroleum oils.

The efficiency of a plasticizer is judged by the amounts of an oil required to improve the processing and/or end use properties of the rubber. There is little information available for comparing the efficiencies of various process oils except in the suppliers' literature [22b]. When compounding on a weight basis, it should be remembered that one requires a larger volume of a low specific gravity than of a high specific gravity plasticizer. The addition of a small amount (say 5 phr) of free processing oil to an oil–black masterbatch during the mixing operation will markedly improve mixing, extrusion, and calendering.

Another important plasticizer problem is staining. In some applications (such as in the vicinity of a white sidewall) a nonstaining oil is needed. There is some variation in staining within a given type of oil, but greatest freedom from staining is possible with paraffinic oils and esters. Some naphthenic oils are only moderately staining.

In industrial products such as molded goods and diaphragms, low-temperature elasticity is one important property. The low-temperature flexibility of the plasticized elastomer will be improved [22a] if the melting point of the oil or plasticizer is lower than the glass transition temperature T_g of the polymer. Esters are frequently selected as plasticizers in rubber compounding to improve low-temperature properties.

Processing aids may be either completely or partially soluble. Soluble oils impart softness to the raw stock without significant reduction in die swell. In the vulcanizates, they are reported to effectively reduce the stiffness with only modest changes in other physical properties [22c]. Examples are listed in Table IX. Softeners of limited solubility reduce extrusion shrinkage, tack, and stickiness. Such materials may be soluble when the stock is hot and bleed to the surface on cooling, which may have both desirable and undesirable features. Waxes used for ozone resistance fall into this category. They migrate to the surface and form a protective coating which resists ozone attack. Such waxes are useful in static applications, but they are virtually useless if the rubber is flexed in service.

The trade literature on rubber processing oils contains many statements which indicate definite interrelations between oil type and the physical properties of rubber stocks containing them. The differences in tensile strength and in hysteresis properties (heat buildup and resilience) may be surprisingly large. Unfortunately, most reports are based on a single mixing procedure (often not specified), and it is not possible to learn whether effects are due to differences in the nature and rate of dispersion of the ingredients or to deficiencies in the state of cure, or are caused by breakdown of the polymer.

An important study of oil extension has been reported in Dunkel *et al.* [23]. Large differences in tensile strength, extrusion shrinkage, and hysteresis were encountered. The higher the aromaticity of the oil, the higher the values for each of these physical properties for SBR, extended by coagulation of the oil with the latex or by Banbury mixing. The large differences in tensile strength were associated with differences in carbon black dispersion—the best dispersions were found for the oils of highest aromaticity. Further, in one set of experiments, the differences between tensile strengths were reduced by remilling. For a better evaluation of the effects of oils on the physical properties of rubber, more mixing studies are needed, since this is a rather neglected area of rubber compounding.

The lower molecular weight, lighter, lower-viscosity, more volatile oils are generally more compatible than their higher molecular weight analogues, but this advantage is diminished by the fact that they are lost in substantially larger amounts during processing when the stocks are hot and, subsequently, in molding and in service.

F. Nonblack Pigments

Large-volume nonblack pigments are divided into three groups, reinforcing, nonreinforcing, and clays and related materials. Some examples are:

1. Reinforcing
   a. Hi-Sil (hydrated silica)
   b. Silene EF (hydrated calcium silicate)
2. Nonreinforcing
   a. Whiting (calcium carbonate)
   b. Titanium dioxide
3. Clays and related materials

Compounds containing reinforcing pigments like Hi-Sil (large specific surface area) and Silene D (smaller specific surface area) must be processed and compounded differently from gums or from carbon black loaded rubbers. These fillers give dry, boardy stocks. They must also be processed hot to reduce problems associated with moisture. They are light and fluffy and slow to incorporate. Their acidity is responsible for the cure retardation resulting from their use, and additional ingredients (amines or glycols) usually have to be added to overcome this feature. Silene D, with its smaller specific surface area, is less retarding than Hi-Sil.

Nonreinforcing materials include titanium dioxide and ground or precipitated whiting ($CaCO_3$) of large particle size which, by giving virtually no reinforcement, come closer to being true fillers than most other materials commonly used. Titanium dioxide is used for extreme whiteness and for pastel colors, and is also effective in masking the inherent color of the rubber and other ingredients.

Clays may be looked upon as slightly reinforcing pigments which improve abrasion resistance and increase moduli. Hard clays give higher moduli than soft clays. Clays also aid in extrusion and provide smoothness and reduce mold shrinkage more per unit volume than any other known material. Here, hard clays are superior. Thus, when molding to close tolerances with minimum mold shrinkage, clays are helpful. Finally, they are effective in reducing gas or air permeability, possibly due to their "platelike" shape.

Useful comparisons of nonblack fillers in natural rubber and SBR are contained in [22c] and [24]. Figures 5 and 6 in this chapter, which show comparisons of modulus changes by various fillers, are plots of data from [24]. Figures 2 and 3 show similar plots of tensile strength data [24].

G. Other Important Compounding Ingredients

1. Resins
    a. Extending resins
    b. Tackifying resins
    c. Curing resins
2. Retarders
    a. Older types
    b. Cyclohexyl-*n*-thiophthalimide (PVI)
3. Special-purpose ingredients, including flame retarders, odorants, colorants, and lubricants

1. Resins

Resins [25] have been separated from softeners and extenders because, although they may be used as softeners and/or extenders, they can be and often are used in other applications. Some scientists believe that the use of high molecular weight resins actually causes extension of the rubber; in other cases, resins are known merely to soften, much like an oil.

One of the other functions of resins is to increase tack. Actually, it would be more accurate to refer to resins as "stickifiers" rather than "tackifiers," since they make the stocks sticky rather than tacky by our definition of tackiness and stickiness [7]. In many uses of compounded rubber, some stickiness is adequate for building or combining the components of a rubber article prior to final cure. Some resins derived from pine trees used for this purpose retard cure somewhat.

The curing resins have a dual purpose. They serve as plasticizers or softeners in the raw stock during the processing steps, since they are quite thermoplastic. They also act as combined curing and reinforcing agents on cure. Figure 20 shows the effect of increasing amounts of curing resin compared to a like

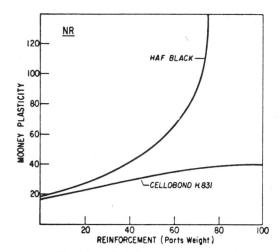

Fig. 20. Effect of reinforcement on processing.

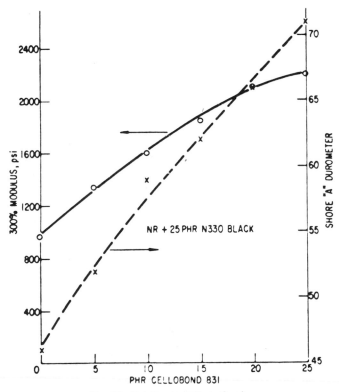

Fig. 21. Reinforcing effect of resin.

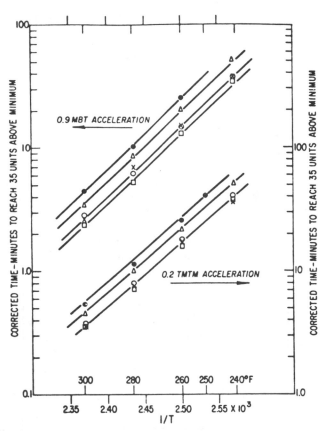

Fig. 22. Retarders that retard both scorch and cure rate: ×, control; ○, 0.8 phr salicylic acid; △, 0.8 phr phthalic anhydride; ●, 0.8 phr diphenyl nitrosamine; □, 0.8 phr benzoic acid.

amount of HAF black on the Mooney viscosity [26]. Figure 21 shows the effect of curing resins on modulus and hardness of the vulcanized rubber compound [26]. In some cases, curing resins are used for both curing and reinforcement, with no added fillers or other curatives.

Many of these curing resins promote adhesion to textile and steel cord. Their chief disadvantage is that they are impart scorch.

2. Retarders

Retarders of vulcanization, or more specifically of scorch, have been used in rubber compounding for many years. In 1957, Craig [27] implied that these materials are involved in the scission of links in the backbone. It is desirable for a retarder to inhibit cross-linking at processing temperatures, but not at curing temperatures. Juve and Shearer [28] showed that these desirable features were not present in retarders of that period. Figure 22 shows their data, obtained

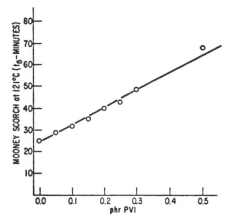

Fig. 23. Effect of PVI on Mooney scorch of natural rubber.

with the Mooney viscometer, for various so-called retarders. They made allowance for the thermal lag in the Mooney machine and measured the time for the viscosity to increase by 35 Mooney units. When these corrected scorch times are plotted against reciprocal temperature in an Arrhenius plot (Fig. 22), a series of parallel straight lines is obtained. These straight lines indicate that the retarders delay scorch and cure to the same relative extents. Thus, when these older conventional retarders were used to reduce scorchiness, the compounder paid a penalty of reduced cure rate, or of state of cure at a fixed cure time.

Recently, a new retarder, cyclohexyl-*n*-thiophthalimide [29a,b], called PVI, has been made available commercially. PVI does not slow the rate of cure,

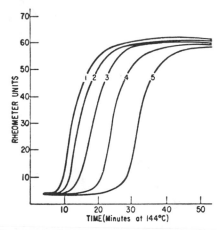

Fig. 24. Effect of PVI concentration on cure rate of natural rubber. Curve 1, none; 2, 0.10; 3, 0.25; 4, 0.50; 5, 1.00 phr PVI.

but works as a true retarder in making a stock much less prone to scorch, thus providing more processing time without affecting cure rate. The cured modulus is not significantly reduced.

Figure 23 shows the reduction in scorch time measured in a Mooney viscometer as a function of (only small amounts of) PVI [29b]. An important application for this material consists in its addition to *slightly* scorched stocks, permitting them to be reworked and returned to the process stream.

Figure 24 shows that increasing the quantities of PVI does not significantly affect cure rate (slope of the ODR curve in the curing region) but that the effect on scorchiness is very pronounced.

III. Compounding Costs

Purchasers of rubber articles—especially mechanical goods—commonly equate tensile strength with quality. There is more than a little justification

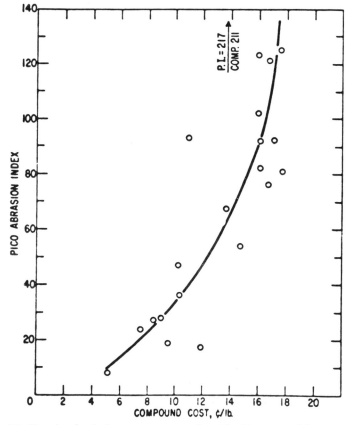

Fig. 25. Pico abrasion index versus compound cost for 21 commercial compounds.

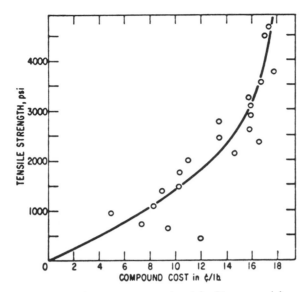

Fig. 26. Tensile strength versus compound cost for 21 commercial compounds.

for this, although exceptions are numerous. Compounding costs are normally reduced through (1) choice of polymer, (2) increasing carbon black and oil levels, and/or (3) by the addition of cheaper, nonreinforcing fillers.

Carbon black increases the hardness; oils decrease it. By adding black and oil together at an appropriate ratio, which will be different for each type of carbon black, the hardness can be kept constant. Abrasion resistance of these black- and oil-extended rubber compounds is often not materially lower than in the unextended compounds. However, other changes in the compound take place. Generally, strength properties suffer.

Cheaper white fillers can be added to reduce cost. Plasticizing oils must be added in quantities dictated by the nature of the filler and its loading.

Generally speaking, these changes, which increase volume loading, decrease cost. For our 21 commercial compounds, the interrelationship between abrasion resistance, tensile strength, and total parts in the compound versus compound cost are shown in Figs. 25–27 (cf. also Figs. 1 and 9).

Some polymers can be extended with large quantities of oil and carbon black, or of other fillers, and still be used to manufacture useful rubber articles. Appliance drain hose made from EPDM with over 500 parts of MT, SRF, FEF, carbon blacks, or combinations is sold commercially. Garden hose with less than 10% elastomer is sold very cheaply. To be extendable with high filler and plasticizer loadings, the polymer must be "stringy." That is, in the raw but milled and broken down state, it must be capable of being extended to extremely high elongations without rupture. Several commercial EPDM's, for example, fulfill these requirements.

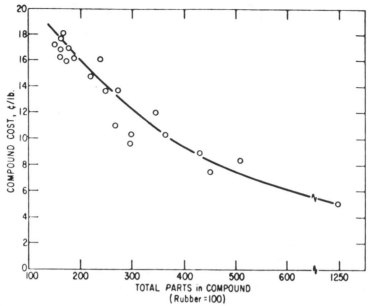

Fig. 27. Compound cost versus total parts in the compound for 21 commercial compounds.

IV. Mixing

The roll mill provides flexibility in operations for use with a variety of stocks of small to modest volume. Roll milling is much slower than mixing in internal mixers, but many high-quality specialty rubber goods are prepared thus. Roll mills are commonly used for cooling stocks when dropped from internal mixers where the intensive high shearing results in hot batches. Roll mills are also used to warm up a stock after a storage period in order to reduce the viscosity of the stock prior to extrusion, calendering, and molding. Roll mills are also commonly employed in sheeting-off a batch to get it into a convenient form for further handling, for blending, for working a stock to get improved pigment dispersion, and for incorporating curatives into the "nonproductive" stock.

In a roll mill (Fig. 28) the shear between the rolls depends on the nip setting and the roll ratio. The nip setting determines the thickness of the sheet that will pass between the rolls. The roll ratio is the relative speed of the back roll to the front roll. Many plants have adopted even-speed rolls under Banburys, etc. to avoid sticking problems; but if one wants a mill to mix well, the higher the roll ratio, the greater the shearing forces and the more intensive the mixing. Figure 29 illustrates the velocity gradient across the mill. Although the literature on mill mixing is scanty, several references are recommended [30, 31, 32].

Fig. 28. Roll mill (26 by 84 in.) mounted on leveling-screw-type vibration isolator pads.

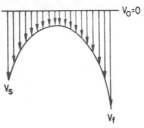

Fig. 29. Velocity gradient across the nip of a roll mill: $V_0 = 0$; V_s, velocity of rubber on face of slow roll; V_f, velocity of rubber on face of fast roll.

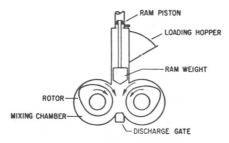

Fig. 30. Internal mixer cross section.

Figure 30 shows a diagram of a typical internal mixer. It is in such devices that most large-volume rubber compounds are mixed at present. There are several manufacturers of these internal mixers, but they all work on the same principle. They knead the pigments into rubber and employ high shear forces to get the rubber to wet the pigment and disperse it.

Continuous mixing has many advantages and may be the mixing of the future. Figure 31 shows one of the continuous mixers, developed by the Farrel Corporation, that is available commercially. Some of the advantages of continuous mixing are uniformity of the product and a continuous supply of stock

Fig. 31. Continuous mixer with mixing rotors shown in open position.

for a calender, tuber, or other continuous operations. This is in contrast to batch mixing, where batch stock sits around and requires additional work on a mill before it will process well on a calender or tuber (unless it is a cold feed tuber).

V. Resistance to Aging

Commercial rubber articles must possess acceptable resistance to changes in physical properties during storage and in service. In practice, this requires protection against oxidation by atmospheric oxygen and ozone. Depending on the application and the nature of the compound, oxidation may be greatly enhanced by heat, light, flexing, and trace metals which catalyze oxidation.

The dominant factors which determine the resistance to oxidative aging are (1) the conditions of service, (2) the nature of the elastomer, (3) type(s) of cross-links, and (4) choice of antidegradant. Hence, in compounding for acceptable resistance to aging, the first consideration must be the anticipated storage and service conditions to which the rubber article is to be submitted. These, along with cost, will govern the choice of polymer. Then the cure system is selected that will provide adequate service life. Next, the aging resistance is "fine tuned" by the selection of antidegradant(s). Finally, it is often necessary to examine the effect of other compounding ingredients on storage and service life. Carbon blacks, in particular, may interact with the antidegradants to render them either more or less effective in a given base formulation. Nonblack pigments may interact with the curing system and/or the antidegradants and often must be used with this in mind for suitable aging resistance.

The moderate changes in physical properties which can be tolerated during prolonged shelf aging are often attainable in either natural or synthetic hydrocarbon rubbers, even without added antidegradants when cured with sulfur donors like TMTD or "efficient sulfur cure systems," i.e., high accelerator/low sulfur cures. The monosulfide cross-links that are produced by these curing systems during prolonged cure times provide excellent resistance to shelf aging for many years at ambient temperature and superior resistance to heat aging as well.

By contrast, pronounced hardening during storage and heat aging is typical of synthetic (hydrocarbon) rubbers, except synthetic polyisoprene, when normally compounded with accelerated sulfur cure systems. When pronounced, this hardening is accompanied by a severe reduction in elongation at break, tensile strength, tear resistance, and resistance to fatigue. These changes in physical properties associated with hardening are believed to be caused by formation of carbon–carbon cross-links; scission (softening) reactions under these conditions are quantitatively of much less significance.

Changes in stress–strain properties and Shore A hardness for two simple SBR 1500 compounds during 8 yr of shelf aging and during 4 days of aging in air at 100°C are tabulated in Table X. The TMTD ("sulfurless") cure produced predominantly monosulfide (R–S–R) cross-links. The NOBS Special accelerated sulfur cure produced a mixture of monosulfide and polysulfide (R–S–S_x–S–R) cross-links with quite appreciable quantities of the S_x-type sulfur in the R–S–S_x–S–R polysulfides. The resistance to change in physical properties of the TMTD (sulfurless) cure is considerably superior to that of the conventional accelerated sulfur cure under both aging conditions. No antidegradants were present in these compounds except the stabilizer present in the raw polymer.

Completely analogous results were observed in carbon black (N330) loaded natural rubber compounds. However, in natural rubber and synthetic polyisoprene, cross-linking and scission reactions normally occur simultaneously in quantitatively significant amounts. Their balance determines whether oxidative hardening or softening takes place. This balance of scission and cross-linking reactions is dependent on the compounding ingredients, the state of cure, and the conditions of oxidative exposure. Many natural rubber gum stocks soften under conditions of oxidative exposure that produce hardening of similarly vulcanized, well-reinforced carbon-black-loaded natural rubber. However, whether hardening or softening predominates, the aging of natural rubber compounds, like that of their synthetic counterparts, normally produces a reduction in elongation and a reduction in resistance to tear, cut growth, and fatigue.

Although much is known about the mechanisms of oxidation of hydrocarbons, including many large-volume plastics, the vulcanization of ordinary

TABLE X

Comparison of Changes in Physical Properties of a TMTD
Sulfurless Cured Compound with a Conventional Accelerated
Sulfur Cure when Aged at Room Temperature (R.T.) and at 100°C

	TMTD sulfurless cure	Accelerated sulfur cure
Ingredient		
SBR 1500	100.	100.
N330 (HAF) black	40.	40.
Zinc oxide	5.00	5.00
Stearic acid	1.00	1.00
TMTD	3.50	—
NOBS Special	—	1.00
Sulfur	—	2.00
Shore A hardness		
Original	62	65
After 8 yr at R.T.	62	70
After 4 days at 100°C	65	72
300% Modulus (psi)		
Original	1600	1950
After 8 yr at R.T.	2340	—
After 4 days at 100°C	2250	—
Tensile strength (psi)		
Original	3800	3700
After 8 yr at R.T.	4025	3165
After 4 days at 100°C	3540	3000
Elongation at break (%)		
Original	480	540
After 8 yr at R.T.	445	270
After 4 days at 100°C	390	270
Increase in oxygen content[a] (%)		
4 days at 100°C	0.10	1.25

[a] Aged in original unextracted state, then extracted with $CHCl_3$ for 24 hr, vacuum dried, and analyzed for total oxygen by a modified Untersaucher procedure. Unaged samples were also extracted prior to direct oxygen analyses.

elastomers greatly complicates the picture. With the exception of peroxide and radiation cure, the extensive mechanistic studies of oxidation of hydrocarbons and its inhibition through antioxidants are of little assistance to the rubber chemist. The introduction of sulfur-containing cross-links and the by-products of vulcanization changes the oxidative behavior of rubber and forces the rubber compounder to use a trial-and-error approach to protecting his compounds against aging. Although it is true that the same general types of inhibitors can often be used in protecting a hydrocarbon plastic and a rubber

TABLE XI

General Classes of Antioxidants and Antiozonants Used in Rubber Compounding:
Uses, and Comparative Ratings

Class of antioxidant, chemical name, (trade name) of example	General properties	Recommended uses	Comparative rating of antioxidant properties		
			Antioxidant activity	Flex-cracking activity	Antiozonant activity
Monohydric hindered phenols 2,6,Di-*t*-butyl-4-methylphenol (Ionol)	Low cost Minimum discoloration and staining Selected ones FDA approved High solubility in rubber No effect on cure rate	Polymer stabilization Carcass and sidewalls Light-colored mechanical goods	Good to poor	Moderate to poor	None
Bis-phenols 2,2′-Methylenebis(4-methyl-6-*t*-butylphenol) (AO 2246)	Medium cost Some discoloration and staining Selected ones FDA approved Low volatility No effect on cure rate	Latex stabilization Light-colored mechanical goods Polymer stabilization	Very good to moderate	Moderate to poor	Negligible
Polyphenols Butylated reaction product of *p*-cresol and dicyclopentadiene (Wing-Stay L)	Medium to high cost Slight discoloration and staining Low volatility Very persistent Some are FDA approved	Latex stabilization Latex compounding Rug backing White sidewalls	Very good to moderate	Moderate to poor	Negligible
Phenolic sulfides 4,4′-Thiobis(6-*t*-butyl-3-methylphenol) (Santowhite crystals)	Medium to high cost Some color problems Low volatility	Latex compounding Foam Rug backing	Good to moderate	Moderate to poor	Negligible

TABLE XI (*Continued*)

Class of antioxidant, chemical name (trade name) of example	General properties	Recommended uses	Comparative rating of antioxidant properties		
			Antioxidant activity	Flex-cracking activity	Antiozonant activity
Mixed phenol phosphites (Age Rite Geltrol)	Medium cost Combined properties of each Autosynergistic	Polymer stabilization Latex compounding	Good to moderate	Moderate to poor	None
Alkylated diphenylamines 4,4'-Dioctyldiphenylamine (Octamine)	Low cost Slight discoloration and staining Negligible effect on cure Better than some phenols	NBR stabilization Latex compounding Limited use in rubber compounding	Good to moderate	Moderate	Negligible
Aryl naphthylamines Phenyl-β-naphthylamine (Neozone D)	Medium cost Active antioxidants Moderate volatility Negligible effect on cure Severe discoloration Toxicity questionable	Polymer stabilization Mechanical goods Tires	Good	Good	Negligible
Polymerized dihydroquinolines Polymerized 1,2-dihydro-2,2,4-trimethylquinoline (Age Rite Resin D)	Medium cost Very low volatility Good heat resistance Moderate discoloration	Heat-resistant stocks of all types	Moderate	Moderate	Moderate

Acetone–diphenylamine reaction products (BLE-25)	Medium cost Active antioxidants Good flex-cracking agents Partially volatile	Polymer stabilization Antioxidant for general compounding	Good	Very good to good	Moderate

N,N′-Dialkyl-p-phenylenediamines
  N,N′-Di(1-methylheptyl)-p-phenylenediamine
  (Eastozone 30)
	Medium cost
	Low volatility
	Discoloring and staining
	Increase cure rates
	Toxicity varies with molecular weight
	Easily oxidized
		Potent antiozonants
		Excellent for short-term ozone resistance
			Good	Good to moderate	Very good to good

N-Alkyl-N′-phenyl-p-phenylenediamines
  N-(1,3-Dimethylbutyl)-N′-phenyl-p-phenylenediamine
  (UOP-588)
	Medium cost
	Voltaility and toxicity depend on size of alkyl groups
	Excellent antioxidants
	Excellent antiflex-cracking agents
	Discoloring and staining
	Activate the cure
		Antiflex-cracking agents
		Metal deactivators
		Treads, sidewalls, and mechanical goods
		Polymer stabilization
			Very good	Very good	Very good to good

N,N′-Diaryl-p-phenylenediamines
  N-Phenyl-N′-o-tolyl-p-phenylenediamine (one constituent)
  (Wing-Stay 100)
	Medium cost
	Low volatility and toxicity
	Limited solubility
	Excellent antioxidants
	Excellent antiflex-cracking agents
	Discoloring
	Low staining
	Negligible effect on cure
		Polymer stabilization
		Treads, sidewalls, and mechanical goods
			Very good	Very good	Good to moderate

vulcanizate, the presence of sulfur in the cross-links, the by-products of vulcanization (many of which have antioxidant properties), and the presence of other compounding ingredients affect the oxidative behavior of the polymer and the action of antidegradants. Only the first steps in understanding the oxidation of real rubber compounds have been taken—but it is a very long road.

The principal types of antidegradants used in the rubber industry are tabulated in Table XI with specific examples and types of protection to be expected based on suppliers' recommendations. This table is abstracted from Tables I–III of an excellent elementary survey of compounding for protection against aging by Parks and Spacht [33].

Ozone cracking is characterized by the formation of cracks that are perpendicular to the direction of strain. This causes rapid deterioration of articles like tire sidewalls in some areas. Paraphenylenediamines are the only effective class of antiozonants for rubbers that must be flexed during service. Microcrystalline paraffin wax provides good protection for rubber in static service. The wax must be sufficiently soluble in the rubber to bloom to the surface and form a protective layer.

In plastics, additives are usually required to reduce *oxidation promoted by ultraviolet light*. These are typically uv absorbers. However, these compounds are not commonly used in rubber compounding. Only one, Irgastab 2002 (nickel salt of 3,5-di-*tert*-butyl-4-hydroxybenzylphosphonic acid monoethyl ester) is listed as a light stabilizer in the section on antidegradants in van Alphen [34]. Fortunately, the best uv absorber, carbon black, is used in substantial quantities in most high-quality rubber articles. When carbon black is not used, the rubber compounder commonly doubles the quantity of antioxidant he would otherwise use. This practice is even followed by some compounders in high-quality black carcass compounds for tires to increase resistance to uv-induced oxidation.

Protection against metal-catalyzed oxidation, particularly by iron, copper, and manganese, is required in the manufacture of synthetic rubber. However, during vulcanization, most of these metal ions are apparently converted to harmless insoluble sulfides. Some multipurpose antidegradants, like the paraphenylenediamines, are active against oxidation catalyzed by metals.

Antidegradants cannot be used indiscriminately. Each has its specific applications. Although some, like the paraphenylenediamines, may protect against more than one type of degradation, instances can be observed where an antidegradant under one type of service is without effect or may even promote oxidative degradation in another service.

VI. Summary

The nature, composition, and preparation of rubber compounds have been described. Factors that affect the selection of compounding ingredients, such as polymer, blends of polymers, curing systems, black selection (as to quality and type), scorch, the effect of processing aids or extenders, nonblack pigments, resins, retarders, and some of the variables involved in mixing have been discussed.

The most important compounding ingredient in a high-quality rubber article is the polymer or blend of polymers. The next most important is the curing system, followed by the carbon black. Selection of the remainder of the components is important, but not as critical as the main ingredients just mentioned. Costs will in all cases have to be weighed against the product and marketing advantages obtained.

REFERENCES

1a. J. R. Beatty and M. L. Studebaker, *Rubber Age* **108**, 27 (1976).

1b. M. L. Studebaker and J. R. Beatty, *Rubber Chem. Technol.* **47**, 803 (1974).

1c. D. V. Rosato, *Rubber World* **166**, 49 (1972).

2. J. R. Beatty, *Rubber Chem. Technol.* **37**, 1341 (1964).

3. E. B. Newton, H. W. Grinter, and D. S. Sears, *Rubber Chem. Technol.* **34**, 1 (1961).

4. A. N. Gent, *in* "Polymer Science and Materials" (A. V. Tobolsky and H. F. Mark, eds.), p. 275. Wiley (Interscience), New York, 1971.

5. J. R. Beatty and A. E. Juve, *Rubber Chem. Technol.* **38**, 719 (1965).

5a. H. E. Railsback, N. A. Stumpe, Jr., and C. R. Wilder, *Rubber World* **160**(6), 63 (1969).

5b. Phillips Petroleum Co., Technical Service Bulletin No. 110, Phillips Petroleum Co., Akron, Ohio, March, 1968.

6. R. P. Zelinski, Structure and Properties of Rubbery Polymers, Advanced Compounding Lecture Series, sponsored by The Akron Rubber Group, February 13, 1973.

7. J. R. Beatty, *Rubber Chem. Technol.* **42**, 1040 (1969).

8. J. W. Lyons, "The Chemistry and Uses of Fire Retardants." Wiley, New York, 1970.

9. R. R. Hindersinn, *in* "Encyclopedia of Polymer Science and Technology," Vol. 7 (H. F. Mark *et al.*, eds.), pp. 1–64. Wiley (Interscience), New York, 1967.

10. ASTM Glossary of Terms Related to Rubber and Rubber-like Materials, ASTM Special Technical Publication No. 184, American Society for Testing and Materials, Philadelphia, Pennsylvania, 1956.

11. G. E. Decker, R. W. Wise, and D. Guerry, *Rubber Chem. Technol.* **36**, 451 (1963).

12. A. E. Juve, P. W. Karper, L. O. Schroyer, and A. G. Veith, *Rubber Chem. Technol.* **37**, 434 (1964).

13. M. L. Studebaker, *Rubber Chem. Technol.* **39**, 1359 (1966).

14. M. L. Studebaker, *Am. Chem. Soc., Div. Rubber Chem. Meeting, Cleveland, Ohio, 1962*.

15. "'Sharples' Brand ETHYLAC as a Delayed Action Activator," Bull. S-130. Pennsalt Chem. Corp., King of Prussia, Pennsylvania, 1957.

16. Sveriges Gummitetniska Forening (Swedish Technical Rubber Association), *Symp. Mixing Tech. Spec. Emphasis Dispersion, Falsterbo, Sweden, May 28–29, 1964*.

17. M. L. Studebaker, *in* "Reinforcement of Elastomers" (G. Kraus, ed.), p. 319. Wiley (Interscience), New York, 1964.

18. M. L. Studebaker, *Rubber Chem. Technol.* **41**, 373 (1968).

19. M. L. Studebaker, *Indian Rubber Bull.* **266**, 4 (Feb) (1971).

20. R. E. Dollinger, R. H. Kallenberger, and M. L. Studebaker, *Rubber Chem. Technol.* **40**, 1311 (1967).

21. R. A. Klyne, B. D. Simpson, and M. L. Studebaker, *Rubber Chem. Technol.* **46**, 192 (1973).

22a. R. D. Deanin, "Polymer Structure, Properties and Applications." Cahners Books, Boston, Massachusetts, 1972. (An excellent recent treatment; contains adequate references.)

22b. S. S. Kurtz, Jr., J. S. Sweely, and W. J. Stout, *in* "Plasticizers Technology" (P. F. Bruins, ed.), Vol. 1, p. 21. Reinhold, New York, 1965. (More specifically directed toward rubber.)

22c. R. R. Barnhart, *in* "Encyclopedia of Chemical Technology" (R. E. Kirk and D. F. Othmer, eds.), 2nd Ed., Vol. 17, p. 543. Wiley, New York, 1968. (An excellent survey.)

23. W. L. Dunkel, F. P. Ford, and J. H. McAteer, *Ind. Eng. Chem.* **46**, 578 (1954).

24. G. G. Winspear, ed., "Vanderbilt Rubber Handbook," pp. 336–337, 356–357. R. T. Vanderbilt Co., New York, 1968.

25. B. Pickup, *in* "Rubber Technology and Manufacture" (C. M. Blow, ed.), pp. 201–203. CRC Press, Cleveland, Ohio, 1971.

26. "Cellobond Rubber Reinforcing Resins," Tech. Manual No. 11. Westerhum Press, London, 1961.

27. D. Craig, *Rubber Chem. Technol.* **30**, 1291 (1957).

28. A. E. Juve and R. Shearer, cited in Craig [27]; see also R. Shearer, A. E. Juve, and J. Musch, *India Rubber World* **117**, 216 (1947).

29a. S. J. Stewart, R. I. Leib, and J. E. Kerwood, *Rubber Age* (*N.Y.*) **102**, 56 (1970).

29b. R. I. Leib, A. B. Sullivan, and C. D. Trivette, Jr., *Rubber Chem. Technol.* **43**, 1188 (1970).

30. B. G. Crowther and H. M. Edmondson, *in* "Rubber Technology and Manufacture" (C. M. Blow, ed.), p. 262. CRC Press, Cleveland, Ohio, 1971.

31. J. R. Scott, *in* "The Applied Science of Rubber" (W. J. S. Naunton, ed.), p. 285. Arnold, London, 1961.

32. H. Colm, *Rubber World* **158**, 67 (1968).

33. C. R. Parks and R. B. Spacht, *Elastomerics* **109**(5), 25 (1977).

34. J. van Alphen, "Rubber Chemicals" (C. M. van Turnhout, ed., rev. ed.), p. 67. Reidel Publ., Boston, Massachusetts, 1973.

Chapter **10**

Strength of Elastomers

A. N. GENT

INSTITUTE OF POLYMER SCIENCE
THE UNIVERSITY OF AKRON
AKRON, OHIO

I. Introduction

Fracture is a highly localized and selective process—only a small number of those molecules making up a test piece or a component actually undergo rupture, while the great majority are not affected. For example, of the 10^{26}

419

Copyright © 1978 by Academic Press, Inc.
All rights of reproduction in any form reserved.
ISBN 0-12-234360-3

chain molecules per cubic meter in a typical elastomer, only one in 100 million (i.e., those crossing the fracture plane, about 10^{18} per square meter) will definitely be broken. Moreover, these will not all break simultaneously, but will rupture successively as the fracture propagates across the specimen at a finite speed. Thus, the first questions that pose themselves in studying the strength of elastomers (and other materials as well) are, where and in what circumstances does fracture begin? Also, what laws govern the growth of a crack once it has been initiated? This chapter seeks to answer such questions, first in a general way and then with particular reference to important modes of failure of elastomers in service. It does not deal with the rather complex problem of the strength of composite structures, such as a pneumatic tire, which involves failure of adhesive bonds at interfaces between the components as well as fracture of the components themselves.

We consider first the initiation of fracture from flaws or points of weakness where the applied stresses are magnified greatly. Fracture begins at such points. The rate of development of cracks after initiation, which depends on the local stress levels as well as on the way in which these stresses vary with time, is treated next. For example, rapid crack growth may take place if stresses are applied and removed frequently, whereas the crack may grow quite slowly, if at all, when the same stresses are held constant and never removed. This phenomenon of accelerated growth under dynamic stressing is termed "mechanical fatigue" or "dynamic crack growth." It is treated in Sections III.E and IV.F.

Because rubbers are viscoelastic, or more generally inelastic (meaning dissipating mechanical energy), to varying extents and their mechanical properties depend on rate of deformation and temperature, it is not surprising to find that their strength is also dependent upon the *rate* at which stresses are applied and upon the temperature of measurement. These effects are discussed in Sections III.B and IV.A. Other effects of the environment, notably the destructive action of ozone, are discussed in Section IV.G. Finally, a brief survey is given of abrasive wear.

II. Initiation of Fracture

A. FLAWS AND STRESS RAISERS

Every solid body contains flaws as points of weakness due to heterogeneities of composition or structure. In addition, because of the presence of sharp corners, nicks, cuts, scratches, and embedded dirt particles or other sharp inclusions, applied stresses are magnified (concentrated) in certain regions of the body, so that they greatly exceed the mean applied stress. Fracture will begin at such a site where the local stress exceeds a critical level and the small flaw starts to grow as a crack.

The stress concentration factor, i.e., the ratio of the stress σ_t at the tip of a sharp flaw to the applied tensile stress σ is given by Inglis's relation for elastic solids that obey a direct proportionality between stress and strain [1]

$$\sigma_t/\sigma = 1 + 2(l/r)^{1/2} \qquad (1)$$

where l is the depth of an edge flaw and r is the radius of the flaw tip in the unstressed state. If the flaw is totally enclosed, it is equivalent to an edge flaw of depth $l/2$ (Fig. 1). Thus, edge flaws are more serious stress raisers than enclosed flaws of the same size, and therefore they are more usual sources of fracture than inclusions. Not exclusively so, however; heterogeneities of composition have been shown to nucleate fatigue cracks internally [2]. Also, some types of crack cannot form near a free surface; see Section II.F.

When the tip radius is much smaller than the depth of the flaw, as seems probable for the severe stress raisers responsible for fracture, Eq. (1) can be approximated by

$$\sigma_b = (\sigma_t r^{1/2})/2l^{1/2} \qquad (2)$$

Thus, the applied breaking stress, denoted by σ_b, is predicted to vary inversely with the depth of the flaw l, and in proportion to $1/l^{1/2}$. This prediction has been tested for brittle polymers, i.e., in the glassy state [3], and for rubbery materials cracked by ozone (Fig. 2). In both cases the breaking stress σ_b was found to vary, in accordance with Eq. (2), with the depth l of a crack or razor cut made in one edge of the test piece. However, for elastomers broken by mechanical stress alone, elongations at break are generally much too large for the assumption of linear stress–strain behavior to be valid and Eq. (2) becomes a relatively poor approximation. Even so, by extrapolating measured values of the breaking stress for different depths of edge cut to the breaking stress for a test piece having no cuts introduced at all, the depth of flaw characteristic of the material may be inferred. (Actually, the value obtained is the depth of a cut equivalent in stress-raising power to natural flaws, which may be smaller and sharper, or larger and blunter, than cuts of equivalent stress-concentration power.)

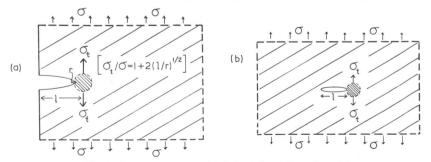

Fig. 1. Stresses near a crack of depth l and tip radius r [1].

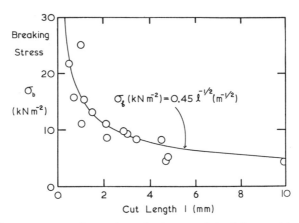

Fig. 2. Fracture stresses for specimens having cuts of depth *l* exposed to ozone [4].

For both rubbers and glasses, the value obtained in this way is about 40 ± 20 μm. The same value is also obtained for ozone cracking (Fig. 2) by extrapolating measured stresses back to that value observed for a test piece having no initial cut in it, and by extrapolating the fatigue lives of test pieces with cuts in them back to the fatigue life for test pieces with no cuts (Fig. 3). In all these cases, substantially similar values are obtained for the natural flaw sizes. Moreover, they are largely independent of the particular elastomer or mix formulation used, even though these factors greatly alter the way in which the breaking stress or fatigue life changes with cut size, as discussed later. Thus, a variety of fracture

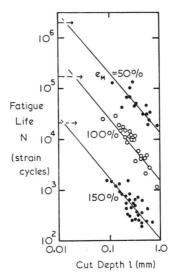

Fig. 3. Fatigue lives *N* for specimens having initial cuts of depth *l* subjected to repeated extensions to the indicated strains.

processes appear to begin from a natural flaw, equivalent to a sharp edge cut, 40 μm in depth.

The exact nature of these failure initiation sites is still not known. They may consist of accidental nicks in molded or cut surfaces, but even if great care is taken in preparing test pieces (e.g., by molding against polished glass surfaces), the breaking stress is not greatly increased. Dust or dirt particles or other heterogeneities nearly as effective as mold flaws seem to be present in a sufficient amount. Only when the sample size is reduced to about 10^{-8} m^3 or less is a significant increase in strength observed, suggesting that the more powerful stress raisers are present only in concentrations of 10^8 m^{-3} or less. Of course, if a way could be found to eliminate them, or at least reduce the effective sharpness of these natural flaws, substantial increases in strength, and even more striking increases in fatigue life, might be achieved, as discussed later. At present, however, they appear to be an inevitable consequence of the processes used in making elastomeric compounds and components.

B. Stress and Energy Criteria for Rupture

Equations (1) and (2) raise several other questions. What is the radius r of a natural flaw? What is the magnitude of the breaking stress σ_t at the tip when the flaw starts to grow as a crack? From a comparison of experimental relations such as that shown in Fig. 2 with the predictions of Eq. (2) only the product $\sigma_t r^{1/2}$ can be determined, and not the two quantities separately. However, the value of r is unlikely to exceed 1 μm for a sharp cut and hence the tip stress σ_t may be inferred to be greater than 200 mega pascal (MPa), taking a value for the product $\sigma_t r^{1/2}$ of 0.2 MN m$^{-3/2}$ as representative of fracture under mechanical stress. For ozone cracking this product takes the value 900 N m$^{-3/2}$ (Fig. 2), and hence the tip stress in this case is presumably 1 MPa or greater.

We must recognize, however, that the tear which begins to propagate from an initial cut or flaw will develop a characteristic tip radius r of its own, independent of the sharpness of the initiating stress raiser [5]. It is therefore more appropriate to treat the product $\sigma_t r^{1/2}$ as a characteristic fracture property of the material. Indeed, Irwin [6, 7] proposed that fracture occurs for different shapes of specimen and under varied loading conditions at a characteristic critical value of the stress intensity factor, K_c, where K_c is defined by

$$K_c = (\pi^{1/2}/2)\sigma_t r^{1/2}$$

or

$$K_c = \pi^{1/2}\sigma_b l^{1/2} \tag{3}$$

when expressed in terms of the applied stress σ_b by means of Eq. (2).

An alternative but equivalent view of the stress criterion for fracture was proposed by Griffith [8, 9], and applied by Irwin [6, 7] and Orowan [10] to elastic solids. Griffith suggested that a flaw would propagate in a stressed material only when, by doing so, it brought about a reduction in elastically stored energy W more than sufficient to meet the surface free energy requirements γ of the newly formed fracture surfaces. Irwin and Orowan recognized that in practice the energy expended in plastic deformation during crack growth generally far exceeds the true surface energy. However, provided that the total energy expended is proportional to the amount of surface created by fracture, Griffith's relation may still be employed.

Griffith's fracture criterion takes the form

$$-(\partial W/\partial l) \geq \tfrac{1}{2}G_c(\partial A/\partial l) \tag{4}$$

where l denotes the length or radius of a fracture, A is the surface area of the specimen, and G_c is the amount of energy required to advance the fracture by unit area. (The factor of $\tfrac{1}{2}$ arises on changing from the area of a fracture plane to the area of the two newly formed surfaces.)

In Griffith's original treatment, the surface free energy 2γ per unit area of fracture plane was employed in place of the generalized fracture energy G_c. His results therefore carried the implication of thermodynamic reversibility, true only for an ideally elastic material. In contrast, G_c represents all forms of energy—surface, chemical, viscous, etc.—dissipated during fracture. Nevertheless, provided that the energy is used in the immediate vicinity of the crack tip, and is independent of the overall shape of the test piece and the way in which forces are applied to its edges, the magnitude of G_c can be employed as a characteristic fracture property of the material, independent of the test method. This expectation has been borne out by critical experiments on a variety of materials, including elastomers, using test pieces for which the relation between the breaking stress σ_b and the rate of release of strain energy on reaching G_c can either be calculated or measured experimentally [11, 12]. Two important cases are considered here.

C. TENSILE TEST PIECE

This test piece is shown in Fig. 4. It consists of a thin strip of thickness t with a cut in one edge of depth l. The effect of the cut in diminishing the total stored elastic energy at a given extension may be calculated approximately by considering a small triangular region around the cut (shown shaded in Fig. 4) to be unstrained, and the strained state of the remainder of the test piece to be unaffected by the presence of the cut, with stored strain energy W' per unit volume (energy density). The reduction in elastic energy of the total test piece due to the presence of the cut is thus $\alpha l^2 t W'$, where α is a numerical constant whose value is given by $\pi(1 + e)^{-1/2}$ and e is the relative extension or strain of the test piece [13]. α thus varies between π and about 1 as the strain e increases.

Hence,

$$-(\partial W/\partial l) = 2\alpha l t W'$$

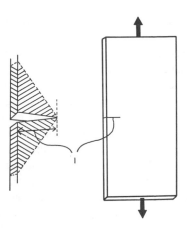

Fig. 4. Tensile test piece.

For this test piece, $(\partial A/\partial l) = 2t$, and the fracture criterion, Eq. (4), becomes [11]

$$2\alpha l W' \geq G_c \tag{5}$$

The breaking stress σ_b does not appear explicitly in Eq. (5); it is that stress at which the strain energy W' satisfies Eq. (5), and thus depends on the elastic properties of the material and the length of the initial cut, as well as on the fracture energy G_c. For an elastic material obeying a linear relationship between tensile stress σ and extension e, the stored energy W' is given by $Ee^2/2$, or $\sigma^2/2E$, where E is Young's modulus. The stress and extension at break are therefore given by

$$\sigma_b = (G_c E/\pi l)^{1/2}, \qquad e_b = (G_c/\pi l E)^{1/2} \tag{6}$$

where α has been given the value π appropriate to linearly elastic materials. Equations (6) were obtained by Griffith [9].

On comparing Eqs. (3) and (6), we see that the critical stress intensity factor K_c and the fracture energy, or critical strain energy release G_c are related to each other and to the breaking stress at the crack tip as follows:

$$K_c^2 = EG_c = (\pi/4)\sigma_t^2 r = (\pi/2)W_t' E r$$

where W_t' is the strain energy density at the crack tip. Hence,

$$G_c = (\pi/2)W_t' r \tag{7}$$

The fracture energy G_c is thus a product of the energy required to break a unit volume of material at the crack tip (i.e., in the absence of nicks or external flaws) and the effective diameter of the tip, as pointed out by Thomas [14]. These two quantities can be regarded as independent factors of the fracture energy: an "intrinsic" strength W_t', and a characteristic roughness or bluntness of a developing crack, represented by r. Because K_c involves also the tensile elastic modulus E, it is not considered as suitable as G_c for a measure of the fracture strength of materials, like elastomers, of widely different moduli.

Equation (5) is more generally applicable than Eqs. (6) because it is not restricted to linearly elastic materials. It constitutes a criterion for tensile rupture of a highly elastic material having a cut in one edge of length l, expressed in terms of the fracture energy G_c. Two important examples of test pieces of this type are (i) ASTM "tear" test pieces for vulcanized rubbers (ASTM D624-54), and (ii) a typical tensile test piece containing small nicks due, for example, to imperfections in the surface of the mold or die used to prepare it.

Several features of Eqs. (5) and (6) are noteworthy. For a given value of fracture energy G_c, stiffer materials with higher values of Young's modulus E will have higher breaking stresses and lower extensions at break than softer materials. These correlations are well known in the rubber industry. Less well known is the direct effect of the size of an initial cut or flaw in reducing both breaking stress and elongation. Finally, if the fracture criterion, Eq. (5), is met for an initial flaw of depth l, it will be increasingly exceeded as l grows, i.e., as fracture proceeds. In consequence, a fracture will accelerate across the specimen.

D. TEAR TEST PIECE

This test piece, shown in Fig. 5, has dimensions such that there are regions I in the arms in simple extension and a region II virtually undeformed. If the arms are sufficiently wide, or if they are reinforced with inextensible tapes, their extension under the tear force F will be negligibly small. In these circumstances the work of fracture $\frac{1}{2}G_c \, \Delta A$ is provided directly by the applied force F acting through a distance $2 \, \Delta l$ where Δl is the distance torn. The torn surface is $2d \, \Delta l$ where d is the thickness of the sheet.* Thus, the fracture criterion, Eq. (4) becomes [11]

$$F \geq G_c d/2 \tag{8}$$

Because the tear force in this case is a direct measure of the fracture energy G_c, and is independent of the elastic constants of the material and of the length of the tear, this test piece is particularly suitable for studying the effects of composition and test conditions upon G_c [16–19].

It is important to remember that the fracture energy G_c is not a constant value for a particular material; it depends strongly on the temperature and rate of tearing, i.e., the rate at which material is deformed to rupture at the tear tip. Hence, several specific values of G_c may be distinguished. The smallest

---

* Actually, the tear tends to run at 45° to the thickness direction, i.e., at right angles to the principal tensile stress, and thus the tear path has a width of about $\sqrt{2}d$ instead of d [15]; however, see Section III.D for further discussion of the tear path.

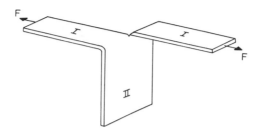

Fig. 5. Tear test piece.

possible value is, of course, twice the surface free energy, about 30–50 mJ m^{-2} for hydrocarbon liquids and polymers [20]. Values of this order of magnitude are indeed observed for fracture induced by ozone, when the function of the applied forces is merely to separate molecules already broken by chemical reaction. This type of "fracture" is discussed further in Section IV.G.

Another special value is that necessary to break all the molecules crossing a plane, in the absence of any other energy-absorbing process. This minimum energy requirement for mechanical rupture is found to be about 50 J m^{-2}; it is treated in the following subsection. Finally, there are the considerably larger values found in normal fracture experiments, ranging from 100 to 100,000 J m^{-2}. They are described in Section III.

E. THRESHOLD STRENGTHS AND EXTENSIBILITIES

A threshold value for the fracture energy of elastomers was first pointed out by Lake and Lindley from studies of fatigue crack growth [21]. By extrapolation they found that a minimum amount of mechanical energy, about 50 J m^{-2} of torn surface, was required for a crack to propagate at all. Mueller and Knauss measured extremely low tearing energies directly, by employing low rates of tear, high temperatures, and a urethane elastomer composition swollen highly with a mobile fluid [22]. Under these near-equilibrium conditions, they obtained a lower limit of about 50 J m^{-2} for the tear energy of their elastomer, in good agreement with Lake and Lindley's extrapolated value. More recently, Ahagon and Gent have obtained threshold tear strengths of 40–80 J m^{-2} for polybutadiene elastomers cross-linked to varying degrees [19]. These threshold values are much smaller than the tear energies determined from normal tearing experiments, which range generally from about 10^2 to about 10^5 J m^{-2}, depending on the rate of tearing, test temperature, and elastomer composition [23]. Indeed, they amount to only about 1 lb of force to tear through a sheet 4 in. thick. Nevertheless, they are much larger than would be expected on the basis of C–C bond strengths alone. For example, about 2×10^{18} molecules m^{-2} cross a randomly chosen fracture plane; the dissociation energy of the C–C bond is about 5×10^{-19} J. Thus, a fracture energy of only about 1 J m^{-2}

(instead of the observed value of 50 J m^{-2}) would be expected if bond rupture were the only mechanism.

This large discrepancy has been attributed by Lake and Thomas [24] to the chain structure of elastomers: many bonds in a molecular chain must be stressed in order to break one of them. Thus, the greater the molecular length between points of cross-linking, the larger the number of bonds which must be stressed in order to break a molecular chain. On the other hand, when the chains are long there will be a smaller number of them crossing a randomly chosen fracture plane. These two factors do not cancel out; their net effect is a predicted dependence of the threshold fracture energy G_0 on the average molecular weight M_c of chains between points of cross-linking, of the form [24]

$$G_0 = k M_c^{1/2} \tag{9}$$

When the presence of physical entanglements between chains at a characteristic spacing along each chain of molecular weight M_e is taken into account, and a correction is also made for the ends of the original molecules which do not enter into the load-bearing network, Eq. (9) becomes

$$G_0 = k(M_c^{-1} + M_e^{-1})^{-1/2}(1 - 2M_c M^{-1}) \tag{10}$$

where M is the molecular weight of the polymer before cross-linking [24]. The constant k involves the density of the polymer; the mass, length, and effective flexibility of a monomer unit; and the dissociation energy of a C–C bond, assumed to be the weakest link in the molecular chain. Taking reasonable values for these quantities [24], k is found to be about 0.3 J m^{-2} (g/g-mole)$^{-1/2}$. Thus, for a representative molecular network, taking $M_c = M_e = 15,000$ and $M = 300,000$, the threshold fracture energy is obtained as 26 J m^{-2}, in reasonable agreement with experiment in view of uncertainties and approximations in the theory. Moreover, the predicted slight increase in fracture energy with molecular weight M_c between cross-links appears to be correct; increased density of cross-linking leads to lower threshold fracture energies [19]. However, because the tensile strength σ_b also involves the elastic modulus E [Eq. (6)] and E is increased by cross-linking, the threshold tensile strength shows a net small increase with increased cross-linking.

Threshold values of tensile strength and extensibility may be calculated from the observed threshold fracture energy, 50 J m^{-2}, the "natural" flaw size, 40 μm (assumed independent of composition), and Young's modulus E for the rubber, by means of Eq. (6). Taking a representative value of E for soft elastomers of 2 MN m^{-2} (corresponding to a Shore A hardness of about 48 degrees), we obtain $\sigma_{b,0} = 0.9$ MN m^{-2}, $e_{b,0} = 0.45$. These values are indeed close to experimental "fatigue limits," i.e., stresses and strains below which the fatigue life is effectively infinite in the absence of chemical attack [21].

F. FRACTURE UNDER MULTIAXIAL STRESSES

Although relatively few studies have been made of the fracture of elastomers under complex stress conditions, some general conclusions can be drawn regarding fracture under specific combined stress states, as follows.

Compression and Shear. Elastomers do not fail along simple shear planes. Instead, fractures develop at 45° to the direction of shear (Fig. 6), i.e., at right angles to the corresponding principal tensile stress [15, 25], at a shear stress theoretically equal to the tensile strength [9]. Indeed, the general condition for rupture appears to be the attainment of a specific tensile stress σ_t at the tip of an existing flaw, and this circumstance can arise even when both applied stresses are compressive, provided that they are unequal [9]. When all the compressive stresses are equal, i.e., under a uniform triaxial compression, the elastomer will merely decrease in volume. No case of fracture under such a loading condition is known.

Under a uniaxial compressive stress, the theory of brittle fracture predicts a breaking stress eight times as large as in tension, Eq. (6), by growth of a crack in an oblique direction [9]. However, a *uniform* compressive stress is not readily achieved. Instead, friction at the loaded surfaces of a thin compressed block generally prevents the elastomer from expanding freely in a lateral direction, and a complex stress condition is set up. The outwardly bulging surfaces may split open when the local tensile stress is sufficiently high but, even so, this local fracture will not propagate inward because the interior is largely under triaxial compression. Rubber in compression is thus "unbreakable" in normal circumstances.

Equibiaxial Tension. Quite surprisingly, the breaking stresses and extensions in equibiaxial tension have been found to be significantly greater than in uniaxial tension [26]. The stored elastic energy at fracture is thus much greater, more than twice as large. Apparently, growth of a crack from an existing flaw does not lead to a strain energy release rate sufficiently large to meet the energy requirements for fracture until a much higher strain energy level is reached than in uniaxial tension.

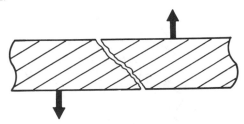

Fig. 6. Tearing under shear stresses (schematic) [15, 25].

There is clearly no preferred direction for crack growth under biaxial stressing, and this feature may be connected in some way with the unexpectedly high strength. However, it must be noted that test sheets put into a state of biaxial extension (by inflation, for example) do not have a cut edge at the point of failure in the central region of the sheet, whereas specimens for uniaxial tests are usually cut out of sheets in the form of thin strips. Stress raisers associated with the cutting process will therefore be present only in the latter case. If they are particularly severe, the measured strength in uniaxial tension will be correspondingly small in comparison with that in biaxial extension. This may account for the observed phenomenon, at least in part.

Triaxial Tension. A small spherical cavity within a block of rubber will expand elastically from its original radius r_0 to a new radius λr_0 under the action of an inflating pressure P. In the same way, it will expand to an equal degree when the faces of the block are subjected to a uniform triaxial tension, i.e., to a negative hydrostatic pressure $-P$ (Fig. 7), provided that the rubber is itself undilatable. When the expansion is small, it is proportional to $-P$ and given by

$$\lambda = 1 + 3P/4E$$

where λ is the expansion ratio and E is Young's modulus of the elastomer; when the expansion is large, it increases more rapidly than in direct proportion with P. Indeed, for rubbers obeying the kinetic theory of rubberlike elasticity (Chapter 1), the expansion becomes indefinitely large at a finite value of the applied tension, given by [27]

$$-P_c = 5E/6 \tag{11}$$

Of course, the original cavity will burst when the expansion of its wall reaches the breaking extension of the rubber, and a large tear will form. Even before this point is reached, the rubber around the cavity will cease to follow elasticity relations valid only for low and moderate strains, so that Eq. (11) may be questioned on this ground. But in any given sample most of the rubber is in

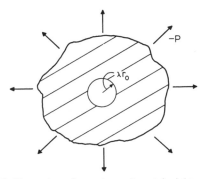

Fig. 7. Expansion of a cavity under a triaxial tension.

fact subjected to small or moderate strains, and hence the relatively small fraction that is highly strained does not affect the observed response appreciably, right up to the point of rupture.

Rubber is commonly found to undergo internal cavitation at triaxial tensions given by Eq. (11). This phenomenon must therefore be regarded as the consequence of an elastic instability (namely, as an unbounded elastic expansion, in accordance with the theory of large elastic deformations) of preexisting cavities too small to be readily detected. It does not involve the fracture energy because it is principally a transformation of potential energy from one strained state to another. In consequence, the critical stress given by Eq. (11) does not depend on the strength of the elastomer but only on its elastic modulus. Cavitation stresses in bonded rubber blocks under tension (Figs. 8 and 9) and near rigid inclusions at points where a triaxial tension is set up (Figs. 10 and 11) are found to be accurately proportional to E and largely independent of the tear strength of the elastomer, in accordance with the dominant role of an elastic rather than a rupture criterion for failure.

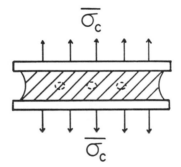

Fig. 8. Cavitation in a bonded block (schematic).

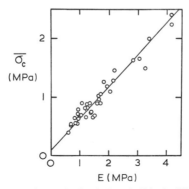

Fig. 9. Critical applied stress $\bar{\sigma}_c$ for cavitation in bonded blocks (Fig. 8) versus Young's modulus E of the elastomer [27].

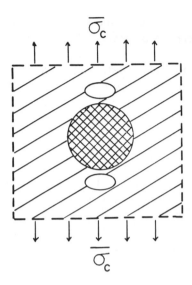

Fig. 10. Cavitation near a rigid inclusion (schematic).

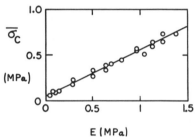

Fig. 11. Critical applied stress $\bar{\sigma}_c$ for cavitation near rigid inclusions (Fig. 10) versus Young's modulus E of the elastomer [28].

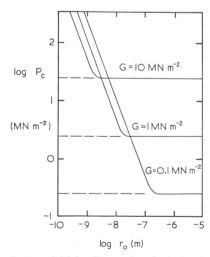

Fig. 12. Cavitation stress P_c versus initial radius r_0 of a spherical cavity for elastomers of varied shear modulus G, assuming a surface free energy of 25 mJ m^{-2} [29].

If the precursor cavities are extremely small, less than about 0.1 μm in radius, their surface energy becomes a significant additional restraint on expansion. As shown in Fig. 12, the triaxial tension necessary for cavity expansion then becomes considerably greater. Thus, if elastomers could be prepared without any microcavities greater than, say, 10 nm in radius, they would be much more resistant to cavitation. This seems an unlikely development, however, so that Eq. (11) remains an important general failure criterion for elastomers. It predicts for soft low-modulus elastomers a surprisingly low critical triaxial tension, of the order of only a few atmospheres. Conditions of triaxial loading should probably be avoided altogether in these cases.

III. Crack Propagation

A. INTRODUCTION

Whereas the initiation of fracture appears to be a similar process for all elastomers, the *propagation* of cracks is widely different. Three basic patterns of crack propagation, or tearing, can be distinguished, corresponding to three characteristic types of elastomeric compound: (i) amorphous elastomers, like SBR; (ii) elastomers, like natural rubber, that crystallize on stretching, even if only at the crack tip where the local stresses are particularly high; and (iii) reinforced elastomers containing large quantities, about 30% by volume, of a finely divided reinforcing particulate filler such as carbon black.

Elastomers in the first category show the simplest tearing behavior and will therefore be described first. For these materials, once fracture has been initiated, a tear propagates at a rate dependent on two principal factors: the strain energy release G and the temperature T. The former quantity represents the rate at which strain energy is converted into fracture energy as the crack advances. It is defined by a relation analogous to Eq. (4):

$$G = -2(\partial W/\partial A) \tag{12}$$

where W denotes the total strain energy of the specimen and A denotes the surface area (which, of course, increases as the crack advances). Even if a crack is stationary, because the critical value G_c at which fracture takes place has not yet been attained, Eq. (12) is still a useful definition of the energy G *available* from the strained specimen for fracture. For a tensile strip with an edge cut, one obtains

$$G = 2\alpha l W' \tag{13}$$

by analogy with Eq. (5), and for a tear test piece

$$G = 2F/d \tag{14}$$

from Eq. (8).

B. VISCOELASTIC ELASTOMERS

Experimental relations for the fracture energy G of an SBR material as a function of the rate of tearing and of the temperature of testing are shown as a three-dimensional diagram in Fig. 13. The fracture energy is seen to be high at high rates of tearing and at low temperatures, and vice versa, in a manner reminiscent of the dependence of energy dissipation in a viscous material on rate of deformation and temperature. Indeed, when the rate of tear is divided by the corresponding molecular segmental mobility ϕ_T [30] at the temperature of test, the tear energy relations at different temperatures superpose to form a single master curve, as shown in Fig. 14 [31]. In this figure, the rates have been multiplied by the factor

$$a_T \equiv \phi_{T_g}/\phi_T \tag{15}$$

in order to convert them into equivalent rates of tearing at the glass transition temperature T_g of the polymer, $-57°C$ for the SBR material of Fig. 13.

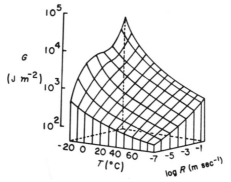

Fig. 13. Fracture energy G for an unfilled SBR material as a function of temperature T and rate of tearing R [16].

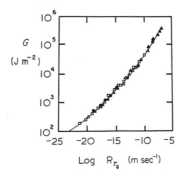

Fig. 14. Fracture energy G versus rate of tearing R reduced to T_g for six unfilled amorphous elastomers [31].

Furthermore, values of tear energy G for five other amorphous elastomers, two butadiene–styrene copolymers of lower styrene content ($T_g = -72°C$ and $-78°C$) and three butadiene–acrylonitrile copolymers having values of T_g of $-30°C$, $-38°C$, and $-56°C$, are all seen to fall on a single curve in this representation, increasing with rate of tearing in accordance with the dissipation of energy by a viscous process [31]. We conclude that the fracture energy G is the same for all unfilled amorphous elastomers under conditions of equal segmental mobility, and that the dependence of tear strength on temperature arises solely from corresponding changes in segmental mobility. Thus, internal energy dissipation determines the tear resistance of such elastomers: the greater the dissipation, the greater the tear strength. This point will emerge also in connection with variations in tensile strength of elastomers (Section IV). It is demonstrated strikingly in the present case by a proportionality between the tear energy G and a direct measure of energy dissipation, namely, the shear loss modulus G'', for the same six elastomers (Fig. 15) [31].

It should be noted that the reduction factors a_T used to transform the results of Fig. 13 to yield the master curve of Fig. 14 were not obtained from the universal form of the WLF rate–temperature equivalence relation [30], i.e.,

$$\log a_T = -17.5(T - T_g)/(52 + T - T_g) \tag{16}$$

but were determined experimentally and found to be somewhat larger [31]. This discrepancy has been traced to changes in effective sharpness of the tear tip when the strength becomes high [32]. Thus, the tear strength reflects both internal energy dissipation and changes in tear tip radius with increasing rate.

At the highest rates of tear and lowest temperatures employed, the tear strength is extremely high, approaching 10^6 J m^{-2}. Simultaneously, the elastomers become first leathery, and eventually glassy. Indeed, at still higher rates and lower temperatures they fracture, like typical polymeric glasses, by a failure process in which a narrow craze band forms and propagates, followed by a running crack [33]. The fracture energy for this process is relatively small, about 100–1000 J m^{-2} [33], in comparison with that of highly viscous but highly deformable elastomers, so that the curve shown in Fig. 14 turns sharply down at higher rates to level off at this value.

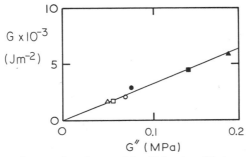

Fig. 15. Fracture energy G versus shear loss modulus G'' for six unfilled amorphous elastomers [31].

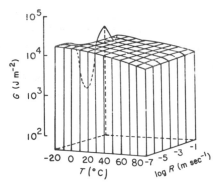

Fig. 16. Fracture energy G for a strain-crystallizing elastomer, natural rubber, as a function of temperature T and rate of tearing R [16].

C. Strain-Crystallizing Elastomers

As shown in Fig. 16, the tear strength of strain-crystallizing elastomers is greatly enhanced over the range of tear rates and temperatures at which crystallization occurs on stretching, i.e., at the tear tip. However, at sufficiently high temperatures crystallization fails to occur. Conversely, at low temperatures and high rates of tear, molecular reorganization into crystallites cannot take place in the short times for which the elastomer is stretched as the crack tip advances. Thus, the strengthening effect of strain-induced crystallization is limited to a particular range of tear rates and temperatures, as seen in Fig. 16. Outside this range, the material has only the strength associated with its viscous characteristics, dependent on $T - T_g$, and with its molecular structure.

The high strength of strain-crystallizing materials has been attributed to pronounced energy dissipation on stretching and retraction, associated with the formation and melting of crystallites under nonequilibrium conditions [34]. Reinforcing particulate fillers have a similar strengthening action, as discussed later; they also cause a marked increase in energy dissipation. Whether this is the sole reason for the strengthening effect of crystallites and fillers, and for other strengthening inclusions (such as hard regions in block copolymers, and hydrogen bonding) is not clear. Even if it is accepted, the exact mechanism whereby irreversible behavior and consequent energy dissipation lead to such pronounced strengthening is not well understood. The phenomenon of "reinforcement" thus requires further investigation.

D. Reinforcement with Fillers

The extent of reinforcement by fine particle fillers is quite remarkable. The tear strength and tensile strength of an amorphous elastomer are increased as much as tenfold when, for example, 20–40% by weight of carbon black is included in the mix formulation. However, as shown in Fig. 17, this strengthen-

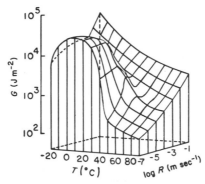

Fig. 17. Fracture energy G for an amorphous elastomer (SBR) reinforced with 30% by weight of FT carbon black [17].

ing action is restricted to a specific range of rates of tearing and temperatures of test, ranges that depend on both the type of filler and the elastomer [17]. Outside this range of effectiveness, the filler does not enhance the observed strength to a comparable degree.

The marked enhancement of tear strength under certain conditions is associated with a pronounced change in the character of the tear process, from relatively smooth tearing with a roughness of the torn surface of the order of 0.1–0.5 mm, to a discontinuous stick–slip tearing process in which the tear deviates from a straight path and even turns into a direction parallel to the applied stress, until a new tear breaks through. This form of tearing has been termed "knotty" tearing [17]; an example is shown in Fig. 18. A typical tear force relation is shown in Fig. 19a; it may be compared with the corresponding relation for an unfilled material in Fig. 19c. The peak tearing force at the stick position reaches high values, but the force during catastrophic slip tearing drops to a much lower level, only about twice as large as that for continuous tearing of the unfilled elastomer. Indeed, when the tear is prevented from deviating from a linear path by closely spaced metal guides, or is made to propagate in a straight line by stretching [32] or prestretching [35] the sample

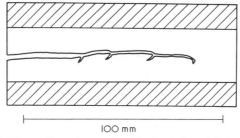

100 mm

Fig. 18. "Knotty" tear in a carbon-black-reinforced elastomer [32].

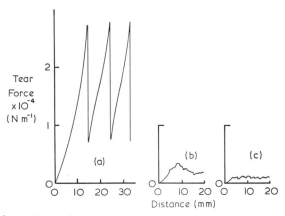

Fig. 19. Tear force relations for (a) a filled elastomer without constraints; (b) the same material with the tear confined to a linear path; (c) the corresponding unfilled elastomer, with and without constraints [32].

in the tearing direction, then the tear force is much smaller, only two to three times that for the corresponding unfilled material (see Fig. 19b).

Thus, the increase in tear strength by fillers is of two kinds: an up to threefold increase in intrinsic strength and a major increase due to a deviation of the tear path (on a scale of several millimeters) under special conditions of rate of tearing, temperature, and molecular orientation. The first effect may be attributed to enhanced energy dissipation as discussed in the previous section. The second is more difficult to explain. It may reflect anisotropy of *stress* around the crack tip, in accord with Andrews' frozen stress hypothesis [36] or anisotropy of *strength*.

Dannenberg [37] has recently surveyed the many molecular and mechanistic processes which have been proposed as explanations for filler reinforcement, and concludes: "There is still no satisfactory general theory" (but c.f. Chap. 8).

E. REPEATED STRESSING: DYNAMIC CRACK PROPAGATION

In contrast to amorphous elastomers, which tear steadily at rates controlled by the available energy for fracture G (as shown in Fig. 14), strain-crystallizing elastomers do not tear continuously under small values of G—values less than about 10^4 J m^{-2} for natural rubber, for example (Fig. 16). Nevertheless, when small stresses are applied repeatedly, a crack will grow in a stepwise manner by an amount Δl per stress application, even though the corresponding value of G is much below the critical level [5].

Experimentally, four distinct growth laws have been observed (Fig. 20), corresponding to four levels of stressing [38, 39]:

(i) $G < G_0$, no crack growth occurs by tearing, but only by chemical (ozone) attack;

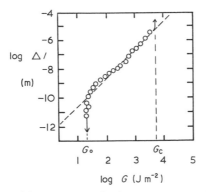

Fig. 20. Crack growth step Δl per stress application versus energy G available for fracture of a natural rubber vulcanizate [21].

(ii) $G_0 < G < G_1$, the growth step Δl is proportional to $G - G_0$;
(iii) $G_1 < G < G_c$, the growth step Δl is proportional to G^2;
(iv) $G \simeq G_c$, catastrophic tearing.

The transitional value of G between one crack growth law and another, denoted G_1 in the preceding list, is found to be about 400 J m^{-2}. No explanation has yet been advanced either for the form of these experimental growth laws or for the transition between them. They must therefore be regarded for the present as empirical relations for the growth step Δl per stress application.

In practice, it is usually possible to approximate the crack growth over the whole range of G values by the relation for region (iii):

$$\Delta l = BG^2 \qquad (17)$$

represented by the broken line in Fig. 20, where the crack growth constant B is found to be about 10^{-14} m(J m^{-2})$^{-2}$ per stress application [5].

It is important to recognize that crack growth is brought about only by *imposing* an adequate deformation; if the deformation is maintained, the crack does not grow further under forces insufficient to cause catastrophic tearing. The reason for this is that a crystalline region develops in the highly stressed material at the crack tip and effectively precludes further tearing. This explains a striking feature of crack growth in strain-crystallizing elastomers: the growth steps under repeated stressing become extremely small if the test piece is not relaxed completely between each stress application [39]. Under these conditions, the crystallized region does not melt; it remains intact to prevent further crack growth when the stresses are reimposed. As a result, the mechanical fatigue life (discussed in the following section) becomes remarkably prolonged if the component is never relaxed to the zero-stress state (Fig. 34). Indeed, failure in these circumstances is normally a consequence of chemical attack, usually by atmospheric ozone [39], rather than mechanical rupture.

Amorphous elastomers show more crack growth under intermittent stressing than under a steady stress, and the additional growth step per stress cycle is found to depend on the available energy G for fracture in substantially the same way as for natural rubber. The principal difference is that over region (iii) the relation obtained is

$$\Delta l = DG^4 \tag{18}$$

in place of Eq. (17). This relation is again a reasonably satisfactory approximation over the entire range of G values for which crack growth occurs, the crack growth coefficient D being about 10^{-18} m$(J\ m^{-2})^{-4}$ per stress application for an SBR vulcanizate [39].

Andrews has put forward a general explanation for the slowing down of a crack in an amorphous elastomer (and the complete cessation of tearing in a strain-crystallizing elastomer) after the stresses have been applied, in terms of time-dependent stress changes at the tip of the crack [36]. As a result, the stress concentration at the growing tip is smaller for a viscoelastic material, or for one that is energy dissipating, than would be expected from purely elastic considerations. Crack growth is correspondingly slowed. Local crystallinity is particularly effective in preventing stress changes, so that further crack growth does not occur at all until the crystalline region melts.

The presence of oxygen in the surrounding atmosphere is found to increase crack growth, presumably by an oxidative chain scission reaction catalyzed by the mechanical rupture. For experiments carried out *in vacuo* the minimum energy G_0 is found to be somewhat larger, and the growth constants B and D appreciably smaller. When antioxidants are included in the elastomer formulation, the results in an oxygen-containing atmosphere approach those obtained *in vacuo*.

IV. Tensile Rupture

A. Effects of Rate and Temperature

In Fig. 21 several relations are shown for the breaking stress σ_b of an unfilled vulcanizate of SBR as a function of the rate of elongation at different temperatures [40]. A small correction factor (T_g/T) has been applied to the measured values to allow for changes in the elastic modulus with temperature. The corrected values are denoted σ_b'.

It is seen that the experimental data form parallel curves, superimposable by horizontal displacements. The strength at a given temperature is thus equal to that at another temperature provided that the rate is adjusted appropriately by a factor depending on the temperature difference. (Using a logarithmic scale for rate of elongation, a constant multiplying factor is equivalent to a constant

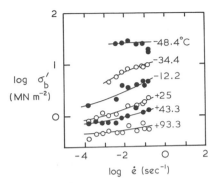

Fig. 21. Breaking stress σ_b' for an SBR vulcanizate versus rate of elongation $\dot{e}$ [40].

horizontal displacement.) This factor is found to be the ratio ϕ_{T_1}/ϕ_{T_2} of segmental mobilities at the two temperatures [40]. It is readily calculated from the WLF relation, Eq. (16). A master curve may thus be constructed for a reference temperature T_s, chosen here for convenience as T_g, by applying the appropriate shift factors to relations determined at other temperatures. The master curve for tensile strength, obtained from the relations shown in Fig. 21, is given in Fig. 22.

The variation of tensile strength with temperature is thus primarily due to a change in segmental mobility. Moreover, the master curve has the form expected of a viscosity-controlled quantity; it rises sharply with increased rate of elongation to a maximum value at high rates when the segments do not move and the material breaks as a brittle glass [41]. The breaking elongation at first rises with increasing rate of elongation, reflecting the enhanced strength, and then falls at higher rates as the segments become unable to respond sufficiently rapidly (Fig. 23).

Rupture of a tensile test piece may be regarded as catastrophic tearing at the tip of a chance flaw. The success of the WLF reduction principle for fracture energy G in tearing thus implies that it will also hold for tensile rupture properties. Indeed, σ_b and e_b may be calculated from the appropriate value of G at

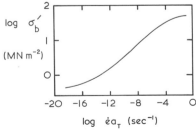

Fig. 22. Master relation for breaking stress σ_b' as a function of rate of elongation $\dot{e}$, reduced to T_g ($-60°$C) by means of the WLF relation, Eq. (16) [40].

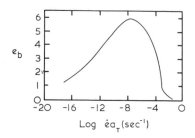

Fig. 23. Master relation for breaking elongation e_b as a function of rate of elongation $\dot{e}$ reduced to T_g ($-60°C$) [40].

each rate and temperature, using relations analogous to Eq. (6). However, the rate of extension at the crack tip will be much greater than the rate of extension of the whole test piece and this discrepancy in rates must be taken into account [42].

In addition, it is clear from the derivation of Eq. (5) that W represents the energy *obtainable* from the deformed material rather than the energy put into deforming it. For a material with energy-dissipating properties, the energy available for fracture is only a fraction of that supplied. Such a material will therefore appear doubly strong in a tensile test, or in any other fracture process in which the tear energy is supplied indirectly by the relief of deformations elsewhere.

B. The Failure Envelope

An alternative representation of tensile rupture data over wide ranges of temperature and rate of elongation is obtained by plotting the breaking stress σ_b against the corresponding breaking extension e_b [43]. Tensile results on which Figs. 21–23 are based are replotted in this way in Fig. 24.

They yield a single curve, termed the "failure envelope," which has a characteristic parabolic shape. Following around the curve in an anticlockwise sense corresponds to increasing the rate of extension or to decreasing the tem-

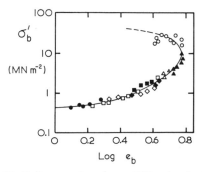

Fig. 24. Failure envelope for an SBR vulcanizate [43].

perature, although these two variables do not appear explicitly. Thus, at the lower extreme, the breaking stress and elongation are both small. These conditions are found at low rates of strain and at high temperatures. Conversely, the upper extreme corresponds to a high breaking stress and low extensibility. These conditions obtain at high rates of strain and low temperatures, when the material responds in a glasslike way.

The principal advantages of the failure envelope representation of data are twofold. First, it clearly indicates the maximum possible breaking elongation $e_{b,max}$ for the material. This is found to be well correlated with the degree of cross-linking, specifically with the molecular weight M_c between cross-links, as predicted by elasticity theory:

$$1 + e_{b,max} \propto M_c^{1/2} \tag{19}$$

Second, the failure envelope can be generalized to deal with different degrees of cross-linking, as discussed later. It has therefore been employed to distinguish between changes in cross-linking and other changes that affect the response to rate of elongation and temperature, but do not necessarily affect M_c. Examples of this are plasticization and adding reinforcing fillers [44].

C. EFFECT OF DEGREE OF CROSS-LINKING

The breaking stress is usually found to pass through a sharp maximum as the degree of cross-linking is increased from zero. An example is shown in Fig. 25. This maximum is primarily due to changes in viscoelastic properties with cross-linking and not to changes in intrinsic strength. For example, it is much less pronounced at lower rates of extension [45] and it is not shown at all by swollen specimens [46]. Bueche and Dudek [45] and Smith and Chu [47] therefore conclude that it would not exist under conditions of elastic equilibrium.

Two different reduction schemes have been employed to construct failure

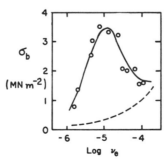

Fig. 25. Tensile strength of SBR vulcanizates versus degree of cross-linking, represented by v_e [45]. Broken curve represents authors' estimate of threshold strengths under nondissipative conditions.

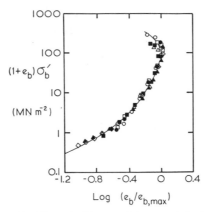

Fig. 26. Failure envelope for Viton A-HV materials cross-linked to various extents ($e_{b,max}$ ranging from 2.5 to 18) [44].

envelopes for materials having different degrees of cross-linking. The first, shown in Fig. 26, consists of scaling the breaking elongation e_b in terms of its maximum value, which is, of course, dependent on the degree of cross-linking, Eq. (19). Also, the breaking stress σ_b is converted into a true stress at break, rather than the nominal stress employed up till now. (The nominal tensile stress is given by the tensile force divided by the unstrained cross-sectional area of the specimen. It has been commonly used in the literature dealing with deformation and fracture of elastomers.) This reduction scheme is clearly quite successful in dealing with a wide range of cross-linking (Fig. 26) [44].

The second method consists of scaling the stress axis by dividing the nominal stress at break by a measure of the density of cross-linking [48]. This method also appears to bring data from differently cross-linked materials into a common relationship. Further work is necessary to decide between these two methods or to combine them into a more comprehensive reduction system.

D. STRAIN-CRYSTALLIZING ELASTOMERS

Whereas amorphous elastomers show a steady fall in tensile strength as the temperature is raised (Fig. 21), strain-crystallizing elastomers show a rather sudden drop at a critical temperature T_c (Fig. 27) [48–51]. This temperature depends strongly on the degree of cross-linking, as shown in Fig. 27. It is clearly associated with failure to crystallize at high temperatures. However, Thomas and Whittle [51] have adduced reasons for believing that whereas the bulk of the sample remains amorphous above T_c, the highly strained material at the flaw tip probably still continues to crystallize. They draw a parallel between the drop in strength at the critical temperature T_c and the similar sharp drop at a critical depth l_c of an edge cut (Fig. 28) for strength measurements made at room temperature.

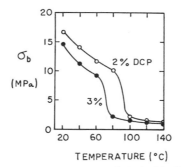

Fig. 27. Tensile strength of natural rubber cross-linked with dicumylperoxide (DCP) versus temperature [51].

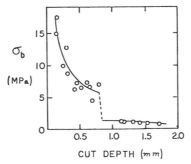

Fig. 28. Tensile strength of natural rubber cross-linked with 2% dicumylperoxide versus depth of initial edge cut [51].

Two other aspects of the critical temperature are noteworthy. First, it is substantially independent of reinforcement with fillers. Second, it depends strongly on the type of cross-linking, being highest for long polysulfidic cross-links and lowest for carbon–carbon cross-links. Apparently, cross-link slippage is an important factor in promoting crystallization.

E. ENERGY DISSIPATION AND STRENGTH

A general correlation between tensile strength and the temperature interval $T - T_g$ between the test temperature T and the glass transition temperature T_g has been recognized for many years [52]. A recent example is shown in Fig. 29, where the strengths of polyurethane elastomers with values of T_g ranging from $-67°C$ to $-17°C$ are plotted against $T - T_g$ [44]. In this representation all the results fall on a single curve, indicating once more that segmental viscosity governs the observed strength.

A more striking demonstration of the close connection between energy dissipation and strength has been given by Grosch et al. [53]. They showed that a direct relationship exists between the energy W_b per unit volume required

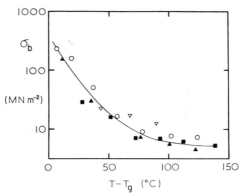

Fig. 29. Tensile strength of polyurethane elastomers versus $T - T_g$ (T_g ranging from $-67°C$ to $-17°C$) [44].

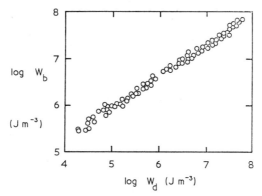

Fig. 30. Work-to-break W_b versus energy W_d dissipated on stretching almost to the breaking elongation [53].

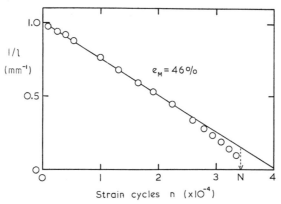

Fig. 31. Growth of an edge crack in a test piece of a natural rubber vulcanizate stretched repeatedly to 46% extension [23].

to break elastomers and the energy W_d dissipated on stretching them almost to the breaking elongation. This relationship held irrespective of the mechanism of energy loss, i.e., for filled and unfilled strain-crystallizing and amorphous elastomers (Fig. 30). Their empirical relation is

$$W_b = 410W_d^{2/3}$$

W_b and W_d being measured in joules per cubic meter. Those materials that require the most energy to bring about rupture (i.e., the strongest elastomers) are precisely those in which the major part of the energy is dissipated before rupture.

F. REPEATED STRESSING: MECHANICAL FATIGUE

Under repeated tensile deformations cracks appear, generally in the edges of the specimen, and grow across it in an accelerating way. This process is known as fatigue failure. It has recently been treated quantitatively in terms of stepwise tearing from an initial nick or flaw [54, 55], as follows.

Every time a deformation is imposed, energy G becomes available in the form of strain energy to cause growth by tearing of a small nick in one edge of the specimen. For tensile test pieces the value of G is given by Eq. (13). The corresponding growth step Δl is assumed to obey Eq. (17), i.e., to be proportional to G^2, so that the law of crack growth becomes

$$\Delta l/l^2 = (4\alpha^2 BW^2) \Delta n$$

where n is the number of times the deformation is imposed and α is a numerical constant, about 2 (see Section II.C). The depth of the crack after n strain cycles is then obtained by integration,

$$l_0^{-1} - l^{-1} = 4\alpha^2 BW^2 n \tag{20}$$

where l_0 is the initial depth of the nick. An example of crack growth is shown in Fig. 31; it conforms closely to Eq. (20).

If the crack grows to many times its original depth, so that $l \gg l_0$ before fracture ensures, the corresponding fatigue life N (the number of cycles up to failure) may be obtained by putting $l = \infty$ in Eq. (20), which yields

$$N = (4\alpha^2 BW^2 l_0)^{-1} \tag{21}$$

This is a quantitative prediction for the fatigue life in terms of one material property, the crack growth constant B, which can be determined in a separate experiment as described earlier (Section III.E). Measured fatigue lives for specimens with initial cuts of different lengths (Fig. 3) and for imposed deformations of different magnitudes have been found to be in good agreement with the predictions of Eq. (21) [38, 54, 55].

Examples of the dependence of fatigue life on initial cut size are shown in Figs. 3 and 32. The lives for specimens that contain no deliberately introduced

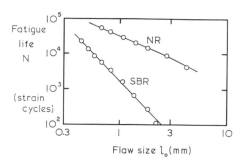

Fig. 32. Fatigue life versus depth of initial cut for test pieces of natural rubber and SBR stretched repeatedly to 50% extension [38].

cuts, represented by horizontal broken lines in Fig. 3, may be interpreted as the result of stepwise tearing from a hypothetical nick or flaw about 20 μm in depth, as discussed previously. It is particularly noteworthy that closely similar sizes for natural flaws are deduced for both strain-crystallizing and noncrystallizing elastomers by such extrapolations. For a noncrystallizing elastomer (SBR) the crack growth law is quite different over the main tearing region [Eq. (18)], and the fatigue life relation becomes

$$N = (48\alpha^4 D W^4 l_0^3)^{-1} \tag{22}$$

where D is the crack growth coefficient, in place of Eq. (21). Measured fatigue lives for an SBR elastomer have been found to be in good accord with this relation for a wide range of initial cut sizes l_0 and deformation amplitudes [38].

Thus, the different crack growth laws for strain-crystallizing and noncrystallizing elastomers lead to quite different fatigue life relations, Eqs. (21) and (22). The fatigue life of a noncrystallizing elastomer is much more dependent on the size of the initial flaw (Fig. 32) and the magnitude of the imposed deformation, so that such elastomers are generally longer lived at small deformations without accidental cuts, but much shorter lived under more severe conditions. The fatigue life is also drastically lowered at high temperatures due to a sharp increase in the cut growth coefficient D as the internal viscosity is decreased (Fig. 33). In contrast, the hysteresis associated with strain-induced crystallization is retained, provided that the temperature does not become so high (about 100°C for natural rubber) that crystallization no longer occurs. The fatigue life for natural rubber is therefore not greatly affected by a moderate rise in temperature.

A more striking difference is found between strain-crystallizing and noncrystallizing elastomers when the stress is not relaxed to zero during each cycle. As shown in Fig. 34, the fatigue life of a natural rubber vulcanizate is greatly increased when the minimum strain is raised from zero to, say, 100%, because the crystalline barrier to tearing at the tips of chance flaws then does not disappear in the minimum strain state. As a result, the growth of flaws is virtually

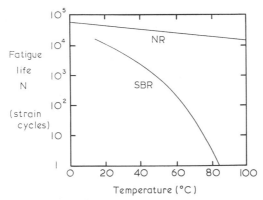

Fig. 33. Fatigue life versus temperature for test pieces of natural rubber and SBR stretched repeatedly to 175% extension [23].

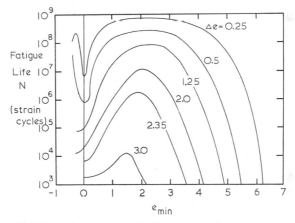

Fig. 34. Fatigue life for test pieces of natural rubber versus minimum extension e_{min}; Δe denotes the additional strain imposed repeatedly [56].

stopped unless the total applied strain is very large, about 400–500%. No comparable strengthening effect is found for noncrystallizing elastomers on raising the minimum strain level.

Corresponding to the threshold value G_0 of tearing energy, below which no crack growth occurs by mechanical rupture, there is a minimum tensile strain e_0 below which normal sized flaws do not grow under fatigue conditions. For typical elastomers, this mechanical fatigue limit is found to be about 50–100% extension, by calculation from Eq. (6) and by direct observation [21]. At extensions below this level, the fatigue life is infinite in the absence of chemical attack.

Reinforcing fillers greatly enhance the tear strength and tensile strength of elastomers but do not cause an equivalent improvement in the crack growth and fatigue properties. At a given strain energy input the measured lives are appreciably larger but if compared at equal *available* energy levels, they would probably be quite similar. The initial flaw size and threshold tear energy G_0 are therefore deduced to be similar to those for unfilled materials. The growth steps are apparently too small for pronounced deviation of the tear, and hence "reinforcement" against fatigue failure by this mechanism does not seem to occur.

G. Surface Cracking by Ozone

In an atmosphere containing ozone, stretched samples of unsaturated elastomers develop surface cracks that grow in length and depth until they eventually sever the test piece. Even when they are quite small, they can cause a serious reduction in strength and fatigue life.

The applied tensile stress necessary for such cracks to grow may be calculated approximately from Eq. (6). The fracture energy G is only about 0.1 J m^{-2} in the present case [4], representing the small amount of energy needed for "fracture" in a liquid medium, i.e., about twice the surface energy for a hydrocarbon liquid [20]. (Molecular scission apparently occurs readily by reaction with ozone, and does not require mechanical energy to induce it.) Taking for a soft rubber a representative value for E of 2 MPa, and a value of 20 μm for the effective depth l of a chance surface flaw, Eq. (6) yields a critical tensile stress for ozone cracking of about 0.06 MPa and a critical tensile strain of about 3%. These predictions are in reasonably good agreement with experimentally observed minimum values for ozone attack.

As the stress level is raised above the minimum value, numerous weaker stress raisers become effective and more cracks form. Actually, the presence of a large number of small, mutually interfering cracks is less harmful than that of a few widely separated cracks which develop into deep cuts, so that the most harmful condition is just above the critical stress level.

The rate at which a crack grows when the critical energy condition is satisfied depends on two factors: the rate of incidence of ozone at the crack tip and the rate of segmental motion in the tip region. When either of these processes is sufficiently slow it becomes rate-controlling. The overall rate R of crack growth is thus given approximately by

$$R^{-1} \, (\text{m sec}^{-1})^{-1} = 8 \times 10^{13} \phi_T^{-1} + 1.2 \times 10^5 c^{-1} \tag{23}$$

where ϕ_T (sec^{-1}) is the natural frequency of Brownian motion of molecular segments at the temperature T, given by the WLF relation, Eqs. (15) and (16), in terms of ϕ_{T_g} (where $\phi_{T_g} \simeq 0.1$ sec^{-1} [57]), and c (mg liter^{-1}) is the concentration of ozone in the surrounding air. For a typical outdoor atmosphere, c is of the order of 10^{-4} mg liter^{-1}; the second term in Eq. (23) then becomes dominant for values of ϕ_T greater than about 10^4 sec^{-1}, that is, at temperatures more than 25°C above T_g [58].

V. Abrasive Wear

Abrasive wear consists of the rupture of small particles of elastomer under the action of frictional forces when sliding takes place between the elastomer surface and a substrate. A suitable measure of the rate of wear is provided by the ratio A/μ, where A is the volume of rubber abraded per unit normal load and per unit sliding distance, and μ is the coefficient of friction. This ratio, termed the abradability, represents the abraded volume per unit of energy dissipated in sliding. Master curves for the dependence of abradability on the speed of sliding, reduced to a convenient reference temperature by means of the WLF relation [Eqs. (15) and (16)], are shown in Fig. 35. The abradability is seen to decrease with increasing speed, pass through a minimum, and then rise again at high speeds as the material becomes glasslike in response. This behavior resembles the variation of the reciprocal of the breaking energy W_b with rate of deformation (a reciprocal relationship because high abradability corresponds to low strength). Indeed, Grosch and Schallamach [59] found a general parallel between A/μ and $1/W_b$. For this comparison, values of the breaking energy W_b were determined at high rates of extension, about 10,000% per second, to bring them into agreement with measurements of abradability carried out at a sliding speed of 10 mm sec^{-1}. This indicates that the size of the rubber elements involved in deformation and wear were of the order of 0.1 mm, a value comparable to the size of the abrasive asperities on the particular track employed in the experiments.

Moreover, the coefficient of proportionality C between abradability and reciprocal breaking energy was found to be similar, about 10^{-3}, for all the unfilled elastomers examined. This coefficient represents the volume of rubber abraded per unit energy applied frictionally to a material for which the energy

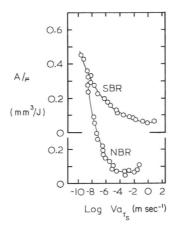

Fig. 35. Abradability A/μ versus speed of sliding V, reduced to 20°C, for an SBR and an ABR vulcanizate [59].

per unit volume, W_b, necessary to cause tensile rupture is itself unity. Thus C may be regarded as a measure of the *inefficiency* of rupture by tangential surface tractions; large volumes are deformed but only small volumes are removed. Apparently the ratio remains similar throughout the rubber-to-glass transition and for a variety of elastomers. It probably reflects a geometrical feature of the rupture process.

The abradabilities A/μ were found to be generally about twice as large for carbon-black-filled elastomers as for corresponding unfilled materials. This surprising observation, that "reinforced" materials wear away faster, can be partially accounted for in terms of the tear strength measurements referred to in a previous section. Under conditions of relatively smooth tearing it was concluded that the intrinsic strength of reinforced materials is not particularly high; instead, it was found to be comparable to that of unfilled elastomers. The measurements of abradability considered here suggest that it is actually somewhat lower under abrasion conditions.

ACKNOWLEDGMENTS

This review was prepared in the course of a program of research on fracture supported by a research grant from the Engineering Division of the National Science Foundation. It is largely based on three earlier reviews [60–62]. The author is also indebted to E. H. Andrews, T. L. Smith, A. G. Thomas, and other authors cited in the text for many helpful discussions, and to R. A. Paden for preparing the diagrams.

REFERENCES

1. C. E. Inglis, *Trans. Inst. Nav. Archit.* (*London*) **55**, 219 (1913).
2. R. J. Eldred, *J. Polym. Sci., Part B* **10**, 391 (1972).
3. J. P. Berry, *in* "Fracture of Non-Metals and Composites" (H. Liebowitz, ed.), Fracture: An Advanced Treatise, Vol. 7, Ch. 2. Academic Press, New York, 1972.
4. M. Braden and A. N. Gent, *Kautsch. Gummi* **14**, WT157 (1961); E. H. Andrews, D. Barnard, M. Braden, and A. N. Gent, *in* "The Chemistry and Physics of Rubberlike Substances" (L. Bateman, ed.), Ch. 12. Wiley, New York, 1963.
5. A. G. Thomas, *J. Polym. Sci.* **31**, 467 (1958).
6. G. R. Irwin, "Fracturing of Metals." Am. Soc. Met., Cleveland, Ohio, 1948.
7. G. R. Irwin, *J. Appl. Mech.* **24**, 361 (1957).
8. A. A. Griffith, *Philos. Trans. R. Soc. London, Ser. A* **221**, 163 (1921).
9. A. A. Griffith, *Proc. Int. Congr. Appl. Mech., 1st, Delft* pp. 55–63 (1924).
10. E. Orowan, *Rep. Prog. Phys.* **12**, 185 (1949).
11. R. S. Rivlin and A. G. Thomas, *J. Polym. Sci.* **10**, 291 (1953).
12. A. G. Thomas, *J. Appl. Polym. Sci.* **3**, 168 (1960).
13. H. W. Greensmith, *J. Appl. Polym. Sci.* **7**, 993 (1963); G. J. Lake, personal communication (1970).

14. A. G. Thomas, *J. Polym. Sci.* **18,** 177 (1955).
15. A. Ahagon, A. N. Gent, H.-W. Kim, and Y. Kumagai, *Rubber Chem. Technol.* **48,** 896 (1975).
16. H. W. Greensmith and A. G. Thomas, *J. Polym. Sci.* **18,** 189 (1955).
17. H. W. Greensmith, *J. Polym. Sci.* **21,** 175 (1956).
18. A. G. Veith, *Rubber Chem. Technol.* **38,** 700 (1965).
19. A. Ahagon and A. N. Gent, *J. Polym. Sci., Polym. Phys. Ed.* **13,** 1903 (1975).
20. H. Tarkow, *J. Polym. Sci.* **28,** 35 (1958).
21. G. J. Lake and P. B. Lindley, *J. Appl. Polym. Sci.* **9,** 1233 (1965).
22. H. K. Mueller and W. G. Knauss, *Trans. Soc. Rheol.* **15,** 217 (1971).
23. H. W. Greensmith, L. Mullins, and A. G. Thomas, *in* "The Chemistry and Physics of Rubber-like Substances" (L. Bateman, ed.), Ch. 10. Wiley, New York, 1963.
24. G. J. Lake and A. G. Thomas, *Proc. R. Soc., Ser. A* **300,** 108 (1967).
25. W. G. Knauss, *Int. J. Fract. Mech.* **6,** 183 (1970).
26. R. A. Dickie and T. L. Smith, *J. Polym. Sci., Part A-2* **7,** 687 (1969).
27. A. N. Gent and P. B. Lindley, *Proc. R. Soc., Ser. A* **249,** 195 (1958); G. H. Lindsay, *J. Appl. Phys.* **38,** 4843 (1967).
28. A. E. Oberth and R. S. Bruenner, *Trans. Soc. Rheol.* **9**(2), 165 (1965).
29. A. N. Gent and D. A. Tompkins, *J. Polym. Sci., Part A-2* **7,** 1483 (1969).
30. J. D. Ferry, "Viscoelastic Properties of Polymers," 2nd Ed. Wiley, New York, 1970.
31. L. Mullins, *Trans. Inst. Rubber Ind.* **35,** 213 (1959).
32. A. N. Gent and A. W. Henry, *Proc. Int. Rubber Conf., 1967* pp. 193–204 (1968).
33. R. P. Kambour, *J. Polym. Sci., Part D* **7,** 1 (1973).
34. E. H. Andrews, *J. Appl. Phys.* **32,** 542 (1961).
35. R. Houwink and H. H. J. Janssen, *Rubber Chem. Technol.* **29,** 409 (1956).
36. E. H. Andrews, *J. Mech. Phys. Solids* **11,** 231 (1963).
37. E. M. Dannenberg, *Rubber Chem. Technol.* **48,** 410 (1975).
38. G. J. Lake and P. B. Lindley, *Rubber J. Int. Plast.* **146**(10), 24; **146**(11), 30 (1964).
39. G. J. Lake and P. B. Lindley, *J. Appl. Polym. Sci.* **10,** 343 (1966).
40. T. L. Smith, *J. Polym. Sci.* **32,** 99 (1958).
41. F. Bueche, *J. Appl. Phys.* **26,** 1133 (1955).
42. F. Bueche and J. C. Halpin, *J. Appl. Phys.* **35,** 36 (1964); J. C. Halpin, *Rubber Chem. Technol.* **38,** 1007 (1965).
43. T. L. Smith, *J. Polym. Sci., Part A* **1,** 3597 (1963).
44. T. L. Smith, *in* "Rheology" (by F. R. Eirich, ed.), Vol. 5, Ch. 4. Academic Press, New York, 1969.
45. F. Bueche and T. J. Dudek, *Rubber Chem. Technol.* **36,** 1 (1963).
46. L. M. Epstein and R. P. Smith, *Trans. Soc. Rheol.* **2,** 219 (1958).
47. T. L. Smith and W. H. Chu, *J. Polym. Sci., Part A-2* **10,** 133 (1972).
48. R. F. Landel and R. F. Fedors, *Proc. Int. Conf. Fract., 1st, Sendai* **2,** 1247 (1966).
49. B. B. S. T. Boonstra, *India Rubber World* **121,** 299 (1949).
50. J. A. C. Harwood, A. R. Payne, and R. E. Whittaker, *J. Appl. Polym. Sci.* **14,** 2183 (1970).
51. A. G. Thomas and J. M. Whittle, *Rubber Chem. Technol.* **43,** 222 (1970).
52. A. M. Borders and R. D. Juve, *Ind. Eng. Chem.* **38,** 1066 (1946).
53. K. A. Grosch, J. A. C. Harwood, and A. R. Payne, *Nature (London)* **212,** 497 (1966).
54. P. B. Lindley and A. G. Thomas, *Proc. Int. Rubber Conf., 4th, London* pp. 428–442 (1962).
55. A. N. Gent, P. B. Lindley, and A. G. Thomas, *J. Appl. Polym. Sci.* **8,** 455 (1964).
56. S. M. Cadwell, R. A. Merrill, C. M. Sloman, and F. L. Yost, *Ind. Eng. Chem., Anal Ed.* **12,** 19 (1940).
57. F. Bueche, "Physical Properties of Polymers." Wiley (Interscience), New York, 1962.

58. A. N. Gent and J. E. McGrath, *J. Polym. Sci.*, *Part A* **3,** 1473 (1965).
59. K. A. Grosch and A. Schallamach, *Trans. Inst. Rubber Ind.* **41,** 80 (1965).
60. E. H. Andrews, "Fracture in Polymers." Am. Elsevier, New York, 1968.
61. A. N. Gent, *in* "Fracture of Non-Metals and Composites" (H. Liebowitz, ed.), Fracture: An Advanced Treatise, Vol. 7, Ch. 6. Academic Press, New York, 1972.
62. F. R. Eirich and T. L. Smith, *in* "Fracture of Non-Metals and Composites" (H. Liebowitz, ed.), Fracture: An Advanced Treatise, Vol. 7, Ch. 7. Academic Press, New York, 1972.

Chapter **11**

The Chemical Modification of Polymers

R. J. CERESA

CHEMISTRY & POLYMER TECHNOLOGY DEPARTMENT
POLYTECHNIC OF SOUTH BANK
LONDON, ENGLAND

I. Introduction

The terms "rubber" and "elastomer" embrace those polymers which have useful rubberlike, highly elastic properties at ambient temperatures. However, many polymers which are nonrubbers by themselves can be chemically modified to a relatively small extent to give products with very useful viscoelastic properties. For example, the introduction of a few chlorine atoms and sulfoxide groups into polyethylene changes the macromolecules so that they no longer tend to form crystalline regions, but become elastomers over a wide temperature range. These groups also allow vulcanization, so that reversibly deforming,

Copyright © 1978 by Academic Press, Inc.
All rights of reproduction in any form reserved.
ISBN 0-12-234360-3

solvent-resistant products can be formulated. On the other hand, starting with a conventional rubber, it is possible, by means of chemical reactions, to convert the macromolecular chains so that they lose their viscoelasticity and become thermoplastic at ambient temperatures (e.g., the cyclization of polyisoprenes). Quite apart from the question of the temperature at which observations are carried out, the borderline between the viscoelastic and the plastic states is a relatively ill-defined one. The common factor of rubbers and plastics is, of course, their macromolecular nature. Now that we have a better understanding of their structure at a molecular level (not forgetting rubber-modified plastics and "plasticized" rubbers), we are able to say that, with some exceptions, a modification that can be carried out on one polymer species may, under suitable conditions, be carried out on a polymer of a different species. Whether the same type of chemical modification will give us the properties we are seeking without the loss of properties we would wish to retain is a matter, first, for conjecture and subsequently, for experimental verification.

In the early days of polymer chemistry and technology, a new chemical modification successfully applied to one polymer was quickly evaluated with a whole range of chemically similar reagents (e.g., the esterification of cellulose), and the same type of reaction was attempted with the available range of similar material. Today, the number of homopolymers available runs into hundreds, the number of random copolymers runs into thousands, and therefore, the number of possible chemical modifications, including block and graft copolymers, runs into astronomical figures. Only a fraction of these potential systems has been evaluated (and then frequently only in a superficial way to obtain patent coverage), so that the field is wide open for research and development. Because of its breadth and depth, the field of the chemical modification of polymers can be treated only in outline in a single chapter and only the more important reactions can be described. To serve the interest of the reader, however, this broad survey will be punctuated by discussions in greater depth of areas that are of interest to rubber chemists, and that, in the opinion of the writer, have considerable potential for further development. The emphasis will be on the principles underlying the chemical modification of polymers, and specific details of reaction conditions will be deliberately omitted. A general bibliography is included for further reading. A number of aspects of the chemical modification of rubbers has been covered in detail elsewhere in this book and the reader is referred to the appropriate chapters on chemistry, vulcanization, characterization, block copolymers, etc.

II. Esterification, Etherification, and Hydrolysis of Polymers

The chemical modifications discussed in this section are historically and scientifically so closely linked to one polymer, cellulose, that although the latter occurs primarily as a fiber and not an elastomer, a discussion of this group of cellulose modifications seems appropriate. Apart from the fact that some cellulose derivatives, like ethyl cellulose, when plasticized, can be quite elastomeric, the effects of modification of a basic polymer are particularly well demonstrable on a substance as stiff and highly crystalline as cellulose. Moreover, in view of the expected hydrocarbon shortage, cellulose may soon gain a new role as a polymeric starting commodity.

Cellulose, identified chemically as beta-1,4-glucan, is the most widely distributed natural polymer, constituting the permanent structure of plant cell walls. For the general properties and chemistry of cellulose itself the reader is referred to the standard textbooks and recent reviews. (See the bibliography.)

Much of the early history of the chemical modification of cellulose is related to the attempts to find a solvent for it, since its macromolecular structure was not understood at that time. In 1844, Mercer discovered and commercialized the interaction of alkali with cellulose fibers, a process which is still in use under the name of mercerization. The initial product of the reaction, alkali cellulose, is not a chemical modification but a physical form in which water and sodium ions penetrate the macromolecular structure and reduce the hydrogen bonding, with consequent swelling of the fibers. The initial product, cellulose I, is converted to cellulose II, a complex physicochemical modification, in the final washing stage. The degree and rate of swelling in this process are dependent upon the source of the cellulose, and if the fibers are stretched prior to and during the reaction, optimum interaction is achieved. Many other inorganic salt solutions swell cellulose [1] and of these, zinc chloride has found the widest application. Aqueous solutions of thiourea, resorcinol, chloral hydrate, and benzenesulfonates also lead to limited swelling of cellulose. In all cases the reduction in physical cross-linking can be followed by a study of the X-ray diffraction diagrams of the crystalline content.

The complete solubility of cellulose in cuprammonium solutions, discovered in 1857 by Schweizer, led to the development of the rayon industry but, as in the case of alkali cellulose, the regenerated polymer is chemically the same as the precursor. Regeneration via cellulose zanthate solutions, invented by Cross and Bevan in 1893, is another process which is still in use; it forms the basis for the manufacture of Cellophane.

The first "chemical" modification of cellulose was achieved by Braconnot in 1833 with the production of cellulose nitrate from a wide range of cellulosic materials. The products were highly inflammable powders which could be dissolved in concentrated acetic acid to give clear tough varnishes. (Note the conversion of a fiber to a film by chemical modification.) In 1847 highly nitrated cellulose, guncotton, was discovered by Bottger and Schonbein, and in 1870 Schutzenberger produced acetylated cellulose using hot acetic anhydride as the reaction medium. These reactions have a common mechanism [2], namely, the esterification of the hydroxyl groups in the basic cellulose moiety (1):

$$\sim \text{cellulose} \sim \qquad (1)$$

Since this early work, a very large range of organic acids has been used to prepare cellulose esters, mixed esters, and ether esters [3]. A typical example of considerable commercial importance is the acetylation of cellulose. As in all esterifications of macromolecular materials, the accessibility of the hydroxyl groups to the esterifying acid is of prime importance. The reaction shown in (1) represents complete esterification, a process which is probably never fully achieved. The identification of the esterified products is, therefore, dependent not only on the content of acetyl groups but also on the location of these groups on the macromolecular backbone. Both factors are affected by the method of preparation and the esterification conditions.

Although many esterification reactions [2] are based on inorganic acids, for insoluble hydroxyl compounds like cellulose, xanthation is more important. Sodium hydroxide is normally used to produce the swollen alkali cellulose, which (after aging) is reacted with carbon disulfide to form the sodium salt of cellulose xanthate:

$$\sim \text{cellulose} \sim + CS_2 \longrightarrow \sim \text{cellulose} \sim \qquad (2)$$

The cellulose can be regenerated by spinning (or extruding as a film) into an acid bath containing salts such as sodium and zinc sulfate [1]. During spinning or extrusion, the macromolecules are oriented in the direction of flow to give high strength to the viscose fiber or the cellophane film. The occurrence of macromolecular orientation during spinning is very important and it is utilized in the chemical modification of many polymers.

The cellulose ethers constitute another important group of cellulose derivatives prepared from alkali cellulose by standard etherification reactions between the hydroxyl groups and an alkyl halide. The properties of the ethers depend upon the extent of the reaction, i.e., the degree of etherification. In general, the ethyl celluloses are water-insoluble thermoplastic materials, whereas the methyl ether, ethyl hydroxyethyl cellulose, and carboxymethyl cellulose are soluble in cold water and are used as viscoelastic thickeners and adhesives.

For the preparation of synthetic hydroxy polymers, hydroxyl groups can be introduced by copolymerization of the base monomer with a hydroxy monomer. These groups can then be used for esterification or etherification [4], but the relatively high cost of hydroxy monomers detracts from the widespread use of direct copolymerization. Instead, one introduces the groups required by the complete or partial hydrolysis of the ester groups in an appropriately hydrolyzable polymer, such as poly(vinyl acetate). Complete hydrolysis yields PVA, poly(vinyl alcohol), a water-soluble polymer with considerable utility as a stabilizer and viscosity modifier for aqueous systems (3):

$$\sim CH_2-\underset{\underset{OH}{|}}{CH}-CH_2-\underset{\underset{OH}{|}}{CH}\sim \tag{3}$$

PVA has a unique use as a strengthening fiber in conjunction with weaker materials such as merino wools in the weaving of delicate fabrics, from which it can afterward be removed by water washing. A major portion of the polymer produced is reacted with aldehydes to form the corresponding poly(vinyl formal), poly(vinyl acetal), and poly(vinyl butyral):

$$\underset{\sim\sim\sim\cdot}{\overset{\overset{OH}{|}}{}} + RCHO \longrightarrow \sim\sim\sim\cdot \overset{\underset{|}{\overset{R}{\diagdown}C\overset{\diagup O}{}}}{} + H_2O \tag{4}$$

There are different grades of each of these materials according to the overall molecular weight and the degree of substitution. These polymers are used as components of systems with unique adhesive properties, e.g., in the manufacture of safety glass laminates [poly(vinyl butyral) and mixed derivatives] and of metal-to-metal adhesive [poly(vinyl formal) cured with phenolics and other resins]. Reactions of poly(vinyl alcohol) with acids or anhydrides occur as normal esterifications, a route used to synthesize polymers and copolymers which cannot be readily formed by conventional polymerization (e.g., when the reactivity ratios of the monomers are not suitable).

Natural rubber and synthetic rubbers in general do not have hydroxyl groups in sufficient numbers for them to be used for esterification reactions. Terminal hydroxyl groups may be introduced into synthetic rubbers as terminal catalyst or initiator fragments and used for coupling or extension reactions.

III. The Hydrogenation of Polymers

The hydrogenation of polymers formed an important part of the early work on the identification of the structure of polymers, but it was not until the introduction of highly selective catalysts, which eliminated many of the degradative side reactions, that extensive commercialization began. Any polymer with unsaturated hydrocarbon groups, either in the main chain or as side groups, can be hydrogenated. Natural rubber was the first polymer to be modified in this way, although the interpretation of Staudinger's data as confirmation of the chainlike nature of the rubber molecule was fiercely debated for many years. Hydrogenation catalysts and reaction conditions giving high yields of low boiling hydrocarbons from natural and synthetic rubbers, a possible answer to the pollution by disused automobile tires in line with energy conservation, are beyond the scope of this chapter, since they involve destructive, rather than nondestructive, hydrogenation, with the consequent loss of the macromolecular structure.

Besides the temperature of reaction and the nature of the catalyst system, the initial molecular weight and degree of branching of the polymer, and the presence of polymerization residues within or associated with the macromolecules (e.g., soaps, catalyst fragments, and residual initiator) that affect the course of hydrogenation reactions are equally important variables. The influence of the additives just mentioned expresses itself in the variations in the final molecular weight of the hydrogenation products and in the resulting differences in tensile strength and other properties.

Polymers and copolymers of butadiene are readily hydrogenated using a nickel–kieselguhr catalyst system in the temperature range 200–250°C at hydrogen pressures of about 500 psi in methylcyclohexane or decalin as the reaction solvent. The degree of residual unsaturation can be controlled by the amount and type of catalyst. Continuous as well as batch processes are used commercially to produce a wide range of butadiene resins and hydrogenated copolymers. The high molecular weight of the hydrogenation products makes recovery of the catalyst systems difficult on an industrial scale. Often, the use of solvent dilution is necessary as a first step: magnetic separation can be employed as a direct method following hydrogenation with nickel or other ferromagnetic catalysts. From a knowledge of the general structure of the polybutadiene molecule, a fully hydrogenated polymer would be expected to behave similarly to polyethylene (low or high density, depending upon the degree of branching in the initial polymer):

$$\sim\!CH_2\!-\!CH\!=\!CH\!-\!CH_2\!\sim \xrightarrow{H_2} \sim\!CH_2\!-\!CH_2\!-\!CH_2\!-\!CH_2\!-\!CH_2\!\sim \qquad (5)$$

and this is, in fact, more or less borne out by such physical properties as tensile strength, elongation at break, and impact strength.

The ease of hydrogenation and the resultant degree of saturation achieved under fixed reaction conditions also reflect the microstructure of the polymer. "External" double bonds such as those of poly-1,2-butadiene are more readily hydrogenated than the "internal" double bonds of reaction (5) [5]. The residual unsaturation in synthetic poly-cis-1,4-isoprene) is consistantly less than that for natural rubber; balata and gutta-percha, too, show rates of reaction markedly different from that of natural rubber, and copolymerization with vinyl monomers, which introduces more saturated groupings into the polymer chain (e.g., styrene, ethyl acrylate, vinylidene chloride) always reduces the hydrogenation of olefinic bonds. Even for the homopolymers polybutadiene and polyisoprene, complete hydrogenation is rarely achieved; the residual 0.5–5.0% unsaturation obtained commercially becomes apparent in pronounced differences in crystallinity, T_g, solubility, and dynamic-mechanical loss spectra of these so-called fully hydrogenated products. Residual unsaturation reveals itself also by poor color retention of white or pastel shades and transparent finishes which yellow by oxidation of the unsaturated components.

IV. Dehalogenation, Elimination, and Halogenation Reactions in Polymers

A. DEHYDROCHLORINATION OF POLY(VINYL CHLORIDE)

The dehydrochlorination of poly(vinyl chloride) has been the subject of much investigation, particularly with the view of developing greater stability in PVC polymers and copolymers. Like many polymeric reactions, dehydrochlorination is a complex process. The vinylene groups, created by the elimination of HCl from adjacent carbon atoms in the chain:

$$\sim CH_2-CHCl-CH_2-CHCl\sim \rightarrow HCl + \sim CH=CH-CH_2CHCl\sim \qquad (6)$$

may be the result of free radical, ionic, or ion–radical steps. The presence of a small proportion of head-to-head, tail-to-tail, and other configurational irregularities in the backbone structure of poly(vinyl chloride) leads to more complex elimination steps by thermal degradation (alone or in the presence of catalysts such as aluminum chloride). The introduction of ring structures, a major process during dehydrochlorination, is likewise affected by the distribution of the chlorine atoms along the polymeric backbone. Hydrogen bromide can be effectively eliminated thermally from poly(vinyl bromide), since dehydrohalogenation is a universal thermal reaction the complexities of which increase from chloride to iodide.

B. THERMAL ELIMINATION

The thermal elimination process can be applied to most "substituted" groups in vinyl polymers by controlled pyrolysis at 600–700°C, producing polyvinylene compounds, e.g., by the splitting off of acetic acid from poly(vinyl acetate). By careful temperature control one can achieve bifunctional reactions and/or intramolecular cyclizations. This has been developed commercially at relatively high temperatures, in the case of the polymerization of methacrylamide above 65°C, to yield a polymer with a substantial proportion of imide groups:

$$+ NH_3 \qquad (7)$$

Polymers and copolymers of multifunctional vinyl monomers (using the term to cover the presence of a halogen or other reactive group in addition to the vinyl group, rather than in the sense of more than one polymerizable group), such as α-chloroacrylic acid, often undergo partial lactonization and hydrolysis during polymerization. Heating in alcohol solution, or electrolyzing alcoholic solutions, one obtains, e.g., the introduction of double lactam rings during the acid hydrolysis of poly(α-acetamino acrylic acid)

$$
\begin{array}{cc}
\text{NHAc} \quad \text{COOH} \\
\text{C—CH}_2\text{—C} \\
\text{COOH} \quad \text{NHAc}
\end{array}
\longrightarrow
\begin{array}{c}
\text{NH———C=O} \\
\text{C—CH}_2\text{—C} \\
\text{COOH} \quad \text{NHAc}
\end{array}
+ \text{AcOH} \longrightarrow
$$

$$
\begin{array}{c}
\text{NH———C=O} \\
\text{C—CH}_2\text{—C} + \text{AcOH} \\
\text{CO———NH}
\end{array}
\tag{8}
$$

The discoloration of polyacrylonitrile is due to a similar type of elimination reaction, which in this case occurs intra- as well as intermolecularly to give cross-linked insoluble ring products,

$$
\begin{array}{c}
\text{CH}_2\text{—CH}\text{CH}_2\text{—CH}
\end{array}
\longrightarrow
$$

(9)

The controlled heating of polyacrylonitrile fibers under tension also causes an elimination of nitrogenous products to leave a "carbon fiber" of high tensile strength which can be considered as the end product of the line of chemical elimination reactions. Carbon fibers from cellulosic materials, lignin, and various interpolymers and blends have been developed. The structures of these products consist largely of three-dimensional carbon networks, partially crystalline and partially graphitic or amorphous.

C. Halogenation of Polymers

The halogenation of polymers can conveniently be divided into the halogenation of halogen-free polymers, such as the polyolefins, and the further halogenation of halogen-containing polymers, such as poly(vinyl chloride). Fluorine reacts with polyethylene in the dark; the reaction can be controlled by nitrogen as diluent but, in general, flourination, like nitration, of hydrocarbon polymers is too energetic to allow the preparation of uniform products under controlled conditions on a commercial scale.

In the absence of oxygen, the chlorination of polyethylene, with or without a catalyst, can be controlled to provide a range of products with varying chlorine contents. The chlorination process is a statistical random reaction, so that chlorination to the same chlorine content as poly(vinyl chloride), say 60%, gives a product which is chemically different from poly(vinyl chloride), yet one which is fully compatible with it. Both polypropylene and polyisobutylene tend to degrade during chlorination, with the loss of many useful properties. The degradation of butyl rubber poly(isobutylene-*co*-isoprene) during chlorination can be avoided at low temperatures by limiting the reaction to a maximum of about 2%, i.e., to a level of chlorination comparable to the unsaturation of the copolymer. The bromination of butyl rubber has also been developed commercially, again to relatively low bromine contents.

The random chlorination of polyethylene destroys the crystallinity of the initial polymer; at a degree of chlorination corresponding to the loss of all crystallinity, the chlorinated product becomes soluble at room temperature. The bromination of polyethylene follows a similar course to yield a rubberlike polymer at 55% bromine content. As in the chlorination process, there is a measurable degree of main-chain degradation; the latter, however, is not as extensive as for polypropylene, where an eightfold decrease in molecular weight has been reported, including dehydrobromination at room temperature. The crystalline, tactic polymers are less readily halogenated than the atactic forms, indicating that halogenation is limited to the more readily accessible amorphous regions. It is not surprising, therefore, that the chlorination or bromination of polyisobutylene leads to severe degradation, while the copolymerization with small proportions of isoprene, as in butyl rubber, confers stability on the halogenated material.

Ethylene–propylene copolymers are, by their random copolymerization, amorphous in structure and therefore easily halogenated. Chlorinated ethylene–propylene copolymers have extensive commercial application due to their compatibility with a wide range of polymers, and have found particular use as blends with poly(vinyl chloride) and poly(vinyl chloride) graft copolymers.

As a general rule, while the direct halogen addition to the double bond by an ionic mechanism does take place in many organic solvents, side reactions, including cyclization and cross-linking, also take place, so that the products of halogenation frequently precipitate from solution at an early stage of the reaction. Natural and synthetic polyisoprenes [6] can be chlorinated, but there is a strong tendency toward cyclization. Polybutadiene, on the other hand, tends to cross-link. Even when all the unsaturated groups available to the reaction have been consumed, halogenation continues, with the evolution of halogen hydride. Copolymers of butadiene, including the whole range of butadiene–styrene copolymers, follow the pattern of polybutadiene itself, but by careful control of the reaction conditions completely soluble products can be obtained which have good film-forming properties. The solubility of chlorinated poly-

(butadiene-*co*-styrene) has led to its use in surface coatings and adhesives. The chlorination of poly(vinyl chloride) to products of increased chlorine content proceeds largely by the abstraction of hydrogen atoms from carbon atoms which are not already "chlorinated," but side reactions and the formation of HCl also occur. The second-order transition temperature of poly(vinyl chloride) is increased by chlorination and the products so formed have considerable utility as PVC modifiers.

Hydrohalogenation of polyisoprene [6] occurs readily by the ionic addition of hydrohalogen to the double bond, but again cyclization tends to take place simultaneously. Carrying out the hydrochlorination of natural rubber in chloroform solution, we obtain a completely soluble rubber hydrochloride that, when stabilized and plasticized, gives a tough flexible transparent film with many uses in the packaging industry.

The chlorination of poly(vinyl aromatic) polymers, such as polystyrene, has been attempted as a way of introducing water-solubilizing groups by subsequent hydrolysis. Unfortunately, the chlorination of the phenyl group takes precedence over that on the main-chain carbon atoms. Poly(vinyl toluene) is more readily attacked at the methyl group; the chlorinated products can be aminated with trimethylamine to give water-soluble derivatives.

V. Other Addition Reactions to Double Bonds

A. ETHYLENE DERIVATIVES

Besides the addition of halogens and hydrohalogens "across the double bond" just covered, there are many other reagents which will react similarly with unsaturated polymers by free radical, ionic, or radical–ion mechanisms. Of prime importance is the addition of ethylene derivatives to polydienes. One of the earliest reactions of natural rubber to be studied in detail was the combination with maleic anhydride [6]. Depending upon the reaction conditions and the presence or absence of free radical initiators, one or more of the four basic reactions may take place, with the products shown (the arrows indicate where the addition has taken place and the new bonds formed):

(a)   Intramolecular addition to the double bond within polyisoprene chains (10);

$$(10)$$

(b)   Intermolecular addition to double bonds in different polymer chains (in this group should be included the statistically possible reaction between widely separated double bonds within the same molecule) (11);

$$(11)$$

(c)   Addition to α-methylenic carbon atoms of a polyisoprene chain (12);

$$(12)$$

(d)   Intermolecular addition to α-methylenic carbon atoms in adjacent chains (or widely spaced α-methylenic carbon atoms in the same macromolecule) (13);

$$(13)$$

In general, the overall reaction rates increase with rising temperature and in the presence of oxygen or free radical initiators, but these same conditions promote intermolecular reactions leading to gel formation. Similar reactions take place with gutta-percha, synthetic poly(cis-1,4-isoprene), and poly(cis-1,4-butadiene). Many workers used two-roll mills and other mastication techniques as convenient ways of blending the maleic anhydride with rubbers at elevated temperatures, but where these techniques have been used mechanochemical reactions have complicated the overall process. Reaction products of natural rubber containing 5–10% of combined maleic anhydride can be vulcanized by conventional sulfur cures; of greater interest is the possibility of creating cross-linking by the use of oxides of calcium, magnesium, and zinc.

Other compounds reacting similarly via activated double bonds (excluding here block or graft copolymerization) include maleic acid, N-methylmaleimide, chloromaleic anhydride, fumaric acid, γ-crotonolactone, p-benzoquinone, and acrylonitrile. Other polymers with unsaturated backbones, such as polybutadiene, copolymers of butadiene with styrene and with acrylonitrile, and butyl rubber, react in similar ways, but the recorded reaction with poly(vinyl chloride) is largely mechanochemical in nature (see later).

B. THE PRINS REACTION

Another addition to polymers with main-chain unsaturation is the Prins reaction between ethylenic hydrocarbons and compounds containing aldehydic carbonyl groups. Kirchof, in 1923, described the reaction of natural rubber in benzene solution with aqueous formaldehyde in the presence of concentrated sulfuric acid. The general reaction of an aldehyde, $RCHO$, with a polyisoprene in the presence of an inorganic or organic acid or an anhydrous metal salt is represented by

$$(14)$$

In the absence of such catalysts, the reaction leads to a shift in the double bond rather than its elimination:

$$(15)$$

These reactions can be carried out in solution or in dispersion, or by reaction in the solid phase [6]; in the last case it is again difficult to differentiate the Prins reaction from mechanochemical reactions initiated by chain rupture during mastication.

Other aldehydic compounds, such as glyoxal and chloral, also react in a similar way with polyisoprenes and unsaturated rubbers (e.g., poly(cis-1,4-isoprenes), poly(cis-1,4-butadiene), and copolymers of isobutylene and isoprene). The use of strong acids, or Lewis acids, causes complications, since the acids themselves, under suitable conditions, catalyze cyclization and cis–trans isomerization, and these reactions may occur simultaneously with the addition reactions.

VI. Oxidation Reactions of Polymers

Oxidation creates such marked changes in properties that it must be considered as a chemical modification of polymers in its own right. Oxidation reactions take place readily at the unsaturated groupings in macromolecules and are often referred to collectively as epoxidation in many textbooks. Oxidation in air is a slow process at ambient temperatures around 25°C or lower, but over long periods of time the effects of the chemical changes are cumulative. It is this process which is responsible for the resinification of natural and synthetic rubbers at the surface of rubber bales after long-term storage. Oxidation is, of course, speeded up at higher temperatures unless the polymers are thermally stabilized (see use of antioxidants. Chapter 7), and particularly in the presence of ozone, which enhances not only oxidation steps [see (16)–(18) and (36)] but also autocatalytic oxidation.

A very fast reaction occurs between unsaturated rubbers and strong oxidizers, such as hydrogen peroxide or peracetic acids, at low temperatures in an appropriate compatibilizing solvent. The main reaction is that of the formation of oxirane (epoxy) ring structures

$$\text{~C}\underset{\underset{H}{|}}{\overset{\overset{R}{|}}{=}}\text{C~} \xrightarrow{\text{H}_2\text{O}_2} \text{~C}\underset{\text{O}}{\overset{\overset{R}{|}}{\diagdown}}\text{C~} \tag{16}$$

but ring opening, isomerization of the epoxide group, and polymerization of the epoxide can also occur to further complicate the overall reaction under any given set of conditions:

$$\text{~C}\underset{\text{O}}{\diagup\diagdown}\text{C~} \xrightarrow{\text{H-OOCR}} \text{~C}\underset{\overset{|}{\underset{H}{O}}}{\overset{\overset{H}{|}}{—}}\text{C}\underset{\overset{|}{\underset{\overset{|}{R}}{C=O}}}{\overset{\overset{H}{|}}{—}}\text{~} \tag{17}$$

Epoxidized polymers are highly reactive with a variety of acid or basic compounds. Cured products can be obtained by reaction with anhydrides, polybasic acids, polyamines, phenolic resins, and organic peroxides. Epoxide groups introduced into polymers allow them to be used as macromolecular initiators of block and graft copolymerization, or for the formation of interpolymers by reaction with unsaturated polyesters.

If mechanochemical breakdown during milling is minimized by operating at temperatures at which the rubbers are thermally softened or by reducing the shearing action of tight nips, many rubbers and some nonrubber plastics are oxidized spontaneously (thermally) in air by way of formation of hydroperoxide and peroxide groups on double bonds, and even along the saturated backbone, by steps similar to the epoxidation reaction

$$\text{cross-links} \qquad \text{chain scission} \qquad (18)$$

This is very much an oversimplification of the reactions that take place, but nevertheless, the presence of hydroperoxide and peroxide groups as well as of epoxy, carbonyl, ether, and aldehyde groups can be readily detected by appropriate chemical tests and by spectrographic analysis. The reactions are largely free radical in character, so that it is difficult to completely differentiate them from similar reactions induced mechanochemically.

VII. Miscellaneous Chemical Reactions of Polymers

The reactivity of the double bond tends to overshadow substitution reactions on saturated hydrocarbons (or saturated portions of macromolecular backbones), but such reactions do also occur and have many commercial applications. Sulfonation of polyethylene with chlorosulfonic acid [7] takes place on the polymer backbone, with direct substitution and the elimination of HCl; but with polystyrene the substitution takes place on the phenyl groups:

$$(19)$$

$$(20)$$

Some cross-linking which occurs as a result of side reactions causes an appreciable gel content in the products. Improved chlorosulfonation in the case of polyethylene led to the development of Hypalon®, a vulcanizable rubber with very good solvent resistance [8].

The nitration of polystyrene by fuming nitric acid or nitric–sulfuric acid mixtures also takes place on the aromatic rings. The same is true of other polymers with aromatic substituents. The main use of this reaction in the case of polystyrene is to serve as a first step toward the synthesis of polyaminostyrene by reduction of the nitro groups. Polyaminostyrene, its copolymers, and similar polymeric amino compounds are used in the manufacture of ion exchange resins. They have also been used via diazotization and coupling reactions in the production of polymeric dyes which can be insolubilized within the dyed fabric.

The direct replacement of the hydrogen atoms of aromatic rings in polymers, such as polystyrene, by metals (e.g., sodium or potassium) can be carried out, but yields tend to be rather low and side reactions are difficult to control. The products are highly colored and usually insoluble, at least in part. Polyiodo-styrene can be prepared by reacting polystyrene in nitrobenzene solution with iodine and iodic acid in high yield. Further reaction with an alkyllithium, such as butyllithium, leads to the formation of polylithiostyrene. This two-stage synthesis is relatively free of side reactions. This polymer is used mainly as a highly reactive intermediary in the synthesis of a wide range of nuclear sub-stituted polystyrenes having groups that are difficult to introduce by more direct methods or without competing side reactions. Typical of such reactions is carbonation—the introduction of carboxyl groups on aromatic rings. Once carboxyl groups have been attached to a polymer, metal ions may be used to form labile cross-links. In the Ionomer® series [9], polyethylene has been co-polymerized with acrylic acid, and the salts formed act as cross-links.

A reaction which can readily be applied to polymers, but one which has not received a great deal of attention, is the Meerheim–Ponsdorf reaction between carbonyl or aldehyde groups and metal alkoxides. Either during the synthesis of polyethylene, or later as a result of oxidative processes, a few carbonyl groups are inadvertently introduced into the polymeric structure of polyethylene. Re-action in the melt with aluminum isopropoxide or aluminum secondary butoxide leads to the formation of a loosely gelled network structure (contain-ing most of the polyethylene molecules) with trifunctional aluminum oxide cross-links (21):

$$3 \quad \underset{O}{\overset{|}{\underset{\diagdown R}{C}}} + \text{Al(OR}^1)_3 \longrightarrow \quad \text{(structure)} \tag{21}$$

These cross-links are not stable against hydrolysis but can be used to introduce other structural modifications [10].

VIII. Block and Graft Copolymerization

A. EFFECTS ON STRUCTURE AND PROPERTIES OF POLYMERS

Some of the most significant changes in structure and properties of polymers can be brought about by either block or graft copolymerization (see Chapter 13). The term *block copolymer* is applied to macromolecules made up of sequences with different chemical (or physical, i.e., tactic) structures where the sequences are of a molecular weight which would give them polymeric features even if separated. They may be linear or branched; the linear structures are called block copolymers (22) and the branched structures graft copolymers (23).

$$\text{\textasciitilde\textasciitilde AAAAAA} \cdot \text{BBBB} \cdots \text{BBBBB} \cdot \text{AAAAA} \text{\textasciitilde\textasciitilde} \tag{22}$$

$$
\begin{array}{cccc}
\text{\textasciitilde\textasciitilde AAA} \cdots \text{AAA\textasciitilde\textasciitilde} & & \text{\textasciitilde\textasciitilde BBBBB} \cdots \text{BBB\textasciitilde\textasciitilde} & \\
\quad \text{B} \qquad\qquad \text{B} & & \text{A} \qquad\qquad \text{A} & \\
\quad \text{B} \qquad\qquad \text{B} & & \text{A} \qquad\qquad \text{A} & \\
\quad \text{B} \quad (a) \quad \text{B} & \text{or} & \text{A} \quad (b) \quad \text{A} & \\
\quad \text{B} \qquad\qquad \text{B} & & \text{A} \qquad\qquad : & \\
& & : \qquad\qquad \text{A} & \\
& & \text{A} \qquad\qquad \text{A} & \\
\end{array}
\tag{23}
$$

Thus the same polymeric sequences may be put together as block or as graft copolymers, with differing properties, though in the author's experience, the major differences between the properties of block and graft copolymers of the same constituent polymers are pronounced only in solution or in the melt. For example, natural rubber may be block copolymerized with poly(methyl methacrylate), or methyl methacrylate monomer may be grafted onto natural rubber. In an attempt to distinguish by nomenclature one structure from the other, the concept of inserting the letters *b* and *g*, for block and graft, respectively, between the names of the specific sequences has been introduced, e.g., natural rubber-*b*-poly(methyl methacrylate) and natural rubber-*g*-poly(methyl methacrylate) in the case of the examples cited. The structure of these two macromolecules would be represented by (22) and (23a), where the $\cdots$ AAAA $\cdots$ sequences represent natural rubber and the $\cdots$ BBBBBB $\cdots$ sequences represent poly(methyl methacrylate).

The number and order of sequences may be more complicated. When the structure is such that it cannot be readily categorized as block or as graft, the term *interpolymer* is frequently used. When the sequences making up the segments are random copolymers, the term "*co*" may be introduced, with the major component monomer preceding the minor constituent. A backbone polymer of butadiene–styrene rubber grafted with styrene containing a small percentage of acrylic acid would be described as poly[(butadiene-*co*-styrene)-*g*-(styrene-*co*-acrylic acid)] and could be schematically represented as in (24) where A represents butadiene, B styrene, and C acrylic acid:

$$\sim\sim\text{AABABAAABBABAAAABBA}\sim\sim \qquad (24)$$

```
         B                    B
         B                    B
         B                    C
         C                    B
         B                    B
         B                    C
         C                    B
         B                    B
                              B
```

The differentiation of tactic sequences can also be accommodated in the nomenclature scheme, as can alternating copolymers, by the use of additional prefixes, but it does add to the complexity. Very complex structures cannot be described in simple terms.

B. BLOCK COPOLYMER SYNTHESIS

The principal method of synthesizing block copolymers to be discussed here consists in allowing a polymeric free radical to initiate the polymerization of a second monomer or combination of monomers. The polymeric radical may be formed by (i) interrupted polymerization (i.e., utilizing the fact that in the final stages of a polymerization reaction the polymeric radicals are still active, although the concentration of monomer is low); (ii) weak linkage initiation (e.g., copolymerization into the backbone polymer a small percentage of a monomer containing, say, an azo grouping which can be thermally degraded in a second-stage polymerization at a higher temperature); (iii) irradiation degradation, when main-chain scission is the primary effect of irradiation; (iv) mechanochemical initiation (e.g., mastication or vibromilling techniques).

(i) Interrupted polymerization was one of the first methods of preparing block copolymers, viz., using photoinitiation to polymerize chloroprene, which then acted as the polymeric initiator of methyl methacrylate. Another important method is to be found in the many examples in the patent literature of latex polymerizations where a second monomer is added to the system without further initiator after the first monomer has been largely consumed. This is one of the many forms of "seed" polymerization. Frequently, the growing radical is preserved by having been trapped within the seed particle, so that the second monomer has to reach the radical site by diffusion to become initiated in the second step of the polymerization. This becomes block copolymerization when the second step involves a monomer of a different chemical type. With some polymers, the lifetime of the polymeric radicals is such that the latex particles may be separated from the aqueous phase and added to the second monomer system to initiate block copolymerization; e.g., block copolymers of poly(vinyl chloride) are prepared commercially in this manner [11]. Interrupted polymerizations can be summarized schematically by

$$\sim\text{AAAA}\cdots\text{AAA}\cdot + n(\text{B}) \longrightarrow \sim\text{AA}\cdots\text{AAABBB}\cdots\text{BB}\sim \qquad (25)$$

(ii) An example of weak-linkage initiation applicable to rubbers is the introduction of peroxide linkages into natural rubber by hot mastication in air at a temperature above 135°C; see reaction (18). The thermal softening of the rubber at this temperature allows very little mechanochemical degradation to take place, so that the reaction with oxygen is largely a thermal oxidation. Although some pendant hydroperoxy groups are introduced, the bulk of the peroxy groups are either main-chain peroxy linkages or terminal hydroperoxy groups (26), and both types of groups can be thermally split. In the presence of a vinyl monomer, such as methyl methacrylate, block copolymerization is then initiated [10]:

$$\sim\text{CH}_2-\underset{\underset{\text{CH}_3}{|}}{\text{C}}=\text{CH}-\text{CH}_2-\text{O}-\text{O}-\text{CH}_2-\underset{\underset{\text{CH}_3}{|}}{\text{C}}=\text{CH}\sim\text{CH}_2-\underset{\underset{\text{CH}_3}{|}}{\text{C}}=\text{CH}-\text{CH}_2-\text{O}-\text{OH} \quad (26)$$

The presence of terminal and pendant hydroperoxy groups will lead to production of block copolymers plus some corresponding grafted copolymer.

(iii) Natural rubber and many synthetic rubbers degrade during irradiation by the random loss of hydrogen atoms and/or side groups, so that irradiation syntheses lead to the formation of graft copolymers.

(iv) Mechanochemical degradation provides a very useful and simple way of producing polymeric free radicals. When a rubber is mechanically sheared [12], as during mastication, a reduction in molecular weight occurs due to the physical pulling apart of macromolecules:

$$\sim\underset{\underset{\text{CH}_3}{|}}{\text{C}}=\text{CH}-\text{CH}_2-\text{CH}_2-\underset{\underset{\text{CH}_3}{|}}{\text{C}}=\text{CH}\sim$$

$$\downarrow$$

$$\sim\underset{\underset{\text{CH}_3}{|}}{\text{C}}=\text{CH}-\dot{\text{C}}\text{H}_2 + \cdot\text{CH}_2-\underset{\underset{\text{CH}_3}{|}}{\text{C}}=\text{CH}\sim \qquad (27)$$

$$\sim\underset{\underset{\text{CH}_3}{|}}{\dot{\text{C}}}-\text{CH}=\text{CH}_2 \qquad \text{CH}_2=\underset{\underset{\text{CH}_3}{|}}{\text{C}}-\dot{\text{C}}\text{H}\sim$$

This rupture mechanism is visualized as occurring by the concentration of the shearing couple on a single bond; thus situation (a) in (I) leads merely to molecular rotation, situation (b) to rupture. It is seen that when either the

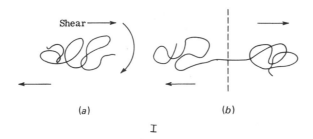

(a)   (b)

I

shearing forces drop too low for bond rupture or the MW is too low to permit large dumbbell-shaped conformations and consequent shearing forces on the end lobes, molecular rupture will no longer occur.

Rupture of the polymer chains tends to take place at the weakest linkage. In the case of natural rubber this is the CH_2—CH_2 bond, due to resonance stabilization of the polyisoprenyl radical formed by its rupture (27). Block copolymerization can be carried out mechanochemically by (i) degrading a rubber in the presence of a monomer polymerizable by free radicals; (ii) co-masticating two rubbers, or a rubber with a polymer, of sufficient compatibility; or (iii) carrying out the mechanochemical degradation in the presence of oxygen, forming peroxy and hydroperoxy groups which can then be used as the active sites in a second-stage copolymerization.

A wide range of monomers is available which can be used to achieve compatible viscoelastic blends that can be masticated by extrusion or by similar techniques under strict exclusion of oxygen (which itself combines with free radicals and thus acts as a polymerization inhibitor) to bring about block copolymerization. Natural rubber, purified by acetone extraction to remove natural or adventitious antioxidants which act as inhibitors, has been extruded with up to 40% of methyl methacrylate to produce natural rubber-b-poly(methyl methacrylate) in semicommercial quantities.

Table I lists the monomers which have been block copolymerized with natural rubber at 15°C using a Baker Perkins type single rotor masticator.

Most butadiene copolymer rubbers tend to form cross-linked gels on mastication because of the interaction of the polymeric radicals with the double bonds within the macromolecule. When such synthetic rubbers are masticated with polyisoprenes, the interpolymeric gel contains a high proportion of polyisoprene. On the other hand, when natural or synthetic polyisoprenes are masticated with nongelling saturated rubbers, such as the polyester rubbers with which they are compatible, the mastication product contains significant amounts of linear block copolymer containing both segment types. The essential feature of all these mechanochemical reactions is the shearing of an intimate mixture of polymer chains in a viscoelastic state. If only one component of a blend of polymers is rubbery under the conditions of mastication, then the polymeric radicals of that component determine the nature of the reactions.

TABLE I

Monomer	Mastication time (min)	Polymerization (%)
Ethyl methacrylate	10	89
Allyl methacrylate	15	98[a]
n-Butyl methacrylate	20	97
Isobutyl methacrylate	15	74
Nonyl methacrylate	15	47
Lauryl methacrylate	5	85
2-Chloroethyl methacrylate	10	93[b]
β-Ethoxyethyl methacrylate	10	79
Ethylene dimethacrylate	20	81[c]
Butyl acrylate	5	74[d]
2,5-Dichlorostyrene	15	83
Methyl vinyl ketone	25	80[e]
Methyl isopropenyl ketone	5	74
2-Vinyl pyridine	10	94
N-Vinyl pyrrolidone	30	47

[a] 53% gel; [b] 11% gel (highly swollen); [c] 71% gel; [d] 23% gel (highly swollen); [e] 78% gel.

Block or graft copolymer reactions depend upon the stability of the free radicals, the presence or absence of reactive groups in the nonrubber, and the availability of such groups for reaction with polymeric free radicals by diffusion of the latter to the reactive sites. A high temperature may be necessary to render both components of a blend viscoelastic, but at such a temperature the viscosity of the blend may be too low to develop effective shearing forces. An example is the case when natural rubber is masticated with polystyrene. At low temperatures, the polymers are incompatible and mastication leads only to breakdown of the rubber. At the temperature at which polystyrene softens, the blend is partially compatible, but too fluid for sufficient shearing forces to develop; thus, thermal oxidative reactions predominate. Small amounts of solvents or plasticizers may be added to lower the temperature at which compatible blends can be achieved to bring about block copolymerization.

The principle of adding plasticizing solvents to create conditions suitable for mechanochemical shearing can be extended to primarily plastic systems [13]. In the absence of a polymerizable monomer, plastics suffer a reduction in molecular weight which continues until a steady state is reached when the viscosity of the system has dropped to the point where no further chain scission occurs. (Lowering the temperature at this point would, in general, increase the viscoelasticity and allow further mechanical scission to occur.) When the plasticizing solvent is also a free-radical-polymerizing monomer, block copolymerization takes place:

$$\sim\sim \cdots AAAAA \cdots \sim\sim \xrightarrow{nB} \sim\sim \cdots AAAABBBBB \cdots \sim\sim \qquad (28)$$

Poly(methyl methacrylate), polystyrene, poly(vinyl acetate), poly(vinyl chloride), poly(vinylidene chloride), and poly(vinyl pyrrolidone) are among the amorphous polymers which have been used for block copolymerization by this technique. Polymers with a high crystallinity, such as polyethylene, polypropylene, and the nylons, are not readily swollen by vinyl monomers and the temperature range for forming viscoelastic systems is very limited. However, acenaphthalene, a solid monomer, can be block copolymerized with polyethylene to a low conversion (25–35%) by masticating just below the melt temperature of the blend, and nylon is readily block copolymerized with acrylic acid at room temperature. The important criterion in all these systems is the existence of a viscoelastic state so as to allow the exertion of sufficient shearing forces on to the polymer chains.

The required vicous state, finally, may be obtained by the addition of a nonreactive plasticizer, which is extracted or evaporated off after the mechanochemical reaction has been carried out. Thus, provided the monomer polymerizes by a free radical reaction,* and oxygen is kept out of the system, block copolymerization can be achieved with practically any polymer–monomer system. More than 200 different block copolymers with rubberlike properties have been synthesized in this way. Only a very few, however, have been fully evaluated and the block copolymers separated and characterized.

C. EXAMPLES

As major examples, let us consider the three monomers butadiene, styrene, and acrylonitrile, and see how they can be block copolymerized together by mechanochemical means. From the large number of theoretical possibilities, eleven have been selected for discussion; these may be prepared by the mastication of:

1. a butadiene–styrene copolymer rubber with acrylonitrile monomer;
2. polyacrylonitrile (plasticized) with a mixture of butadiene and styrene monomers;
3. a butadiene–acrylonitrile copolymer rubber with styrene monomer;
4. polystyrene with a mixture of butadiene and acrylonitrile monomers;
5. a styrene–acrylonitrile resin with a mixture of styrene and butadiene monomers;
6. polybutadiene with a mixture of styrene and acrylonitrile monomers;
7. a butadiene–styrene rubber with polyacrylonitrile (best plasticized);
8. a styrene–acrylonitrile resin with polybutadiene;
9. a butadiene–acrylonitrile rubber with polystyrene;
10. a high styrene–butadiene resin with acrylonitrile monomer;
11. a high styrene–butadiene resin with polyacrylonitrile (plasticized).

* It has been found in practice that a number of monomers that normally do not polymerize by free radical processes in the temperature range 10–50°C can be block copolymerized by cold mastication techniques, indicating ionic initiation via heterolytic scission.

(All of the foregoing reactions except 2 and 11 have been reported in the patent literature or are known to have been commercially evaluated.) In each of these examples, the products would be chemically and physically different in terms of the makeup of the structural sequences, and all properties would also depend upon the relative proportions of the initial components. In all mechanochemical reactions, some of the starting polymer or copolymer remains unchanged, mainly the low molecular weight fraction, which is not effectively sheared, and some homopolymer may be formed from the polymerizing monomers by chain transfer reactions. Varying the mastication conditions greatly influences the yield and rate of reaction; the chemical nature of the products is less affected, except that the presence of butadiene as one of the constituents (either polymer or monomer) will cause increasing gel contents with continued mastication. Processes 3, 5, 6, 8, and 9 are known to give products in which a rubber phase is dispersed in a resinous matrix; i.e., they are alternative methods for producing an A-B-S type copolymer.* The presence of a proportion of block or graft copolymer in the system assists in stabilizing the dispersion of the rubber phase in the resin matrix by acting at the phase boundary as a "soap," i.e., a compatibilizing agent at the phase boundary.

D. OTHER METHODS OF EFFECTING MECHANICOCHEMICAL REACTIONS

Although mastication is the simplest procedure for bringing about mechanochemical reactions with rubbers and rubberlike systems, other methods have been used. Comminution and vibromilling have largely been applied to polymers that are glassy at ambient temperatures such as polystyrene and poly-(methyl methacrylate). Mechanochemical degradation can be carried out in an agate vibromill or similar equipment, and the breakdown followed by viscometry. Block copolymerization can often be achieved by vibromilling (or ball-milling) the polymer in the presence of a vinyl monomer. High yields of material in sufficient quantity for evaluation are not readily obtained. Vibromilling below the glass transition temperature has enabled block copolymerization of natural rubber with methyl methacrylate to be carried out on a small scale, with conversions as high as 86% and with up to 70% of block copolymer in the product. Similar results were achieved with styrene and with acrylonitrile, but in the latter case no free polyacrylonitrile was detected.

Ultrasonic irradiation [14], high-speed stirring, freeze–thaw cycling [15], and spark discharging of polymer–monomer systems are all mechanochemical degradative mechanisms, but they have not been applied with any great success

---

* ABS copolymers, as generally produced commercially, are *graft* copolymers of styrene–acrylonitrile on polybutadiene, and each of the listed reactions could be carried out as a grafting reaction to give products with closely related properties.

to rubbers as methods of synthesizing block copolymers. Vapor and liquid phase swelling techniques are, however, directly applicable to the block copolymerization of many monomers onto uncross-linked natural and synthetic rubbers. Swelling takes place when polymers of a high molecular weight, in which the chain entanglements act as pseudo-cross-links,* are allowed to come in contact with the vapors of a wide range of organic liquids, e.g., poly(methyl methacrylate) (Plexiglas) with benzene, toluene, or chloroform. The swelling curves exhibit typical swelling behavior with a distinct and reproduceable change in swelling rate at a critical degree of volume swelling. It has been postulated that if the swelling process develops sufficiently rapidly, the entangled chains resisting the swelling cannot part quickly enough, so that the forces become sufficient to bring about mechanochemical scission at the critical degree of swelling. Thus, if a polymer is allowed to swell in the vapor of a polymerizable vinyl monomer, such as acrylonitrile, then prior to the critical swelling level the system is completely reversible and the monomer can be recovered unchanged. Once a critical degree of swelling is reached, block copolymerization takes place [16] and much of the monomer can no longer be recovered.

Similar discontinuities have been detected in the liquid phase swelling of natural rubber and have again been attributed to mechanochemical breakdown of entangled chains. Low yields of block copolymers have been obtained with methyl methacrylate, styrene, and acrylonitrile.

E. Ionic Mechanisms

So far, discussion of block copolymerization has been restricted to free radical processes, but ionic mechanisms are of equal importance. As a special feature of many homogeneous anionic polymerizations in solution, spontaneous termination can be avoided by a careful choice of experimental conditions. In fact, an infinite life of the active chain end is theoretically possible, and this has led to the term "living" polymers. Thus, polymer carbanions can resume growth after the further addition of monomer; by changing the monomer composition, block copolymerization is readily initiated, and this process can be repeated. A major advantage of this type of synthesis over most free radical processes is the ability to control the chain length of the sequences by adjusting the concentrations of initiating sites and of monomer at each stage of the block copolymerization. It is this convenience which has led to the commercial acceptance of this intrinsically difficult process and to the development of the many products which have recently been obtained. The reaction can be repre-

---

* Bond rupture occurs to a similar extent in chemically cross-linked rubber gels, too, but the utility of these systems is not as great. The surprising discovery was that high molecular weight uncross-linked rubbers also suffer chain rupture during swelling [10].

sented by equations (29)–(32), representing the block copolymerization of

$$\left[\bigcirc\bigcirc\right]^{-}Na^{+} + PhCH{=}CH_2 \rightleftharpoons [PhCH{-}CH_2]^{-}\cdot Na^{+} + \bigcirc\bigcirc \tag{29}$$

$$2[PhCH{-}CH_2]^{-}\cdot Na^{+} \longrightarrow [PhCH{-}CH_2{-}CH_2{-}CHPh]^{2-}2Na^{+} \tag{30}$$

$$[PhCH{-}CH_2{-}CH_2{-}CHPh]^{2-}2Na^{+} + 2(n-1)(PhCH{=}CH_2) \longrightarrow \tag{31}$$
$$[(PhCH{-}CH_2)_n{-}(PhCH{-}CH_2)_n]^{2-}2Na^{+}$$

$$[(PhCH{-}CH_2)_{2n}]^{2-}2Na^{+} + 2m(CH_2{=}\overset{\overset{\displaystyle CH_3}{|}}{C}{-}CH{=}CH_2 \longrightarrow \tag{32}$$
$$(CH_3{-}\overset{\overset{\displaystyle CH_3}{|}}{C}{=}C{-}CH_2)_m(PS)_{2n}(CH_3{-}\overset{\overset{\displaystyle CH_3}{|}}{C}{=}C{-}CH_2)_m$$

isoprene by polystyryl anions initiated by a sodium naphthalene complex in ether [17]. The styryl ion-radical dimerizes to the dianion initiator and each initiator species forms one chain which remains unterminated at the end of the fast propagation step during which all the monomer is consumed. The length of the sequence thus depends upon the concentration of initiator and the concentration of monomer. The addition of a further quantity of the monomer extends the chain lengths without increasing the number of chains. Since the initiator concentration is fixed when the monomer is added, all polymer chains are about the same length. Addition of a second monomer adds sequences of this, and further additions of the first or second monomer can be made to build up multisegment block copolymers. When the desired structure has been built, the reaction can be terminated by the addition of a polar compound, usually an alcohol.

A second type of anionic block copolymerization employs organolithium initiators, which have wider use because of their extended range of solubility, which includes hydrocarbons [18]. Organolithium compounds act as initiators by direct attack of the organic anion on the monomer species, again a fast reaction and, in the absence of compound with active hydrogen atoms, without transfer or termination steps. If carefully executed, the reaction permits one to have precise control over molecular weight and (the narrow) molecular weight distribution.

Four variations of the preparation of these so-called A-B-A block copolymers are used in practice, typified by the following examples.

1. Three-Stage Process with Monofunctional Initiators

In this technique, used, e.g., for the synthesis of block copolymers of [poly(styrene-*b*-butadiene-*b*-styrene)] (S-B-S), a polystyrene block is first formed by using *n*-butyllithium as the initiator in an aromatic solvent, such as benzene, followed by the polymerization of the diene monomer added to the solution (the red color of the styryllithium immediately disappears); the third step is the readdition of styrene monomer.

2. Two-Stage Process with Difunctional Initiators

The use of sodium naphthalene or of a dilithium complex as the initiator is potentially the best method, since there are only two monomer addition steps. It is also very useful in the case of A-B-A block copolymers, where B can initiate A, but A cannot initiate B. This method is used to make poly(methacrylonitrile-b-isoprene-b-methacrylonitrile). The solubility restrictions with respect to the initiator are a disadvantage of this process.

3. Monofunctional Initiation and Coupling

This is a two-stage process whereby B is sequentially polymerized onto A, and then the two chains are coupled to yield an A-B–B-A block copolymer. In this process, a monomer pair can be used such that the chain end of A will initiate B, while the chain end of B is incapable of initiating the polymerization of A. Poly(styrene-b-propylene sulfide-b-styrene) is usually made by this process.

4. Two-Stage Tapered Block Copolymerization

This method is used to form a block copolymer which consists of two segments of essentially homopolymeric structure separated by a block of a "tapered" segment of random copolymer composition. Thus, a first addition of styrene to an organolithium hydrocarbon system produces the homopolymeric blocks, while at the second stage a mixture of styrene and butadiene is added, followed by pure styrene in the end.

For details of the syntheses of linear block copolymer rubbers, the reader is directed to Chapter 13 and to the bibliography at the end of this chapter.

F. GRAFT COPOLYMER SYNTHESIS

The syntheses of graft copolymers are much more diverse, but they can nevertheless be divided into groups of related processes, of which (i) polymer transfer; (ii) copolymerization via unsaturated groups; (iii) redox polymerizations; (iv) high-energy irradiation techniques; and (v) photochemical syntheses are perhaps the most important.

(i) During the polymerization of a monomer by free radical initiation, chain transfer to initiator; transfer to monomer, solvent, mercaptan, or other added transfer agents; and transfer to polymer take place (see again Chapter 2). It is the latter that is of interest as a potential grafting technique:

$$
\begin{array}{l}
P\cdot + P \longrightarrow PH + P\cdot \\
R\cdot + P \longrightarrow RH + P\cdot \\
P\cdot + P\cdot \longrightarrow PH + P\cdot
\end{array}
\qquad (33)
$$

The reaction proceeds by the transfer of a hydrogen, or halogen (in the case of halogenated polymers), atom from a macromolecule P to the growing chains

$$\text{ⱮⱮ} \cdot + \Big| \;\; \longrightarrow \;\; \text{ⱮⱮH} + \Big| \cdot$$

$$\text{monomer} \tag{34}$$

$$\Big|_{\text{ⱮⱮ}}$$

P· (or to an excess initiator free radical R·, thereby "terminating" them). The reactivity is now located on the transfer molecule, which in turn initiates co-polymerization, i.e., the growth of a grafted side chain, of a newly introduced second monomer. A measure of grafting occurs with most monomer–polymer systems, especially those initiated by benzoyl peroxide, if the concentrations of polymer and initiator are high. Interestingly, in many latex and suspension polymerizations it is customary to add a water-soluble polymer such as poly-(vinyl alcohol) to aid in dispersing the monomer and stabilizing the dispersion. A significant proportion of this added polymer cannot be extracted and must be presumed to have also been grafted to the base polymer.

The simplest technique is to dissolve the polymer in the desired monomer at as high a concentration as possible while avoiding phase separation; add the peroxide or other initiator to the cooled solution; and raise the temperature to obtain a controllable rate of polymerization. Methyl methacrylate and styrene, as well as various copolymer mixtures, have been grafted to natural and synthetic polyisoprenes in this way, using benzoyl peroxide as initiator. The method is particularly suitable for lower molecular weight fractions of polyisoprenes, polybutadienes, etc., and for low molecular weight polymers such as the polyethylene oxides [19].

In a bulk variant of this process one swells the rubber with the monomer–initiator system, homogenizes, and reacts the swollen product in an extruder possessing heating, followed by cooling, zones. Some mechanochemical reaction may take place simultaneously but the bulk of the grafting is by thermally initiated transfer via the peroxide or other added initiator. The effects of concentration variables on the extent of grafting, the number of grafts per macromolecule, and the lengths of the grafted chains have been studied for a number of monomers grafted to polyisoprenes, and the systems have been extensively characterized [20].

Since chlorine atoms are more susceptible to transfer processes than hydrogen atoms, it is surprising that so little attention has been devoted to grafting to chlorinated and hydrochlorinated polyisoprenes. The hydrochloride of natural rubber has been used, fairly recently, as the backbone polymer for grafting. Reasonably good yields were obtained by using acrylonitrile as the monomer and either benzoyl peroxide or azobis(isobutyronitrile) as the initiators.

(ii) Other methods (e.g., most car body repairs) are based on the polyester–styrene copolymerization process (reinforced with various types of inert mesh or glass fabric), a graft copolymerization of styrene onto backbone unsaturation of a polymer of relatively low molecular weight. In general, for high grafting yields, a reasonably high concentration of pendant vinyl groups is required on the backbone polymer. For glass-reinforced plastics the polyester resins are selected with this in view. In natural rubber, a few such groups per molecule* are always present and these undoubtedly participate during normal grafting. The content of pendant vinyl groups can be increased by mastication of unsaturated rubbers under nitrogen, since the resonance structures recombine as

$$
\begin{array}{ccc}
& \overset{\displaystyle CH_3}{\underset{\displaystyle |}{}} & \\
\text{\textasciitilde} C{=}CH{-}CH_2\cdot & & \cdot CH_2{-}\overset{\displaystyle CH_3}{\underset{\displaystyle |}{C}}{=}CH\text{\textasciitilde} \\
\updownarrow & + & \updownarrow \\
\text{\textasciitilde}\overset{\displaystyle CH_3}{\underset{\displaystyle |}{C}}{-}CH{=}CH_2 & & CH_2{=}\overset{\displaystyle CH_3}{\underset{\displaystyle |}{C}}{-}CH\text{\textasciitilde} \\
\end{array}
\qquad (35)
$$

$$
\begin{array}{ccc}
\overset{\displaystyle CH_3}{\underset{\displaystyle |}{}}\quad\overset{\displaystyle CH_3}{\underset{\displaystyle |}{}} & & \overset{\displaystyle CH_3}{\underset{\displaystyle |}{}} \\
\text{\textasciitilde} C{-}CH_2{-}C{=}CH\text{\textasciitilde} & \text{and} & \text{\textasciitilde} C{=}CH{-}CH_2{-}CH\text{\textasciitilde} \\
\underset{\displaystyle CH_2}{\underset{\displaystyle \|}{CH}} & & \underset{\displaystyle CH_2}{\underset{\displaystyle \|}{C{-}CH_3}}\ \text{etc.}
\end{array}
$$

The direct introduction of peroxide groups into the backbone of polymers, such as poly(methyl methacrylate), has been used to produce macromolecular initiators for the synthesis of block copolymers, e.g., poly(methyl methacrylate-*b*-acrylonitrile) and poly(methyl methacrylate-*b*-styrene). Ozonization can also be used, with careful control of the degree of ozonolysis, to introduce epoxy ring structures into natural rubber:

$$
\text{\textasciitilde}CH_2{-}\overset{\displaystyle CH_3}{\underset{\displaystyle |}{C}}{=}CH{-}CH_2\text{\textasciitilde} \xrightarrow{O_3} \underset{\text{\textasciitilde}CH_2}{\overset{CH_3}{}}\!\!\diagdown\!\!\underset{O{-}O}{\overset{O}{}}\!\!\diagup\!\! CH{-}CH_2\text{\textasciitilde} \qquad (36)
$$

By carrying out the reaction to about 4% of the available double bonds in a solvent such as toluene at a low temperature followed by a nitrogen purge, grafting can be effected by addition under nitrogen of methyl methacrylate monomer (reacting at 80°C in sealed ampules) and formation of two MMA chains attached to the oxygens of the opened —O—O— bridge. This technique should be applicable to isoprene and butadiene copolymers.

---

* About 0.4% of the unsaturated groups are pendant vinyl groups in an average sample of acetone-extracted pale crepe rubber.

(iii) Redox polymerizations are among the most popular techniques for grafting reactions and, of the possible initiator systems, ferrous ion oxidation and those based on ceric ion reduction are widely used. In a redox polymerization, a hydroperoxide or similar group is reduced to a free radical plus an anion, while the metal ion is oxidized to a higher valency state, and at the same time a monomer is added. When the reducible group is attached to a polymeric chain, the free radical grafting sites thus formed on the macromolecular backbone act as initiators for graft copolymerization:

$$\sim\!CH_2\!-\!\underset{\underset{O-OH}{|}}{CH}\!\sim + Fe^{2+} \xrightarrow[\substack{amine\ or\\aldehyde}]{} \sim\!CH_2\!-\!\underset{\underset{O}{|}}{CH}\!\sim + Fe^{3+} + OH^- \qquad (37)$$

This method has been used to graft methyl methacrylate to natural rubber latex. (Actually, fresh latex contains a few hydroperoxide groups per macromolecule, which can take part in grafting reactions.) Recentrifuged latex concentrate is mixed with methyl methacrylate and a solution of tetraethylenepentamine is added, followed by a small quantity of ferrous sulfate solution. The homogenized blend is allowed to stand, often overnight [20]. The graft copolymer is isolated by coagulation. Since practically all free radical sites are formed on the rubber backbone, there is very little free poly(methyl methacrylate) in the grafted system; on the other hand, some rubber chains are without grafts, since not all chains have hydroperoxy groups. Higher yields of graft copolymer are obtained by allowing the monomer to dissolve in, and equilibrate with, the latex particles before adding the amine and ferrous ion initiator. Passing oxygen (air) through the latex for several hours has been claimed to reduce the free rubber content of the polymerization product, but nitrogen purging is then necessary to prevent dissolved oxygen from acting as a polymerization inhibitor.

Hydroxy polymers can be grafted by redox polymerization by using a water-soluble peroxide, such as hydrogen peroxide in conjunction with ferrous ions. The OH radicals thus produced abstract H atoms from the hydroxy groups in the polymer, giving free radical grafting sites on the backbone. This method has been used with starch and cellulose derivatives, but considerable quantities of homopolymer are formed from the initial hydroxyl radicals in parallel with the H abstraction. By introducing a few hydroxyl groups into a copolymeric synthetic rubber, grafting can be effected provided the presence of homopolymer can be tolerated. Mixtures of ferrous ammonium sulfate and ascorbic acid are suitable redox initiation systems. Many patents claim the preferred use of ceric ions, which easily oxidizes hydroxyl groups by a radical–ion reaction:

$$R\!-\!OH + Ce^{3+} \rightarrow R\!-\!\overset{+}{\underset{\cdot}{O}}H + Ce^{2+}$$

$$R\!-\!\overset{+}{\underset{\cdot}{O}}H + OH^- \rightarrow R\!-\!O\cdot + H_2O \qquad (38)$$

The advantage of this reaction lies in the fact that only hydroxyls on the polymer are converted into R—O· free radicals, so that no homopolymer can be produced and pure graft is obtained.

   (iv) During high-energy irradiation in vacuum, e.g., from a ^{60}Co source, some main-chain degradation of natural rubber and other polyisoprenes occurs, according to Eq. (39). However, much of the irradiation energy is also adsorbed by the removal of hydrogen atoms (40):

$$\text{CH}_3 \qquad\qquad \text{CH}_3$$
$$\text{\char126CH}_2\text{—C}=\text{CH—CH}_2\text{—CH}_2\text{—C}=\text{CH—CH}_2\text{\char126} \longrightarrow$$

$$\text{CH}_3 \qquad\qquad \text{CH}_3 \tag{39}$$
$$\text{\char126CH}_2\text{—C}=\text{CH—CH}_2\text{·} + \text{·CH}_2\text{—C}=\text{CH—CH}_2\text{\char126}$$

$$\text{CH}_3 \qquad\qquad\qquad \text{CH}_3$$
$$\text{\char126CH}_2\text{—C}=\text{CH—CH}_2\text{\char126} \xrightarrow{\text{irrad}} \text{\char126CH}_2\text{—C}=\text{CH—ĊH\char126} \tag{40}$$

The irradiation of natural rubber in the presence of a vinyl monomer thus leads primarily to a synthesis of graft copolymers, but some block copolymer is certainly always present. Irradiation syntheses may be carried out in solution, either in contact with liquid monomer (with or without a diluent) or in contact with monomer in the vapor phase; or in emulsion or suspension. The rubber may be preirradiated in the absence of air to produce free radicals for later monomer addition, but the life of these radicals is short due to mobility within the rubber matrix. Irradiation at very low temperatures makes it possible to use the trapped radicals technique for a variety of natural and synthetic rubbers. Plastics and polymers with a crystalline phase are more readily preirradiated to initiate later grafting by trapped radicals. Irradiation may also be carried out in air to introduce peroxide groupings:

$$\text{CH}_3 \qquad\qquad \text{CH}_3$$
$$\text{\char126CH}_2\text{—C}=\text{CH—CH}_2\text{·} \xrightarrow{\text{O}_2} \text{\char126CH}_2\text{—C}=\text{CH—CH}_2\text{—O—O·}$$

$$\tag{41}$$

$$\text{CH}_3 \qquad\qquad\qquad \text{or} \qquad \text{CH}_3 \qquad\qquad\qquad\qquad \text{CH}_3$$
$$\text{\char126CH}_2\text{—C}=\text{CH—CH}_2\text{—OOH} + \text{R·} \quad \text{\char126CH}_2\text{—C}=\text{CH—CH}_2\text{—O—O—CH}_2\text{—C}=\text{CH\char126}$$

$$\text{CH}_3 \qquad\qquad\qquad \text{CH}_3$$
$$\text{\char126CH}_2\text{—C}=\text{CH—ĊH\char126} \longrightarrow \text{\char126CH}_2\text{—C}=\text{CH—CH\char126}$$
$$\qquad\qquad\qquad\qquad\qquad\qquad\qquad\qquad \text{O—O·}$$

$$\text{CH}_3 \qquad\qquad\qquad \text{or} \qquad \text{CH}_3$$
$$\text{\char126CH}_2\text{—C}=\text{CH—CH\char126} + \text{RH} \qquad \text{\char126CH}_2\text{—C}=\text{CH—CH\char126} \tag{42}$$
$$\qquad\qquad\quad\text{OOH} \qquad\qquad\qquad\qquad\qquad\qquad \text{O}$$

$$\text{CH}_3 \qquad \text{O}$$
$$\text{\char126CH}_2\text{—C}=\text{CH—CH\char126}$$

These groups can then be used to initiate grafting by any of the methods already discussed. Latex phase grafting is generally favored for its simplicity; natural rubber grafts with methyl methacrylate, styrene, acrylonitrile, and vinyl chloride have been made in this way [21].

The irradiation of mixed latices for subsequent combination of the ruptured chains is another approach; it has been carried out with natural rubber and poly(vinyl chloride) latices to prepare graft (and block) copolymers in fairly high yields without the problem of monomer recovery. The same method has been used to graft polychloroprene onto synthetic polyisoprene dispersions and onto polybutadiene latices of various compositions.

(v) Macromolecules containing photosensitive groups which absorb energy from ultraviolet frequencies often degrade by free radical processes. The degradative process as a rule is fairly slow, but by the addition of photosensitizers, such as xanthone, benzyl, benzoin, and 1-chloroanthraquinone, the rate can be speeded up to enable graft copolymerization to take place in the presence of methyl methacrylate or other monomers. This can be done in the case of natural rubber in the latex phase with reasonably high yields of graft copolymer. Natural rubber-*g*-polystyrene, and poly(butadiene-*g*-styrene) have both been prepared by uv irradiation of sensitized latex–monomer dispersions. A combination of photochemical synthesis and redox type initiation can also be carried out—a process known as one-electron oxidation—to achieve grafting with minimal homopolymer formation.

Bromine atoms on the backbone of a polymer can be liberated readily by uv irradiation to give free radical sites for grafting reactions. The bromination can be photochemically induced (43), or a chain transfer agent such as carbon tetrabromide may be used in the polymerization step to introduce the labile groups (44):

$$\sim\!CH_2\!-\!CH\!-\!CH_2\!-\!CH\!\sim \xrightarrow[h\nu]{Br_2} \sim\!CH_2\!-\!\overset{Br}{\underset{|}{C}}\!-\!CH_2\!-\!CH\!\sim + HBr \qquad (43)$$

$$\sim\!CH_2\!-\!CH\cdot + CBr_4 \xrightarrow[h\nu]{Br_2} \sim\!CH_2\!-\!CHBr + \cdot CBr_3 \ etc. \qquad (44)$$

With the aid of suitable sensitizers, polymers such as brominated butyl rubber, valuable because of their flame retardancy, may act as backbone polymers for a variety of grafting reactions.

An early synthesis of block copolymers was based upon the uv irradiation of poly(methyl vinyl ketone) in the presence of acrylonitrile. The initial degradative step is

$$
\begin{array}{ccc}
\overset{H}{\underset{H}{\wedge\!\!\wedge C}} - \overset{H}{\underset{\underset{CH_3}{|}}{\underset{C=O}{C}}} - \overset{H}{\underset{H}{C\wedge\!\!\wedge}} \quad\longrightarrow\quad \wedge\!\!\wedge CH_2\cdot \; + \; \cdot\overset{H}{\underset{\underset{CH_3}{|}}{\underset{C=O}{C}}} - CH_2\wedge\!\!\wedge
\end{array}
\tag{45}
$$

This degradation reaction, supplemented by various subsequent oxidation steps, has found renewed interest in the form of the introduction of photodegradable plastics as part of the campaign to reduce plastic litter from throwaway packaging. Although as yet there has been no demand for photodegradable rubbers, the incorporation of a small percentage of a vinyl ketone into a rubber copolymer or homopolymer would open the way to a useful synthesis of block copolymers.

Many other syntheses of block and graft copolymers have been reported, but enough has been said to indicate the scope of these reactions, and to indicate a potential that has still to be thoroughly explored. Many grafting and block copolymerization systems have only been evaluated for plastic materials, but are capable of extension to rubbers.

G. Base Polymer Properties

Though the properties of block and graft copolymers are discussed in Chapters 3–5 and 13, some properties of the copolymer which are particularly germane to this discussion will be briefly mentioned.

The properties of, say, natural rubber grafted with poly(methyl methacrylate) cannot be evaluated unless the copolymer is isolated from either homopolymer species. The methods used are based upon fractional precipitation, selective solution, or a combination of these basic techniques. For details, the reader is referred to Chapter 3, and to the bibliography at the end of this chapter. In many cases, though, technologists are concerned with the materials as manufactured, so that we will consider in this context also the properties of the block and graft copolymers without homopolymer removed.

The presence of two chemically different polymeric sequences in the same chain causes that macromolecule to act as a soap; i.e., helps to compatibilize two species of homopolymer in a blend by accumulating at their interface, assisting a more gradual transition from one phase to the other, and thus reducing the interfacial energy. Microphase separation, of course, still occurs, the predominant case in practice, but macrophase separation is thereby usually prevented. In high-impact polystyrene and in A-B-S copolymers prepared by grafting reactions, the dispersed rubber phase in the glassy matrix and the dispersed glassy phase within the rubbery particles are both prevented from forming separate phases by the graft copolymer chains, which on a molecular scale have their rubbery segments associated with the rubber particles, and their plastic segments with the glass phase. In this respect there is little difference in properties between a graft and a linear block copolymer—the essential feature is the presence of the two types of sequences in the same macromolecule.

The block copolymeric thermoelastic polymers owe their properties to this very structure, whereby the polystyrene end blocks (along with any homopolymeric polystyrene) form the microphase, which is dispersed within a continuous phase of polybutadiene formed from the polybutadiene segments of the central sequences in S-B-S type block copolymers. (For this to happen, the total volume of the polybutadiene segments must exceed the total volume of the polystyrene segments. When the reverse is the case, the product exhibits the properties of a high-impact polystyrene; see again Chapter 13.) While the polystyrene "structures" act as physical cross-links at low temperature, at processing temperatures above the softening temperature of polystyrene both segment types exhibit viscoelasticity, allowing the material to be extruded, injection molded, etc. On cooling, the polystyrene domains become rigid again and assert their influence on the material properties.

When block and graft copolymers are dispersed in solvents, the solutions have properties which depend upon whether or not the copolymer is eventually fully solvated. If the solvent is a "good" solvent for both sequences [e.g., chloroform in the case of natural rubber graft copolymerized with poly(methyl methacrylate)] [22], then both segment types are expanded and films cast from dilute solutions will usually be intermediate in properties to the two homopolymers (in this example the properties of a reinforced rubber film). If the solvent is a good solvent for the rubber but a poor solvent or nonsolvent for poly(methyl methacrylate), e.g., petroleum ether, then the solutions show the typical turbidity of a block or graft copolymer and the cast film is highly elastic. When the solvent is acetone, a good solvent for poly(methyl methacrylate) but a nonsolvent for rubber, the cast films are plastic with high tear strengths. See again Chapter 10.

When grafting is carried out on a polymer under conditions such that the physical form of the substrate polymer is maintained, then the original properties of the substrate usually predominate while supplementary properties accrue due to the grafting. This is invariably the case when the substrate is in fibrous form, e.g., cellulose, nylons, and terylene grafted with various monomer systems. The nature of the grafting reaction to these fibers is usually such as to form a surface coating over the substrate polymer; the surface characteristics, such as dyeing, are therefore usually those of the grafting system.

It is very doubtful that any blends of two polymers, or of chemically different copolymers, can from a thermodynamic point of view ever be fully compatible. Even most block or graft copolymer systems therefore show microphase separation which will be typical for the properties of a given system. Chemical modification, as discussed in the earlier part of this chapter, will in general lead to the formation of polymeric single phases, provided the reaction has been carried out homogeneously. The choice need not be restricted, however, to just these two approaches, since chemical modification can be carried out after block or graft copolymerization or vice versa. Very little has been published on such a consecutive use of these two physically different ways of modifying polymers chemically, so that there is considerable scope for developing new modifications of long-established rubbers, as well as generally for changing old into new polymers.

REFERENCES

1. V. C. Haskell, in "Encyclopedia of Polymer Science and Technology," Vol. 3 (H. S. Mark et al., eds.), p. 60. Wiley (Interscience), New York, 1965.
2. G. N. Bruxelles and N. R. Grassie, in "Encyclopedia of Polymer Science and Technology," Vol. 3 (H. S. Mark et al., eds.), p. 307. Wiley (Interscience), New York, 1965.
3. B. P. Rouse, in "Encyclopedia of Polymer Science and Technology," Vol. 3 (H. S. Mark et al., eds.), p. 325. Wiley (Interscience), New York, 1965.
4. W. Jarowenko, in "Encyclopedia of Polymer Science and Technology," Vol. 12 (H. S. Mark et al., eds.), p. 787. Wiley (Interscience), New York, 1970.
5. A. I. Yakubchik, B. I. Tikhominov, and V. S. Sumilov, Rubber Chem. Technol. 35, 1063 (1962).
6. J. I. Cunneen and M. Porter, in "Encyclopedia of Polymer Science and Technology," Vol. 12 (H. S. Mark et al., eds.), p. 304. Wiley (Interscience), New York, 1970.
7. C. F. Reed and C. Horn, U.S. Patent 2,046,090 (1936).
8. A. Nerasian and D. E. Anderson, J. Appl. Polym. Sci. 4, 74 (1960).
9. R. W. Rees (to DuPont), Can. Patent 674,595 (1963).
10. R. J. Ceresa, unpublished observations.
11. P. Christen and E. M. Guidal (to Rhone-Poulcc, S.A.), U.S. Patent 3,119,786 (1960).
12. R. J. Ceresa, in "Encyclopedia of Polymer Science and Technology," Vol. 2 (H. S. Mark et al., eds.), p. 502. Wiley (Interscience), New York, 1965.
13. R. J. Ceresa, Ph.D. Thesis, Univ. of London, London, 1959.
14. M. S. Akutin, Rubber J. Int. Plast. 138, 730, 732 (1960).
15. R. J. Ceresa, Polymer 2, 213 (1961).
16. R. J. Ceresa, Polymer 1, 477, 488 (1960).
17. M. Szwarc, Rubber Plast. Age 39, 859 (1958).
18. W. H. Janes, in "Block Copolymers" (D. C. Allport and W. H. Janes, eds.), p. 62. Appl. Sci. Publ., London, 1973.
19. R. J. Ceresa (to Koppers Co.), U.S. Patent 3,110,695 (1963).
20. J. E. Morris and B. C. Sekhar, Proc. Int. Rubber Conf., Washington, D.C. p. 277 (1959).
21. E. G. Cockbain, T. D. Pendle, and D. T. Turner, J. Polym. Sci. 39, 419 (1959).
22. F. M. Merrett, Trans. Faraday Soc. 50, 759 (1954).

BIBLIOGRAPHY

Allen, P. W., ed., "Polymer Characterization." Butterworth, London, 1967.
Allport, D. C., and Janes, W. H., eds., "Block Polymers." Appl. Sci. Publ., London, 1973.

Bateman, L., ed., "The Chemistry and Physics of Rubberlike Substances." Wiley, New York, 1963.

Ceresa, R. J., "Block and Graft Polymers." Butterworth, London, 1962.

Ceresa, R. J., ed., "Block and Graft Copolymerization," Vol. 1. Wiley, New York, 1973.

Ceresa, R. J., ed., "Block and Graft Copolymerization," Vol. 2. Wiley, New York, 1976.

"Encyclopedia of Polymer Science and Technology." H. S. Mark *et al.*, eds., Wiley (Interscience), New York, 1965–1968, with particular reference to:

"Anionic Polymerization," Vol. 2, pp. 95–137; "Block and Graft Copolymerization," Vol. 2, pp. 485–528; "Cellulose," Vol. 3, pp. 131–226; "Cellulose Graft Copolymers," Vol. 3, pp. 242–283; "Cellulose Derivatives," Vol. 3, pp. 291–306; "Cellulose Esters, Inorganic," Vol. 3, pp. 307–325; "Cellulose Esters, Organic," Vol. 3, pp. 325–419; "Cellulose Ethers," Vol. 3, pp. 476–549; "Copolymerization," Vol. 4, pp. 165–243; "Elastomers, Synthetic," Vol. 5, pp. 406–481; "Emulsion Polymerization," Vol. 5, pp. 801–859; "Epoxidation," Vol. 6, pp. 83–102; "Ethylene Sulphonic Acid Polymers," Vol. 6, pp. 455–465.

Fettes, E. M., ed., "Chemical Reactions of Polymers." Wiley (Interscience), New York, 1964.

Ham, G. E., ed., "Kinetics and Mechanisms of Polymerization," 3 vols. Dekker, New York, 1967.

Lenz, R. W., "Organic Chemistry of Synthetic High Polymers." Wiley (Interscience), New York, 1967.

Natta, G., ed., "Stereoregular Polymers and Stereospecific Polymerization," 2 vols. Pergamon, Oxford, 1965.

Stern, H. J., "Rubber, Natural and Synthetic," 2nd Ed. Maclaren, London, 1967.

Wilson, J. E., "Radiation Chemistry of Monomers, Polymers and Plastics." Dekker, New York, 1974.

Chapter 12

Elastomer Blends

P. J. CORISH

DUNLOP RESEARCH CENTRE
BIRMINGHAM, ENGLAND

I. Introduction

Whereas many organic liquids at room temperature are fully or partially miscible, polymers, by virtue of the length of their molecular chains, are not easily miscible. It was demonstrated by Walters and Keyte [1] some years ago that elastomer blends are never truly homogeneous, showing discrete areas of each elastomer varying from ~ 0.5 μm upward, depending on methods of mixing, elastomer viscosity, crystallinity, etc. The type of result obtained is shown in Fig. 1, which illustrates phase-contrast optical microscopy of an elastomer/elastomer blend, viz., a mill-mixed blend of natural rubber with SBR. This early work heralded the beginning of a new look at elastomer blends from both technical and economic points of view.

All rubbers have shortcomings in one or more properties; thus there are technical reasons for blending two rubbers in order to try to obtain the right

489

Copyright © 1978 by Academic Press, Inc.
All rights of reproduction in any form reserved.
ISBN 0-12-234360-3

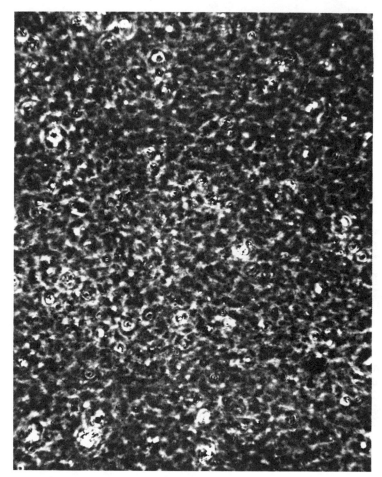

Fig. 1. Mill-mixed blend of natural rubber with SBR showing interlocking zone pattern (magnification: 1400). (From Walters and Keyte [1].)

compromise. Processing difficulties may also be overcome by blending, and appreciable differences in the prices of different rubbers emphasize economic reasons for blends. There is a need to understand which polymers can be successfully blended and what underlying factors influence the results. This chapter will attempt to present an account of the present state of knowledge on elastomer blends and to indicate future development areas where further research would seem desirable.

The first difficulty is that mixtures and blends occur at various scales in the range of materials employed in the rubber industry. A composite article such as a tire is a mixture of wire (metal), textile cord (organic fiber), and rubber compounds. The rubber compound itself is a mixture of elastomer, filler, and usually, extender. The elastomer phase itself may be a mixture of two or more rubbers.

On a smaller scale, the elastomer may contain blocky segments or crystalline portions along the polymer chains which act as reinforcing agents or stiffeners. These may be similar to the repeating units of the elastomeric part, e.g., iso-tactic–atactic block polymers of polypropylene or polypropylene oxide. They may, however, be different from the elastomeric part, e.g., the range of styrene–butadiene thermoplastic copolymers described by Shell and others.

Two types of elastomer may of course, be copolymerized, or one elastomeric type may be grafted onto another. It is possible that selection of catalyst and monomers could cause simultaneous polymerization of two polymers.

All the foregoing systems are different facets of the broad concept of blends, but some selection is obviously needed. In this chapter, the greater emphasis will be placed on elastomer/elastomer blends which will involve their preparation from the constituent elastomers. Reference will also be made to elastomer/plastic blends prepared from separate materials and by block and graft type reactions.

The accompanying tabulation lists abbreviations for many of the substances and a few organizations discussed.

NR	Natural rubber
SBR	Styrene–butadiene rubber
BR	Polybutadiene rubber
CR	Chloroprene rubber
IR	Isoprene rubber, synthetic polyisoprene
IIR	Isobutene–isoprene rubber, butyl rubber
EPM	Ethylene–propylene copolymer
EPDM	Ethylene–propylene–diene terpolymer
NBR	Acrylonitrile–butadiene rubber
Thiokol	Polysulfide rubber
SKD	Polybutadiene rubber (Russian)
SKD 10	Butadiene–methacrylic acid copolymer (Russian)
SKI-3	High-cis synthetic polyisoprene (Russian)
SKI	Lithium-catalyzed synthetic polyisoprene (Russian)
SKN	Acrylonitrile–butadiene rubber (Russian)
Nairit	Chloroprene rubber (Russian)
PVC	Polyvinyl chloride
EVA	Ethylene–vinyl acetate copolymer
ISAF	Intermediate super abrasion furnace black
EPC	Easy processing channel black
TMTD	Tetramethylthiuram disulfide
MBTS	Dibenzthiazyl disulfide
DOTG	Di-ortho-tolylguanidine
JSR	Japan Synthetic Rubber Co. Ltd.
RAPRA	Rubber and Plastics Research Association of Great Britain
ICI	Imperial Chemical Industries
G.D.Ch.	Gesellschaft Deutscher Chemiker

II. Elastomer/Elastomer Blends

A. BLENDING METHODS

From the practical point of view, the first consideration must be the available methods for elastomer blending. These encompass latex blending, solution blending, combinations of these methods, conventional mechanical and mechanochemical mixing methods, and powder mixing in special cases.

1. Latex

Since the coagulation of a mixture of latices may be a random process entirely dependent upon soap concentration [2], latex blending offers the possibility of finer-scale dispersions than solution blending. It also appears to receive greater attention as a commercial manufacturing process [3–5].

Angove [3] has reviewed the field of latex blends with particular attention to processing, properties, and economics. More recently [6, 7], Blackley and Charnock have investigated the viscosity and mechanical stability of NR/SBR latex blends, while King has attempted to improve their compatibility by modification of the stabilizing system [8].

Except for latex foam or thread applications, where use of blends may provide cheaper compounds or advantages in heat resistance or solvent swelling, latex blending would not generally appear to have any advantages over other methods. A JSR patent [9], however, claims that cis BR latex blended and coagulated with SBR or NR latex gives a more homogeneous dispersion of ISAF black. The blending of cross-linked polyurethane–urea and linear polybutadiene–acrylonitrile as aqueous emulsions [10], recently used to effect the formation of an interpenetrating polymer network, is another exception to this generality.

2. Solution

The solution blending of incompatible polymers is hindered to some extent by the high mobility of the solutions prior to recovery, which permits a relatively gross separation of the polymers into their constituent phases. Such blends may be macroheterogeneous rather than microheterogeneous.

Walters and Keyte [1] in an early study of gum rubber blends observed that droplets of SBR solution suspended in natural rubber solution could be broken down by agitation, but rapidly recoalesced on standing. Evaporation of the solvent produced a coarse polymer mixture comprising layers of SBR surrounded by NR. Livingston and Rongone [11] have considered such blends to resemble mill-mixed compositions in the early stages of mixing. Bohn [12] has suggested that the use of a nonsolvent to effect recovery of the elastomer mixture may produce similar macroheterogeneity by virtue of differing concentration limits for the component polymers. Shundo et al. [13], however, have reported satisfactory solution blending of NR and SBR by coprecipitation in methanol with vigorous agitation.

Solution blending of elastomers normally produced by solution polymerization processes is obviously advantageous, since a blended rubber can be produced "on site" and advantages can accrue from the incorporation of black and oil in the blend, producing a masterbatch. It would appear necessary that a certain level of work or power be put into a solution blend, especially one involving carbon black masterbatching, to achieve a satisfactory dispersion [14]. It is claimed that work should be done on the mixture of filler dispersion and polymer solution at a rate of at least 20 W, until the total amount of work is at least 15 J/cc of the mixture. Rapid evaporation or removal (or both) of solvent appears to be a recommended procedure to minimize any heterogeneity, especially that due to black redistribution during drying.

3. Solution and Latex

A number of processes have been developed for preparation of elastomer blends from combinations of polymer solutions and latices [15–17]. Claimed advantages of the processes are simplification of the polymer recovery procedures, and products with better dispersion of fillers than would be obtained from a mechanical mixture of dry rubbers.

Combinations of latex and solution blending techniques to produce filler masterbatches are involved in two other processes. The Columbian Hydrodisperse method [18] is used for masterbatching solutions of elastomers in organic solvents (as produced by solution polymerization) with black dispersions in water. The black passes into the organic solvent phase, thus producing a masterbatch [19].

Another method studied [20] involves the masterbatching of black dispersed in an organic solvent with a rubber latex. Use of oppositely charged soaps is necessary, the black dispersion in solvent being dispersed itself in water using a soap opposite in polarity to that on the latex particle. If a solution of a polymer containing a black dispersion is used instead of the pure solvent and black, a blend masterbatch is produced.

4. Mechanical

It has been recognized [21] that, while two polymers may be virtually mutually insoluble, blends may be produced which are macroscopically homogeneous and have useful properties, provided mechanical mixing is sufficiently intense and provided also that the viscosities after mixing are sufficiently high to prevent gross phase separation. Such blends are termed microheterogeneous. Fan et al. [22] reviewed mixing of solids over the period 1958–1969, covering many aspects of such empiricism.

The high shearing forces required to blend high molecular weight elastomers are such that mechanical blending is confined to open roll mills, internal mixers (e.g., Banbury), and extruders. With regard to open roll mills, attention has been called to the desirability of similar polymer viscosities (at the mixing temperature) for ease of dispersion, and to the mastication of NR prior to blending with synthetic rubbers [23]. Where elastomers of different cure rates are to be blended, a masterbatch process may be desirable [23, 24]; that is, the component polymers are precompounded with vulcanizing agents and additives, and then the individual stocks are blended in the desired proportions. This technique has found general acceptance in mechanical blending technology [11, 13]. However, there are exceptions, such as the blends of chlorobutyl and nitrile rubbers studied by Evans and Partridge [25], who found that mixing compounding ingredients directly into a blend of the elastomers showed advantages over previously prepared masterbatches.

Shundo et al. [13] have compared the use of roll mills and Banbury mixers in the preparation of NR/SBR blends and found the mills to furnish the more uniform compositions. On the other hand, Hess et al. [26] have reported wider distributions of polymer zone size in mill mixtures, according to mixing time, than for Banbury mixtures, confirming the less critical nature of Banbury blending reported elsewhere [23].

Ways of manipulating mechanical blending with regard to filler distributions are discussed in a later section.

5. Mechanochemical

Under certain conditions the mechanical working of a mixture of polymers can lead to interpolymerization. For example, when the tendency of the polymeric radicals produced by mechanical rupture is to recombine rather than to disproportionate, block copolymer formation may be favored [27]. Alternatively, when sites for hydrogen or halogen abstraction are present in one of the polymeric constituents, graft copolymer formation may be favored.

Angier and Watson [28, 29] have studied interpolymers of NR, CR, NBR, SBR, and BR produced by cold mastication under nitrogen in an internal mixer. Evidence for interpolymerization was established by finding differences in gel contents of the products after mastication as compared with those of simple polymer mixtures. Slonimskii and Rečtsova [30], in an investigation of SBR and BR blends, found that in the absence of oxygen and with polymers of high purity (i.e., essentially free from radical inhibitors or acceptors) mechanical blending produced materials whose property–composition relationships differed from those of simple mill mixtures. Benzene solutions of the blends did not separate into discrete phases, and a considerable portion of the products remained insoluble. This was considered to indicate chemical grafting of the constituent polymers.

Reviews of mechanochemical blending have been made by Ceresa [31] and more recently by Allport [32]. The latter author and Burlant and Hoffman [33] emphasize the complexity of mechanochemical interpolymer formation and the difficulty in isolating and identifying pure products.

For filled elastomers, the presence of the filler leads to bound rubber formation, discussed more fully later.

6. Powdered Rubbers

The use of powdered rubbers is increasing and the possibility of blending a powdered rubber with a solid rubber by conventional mixing techniques is obvious. Blending two rubbers—both in powdered form—is also possible.

B. CHARACTERIZATION OF BLENDS BY PHYSICAL AND CHEMICAL TECHNIQUES

Before discussing the structural aspects of blends, it is appropriate to describe their characterization. The methods, which apply generally to both elastomer/elastomer and elastomer/plastic combinations, vary in effectiveness depending on which combination is involved. They range from microscopy and solubility methods, which characterize the compatibility of polymers on a gross scale, to spectroscopic techniques which identify the polymer types and bonds involved.

1. Microscopy

Microscopy methods have been used for many years, literally to see what component parts a polymer blend contains. The optical microscope, for example, was used extensively by Walters and Keyte [1] to study the blending of gum and lightly compounded elastomers involving the use of phase contrast to differentiate the components. When filled elastomer blends were investigated, workers such as Hess et al. [26] and Marsh et al. [34] resorted to electron microscope techniques, including differential swelling of the component phases to provide contrast. Electron microscopy has also been used by Lewis et al. [35] in autoradiographic studies of both gum and black-filled elastomer blends, while a combination of electron microscopy with DTA and GPC has been employed by Callan et al. [36] in compatibility investigations.

Elastomer/plastic blends are somewhat easier to identify than elastomer/elastomer blends, particularly in electron microscopy, since staining techniques (e.g., osmium tetroxide) can be employed. Moore [37] and Keskkula and Traylor [38] have used such methods to study rubber-reinforced polystyrenes, while Miyamoto et al. [39] have used them together with dynamic loss (tan δ) measurements to investigate the structure and properties of styrene–butadiene–styrene (S-B-S) block copolymers.

More recently, Smith and Andries [40] have used two new methods of sample preparation for electron microscopic examination of polymer blends. The first is an ebonite method where the blend is cured and hardened by a sulfur–sulfenamide–zinc stearate mixture and then microtomed. The second is a cryogenic method where the blend is microtomed in the frozen state and then stained with osmium tetroxide, which stains only unsaturated elastomers. Sharp, clear phase differentiation was obtained with all the blends studied.

2. Solubility

Solubility differences between components of polymer blends have been utilized in several ways. Voyutsky [41] employed a compatibility test for two rubbers which consisted simply of mixing and stirring 10% solutions of the rubbers at room temperature. Breitenbach and Wolf [42] studied the mutual influence of polymers of different molecular weight on solubility, while Chen [43] investigated polymer miscibility in mixed organic liquids. Schnecko and Caspary [44] determined the compatibility of some binary rubber mixtures by dilute solution viscosity measurements, while Burrell [45] discussed the challenge of the solubility parameter concept. Although these investigations of solubility differences between polymers give some insight into their blending potential, it should be noted that measurements in dilute solution do not represent the state of matter that exists in practice where two or more bulk polymers are blended together.

Feldman and Rusu [46] studied the compatibility of PVC with other polymers using a viscometric method involving coaxial cylinders. Englert and Tompa [47] used the incompatibility of polymers for countercurrent fractionation, while Menin and Roux [48] used gel permeation chromatography for the determination of the composition of two-polymer mixtures including BR and polyisobutylene.

3. Optical Properties

Optical properties of polymer blends have been used by some authors to characterize blends. Kraus and Rollmann [49] used stress-optical properties in combination with dynamic data to study polyblends. Kuhn et al. [50] used light scattering methods to investigate polymer mixtures, especially their incompatibility.

4. Thermal and Thermomechanical Analysis

Several different types of thermal method have been used to characterize polymer blends, particularly in the plastic/elastomer area where the existence of high-T_g (glass transition temperature) and low-T_g components eases the problems of experimental differentiation. The methods used have varied in their inherent frequency from dilatometric measurements (static) to dynamic tests at high frequencies.

Dilatometry has been used extensively by various authors to correlate T_g with chemical structure: these include Boyer [51], Shen and Eisenberg [52], Weyland et al. [53], and Roller and Gillham [54]—the last-named specifically for high-temperature elastomers. Manabe and Takayanagi [55] have considered the relationship of free volume, which involves T_g determinations, to polymer blends. Chandler and Collins [56] have studied the multiple T_gs in NBR copolymers, while Hoffmann et al. [57] have investigated the molecular structure of anionic butadiene–styrene copolymers using T_g data.

Corish [58] has described both static and dynamic methods, the latter involving rolling ball spectrometer studies (see Fig. 2, which shows the basic instrumental setup [59]); results will be shown later. Dynamic damping measurements have been used by de Decker and Sabatine [60] for elastomer blends, as well as by Keskkula *et al.* [61] to study the significance of the rubber-damping peak in rubber-modified polymers and by Davies [62] to derive a new formula for the elastic constant of two-phase composite materials.

Further techniques include thermal analysis as used by Miller [63] for polymers, and by Klempner and Frisch [64] for interpenetrating networks. Dif-

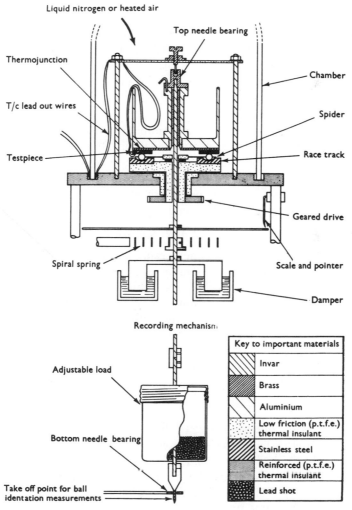

Fig. 2. The rolling ball spectrometer. (From Cheetham [59].)

ferential scanning calorimetry (DSC) also provides useful information and has been used by Landi [65] to study phase stability in heterogeneous compositions of NBR, and by Seymour and Cooper [66] to investigate polyurethane block polymers.

5. Dielectric Relaxation

Dielectric relaxation measurements have provided novel data for characterization of blends. Fujimoto and Yoshimiya [67] have studied blends of cis BR with NR or SBR, while North and Reid [68] have investigated a series of heterophase polyether–urethanes.

6. Infrared Spectroscopy

The complementary nature of evidence from various techniques is illustrated by the work of Gesner [69], in which infrared (ir) spectroscopy, electron microscopy, microphotometry, and solvent elution and precipitation are used to characterize polyblends. More traditional spectroscopic-plus-chemical methods have been used by Clark and Scott [70] to characterize sulfur-cured SBR/BR blends.

Yamagi et al. [71] have heat-treated rubber blends at $200°C$ to obtain carbon disulfide-soluble polymer for infrared analysis. By calibrating against binary blends of known composition, analysis of ternary blends of NR, SBR, and BR has been possible. The use of polarized radiation has enabled Estes et al. [72] to characterize segmented polyurethane elastomers—an example of an elastomer/plastic intramolecular blend.

7. Nuclear Magnetic Resonance (NMR) Spectroscopy

NMR studies of cured elastomers, especially those containing fillers, have been limited due to line broadening, a restriction which reduces the effective information. Thus, Fujimoto [73] has studied heterogeneity in filled rubber systems using NMR in combination with dynamic mechanical measurements and DSC data. Legrand [74] has investigated molecular motion in block copolymers by NMR, a simpler problem.

Some advances may well be expected in this area, perhaps employing spinning at the magic angle to narrow the NMR line widths, as recently described by Schaefer et al. [75] for a polyisoprene compound.

8. X-Ray Analysis

X-ray methods, due to the size of the domains necessary to give characteristic X-ray patterns, have been restricted in the main to rubber/plastic blends, more especially to block polymers. Lewis and Price [76] used X-ray and electron microscope methods to study the morphology of S-B-S block copolymers; Douy and Gallot [77] have investigated the organized structure of B-S-B copolymers by the same two techniques plus DSC. Legrand [78] has made a more general study of domain formation in block copolymers, while McIntyre and Campos-Lopez [79] have restricted their investigation to the macrolattice of a triblock polymer.

9. Gas Chromatography

Gas chromatographic techniques were used by Guillet [80] to study polymer structure; in these studies a pulse of molecules was sent along a narrow tube coated with the polymer under investigation. Interaction in collision with the wall altered the translational velocity along the tube. This method could have potential for the study of blends. More conventional analytical work on three-component blends (NR, SBR, and BR) has been described by Sugiki et al. [81]; it involved pyrolysis and gas phase chromatography (GPC) using an rf induction apparatus. An accuracy of 5% is claimed.

10. Miscellaneous

Studies of the molecular structure of SBR's by infrared measurements, refractive index, viscosity, and light scattering have been reported by Hoffmann et al. [57]. Detailed discussions are also presented of T_g, flow behavior, and structural viscosity, as well as dynamic behavior. The chemical heterogeneity of copolymers, involving experimental investigation of compatibilities, is reported by Kellinsky and Markert [82], and Heinze et al. [83] have carried out similar work on physical characterization of copolymers and polymer mixtures.

C. STRUCTURAL ASPECTS

Microheterogeneity in elastomer blends was originally studied by Walters and Keyte [1] using phase-contrast microscopy. Results from two-component mill and solution mixtures of NR, SBR, BR, CR, IR, and IIR were examined. In no case was mixing on a molecular scale observed. Phase micrographs revealed the presence of discrete domains of each elastomer in the blends, and microphotometer measurements indicated their areas to be in approximately the same proportion as the volume composition of the blends. Generally, disperse zones 1–2 μm in diameter were observed in 50/50 wt % mixtures, although in certain instances interlocking networks 0.5 μm in diameter were detected.

The observations of Walters and Keyte have been confirmed and extended by more recent investigations using phase-contrast and electron microscopy. Callan et al. [84], for example, have demonstrated the phase structure of butyl/ EPDM blends to consist of discrete zones of the minor polymer dispersed in a continuous matrix of the major polymer. Phase inversion occurred at a 50/50 blend ratio; the mixture then consisted roughly of interpenetrating polymer networks. Hess has displayed similar structure relationships in NR/BR blends

[26] and Powell in a series of SBR/EPDM blends [85]. Marsh has reported the absence of interpenetrating networks in many 50/50 blends [34], one component apparently preferring to adopt a discontinuous form. This absence was considered to be related to differences in the response of the elastomers to milling, in that the component with the lower viscosity tends to form the continuous phase. A blend of NR and CR was particularly unusual in that natural rubber remained the continuous phase when it constituted only 25 wt % of the blend. More recently, the morphology of an NR/CR/EPDM triblend has been investigated, and the structure of the components resolved [86].

Marsh et al. [34] have shown that whereas IR/SBR and IR/BR blends are microheterogeneous, SBR/BR blends are homogeneous, at least to the extent that SBR itself is homogeneous. This has been confirmed in an additional study in which the zone sizes observed during phase microscopic examination of NR, SBR, IR, BR, NBR, CR, and EPDM blends were reported. Domain sizes ranged from 0.2 to 30 μm, but of the 11 elastomer combinations examined, only SBR/BR appeared microhomogeneous. The differing observations of Walters and Keyte [1] and Yoshimura and Fujimoto [87] suggests that homogeneity in the SBR/BR system may be intimately related to the molecular characteristics of the particular polymers involved. Such a conclusion may not be unexpected, since the requirements for molecular compatibility in high molecular weight polymers are stringent [21, 88, 89]. The mutual solution of two materials demands that the free energy of mixing be negative, a situation brought about by exothermic mixing and/or a large entropy of mixing. In the great majority of cases, though, the mixing of polymers is endothermic [90], while the entropic contribution is small by virtue of the high molecular weights. In consequence, mutual solution occurs only over a very narrow range of enthalpy changes, corresponding to extremely small cohesive energy density differences [21]. It may be said that, as far as chemically dissimilar high polymers are concerned, incompatibility is the rule and compatibility the exception.

An attractive demonstration of these principles is found in Livingston and Rongone's study of mixtures of styrene–butadiene rubbers of differing styrene contents [11]. Where differences in styrene content were small, the mixtures were homogeneous, whereas where differences were greater than 20%, the mixtures were heterogeneous. Shundo et al. [91] likewise have demonstrated that in binary blends of NBR's, copolymers of similar acrylonitrile content display superior properties to those of differing acrylonitrile content, the total concentration being constant. Satake et al. [92] have studied unvulcanized blends of high- and low-cis polybutadiene at $-40°$C, a temperature at which high-cis polybutadiene crystallizes. A sigmoid dependence of hardness and compression modulus on blend ratio was noted, indicating a heterogeneous nature for this chemically similar mixture. This conclusion was confirmed by the crystallization of the high-cis polybutadiene component, observed by dilatometry and DSC.

Bohn [12] has suggested that equilibrium may not always be reached in mechanical blending processes, particularly in time-limited mixing operations. However, studies by Marsh et al. [34] have indicated that where compatibility exists, this is achieved in a relatively short time, of the order of 5 min. Additionally, Rehner and Wei [24] have demonstrated that the structure of mill-mixed chlorobutyl/polybutadiene rubber blend does not change appreciably with milling times of 20 min to 20 hr.

Further conclusions of Walters and Keyte [1] with regard to elastomer blending were that the fineness of dispersion varied according to the polymer viscosities and the degree of mixing involved. Disparate polymer viscosities were considered to produce coarse heterogeneity by virtue of an aggregation of the stiffer polymeric component in the softer component. Solubility parameters were not considered to be important. Evidence to support this theory has been supplied by Callan et al. [84] and Marsh et al. [34], although inconsistencies in the former work suggested that additional polymer characteristics were also influential. Powell [85] has suggested that chain branching may be a factor in the breakdown of polymer during blending, modifying the effective viscosity and the resultant scale of dispersion accordingly. Gardiner [93] has interpreted similar dispersion relationships in solution blends as a function of surface tension.

Hess et al. [26] have reported a reduction in disperse polymer domain size in NR/BR blends with increased milling time, confirming the observations of Walters and Keyte. Powell [85] has shown that at equal mixing time the zone size decreases as the mill roll separation (nip) decreases. A limiting zone size of ~0.5 μm was reported, in agreement with others. Rehner and Wei [24] have reasoned that the physical processes involved in mill mixing (or similar mechanical blending process) probably prevent breakdown of the elastomer zones beyond this limiting size.

In summary, it may be stated that with few exceptions elastomer blends are microheterogeneous, the continuous phase being either the polymer in highest concentration or the polymer of lowest viscosity. The sizes of the zones of disperse phase vary according to the polymer characteristics, viscosity differential, and manner of blending. Typically, they may be of the order of 1–5 μm, although in favorable circumstances zones of 0.5 μm may be obtained. Phase inversion, the reversal of the continuous and disperse phases, occurs via an interpenetrating network of polymers, often at 50/50 blend ratio, but all such transitions depend very much on polymer viscosity and type.

Little systematic work on the relationship between scale of dispersion and physical properties has been undertaken. Walters and Keyte [1] prepared a series of NR/SBR and NR/BR blends of varying zone size (2–100 μm) by a process of solvent mixing followed by progressive milling. The effect on physical properties was minimal. Tear strength fell as the zone size decreased, but the difference was small and may have been caused by increased breakdown of the polymer under the extended milling conditions required for preparation of the finer dispersion. In similar manner, Livingston and Rongone [11] studied the vulcanizate properties of BR/SBR and BR/IR blends as a function of degree of mixing. For both systems, physical properties varied with degree of mixing up to 100 mill passes, but no microscopic study was undertaken and a correlation of these variations with zone size cannot be made. Powell [85] has demonstrated that the ozone resistance of SBR/EPDM blends increases as the fineness of dispersion increases. On the basis of Andrews' microcrack termination theory [94], this was attributed to an increase in the number of EPDM zones and a consequent reduction in their separation. The effect of reduced zone separation is to restrict ozone cracks in the SBR phase to microscopic dimensions, thereby increasing the resistance of the vulcanizate to macroscopic failure.

The structural aspects of unfilled elastomer blends just discussed, involving considerations of differences in viscosity between the components, continuous versus disperse phase, etc., are likewise relevant in filled elastomers. However, a major and sometimes overriding factor is the dispersion (i.e., distribution and disaggregation) of the filler. Also, the distribution of soluble compounding ingredients, especially curatives, may exert significant influence on the performance of the vulcanized product. These factors are discussed in detail in the following sections.

D. DISTRIBUTION OF SOLUBLE COMPOUNDING INGREDIENTS

Using optical microscopy, attenuated total reflectance infrared spectroscopy, and microinterferometry, Gardiner [95, 96] has demonstrated that diffusion of common vulcanizing ingredients, such as sulfur, TMTD, MBTS, and DOTG occurs from compounded low unsaturated rubbers such as IIR and EPDM to uncompounded high unsaturation rubbers such as NR and SBR; also to a lesser extent from compounded high unsaturation rubbers to uncompounded low unsaturation rubbers. Migration was observed to commence in a very short time (e.g., 3 sec at 307°F), indicating that a diffusion gradient may be produced between dissimilar elastomers considerably before significant vulcanization occurs. Since the solubility of the common curatives is greater in high unsaturation elastomers than in low unsaturation elastomers, migration from the latter to the former may be inevitable, even when concentrations are equal initially [93]. Such migration will be accentuated by higher curative reaction rates in the higher diene elastomers, causing excess material to be drawn from adjacent phases.

Figures 3–5, in which curative concentration is plotted against distance from the interface, are typical of the results obtained. For the case of sulfur diffusing from NR to SBR (Fig. 3) considerable transfer occurs and significant concentrations of sulfur build up in the SBR phase, in the 20 μm next to the interface. Figure 4 shows the completely different character of diffusion of

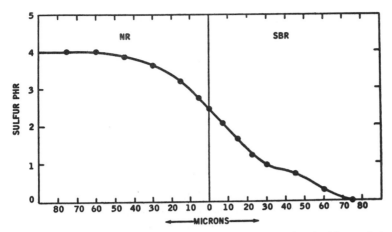

Fig. 3. Diffusion of sulfur from NR to SBR (150°C, 9 sec). Reprinted with permission from Gardiner, *Rubber Chem. Technol.* **42,** 1058 (1969). Copyright by the American Chemical Society.

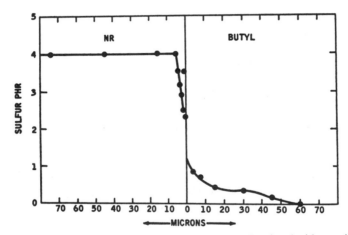

Fig. 4. Diffusion of sulfur from NR to butyl (150°C, 9 sec). Reprinted with permission from Gardiner, *Rubber Chem. Technol.* **42,** 1058 (1969). Copyright by the American Chemical Society.

sulfur between NR and butyl. It would appear that diffusion control by the butyl is taking place and limiting NR diffusion. MBTS diffusion is similar to sulfur diffusion but the diffusion coefficients tend to be an order of magnitude smaller, as might be expected with a larger molecule, having low solubility in each rubber compared to sulfur. Figure 5 typifies the behavior found, in this case relating to diffusion of MBTS from SBR to BR.

The practical consequences of cure rate, solubility, and diffusion differences in elastomer blends are curative imbalance between the component phases and

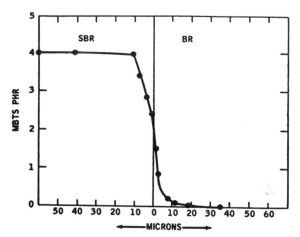

Fig. 5. Diffusion of MBTS from SBR to BR (150°C, 16 sec). Reprinted with permission from Gardiner, *Rubber Chem. Technol.* **42**, 1058 (1969). Copyright by the American Chemical Society.

associated over- and undercure [1, 93, 97]. In such circumstances vulcanizate properties may not approach the levels attainable by either of the elastomers alone. Attention to mixing schedules and masterbatching procedures has been demonstrated to offer advantages in certain instances, according to vulcanization recipes and individual accelerator influence [24]. Also, vulcanizing systems which function independently of polymer unsaturation have been suggested to permit the development of highest mechanical performance in elastomer combinations of differing unsaturation [98].

Interphase cross-linking of blends, particularly those containing EPDM, has received recent attention. The incompatibility of EPDM and high diene elastomers has been considered to preclude the possibility of mutual molecular chain interactions and to make co-cross-linking of the polymers difficult [99]. However, it has been reported that co-cross-linking can be achieved if the cure rates of the polymers are virtually identical and if the cross-link structure is predominantly polysulfidic [99]. In contrast, interfacial bonding in chlorobutyl/polydiene blends has been reported only in the presence of efficient thiuram disulfide and thiuram tetrasulfide vulcanization systems [100]. Chemical probe analysis has established that interfacial bonds are associated with a preponderance of monosulfidic cross-links, formed during the initial stages of vulcanization rather than by the maturation of polysulfidic cross-links. In other investigations [101], co-cross-linking in EPDM/CR blends has been attributed to the independent consumption of differing curatives in each rubber phase. With less disparate combinations of elastomers (i.e., SBR/BR and NR/BR), Yoshimura and Fujimoto [87] and Corish [58] have demonstrated that the development of interphase cross-linking causes a merging of the dynamic mechanical loss peaks characteristic of the individual polymers.

Separate loss peaks, which were observed in uncured blends of the polymers, merged upon progressive vulcanization to provide a pseudo-single-phase system. The formation of a mechanically "compatible" blend from an intrinsically incompatible polymer combination is thus effected. This is illustrated in Fig. 6, which shows results obtained from a rolling ball spectrometer for a 50/50 blend of cis BR with natural rubber.

The use of radiochemical methods to estimate distribution of sulfur in black-loaded rubber blends was reported by Corish and Palmer [102] and the control of such distribution was indicated as a means of varying blend properties. This point was also recognized by Amidon and Gencarelli [103] and Mastromatteo *et al.* [104, 105], who claimed that use of long-chain hydrocarbon dithiocarbamate accelerators is beneficial in sulfur cures of EPDM's with highly unsaturated rubbers. Vulcanizates of high tensile strength were obtained. The problem of covulcanization with EPDM in blends with higher unsaturated elastomers has also received attention from Ishitobi *et al.* [101], as discussed previously, and from Kerrutt *et al.* [106].

Recent evidence [107] suggests that the superiority of long-chain hydrocarbon accelerators (e.g., thiuram sulfides) in the covulcanization of rubber blends derives from their more similar solubilities in high and low unsaturation rubbers than the short-chain hydrocarbon homologs. Cure rate differences may be of secondary importance.

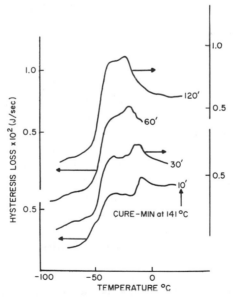

Fig. 6. Hysteresis loss of a blend of polybutadiene and natural rubber as a function of temperature and cure time. Reprinted with permission from Corish, *Rubber Chem. Technol.* **40**, 324 (1967). Copyright by the American Chemical Society.

Some recent work by Baranwal and Son [108] provides an alternative way of co-curing blends of EPDM and diene rubbers for improved properties. By grafting accelerators onto EPDM, excellent cure compatibility between EPDM and natural rubber was obtained, leading to much improved tensile properties and to reduced heat buildup.

Another important compounding ingredient whose migration has been studied extensively is extender oil. Corman *et al.* [109] have investigated the migration of extender oil in natural and synthetic rubbers by radiochemical methods. Their findings were that variations in carbon black type and loading have no detectable effect on migration: the most important factor is the elastomer matrix, which for the oils studied provides the order of diffusion coefficients D in the accompanying tabulation.

	Approximate D (cm^2 sec^{-1})	Temperature (°C)
cis BR	6.4×10^{-7}	100
NR	3.5×10^{-7}	100
EPDM	2.6×10^{-7}	100
SBR	1.9×10^{-7}	100

Activation energies for the diffusion process are also quoted.

Further work by Deviney and Whittington [110] on the dependence of diffusion rates on oil concentration and elastomer blend ratio showed that for two-component blends involving NR, SBR, and BR, diffusion coefficients can be accurately calculated from the values for each elastomer and its relative concentration in the blend.

Antidegradants play an important part in the resistance of elastomers to service under severe environmental conditions of temperature or stressing. It appears from limited unpublished studies [111] that antidegradants are liable to migrate in the same way as curatives and extender oils. In tires which initially contain different antidegradants in treads and casings, the antidegradants diffuse so as to "equilibrate" themselves in the tire. This process is initiated during curing operations—presumably due to the high temperature involved—but continues during service. On a smaller scale, distribution of antidegradants between the components of an elastomeric blend can be expected in the same way. Lewis *et al.* [112], for example, have indicated that the antioxidant/ antiozonant N,N'-diphenyl-p-phenylenediamine migrates preferentially from EPDM to SBR during cure; and similarly from SBR to NR, and from EPDM to NR.

E. DISTRIBUTION OF INSOLUBLE COMPOUNDING INGREDIENTS

The effect of an appreciable volume loading of filler on the properties of cured elastomeric vulcanizates depends on whether the elastomer is stress-crystallizing or not. Dinsmore and Juve [113] reported the following information for comparable loadings (15–25% by volume).*

$$\frac{\text{TS black}}{\text{TS gum}} = 1\text{–}1.5 \text{ for NR vulcanizates}$$

$$= 5\text{–}10 \text{ for SBR, etc. vulcanizates}$$

where TS is the tensile strength. Thus, appreciable improvements of strength properties may be obtained for non-stress-crystallizing rubbers by incorporation of carbon black.

Another important factor is the type of filler: two types can broadly be defined, reinforcing and nonreinforcing. Carbon blacks and silicas of the smaller particle sizes and active surfaces belong to the first class, while clays and whiting are examples of the second category.

The effect of variation of filler loading, including various carbon black types, on cured physical properties has been reviewed by various authors, including Leigh-Dugmore [114]. It is clear from such studies and from more recent work by Corish and Palmer [102] and Bulgin and Walker [115] that every elastomer has its optimum filler, especially carbon black, loadings for particular important properties, such as wear and tear resistance. Since different rubbers have different responses to filler loading for certain properties, control of filler distribution in elastomer blends becomes very important.

Hess and co-workers [26] demonstrated that carbon black normally locates preferentially in the BR component of a 50/50 NR/BR preblend, and that this distribution corresponds with optimum vulcanizate performance, including road wear, tensile and tear strengths, and hysteresis (see Figs. 7 and 8). Electron microscopy was used as the main tool for determining the relative proportions of black in the elastomer phases concerned. Further work [116] indicated that in the preparation of blended systems from gum BR and an NR solution masterbatch, carbon black could migrate from the latter to the former during mixing. Marsh and co-workers [34, 86] recognized the existence of unequal filler distribution in elastomer blends but were unable to verify black transfer during mixing. This discrepancy has been largely resolved by further investigations by Callan et al. [117] using techniques such as DTA and GPC as well as a Quantimet Image Analysing Computer in combination with electron microscopy. The incorporation of carbon black into 50/50 elastomer preblends indicated that black affinity decreased in the order BR, SBR, CR, NBR, NR, EPDM, IIR. In addition, black transfer during blending was observed from a mechanically mixed IIR masterbatch to a high unsaturation gum rubber, but not from an NR masterbatch. Transfer thus appears to be confined to those situations in which the adsorptive capacity of the filler has not been fully

* See Chapter 8 for filler effects.

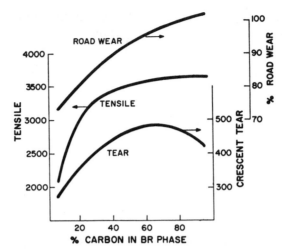

Fig. 7. Effect of carbon black distribution on vulcanizate properties (40 parts per hundred rubber ISAF black; 50 NR/50 BR). Reprinted with permission from Hess *et al.*, *Rubber Chem. Technol.* **40,** 371 (1967). Copyright by the American Chemical Society.

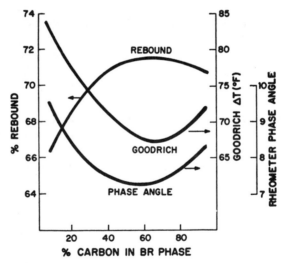

Fig. 8. Effect of carbon black distribution on hysteresis (40 parts per hundred rubber ISAF black; 50 NR/50 BR). Reprinted with permission from Hess *et al.*, *Rubber Chem. Technol.* **40,** 371 (1967). Copyright by the American Chemical Society.

realized. This may occur if the polymer/filler masterbatch has minimum thermal or mechanical history (e.g., solution masterbatch), or involves low molecular weight or low unsaturation elastomers—i.e., situations in which bound rubber formation is minimized. These conclusions are in general agreement with those of Marsh *et al.* [86], and have been further verified by Sircar and Lamond [118], who show in more recent work [119] that hysteresis increases and cut growth resistance decreases by heterogeneous black distribution. The accumulation of black in the BR phase of NR/BR blends has also been confirmed by Vohwinkel [120].

It is apparent from some of those studies [26, 116] that the surface polarity of carbon black influences its distribution in elastomer blends, and that the behavior of inorganic fillers is disparate (e.g., in cis-BR/NR blends, silica tends to accumulate in the NR phase). Filler distribution is also influenced by the point of addition of filler, the viscosity levels of the elastomers, and the blending methods employed. Chemical or thermal promotion may locate a filler permanently in a particular phase [117, 118], allowing a predetermined distribution to be obtained. The practical advantages accompanying preferred filler distribution have been demonstrated by Corish and Palmer [102] in the achievement of superior wear performance in high styrene SBR/cis-BR blends; attention to mixing and blending methods was involved. Marsh has reported a similar study concerning an NR/CR/EPDM triblend and various methods of black incorporation [86].

The degree of dispersion of filler—especially of carbon black—in blends is an important criterion and this is emphasized by Scott and Chirico [121], who related the performance of SBR/cis-BR tread compounds to their black dispersion. Smith [122], too, investigated SBR/BR blends, using electron microscopy, to assess pigment "wetting" characteristics. The ability of rubbers to "accept" black and other fillers is an important factor in attaining good dispersion, which in turn assists reinforcement and ultimate performance properties.

One method of assessing the reinforcing ability of a filler in an elastomer is the determination of bound rubber. Ivan et al. [123] recently described the present state of knowledge in this area. They carried out an experimental study of binary systems of IR, SBR, BR, IIR, CR, and NBR filled with various blacks or active silica. The influence of filler and that of the method of mixing on the competition between elastomers in forming bound rubber is shown. A study of bound rubber in BR/IR blends by Ashida et al. [124] using a pyrolysis–GPC technique also shows up black distribution differences. When HAF black was added to a preblend, the BR component in the bound rubber was slightly larger than the IR component and both components increased with mixing time. When EPC black was added to a preblend, IR formed a relatively large part of the bound rubber at the first stage of mixing and then, as mixing time increased, decreased appreciably to a level lower than the BR component. When IR was added to a BR masterbatch, the resulting bound rubber was mostly BR. At an early stage of mixing, the IR in the bound rubber was higher for EPC than HAF black. Addition of thermally treated HAF black showed no dependence of bound rubber on surface acidity. A similar study has been made using compounds prepared by the solution masterbatch process [125].

F. Effect of Blending on Properties

Most studies of the effect of blending on properties are concerned with general-purpose elastomers. In general, the influence on processing and cured properties is considered—in fact, the achievement of best cured properties inherently depends on successful processing.

1. Polybutadiene (BR)

Polybutadiene blends were discussed in the symposium mentioned previously [23]. One particular advantage conveyed by blending BR with NR is discussed by Glanville and Milner [126]: allowing higher curing temperatures, which yield better physical properties. The heat stability conveyed by cis-BR when blended with NR or SBR is confirmed by Svetlik et al. [127]. Similar advantages are claimed for blends of emulsion BR with NR by McCall and de Decker [128]. Though the addition of cis-BR is said by Grimberg [129] to decrease the tensile strength and modulus of NR vulcanizates, it increases elasticity and abrasion resistance.

The processing properties of cis-BR are improved by blending with liquid BR: excellent properties are claimed [130] for sulfur vulcanizates. Difficulties in the processing of stereoregular polybutadienes can be overcome, as claimed by Snow [131], by use of specific blacks and/or blending with NR or SBR. The processing of car and truck tire tread compounds based on blends of BR prepared with different catalysts and SBR or NR is described by Abagi et al. [132].

Railsback et al. [133, 134] have made detailed studies of blends of BR with SBR (plus oil extension) to obtain improved tire performance and lowest cost. They also showed that properties could be controlled by use of solution polymers in blends: data were presented to show how mixing times can be reduced, or extrusion characteristics modified, by blending solution polymers with NR or emulsion SBR. Tests designed around low-cost tread compositions for passenger car tires were reported by Sarbach et al. [135], who showed that miles of service per compound cost increased when various BR–black masterbatches were blended 50/50 with SBR–black masterbatches. Factors influencing the creep of tire tread compounds based on blends of cis-BR with methylstyrene rubber have been investigated by Russian workers [136].

Work at Chemische Werke Hüls [137] shows that the addition of 10–30% of another elastomer to a vulcanizing mixture containing cis-BR and, optionally, an alkali or alkaline earth salt of a long-chain carboxylic acid, gives a vulcanizate with improved resistance to cutting. Elastomers claimed to be useful include SBR, NR, IR, and even emulsion BR.

The effect of the addition of a liquid polybutadiene SKD10 (containing 10% carboxyl groups) to SKD (cis-BR) synthetic compounds, described by Epshtein et al. [138], led to better technical properties.

2. Styrene–Butadiene Rubbers (SBR)

Flanigan [139], commenting on future consumption of SBR, reports that blends with highly unsaturated EPDM could lead to new applications. The treadwear and wet skid resistance of SB elastomers and blends is dealt with by Kienle [140]. Information on the practical aspects of NR/Buna S blends, including flex cracking, processing, and vulcanizing, was reported by Springer [141]. Hard rubber polyblends of SBR and butyl rubber vulcanized with sulfur have been studied in detail by Meltzer and Dermody [142]. The T_g was found to increase with degree of vulcanization. Reversion occurred on overcure. Dual or stepwise transition from glassy to rubbery modulus behavior, characteristic of polyblends, was found in all mixtures. Temperature coefficients of vulcanization were calculated [143].

Fatigue characteristics of mixtures of SBR, NR, and low molecular weight substances have been investigated by Fujimoto et al. [144]. Cut growth rates of vulcanized pure and black-filled mixtures which can be regarded as compatible systems have been determined. The growth rate of samples obtained by mixing a low molecular weight substance having a higher T_g, and consequently a higher viscosity, is found to be lower. The phenomenon is explained as the reinforcing and retarding effect of the higher-T_g material against the translation or rotation of the polymer chains.

3. Polyisoprene (IR)

The physical properties of mixtures of Cariflex IR305 and IR500 with NR were investigated by Ghircoiasu et al. [145], who found that IR305 had better mechanical properties than the oil-extended IR500, and that on addition of NR, properties of the latter, especially breaking strength, improved more than those of the former. Hoffmann and Sandray [146] also studied Cariflex IR, investigating blends with NR, SBR, and BR from the point of view of processing and technical properties. SKI-3 (high-cis IR) and SKI lithium isoprene rubbers were investigated by Korotkov et al. [147], who concluded that the best blend was 1:1, and that applications in tread and breaker compounds were possible.

Combinations of stereoregular cis-IR and α-methylstyrene BR rubbers were discussed by Levitin et al. [148], who found optimum processing and technical properties of carcass vulcanizates at a 1:1 ratio, including bond strength to cord. Triple blend treads for giant tires based on IR/NR/BR compounds have been reported by Mekel and Bierman [149] to provide advantageous properties.

4. Ethylene–Propylene (EP) Rubbers

EP elastomers are dealt with at some length by Blümel and Kerrutt [150], particularly from the standpoint of development and modification. Blends with other elastomers can be broadly divided into two classes, EP/low unsaturation elastomers and EP/high unsaturation elastomers. These will be discussed in turn.

Callan *et al.* [84] deal with the processing and properties of butyl/EPDM blends, as mentioned in Section II C. A technical information bulletin [151] on EPDM/butyl blends mentions their enhanced heat resistance in black- and mineral-filled stocks. Imoto [152], in studies of blending of EPDM with EPM, used roll blending and three different vulcanizing systems, sulfur, Dicup, and brominated alkyl phenol resin. The physical properties of the blends were proportional to the blend ratio. Further investigations on the peroxide and sulfur cures of EPDM and EPM in the presence of various unsaturated polymers have been made by Imoto *et al.* [153]. For peroxide cures, scorch rate was retarded by addition of liquid diene polymers and promoted by adding liquid polychloroprenes. On addition of liquid Thiokol, the Mooney scorch rate of EPDM was accelerated and that of EPM retarded. Also for peroxide cures of EPDM and EPM, cross-link density, modulus, tensile strength, and hardness are decreased by addition of liquid SBR, NBR, and Thiokol, but increased by 1,2-BR and CR. For sulfur cures, promotion of the cure of EPDM is achieved by adding liquid 1,4-diene polymers such as SBR or NBR, but retarded by addition of liquid 1,2-BR. It was also found that the resistance to hot air degradation of EPM vulcanizates was improved by adding liquid polymers. Loheac and Odam [154] discussed EPDM and chlorobutyl blends in sidewall mixes for radial ply tires: this type of application is also reported by Speranzini and Drost [155]. Improving butyl treadwear by blending with EPM and EPDM has been investigated by Willis and Denecour [156].

Blending of high unsaturation rubbers with EPDM and EPM usually seeks to take advantage of their ozone resistance. This is instanced by Ossefort and Bergstrom [157], who studied blends of EPM and EPDM with diene elastomers using the EP rubbers as ozone shields. Spenadel and Sutphin [158] also discussed ozone protection with EPDM; their data mostly apply to EPDM/CR blends, but are broadly applicable for other rubbers. The use of ultrahigh 1,2-BR's as plasticizing coagents in filled and oil-extended EPM's is reported by Dorman *et al.* [159].

The covulcanization of EP rubbers with dienes is an important consideration. Leibu *et al.* [160] discussed a type of EPM based on 1,4-hexadiene which on blending improves the ozone resistance of SBR and NR. Blending EPDM with dienes is further discussed by Sutton [161], who reports that blends with NR, SBR, NBR, and CR give compounds with good ozone and chemical resistance, and reduced compression set. Satake *et al.* [162] compared laboratory studies in tire tread compounds based on SBR/EPDM and SBR/BR/EPDM blends with field tests. Results of tests of heat aging, groove cracking, abrasion resistance, and cornering characteristics relative to SBR and SBR/BR controls indicate that the EPDM blends can be used for treads without antioxidants, and that the molecular weight distribution, propylene content, and termonomer content of EPDM blends should be investigated further to improve abrasion resistance.

5. Butyl and Chlorobutyl

Two characteristics of butyl rubber which are particularly advantageous are its high mechanical damping, referred to by Dunnom and de Decker [163], and its impermeability, studied by Topcik [164] in blends with NR for the inner liners of tubeless tires.

The difficulty experienced with covulcanization of butyl with diene rubbers is alleviated by use of chlorobutyl. Banks *et al.* [165] referred to blends of NR with chlorobutyl for transport applications. Vitolins [166] discussed the adhesion of chlorobutyl blends to highly unsaturated elastomers and detailed the effects of mineral fillers, oils, and resin. Rehner and Wei [24] also investigated the heterogeneity and cross-linking of elastomer blends of chlorobutyl with highly unsaturated elastomers.

6. Special-Purpose Rubbers

Blends of chlorobutyl with nitrile rubbers were studied by Evans and Partridge [25] in an attempt to develop vulcanizates having the desirable properties of each type as well as lower cost. Comparison of mixing methods showed an advantage for directly compounding ingredients into the blends of elastomers over mixing previously prepared masterbatches. Study of black loadings showed reinforcing furnace blacks to be the best type generally.

The influence of added amounts of SKD (*cis*-BR) on the low-temperature and abrasion properties of SKN (NBR's) was investigated by Devirts and Manvelova [167]. Data on EPDM/NBR blends have also been given in an RAPRA report [168], which considered the possibility of obtaining moderate resistance to oils and ozone plus an acceptable level of general mechanical properties.

Use of mixtures of SKD (*cis*-BR) with Nairit (neoprene) in production of rubber articles from the viewpoint of processing and properties is discussed by Koldunovich *et al.* [169]. The influence of the vulcanization process on the properties of neoprene/*cis*-BR vulcanizate is considered by Orekhov *et al.* [170]. Use of oil-extended emulsion BR blends with neoprene is reported by Wingrove [171] to offer improved low-temperature properties, abrasion resistance, and heat-aging properties. Such blends also have less tendency to soften and stick to mill rolls, fewer calendering problems, faster extrusion rates, and better molding characteristics.

The covulcanization of epichlorohydrin polymers with rubbers, especially sulfur-curable *cis*-BR, is the subject of a Hercules patent [172]. Products with improved low-temperature flexibility and reduced air permeability are obtained by use of an organic accelerator, a metal salt or oxide, and sulfur. The Dow Corning Corporation [173] lists the following advantages for a blend of silicone and fluorosilicone polymers: it bonds easily with fluorosilicone adhesives; it has better physical properties (e.g., compression set); it processes better and is cheaper.

G. CONCLUSIONS

It is clear from the foregoing discussion that the technical reasons for blending fall broadly under two headings: processing and cured physical properties.

The processing characteristics (i.e., the mixing, extruding, and calendering behavior) of many new stereoregular elastomers with narrow molecular weight distributions need modification to render these materials technically acceptable for factory operations. This may be achieved by addition of oil, liquid polymers, or other broader distribution elastomers such as NR or emulsion SBR.

Cured physical properties are affected by two major factors: vulcanization and filler distribution. In order to achieve a single T_g, which controls blend properties such as resilience, some covulcanization (interphase cross-linking) between the different phases in the blends is needed. This explains why the use of the more reactive chlorobutyl in place of butyl rubber for blending with diene elastomers gives better general physical properties, and why monomer variation in blends of the terpolymer EPDM with dienes influences physical properties.

The covulcanization in turn is influenced by the type of curative system used, including its solubility in each phase, and the response of the individual elastomers to the curatives. The systems can be based on sulfur, peroxide, metallic oxide, and radiation cross-linking.

Filler dispersion to an adequate level is needed for optimum properties, as reported by many authors, including Boonstra and Medalia [174], but another potent factor appears to be the distribution of the filler in the phases of the blends. This distribution is clearly influenced by the molecular weight level of each phase, by the method of mixing, and by chemical interaction between the filler surfaces and the chemical groupings in the elastomer.

There are many unanswered questions which could lead to fruitful research in the field of filled elastomer blends. These include the effect of storage and processing (e.g., extruding at high temperatures) on distribution of both soluble and insoluble ingredients. Could use of rubber-bound antioxidants minimize redistribution of antioxidants in elastomer blends? In blends of rubbers which have different affinities for carbon black and silica, would use of carbon black and silica loadings lead to advantageous vulcanizate properties? Can modification of filler surfaces enable control of their location in rubber blends?

III. Elastomer/Plastic Blends

Elastomer/plastic blends, block copolymeric elastomers with thermoplastic sections, and polyurethanes have a general similarity in their biphasic, but elastomeric, behavior. It seems thus justified for comparative purposes to include them in this chapter. Plastic/elastomer blends, in which the former phase is continuous, have not been included, since their properties are essentially plasticlike, though considerations similar to those relating to elastomer/plastic blends hold good.

A. PHYSICAL MIXTURES

The blending of glassy and crystalline plastics, notably polystyrene, polyethylene, polypropylene, and PVC, into IR and BR has been shown by Blondel [175] to provide systematic improvements in hardness and reduction of hysteresis. Polyethylene also improves tear strength. While the mixing process demands relatively high temperatures, particularly in the case of polypropylene, application in footwear, flooring, and damping products is suggested. The studies of other workers have been less extensive and may best be considered according to plastic type.

It is well known that the properties of styrene–butadiene rubbers are influenced by the styrene concentration: as this increases, the polymer becomes less rubbery, more leathery, and eventually resinlike. In order to clarify the difference between copolymers and blends, Shundo et al. [176] have studied a series of butadiene–styrene copolymers and polybutadiene/polystyrene blends of identical styrene contents. The properties of the copolymer vulcanizates differed markedly from those of the blends. In the former, a curvilinear relationship with monomer ratio was displayed, whereas in the latter a linear relationship was shown. These differences were interpreted on the basis of differing morphologies of the polystyrene phase.

Satake [177] has indicated that BR/polystyrene blends prepared by mill mixing are very hard and difficult to process if greater than 20% polystyrene is present. Softening can be effected, however, by remilling at temperatures below 120°C. Microscopic studies indicate that the remill process corresponds with inversion of the phases, the polystyrene component becoming discontinuous. A comparison of physical properties of remilled and unremilled vulcanizates [178] indicates that whereas tensile strength is determined solely by the blend ratio, hardness, elongation, set, and resilience are determined by the nature (i.e., continuity or discontinuity) of the polystyrene phase. Similar studies have been made in SBR/polystyrene blends [179].

The effect of incorporating finely divided polystyrene filler into SBR vulcanizates has been considered mechanistically by Morton and Healey [180, 181]. An increase in tensile strength was shown to depend upon polystyrene concentration and to be related to the viscoelastic properties of the material. It was concluded that polystyrene filler increases the viscous component of the rubbery network, analogous to the effect of decreasing the temperature or increasing the rate of test.

In the case of polyethylene blends, Fischer [182] has reported improved mechanical and electrical properties in EPDM vulcanizates, while Danic *et al.* [183] found good oil and fuel resistance and good ozone and electrical properties in butyl vulcanizates. Improvements in wet skid resistance with SBR/BR tire compounds have also been demonstrated [184]. Nishioka *et al.* [185] have reported that polyethylene acts as a reinforcing agent for IR only when it is chemically linked to the rubber matrix, similar to the hard domains in block copolymers. This is achieved by peroxide vulcanization rather than sulfur. Covulcanization of similar blend systems has been reported to provide a route for the formation of artificial leathers [186]. Satake [187] has studied the mixing state of BR/polyolefin blends, while Imoto [188] has considered the in situ polymerization of vinyl monomer/EPDM blends. In this situation, graft polymerization and homopolymerization of the vinyl monomer occur simultaneously during the peroxide cure of the blend, providing a composition with superior dispersion to a conventional EPDM/vinyl polymer blend.

More recently, Uniroyal has marketed grades of thermoplastic rubbers (TPR's) which appear to be blends of EPDM with polypropylene or polyethylene, or both. This type of thermoplastic rubber is described in a recent US Patent to Uniroyal [189] which relates to dynamically partially cured blends of EPM or EPDM rubbers with polyolefin resins such as polypropylene and polyethylene. The dynamic partial cure of the blend is effected by heating with a curative such as a peroxide while shearing the blend. The blends are elastomeric, have good physical properties without vulcanization, and are reprocessible.

The modification of butadiene–acrylonitrile elastomers with PVC resins has long been recognized to provide excellent ozone resistance and weathering properties, thereby increasing the useful range of these elastomers [190–192]. The principal disadvantages are a reduction in low-temperature flexibility and poor compression set [190, 193]. Adequate fluxing of the polymers during mechanical blending is found to be essential for development of maximum physical properties, processing temperatures of 150–200°C being recommended. These conditions have been shown to correspond to mutual solution of the polymers [194]. Jorgensen and Frazer [195], in a study of component polymer variations, have shown that the molecular weight and acrylonitrile content of NBR influence the blend performance, while the type and molecular weight of PVC do not. Fisher *et al.* [196] have reported that the blending of NBR and PVC in the form of latices overcomes the requirement for high-temperature processing while providing higher tensile strength and ozone resistance. Latex blending has found application in the production of foam coatings for textile floor covering [197].

B. BLOCK COPOLYMERS

While block copolymers of various forms have been known for many years [198–200], the development by the Shell Chemical Company of Kraton (Cari-

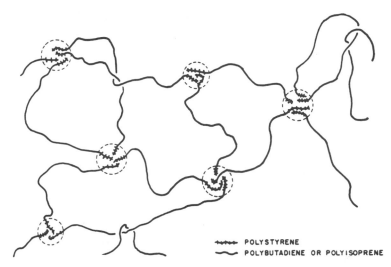

Fig. 9. Phase arrangement in S-B-S and S-I-S block copolymers (schematic). (From Holden et al. [201].)

flex) thermoplastic elastomers heralded a rapid growth in fundamental and technological interest. Kraton elastomers consist of ordered block copolymers of general structure S-B-S where S is polystyrene and B is polybutadiene [201]. Incompatibility of the styrene and butadiene blocks causes the former to aggregate in the bulk state to form glassy particles which act as network junctures and as reinforcing filler [202]. Figure 9 gives a schematic representation of the phase arrangement. In this way, a system is produced which, without vulcanization, displays rubberlike properties (cf. Fig. 10) at ambient temperatures but which flows at temperatures above the glass transition of the polystyrene block, similar to thermoplastics. This topic is dealt with in detail in another chapter* but, as may be seen from the preceding discussion, block copolymers are a special type of blend and some discussion is indicated.

Preparation of styrene–butadiene block copolymers is by homogeneous anionic polymerization, as reviewed by Fetters [203]. The application of this synthesis route to block copolymer formation can be traced to Zwarc *et al.* [204], who observed that polystyrene anions in solution remained stable over extended periods of time and could add further monomer units to increase the homopolymer molecular weight or to form block copolymers. These observations led to the concept of stable or living anions and laid the foundation for subsequent investigations into commercial processes for the production of block copolymers. A further review of synthesis routes has been prepared by Allport and Janes [198].

* See Chapter 13.

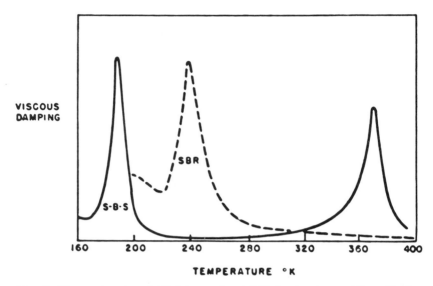

VISCOUS
DAMPING

S-B-S

SBR

160   200   240   280   320   360   400

TEMPERATURE °K

Fig. 10. Viscous damping of S-B-S (———) and SBR (---) polymers. (From Holden *et al.* [201].)

Phase separation occurs in block copolymers [205] as in a mixture of homopolymers, but morphological constraint is introduced by the existence of permanent attachments between the polymer blocks. Krause [206, 207] and Bianchi *et al.* [208] have presented a thermodynamic treatment of phase separation, assuming sharp boundaries between the phases and complete separation, i.e., interfaces composed solely of interblock chemical bonds. A consequence of this arrangement is that for a given block length the phases are restricted in size if their density is to remain uniform. Electron microscopic evidence provided by Matsuo [205] supports this view, but Legrand has shown by small-angle X-ray scattering that the boundaries between matrices and domains in block copolymers are not sharp [78]. This may be especially true under non-equilibrium conditions where domain morphology is often poorly defined [209, 210]. A thermodynamic treatment of phase separation in which regions of mixed polymer blocks are allowed has been given by Leary and Williams [211].

Detailed studies of phase morphology in solvent cast films of styrene–butadiene–styrene block copolymers [210, 212–214] have revealed five fundamental structures: spherical domains of component A (or B) in matrix B (or A), rodlike domains of component A (or B) in matrix B (or A) and alternating lamellae of the two components (Fig. 11). A thermodynamic analysis of the formation of the different types of micelles in solution has been made by Inoue *et al.* [215], while calculations of the free energy changes required to establish each microstructural form have been carried out by Meier [216] and by Leary and Williams [217]. Observed domain sizes are ~10–40 mμ [202, 210], approximately 2 orders of magnitude smaller than in homopolymer

├─ 1 μm ─┤

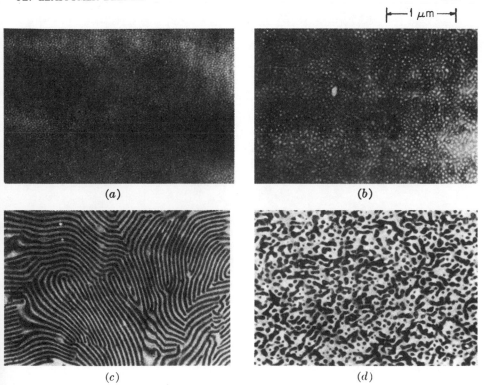

Fig. 11. Electron micrographs of ultrathin sections (∼350 Å thick) cut normal to the surface of films cast from 5% toluene solutions of styrene–isoprene–styrene block copolymers varying in composition from 9.5% to 72% weight fraction of styrene: (a) S-I-S-1; (b) S-I-S-2; (c) S-I-S-3; (d) S-I-S-4. (From Uchida et al. [212].)

blends. Agreement with Meier's theoretical treatment of domain size as a function of block length is reasonable [218].

Investigations of the influence of casting solvent on phase morphology [39, 205, 212, 215, 219] indicate that the use of a selective good solvent for a given block component tends to favor the development of that component as a continuous matrix (see Fig. 12). A nonselective solvent (i.e., a good solvent for both block components) may produce alternating lamellar morphology, but this is subject to copolymer composition [220]. Meier [216] has calculated that the spherical, cylindrical, and lamellar forms of phase structure each have lowest free energy over a particular range of block molecular weight ratios.

The slow evaporation of solvent during solution casting has been demonstrated to produce a high degree of domain ordering, in some instances effecting a three-dimensional array [210, 218]. Small-angle X-ray scattering investigations indicate that the domains can be regarded as participating in a macrolattice in a way comparable to the atoms in a crystal lattice [209, 221–223].

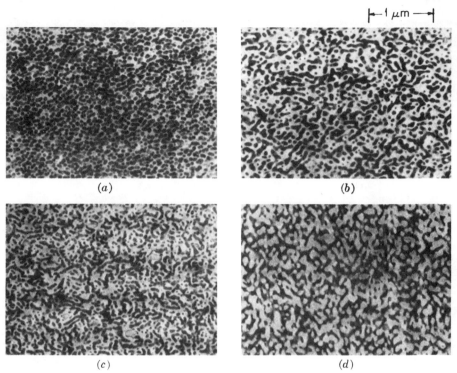

⊢—1 μm —⊣

(a)

(b)

(c)

(d)

Fig. 12. Electron micrographs of ultrathin sections (∼350 Å thick) cut normal to the surface of films cast from 5% solutions of S-I-S-4 block copolymer in various solvents: (a) MEK; (b) toluene; (c) CCl₄; (d) cyclohexane. (From Uchida *et al.* [212].)

Melt processing may produce less ordered structures or, in the case of extrusion and injection molding, oriented cylindrical rather than spherical domains [224, 225]. This leads to strongly anisotropic behavior and pronounced yielding in those directions in which the polystyrene phase is continuous.

The mechanical properties of styrene–butadiene block copolymers have been well studied [201, 219, 226–230]. Generally, tensile properties vary according to the copolymer structure, i.e., the number and length of component blocks [201, 205, 231, 232], the monomer ratio [201], and the morphology [209, 219, 233, 234]. Low styrene concentration produces stress–strain properties similar to uncross-linked conventional rubbers, but as the styrene concentration is increased, the performance approaches that of cross-linked rubbers and, ultimately, of styrene homopolymer [201]. Styrene concentrations of 25–30 wt % may usefully provide elastomeric performance. The nature of the solvent in solvent-cast preparations has a substantial influence on the modulus and associated extension behavior [219, 233, 234]. Films cast from selective poly-butadiene solvents exhibit low modulus, rubbery behavior, whereas films cast from selective polystyrene solvents exhibit high modulus and a plastic yield behavior. Yielding is considered to represent fragmentation of the continuous polystyrene network, thereby effecting an increase in rubbery behavior on further extension cycles.

The modification of styrene–butadiene block copolymers by addition of butadiene and styrene homopolymer, and vice versa, the modification of butadiene and styrene homopolymer mixes by the blocks, has been studied with particular regard to morphology [235–238] and physical properties [239]. Inoue et al. [235] have stated that solubilization of homopolymer by the respective block segments occurs provided that its molecular weight is equal to or less than that of the corresponding block. Modification of block copolymers with butadiene–styrene random copolymers does not provide solubilization [240], although much stronger mutual interaction is reported than with, for example, EPM [241]. An improved balance of flow and hardness properties has been claimed by Van Henten and Stephens [242] through the blending of block copolymers with polystyrene and EVA. Superior wet skid resistance has been claimed for blends of styrene–butadiene block and random copolymers [243], while the addition of minor amounts of high softening point resin has been shown to improve high-temperature performance [244]. Similar results may be achieved with α-methylstyrene as the block hard phase [245–247]. Polydimethylsiloxane may be employed as the block soft phase [245, 248], improving thermooxidative stability.

C. POLYURETHANES

Segmented polyurethane elastomers are essentially alternating block copolymers of flexible segments, such as polyether and polyester, and rigid segments, such as polyurethane or polyurea (formed by the reaction of a diisocyanate with a diol or diamine). The hard segments are associated by interchain hydrogen bonds [249, 250] to form discrete, sometimes crystalline, domains acting as physical cross-links [251]. Frisch [252] has recently reviewed the chemistry of these materials, and Allport and Mohajer [253] the property–structure relationships. DSC and dielectric relaxation measurements by Seymour and Cooper [66] and North and Reid [68] have confirmed the two-phase structure of polyurethanes. Koutsky et al. [254] have used solvent etching and iodine staining to reveal the hard and soft blocks by electron microscopy: these are approximately 30–100 Å wide.

Estes et al. [72], in a study of ir dichroism, have proposed that block polyurethanes comprise two interpenetrating semicontinuous phases, one containing urethane segments and the other prepolymer segments. Characteristic domain dimensions are 50 Å. Harrell [255] has studied the relationship between physical properties and hard segment size, size distribution, and spacing. Results indicate that the hard segments are crystalline with well-defined melting points which increase with segment size. The effect of cross-linking and hard block crystallinity on thermal and mechanical behavior has been studied by Miller [256] and the effect of hard/soft segment ratio has been studied by Ferguson *et al.* [257].

An interesting facet of polyurethanes is their end use as spandex fibers in which properties intermediate between those of natural rubber and conventional nylons or polyester fibers are obtained. This is a further ramification of the use of the alternating flexible and rigid segments of the polyurethane block copolymers, conferring controllable stress–strain properties on an end product.

Block copolymers containing segments of an elastomeric polyurethane and a hard glassy polymer have also been claimed [252] to give materials with desirable mechanical and physical properties. A recent RAPRA Information Report [258] quotes reference to polyurethane blends with other rubbers and plastics.

Another new area involving polyurethanes [252] is that of interpenetrating polymeric networks (IPN), in which a second linear polymer is cross-linked in the presence of a first polymer already cross-linked, resulting in interlocking

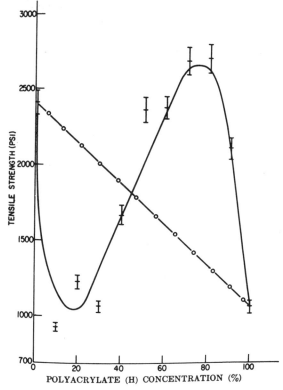

Fig. 13. Tensile strength as a function of composition of interpenetrating polymer network. Reprinted with permission from Frisch, *Rubber Chem. Technol.* **45,** 1442 (1972). Copyright by the American Chemical Society.

rings of polymer chains. Among the IPN's investigated were combinations of urethane–urea with an acrylate copolymer. They were prepared from water emulsions of urethane–urea (made from blends of polyester diols and triols with toluene diisocyanate and chain extended with a diamine) and a polyacrylate copolymer latex which was cross-linked by means of water dispersions of sulfur, zinc oxide, and zinc dibutyldithiocarbamate. After thorough blending of the latex and curatives, films were cast and cured. As may be seen from Figs. 13 and 14, graphs of tensile strength and cross-link density versus concentration of polyacrylate do not follow a straight-line relationship. It is possible to achieve a maximum in tensile strength where the cross-link density exhibits a definite maximum, i.e., 60–80% polyacrylate.

The formation of IPN's is a way of overcoming the mixing difficulty encountered with chemically incompatible polymers and of reducing the problem of interfacial separation. The types of elastomer pairs suitable for preparation of IPN's extend beyond polyurethanes plus polyacrylates to polymethacrylates

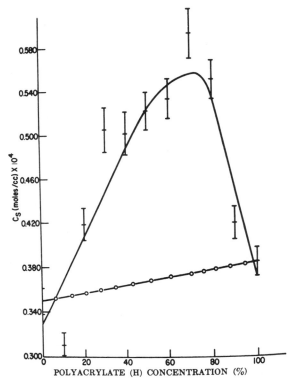

Fig. 14. Cross-link density as a function of composition of interpenetrating polymer network. Reprinted with permission from Frisch, *Rubber Chem. Technol.* **45**, 1442 (1972). Copyright by the American Chemical Society.

with polyacrylates and/or polystyrene [259, 260], or polysiloxane with polystyrene or polymethylmethacrylate [261]. According to the choice of polymerization/vulcanization system in the second component, intermolecular cross-links may be largely present or absent. Evidence indicates that softer, more elastic materials are produced in the former case. Microscopic and mechanical investigations suggest that the network formed first constitutes the more continuous phase [262]. Manipulation of these several variables can be expected to furnish a range of new and useful blend compositions.

IV. General Correlations and Conclusions

The following highlights may be distilled out of this chapter, which like the blends it describes is heterogeneous in character. The method of blending as a means of controlling the structure of the phases is a very important factor in the properties of elastomer/plastic blends, especially impact strength and crack resistance. With unfilled elastomer/elastomer blends, blending methods and blend structure are primarily important, especially for crack resistance (e.g., ozone attack). With filled elastomer/elastomer blends, blend structure remains a determining factor, but filler and curative distribution may be more important.

If it is possible by variation in blending and mixing methods to achieve a heterodistribution of fillers and/or curatives between the phases of the blend, then novel properties may be realized. Co-cross-linking between the phases seems a necessary prerequisite for some properties (e.g., resilience) if the blend performance is to be "intermediate" between that of the components. Not enough is known about this area, nor of the contributory factors of oil and antioxidant migration. Surface modifications of fillers, to change their distribution, also appear worthwhile investigating—if this can be done without seriously interfering with reinforcement.

It is generally found (see, e.g., the conclusions of the recent review article [263] on which this chapter is substantially based) that various authors quote different results from otherwise similar polymers. It has been shown that molecular weight differences have a major effect on blending propensity; thus, it would seem appropriate to suggest that a detailed study of the blending of well-characterized polymers be made, paying attention to micro- and macro-structure, especially of high and low molecular weight "tails" and microgels.

The influence of compatibilizing agents such as A-B and A-B-A block copolymers would seem to be another fruitful area for further research. Another method of "bringing together" two incompatible phases is the fairly recent innovation of interpenetrating networks: more studies of this novel area would seem to be indicated, especially where fillers are incorporated.

In general, elastomer/plastic blends, block and graft copolymers (typified by the thermoplastic elastomers), and polyurethanes have a major characteristic in common: they contain hard and soft phases. The former act in the same way as reinforcing particulate fillers in elastomers. The interconnecting polymer between the hard and soft phases may function in the same way as the adsorbed or bound rubber on particulate fillers. There would appear to be some basic research opportunities also in this area.

REFERENCES

1. M. H. Walters and D. N. Keyte, *Trans. Inst. Rubber Ind.* **38**, 40 (1962).
2. B. S. Gesner, Polyblends *in* "Encyclopaedia of Polymer Science and Technology" (N. M. Bikales, ed.), Vol. 10, p. 697. Wiley (Interscience), New York, 1969.
3. S. N. Angove, *Rubber J.* **149**(3), 37 (1967).
4. Greengate and Irwell Rubber Co., Fr. Patent 1,446,617 (1966).
5. Ashland Oil Refining Co., U.S. Patent 3,294,714 (1966).
6. D. C. Blackley and R. C. Charnock, *J. Inst. Rubber Ind.* **7**, 60 (1973).
7. D. C. Blackley and R. C. Charnock, *J. Inst. Rubber Ind.* **7**, 113 (1973).
8. A. King, *RAPRA Res. Bull.* No. 127 (1964).
9. Japan Synthetic Rubber Co., Ltd., Br. Patent 1,046,215 (1966).
10. H. L. Frisch, D. Klempner, and K. C. Frisch, *Rubber Chem. Technol.* **43**, 883 (1970).
11. D. I. Livingston and R. L. Rongone, *Proc. Int. Rubber Technol. Conf., Brighton*, p. 337 (1968).
12. L. Bohn, *Rubber Chem. Technol.* **41**, 495 (1968).
13. M. Shundo, M. Imoto, and Y. Minoura, *J. Appl. Polym. Sci.* **10**, 939 (1966).
14. M. J. Palmer, J. F. Yardley, and C. H. Leigh-Dugmore (to Dunlop Ltd.), Br. Patent 1,095,776 (1965).
15. Phillips Petroleum Co., U.S. Patent 3,304,281 (1967).
16. Columbian Carbon Co., Br. Patent 980,280 (1965).
17. Japan Synthetic Rubber Co. Ltd., Br. Patent 1,045,980 (1966).
18. K. A. Burgess, S. M. Hirshfield, and C. A. Stokes, *Rubber Age* **97**(9), 85 (1965).
19. Columbian Carbon Co., Br. Patent 980,280 (1965).
20. R. P. Campion and J. F. Yardley (to Dunlop Ltd.), Br. Patent 1,173,171 (1968).
21. T. Pazonyi and M. Dimitrov, *Rubber Chem. Technol.* **40**, 1119 (1967).
22. L. T. Fan, S. J. Chen, and C. A. Watson, *Ind. Eng. Chem.* **62**(7), 53 (1970).
23. *JSR News* **5**(6), 1 (1967).
24. J. Rehner and P. Wei, *Rubber Chem. Technol.* **42**, 985 (1969).
25. L. Evans and E. G. Partridge, *ACS/CIC, Div. Rubber Chem., 83rd Meet., 1963; Rubber Age* **94**(2), 272 (1963).
26. W. M. Hess, C. E. Scott, and J. E. Callan, *Rubber Chem. Technol.* **40**, 371 (1967).
27. R. J. Ceresa, *Rubber Chem. Technol.* **33**, 923 (1960).
28. D. J. Angier and W. F. Watson, *J. Polym. Sci.* **18**, 129 (1955).
29. D. J. Angier and W. F. Watson, *Trans. Inst. Rubber Ind.* **33**, 22 (1957).
30. G. L. Slonimskii and E. V. Reztsova, *Rubber Chem. Technol.* **33**, 457 (1960).
31. R. J. Ceresa, "Block and Graft Copolymers," Ch. 5. Butterworth, London, 1962.
32. D. C. Allport, *in* "Block Copolymers" (D. C. Allport and W. H. Janes, eds.), Ch. 2. Appl. Sci. Publ., London, 1973.
33. W. J. Burlant and A. S. Hoffman, "Block and Graft Copolymers," Ch. 8. Van Nostrand-Reinhold, New York, 1960.
34. P. A. Marsh, A. Voet, and L. D. Price, *Rubber Chem. Technol.* **40**, 359 (1967); **41**, 344 (1968).
35. J. E. Lewis, H. N. Mercer, M. L. Deviney, L. Hughes, and J. E. Jewell, *ACS/CIC, Div. Rubber Chem., 98th Meet., 1970; Rubber Chem. Technol.* **44**, 855 (1971).

36. J. E. Callan, W. M. Hess, and C. E. Scott, *Rev. Gen. Caoutch. Plast.* **47,** 1159 (1970).
37. J. D. Moore, *Polymer* **12,** 478 (1971).
38. H. Keskkula and P. A. Traylor, *J. Appl. Polym. Sci.* **11,** 2361 (1967).
39. T. Miyamoto, K. Kodama, and K, Shibayama, *J. Polym. Sci., Part A-2* **8,** 2095 (1970).
40. R. W. Smith and J. C. Andries, *Rubber Chem. Technol.* **47,** 64 (1974).
41. S. S. Voyutsky, "Autoadhesion and Adhesion of High Polymers," p. 141. Wiley (Interscience), New York, 1953.
42. J. W. Breitenbach and B. A. Wolf, *Makromol. Chem.* **117,** 163 (1968).
43. S. A. Chen. *J. Appl. Polym. Sci.* **16,** 1603 (1972).
44. H. Schnecko and R. Caspary, *Kautsch. Gummi, Kunstst.* **25,** 309 (1972).
45. H. Burrell, *J. Paint Technol.* **40,** 197 (1968).
46. D. Feldman and M. Rusu, *Rev. Gen. Caoutch. Plast.* **48,** 687 (1971).
47. A. Englert and H. Tompa, *Polymer* **11,** 507 (1970).
48. J. P. Menin and R. Roux, *J. Polym. Sci., Part A-1* **10,** 855 (1972).
49. G. Kraus and K. W. Rollmann, *Adv. Chem. Ser.* **99,** 189 (1971).
50. R. Kuhn, S. B. Liang, and H. J. Cantow, *Angew. Makromol. Chem.* **18,** 101 (1971).
51. R. F. Boyer, *Rubber Chem. Technol.* **36,** 1303 (1963).
52. M. C. Shen and A. Eisenberg, *Rubber Chem. Technol.* **43,** 95 (1970).
53. H. G. Weyland, P. J. Hoftyzer, and D. W. Van Krevelen, *Polymer* **11,** 79 (1970).
54. M. B. Roller and J. K. Gillham, *J. Appl. Polym. Sci.* **16,** 3105 (1972).
55. S. Manabe and M. Takayanagi, *Kogyo Kagaku Zasshi* **70,** 525 (1967).
56. L. A. Chandler and E. A. Collins, *J. Appl. Polym. Sci.* **13,** 1585 (1969).
57. M. Hoffmann, G. Pampus, and G. Marwede, *Kautsch. Gummi, Kunstst.* **22,** 691 (1969).
58. P. J. Corish, *Rubber Chem. Technol.* **40,** 324 (1967).
59. I. C. Cheetham, *Trans. Inst. Rubber Ind.* **40,** 156 (1965).
60. H. K. de Decker and D. J. Sabatine, *Rubber Age* **99**(4), 73 (1967).
61. H. Keskkula, S. G. Turley, and R. F. Boyer, *J. Appl. Polym. Sci.* **15,** 351 (1971).
62. W. E. A. Davies, *J. Phys. D* **4,** 1176 (1971).
63. G. W. Miller, *J. Appl. Polym. Sci.* **15,** 39 (1971).
64. D. Klempner and H. L. Frisch, *J. Polym. Sci., Part B* **7,** 525 (1970).
65. V. R. Landi, *Rubber Chem. Technol.* **45,** 222 (1972).
66. R. W. Seymour and S. L. Cooper, *J. Polym. Sci., Part B* **9,** 689 (1971).
67. K. Fujimoto and N. Yoshimiya, *Rubber Chem. Technol.* **41,** 669 (1968).
68. A. M. North and J. C. Reid, *Eur. Polym. J.* **8,** 1129 (1972).
69. B. D. Gesner, *Am. Chem. Soc., Div. Org. Coat. Plast. Chem., 154th Meet., 1967; Appl. Polym. Symp.* **7,** 53 (1968).
70. J. K. Clark and R. A. Scott, *Rubber Chem. Technol.* **43,** 1332 (1970).
71. I. Yamagi, I. Yamaja, and M. Watanabe, *J. Soc. Rubber Ind. Jpn.* **46,** 411 (1973).
72. G. M. Estes, R. W. Seymour, and S. L. Cooper, *Macromolecules* **4,** 452 (1971).
73. K. Fujimoto, *J. Soc. Rubber Ind. Jpn.* **43,** 54 (1970).
74. D. G. Legrand, *Trans. Soc. Rheol.* **15,** 541 (1971).
75. J. Schaefer, S. H. Chin, and S. I. Weissman, *Macromolecules* **5,** 798 (1972).
76. P. R. Lewis and C. Price, *Polymer* **12,** 258 (1971).
77. A. Douy and B. Gallot, *Makromol. Chem.* **156,** 81 (1972).
78. D. G. Legrand, *J. Polym. Sci., Part B* **8,** 195 (1970).
79. D. McIntyre and E. Campos-Lopez, *Macromolecules* **3,** 322 (1970).
80. J. E. Guillet, *J. Macromol. Sci., Chem.* **4,** 1669 (1970).
81. S. Sugiki *et al., J. Soc. Rubber Ind. Jpn.* **45,** 299 (1972).
82. F. Kellinsky and G. Markert, *Makromol. Chem.* **121,** 117 (1969).
83. D. Heinze, W. Ring, S. Tocker, and G. Pampus, G.D. *Ch. Group Lect. Meet., Bad Nauheim, Germany 1966.*
84. J. E. Callan, B. Topcik, and F. P. Ford, *Rubber World* **151**(6), 60 (1965).

85. B. D. W. Powell, M.S. Thesis, Loughborough Univ. of Technol., Loughborough, England, 1970.
86. P. A. Marsh, T. J. Mullens, and L. D. Price, *Rubber Chem. Technol.* **43**, 400 (1970).
87. N. Yoshimura and K. Fujimoto, *Rubber Chem. Technol.* **42**, 1009 (1969).
88. R. L. Scott, *J. Chem. Phys.* **17**, 279 (1949).
89. P. J. Flory, "Principles of Polymer Chemistry," p. 555. Cornell Univ. Press, Ithaca, New York, 1953.
90. G. L. Slonimskii, *J. Polym. Sci.* **30**, 625 (1958).
91. M. Shundo, K. Komuro, K. Goto, M. Imoto, and Y. Minoura, *J. Appl. Polym. Sci.* **12**, 2047 (1968).
92. K. Satake, K. Shinki, T. Teraoka, and S. Ibe, *J. Soc. Rubber Ind. Jpn.* **44**, 54 (1971).
93. J. B. Gardiner, *Rubber Chem. Technol.* **43**, 370 (1970).
94. E. H. Andrews, *Rubber Chem. Technol.* **40**, 635 (1967).
95. J. B. Gardiner, *Rubber Chem. Technol.* **41**, 1312 (1968).
96. J. B. Gardiner, *Rubber Chem. Technol.* **42**, 1058 (1969).
97. K. Fujimoto and N. Yoshimiya, *Rubber Chem. Technol.* **41**, 669 (1968).
98. K. V. Boguslavskii, *Sov. Rubber Technol.* **29**(1), 7 (1970).
99. K. Hashimoto, T. Harada, I. Ando, and N. Okubo, *J. Soc. Rubber Ind. Jpn.* **43**, 652 (1970).
100. R. L. Zapp, *Rubber Chem. Technol.* **46**, 251 (1973).
101. T. Ishitobi *et al.*, *Soc. Rubber Ind. Jpn.*, *38th Annu. Meet.*, *1971*.
102. P. J. Corish and M. J. Palmer, "Advances in Polymer Blends and Reinforcement," p. 1. *Inst. Rubber Ind. Conf.*, *Loughborough, Eng.*, *1969*.
103. R. W. Amidon and R. A. Gencarelli, U.S. Patent 3,674,824 (1972).
104. R. P. Mastromatteo and T. J. Brett, U.S. Patent 3,678,135 (1970).
105. R. P. Mastromatteo, J. M. Mitchell, and T. J. Brett, *Rubber Chem. Technol.* **44**, 1065 (1971).
106. G. Kerrutt, H. Blümel, and H. Weber, *DKG Meet. 1968; Kaut. Gummi Kunstst.* **22**, 413 (1969).
107. Sumitomo Chemical Co., Br. Patent 1,325,064 (1973).
108. K. C. Baranwal and P. N. Son, *Rubber Chem. Technol.* **47**, 88 (1974).
109. B. G. Corman, M. L. Deviney, and L. E. Whittington, *Rubber Chem. Technol.* **43**, 1349 (1970).
110. M. L. Deviney and L. E. Whittington, *Rubber Chem. Technol.* **44**, 87 (1971).
111. P. J. Corish and B. E. Dawson, unpublished observations (1965).
112. J. E. Lewis, M. L. Deviney, and L. E. Whittington, *Rubber Chem. Technol.* **42**, 892 (1969).
113. R. P. Dinsmore and R. D. Juve, *in* "Synthetic Rubber" (G. S. Whitby, ed.), p. 399. Wiley, New York. 1954.
114. C. H. Leigh-Dugmore, *in* "The Applied Science of Rubber" (W. J. S. Naunton, ed.), p. 480. Arnold, London. 1961.
115. D. Bulgin and D. L. Walker (to Dunlop Ltd.), Br. Patent 1,061,017 (1964).
116. C. E. Scott, J. E. Callan, and W. M. Hess, *Natur. Rubber Conf.*, *Kuala Lumpur*, *1968; J. Rubb. Res. Inst. Malaya*, **22**, 242 (1969).
117. J. E. Callan, W. M. Hess, and C. E. Scott, *Rubber Chem. Technol.* **44**, 814 (1971).
118. A. K. Sircar and T. G. Lamond, *Rubber Chem. Technol.* **46**, 178 (1973).
119. A. K. Sircar, T. G. Lamond, and P. E. Pinter, *Rubber Chem. Technol.* **47**, 48 (1974).
120. K. Vohwinkel, *Ned. Rubberind.* **30**(5), 1 (1969).
121. C. E. Scott and V. Chirico, *ACS/CIC, Div. Rubber Chem.*, *94th Meet.*, *1963*.
122. R. W. Smith, *Rubber Chem. Technol.* **40**, 350 (1967).
123. G. Ivan, G. Balan, and M. Giurgiuca, *Ind. Usoara* **19**, 665 (1972).
124. M. Ashida, Y. Inugai, T. Watanabe, and K. Kobayashi, *J. Soc. Rubber Ind. Jpn.* **46**, 241 (1973).
125. K. Abe, M. Ashida, and T. Watanabe, *J. Soc. Rubber Ind. Jpn.* **46**, 397 (1973).
126. L. M. Glanville and P. W. Milner, *Rubber Plast. Age* **48**, 1059 (1967).
127. J. F. Svetlik, E. F. Ross, and D. T. Norman, *ACS/CIC, Div. Rubber Chem.*, *86th Meet.*, *1964; Rubber Age (N.Y.)* **96**(4), 570 (1965).
128. C. A. McCall and H. K. J. de Decker, Fr. Patent 1,442,275 (1966).

129. R. Grimberg, *Ind. Usoara* **13**, 339 (1966).
130. S. M. Hirshfield, U.S. Patent 3,281,289 (1966).
131. C. W. Snow, U.S. Patent 3,280,876 (1966).
132. V. Abagi, C. Dragus, A. Verejan, and M. Resmerita, *Mater. Plast.* (*Bucharest*) **6**(2), 84 (1969).
133. H. E. Railsback, J. R. Haws, W. T. Cooper, and J. H. Tucker, *Rubber World* **148**(1), 40 (1963).
134. H. E. Railsback and G. Porta, *Mater. Plast. Elastomeri* **35**, 63 (1969).
135. D. V. Sarbach, R. W. Hallman, and W. E. Moles, *Rubber Age* (*N.Y.*) **102**(8), 47 (1970).
136. N. G. Chuprina, I. A. Levitin, G. N. Buiko, V. A. Krol, and E. Z. Diner, *Sov. Rubber Technol.* **29**(10), 23 (1970).
137. Chemische Werke Hüls A. G., Fr. Patent 1,449,094 (1966).
138. V. G. Epshtein, O. A. Zakharkin, M. A. Polyak, and S. G. Yukshnovich, *Sov. Rubber Technol.* **26**(1), 10 (1967).
139. W. C. Flanigan, *Rubber Age* **101**(2), 49 (1969).
140. R. N. Kienle, E. S. Dizon, T. J. Brett, and C. F. Eckert, *Rubber Chem. Technol.* **44**, 996 (1971).
141. A. Springer, *RAPRA Transl.* **1170** (1964).
142. T. H. Meltzer and W. J. Dermody, *J. Appl. Polym. Sci.* **8**, 773 (1964).
143. T. H. Meltzer and W. J. Dermody, *J. Appl. Polym. Sci.* **8**, 791 (1964).
144. K. Fujimoto, I. Inomata, and T. Nishi, *J. Soc. Rubber Ind. Jpn.* **46**, 216 (1973).
145. C. Ghircoiasu, D. Mihaila, and G. Petrescu, *Ind. Usoara* **13**, 82 (1966).
146. J. Hoffmann and V. Sandray, *Rev. Gen. Caoutch. Plast.* **47**, 1273 (1970).
147. A. A. Korotkov, V. N. Reikh, and N. F. Kovalev, *Sov. Rubber Technol.* **25**(8), 5 (1966).
148. I. A. Levitin, Y. G. Korablev, and T. A. Kadyrov, *Sov. Rubber Technol.* **26**(12), 18 (1967).
149. A. L. Mekel and H. Bierman, *Rubber J.* **149**(4), 54 (1967).
150. H. Blümel and G. Kerrutt, *Kautsch. Gummi, Kunstst.* **24**, 517 (1971).
151. *ICI Tech. Inf. Bull.* **R165** (1969).
152. M. Imoto, *J. Soc. Rubber Ind. Jpn.* **42**, 439 (1969).
153. M. Imoto *et al.*, *J. Soc. Rubber Ind. Jpn.* **42**, 1016 (1969); **43**, 29 (1970).
154. L. Loheac and N. Odam, *Rev. Gen. Caoutch. Plast.* **48**, 509 (1971).
155. A. H. Speranzini and S. J. Drost, *Rubber Chem. Technol.* **43**, 48 (1970).
156. J. M. Willis and R. L. Denecour, *Rubber Age* (*N.Y.*) **100**(10), 61 (1968).
157. Z. T. Ossefort and E. W. Bergstrom, *Rubber Age* (*N.Y.*) **101**(9), 4 (1969).
158. L. Spenadel and R. L. Sutphin, *Rubber Age* (*N.Y.*) **102**(12), 55 (1970).
159. E. N. Dorman, E. K. Hellstrom, and J. W. Prane, *Rubber Age* (*N.Y.*) **104**(1), 49 (1972).
160. H. J. Leibu, S. W. Caywood, and L. H. Knabeschuh, *Rubber World* **165**(3), 52 (1971).
161. M. S. Sutton, *Rubber World* **149**(5), 62 (1964).
162. K. Satake, T. Wada, T. Sone, and M. Hamada, *J. Inst. Rubber Ind.* **4**, 102 (1970).
163. D. D. Dunnom and H. K. de Decker, *Rubber World* **151**(6), 108 (1965).
164. B. Topcik, *Mater. Plast. Elastomeri* **35**, 201 (1969).
165. S. A. Banks, R. H. Dudley, D. G. Onken, and J. A. Rae, *Rubber World* **151**(3), 62 (1964).
166. G. Vitolins, *ACS/CIC, Div. Rubber Chem., 94th Meet., 1968; Rubber Chem. Technol.* **42**, 971 (1969).
167. E. Y. Devirts and I. G. Manvelova, *Sov. Rubber Technol.* **28**(11), 31 (1969).
168. J. R. Pyne, *RAPRA Mater. Eval. Rep.* No. 5, Sheets 9, 10 (1968).
169. E. B. Koldunovich *et al.*, *Sov. Rubber Technol.* **25**(12), 15 (1966).
170. S. V. Orekhov, N. D. Zakharov, M. A. Polyak, and V. G. Epshtein, *Sov. Rubber Technol.* **27**(6), 14 (1968).
171. D. E. Wingrove, *Rubber Age* (*N.Y.*) **102**(4), 74 (1970).
172. Hercules Inc., Br. Patent 1,011,847 (1965).
173. Dow Corning Corp., *Adhes. Age* **14**(6), 33 (1971).
174. B. B. Boonstra and A. A. Medalia, *Rubber Age* (*N.Y.*) **92**(6), 892 (1963); **93**(1), 82 (1963).
175. J. C. Blondel, *Rev. Gen. Caoutch. Plast.* **44**, 1011 (1967).

176. M. Shundo, T. Hidaka, K. Goto, M. Imoto, and Y. Minoura, *J. Appl. Polym. Sci.* **12,** 975 (1968).
177. K. Satake, *Rev. Gen. Caoutch. Plast.* **46,** 63 (1969).
178. K. Satake, *J. Appl. Polym. Sci.* **14,** 1007 (1970).
179. K. Satake, K. Shinki, T. Teraoka, and S. Ibe, *J. Appl. Polym. Sci.* **15,** 2775 (1971); **15,** 2807 (1971).
180. M. Morton, *J. Elastoplast.* **3,** 112 (1971).
181. M. Morton and J. C. Healey, *ACS/CIC. Div. Rubber Chem., 92nd Meet., 1967; Rubber Chem. Technol.* **41,** 773 (1968).
182. W. F. Fischer, *ACS/CIC, Div. Rubber Chem., 86th Meet., 1964.*
183. J. J. Danic, D. C. Edwards, and J. Walker, *Rubber World* **149**(5), 36 (1964).
184. E. W. Duck and P. W. Milner, *Proc. Int. Rubber Technol. Conf., Brighton* p. 359 (1967).
185. A. Nishioka, J. Furukawa, S. Yamashita, and T. Kotani, *J. Appl. Polym. Sci.* **14,** 1183 (1970).
186. G. A. Blokh *et al., Radiats. Khim. Polim., Mater. Simp., Moscow, 1964* p. 321 (1966).
187. K. Satake, *J. Appl. Polym. Sci.* **15,** 1819 (1971).
188. M. Imoto, *J. Soc. Rubber Ind. Jpn.* **42,** 921 (1969).
189. W. K. Fischer (to Uniroyal Ltd.), U.S. Patent 3,806,558 (1971).
190. T. J. Sharp and J. A. Ross, *Trans. Inst. Rubber Ind.* **37,** 157 (1961).
191. W. J. Abrams, *Rubber Age (N.Y.)* **91**(2), 255 (1962).
192. W. A. Wilson, *Rubber Age (N.Y.)* **90**(1), 85 (1961).
193. E. A. Devirts and A. S. Novikov, *Sov. Rubber Technol.* **27**(5), 11 (1968).
194. J. W. Horvath, W. A. Wilson, H. S. Lundstrom, and J. R. Purdon, *ACS/CIC. Div. Rubber Chem., 92nd Meet., 1967; Rubber Chem. Technol.* **41,** 770 (1968).
195. A. H. Jorgensen and D. G. Frazer, *Rubber World* **157**(6), 57 (1968).
196. S. L. Fisher, A. M. Perminov, I. I. Radchenko, I. Y. Ezerets, and T. E. Yurchuk, *Sov. Rubber Technol.* **29**(6), 7 (1970).
197. T. Schuette and G. Jacobs, U.S. Patent 3,296,009 (1967).
198. D. C. Allport and W. H. Janes, eds., "Block Copolymers." Appl. Sci. Publ., London, 1973.
199. R. J. Ceresa, "Block and Graft Copolymers." Butterworth, London, 1962.
200. R. J. Ceresa, *in* "Encyclopaedia of Polymer Science and Technology" (N. M. Bikales, ed.), Vol. 2, p. 485. Wiley (Interscience), New York, 1965.
201. G. Holden, E. T. Bishop, and N. R. Legge, *J. Polym. Sci., Part C* **26,** 37 (1969).
202. D. J. Meier, *J. Polym. Sci., Part C* **26,** 81 (1969).
203. L. J. Fetters, *J. Polym. Sci., Part C* **26,** 1 (1969).
204. M. Zwarc, M. Levy, and R. Milcovich, *J. Am. Chem. Soc.* **78,** 2656 (1956).
205. M. Matsuo, *Jpn. Plast.* **2**(3), 6 (1968).
206. S. Krause, *J. Polym. Sci., Part A-2* **7,** 249 (1969).
207. S. Krause, *Macromolecules* **3,** 84 (1970).
208. U. Bianchi, E. Pedermonte, and A. Turturro, *Polymer* **11,** 268 (1970).
209. P. R. Lewis and C. Price, *Polymer* **12,** 258 (1971).
210. P. R. Lewis and C. Price, *Polymer* **13,** 20 (1972).
211. D. F. Leary and M. C. Williams, *J. Polym. Sci., Part B* **8,** 335 (1970).
212. T. Uchida, T. Soen, T. Inoue, and H. Kawai, *J. Polym. Sci., Part A-2* **10,** 101 (1972).
213. T. Soen, T. Inoue, K. Miyoshi, and H. Kawai, *J. Polym. Sci., Part A-2* **10,** 1757 (1972).
214. J. Kohler, A. Banderet, G. Riess, and C. Job, *Rev. Gen. Caoutch. Plast.* **46,** 1317 (1969).
215. T. Inoue, T. Soen, T. Hashimoto, and H. Kawai, *J. Polym. Sci., Part A-2* **7,** 1283 (1969).
216. D. J. Meier, *Polym. Prepr., Am. Chem. Soc., Div. Polym. Chem.* **11**(2), 400 (1970).
217. D. F. Leary and M. C. Williams, *J. Polym. Sci., Part A-2* **11,** 345 (1973).
218. C. Price, A. G. Watson, and M. T. Chow, *Polymer* **13,** 333 (1972).
219. J. F. Beecher, L. Marker, R. D. Bradford, and S. L. Aggarwal, *J. Polym. Sci., Part C* **26,** 117 (1969).

220. M. Matsuo, S. Sagae, and H. Asai, *Polymer* **10**, 79 (1969).
221. D. McIntyre and E. Campos-Lopez, *Macromolecules* **3**, 322 (1970).
222. G. Kämpf, H. Krömen, and M. Hoffman, *Rubber Chem. Technol.* **45**, 1564 (1972).
223. W. R. Wrigbaum, S. Yazgan, and W. R. Tolbert, *J. Polym. Sci., Part A-2* **11**, 511 (1973).
224. H. J. M. A. Mieras and E. A. Wilson, *J. Inst. Rubber Ind.* **7**, 72 (1973).
225. R. G. C. Arridge and M. J. Folkes, *J. Phys. D* **5**, 344 (1972).
226. J. T. Bailey, E. T. Bishop, W. R. Hendricks, G. Holden, and N. R. Legge, *Rubber Age* (*N.Y.*) **98**(10), 69 (1966).
227. T. L. Smith and R. A. Dickie, *J. Polym. Sci., Part C* **26**, 163 (1969).
228. D. M. Brunwin, E. Fischer, and J. F. Henderson, *J. Polym. Sci., Part C* **26**, 135 (1969).
229. C. W. Childers and G. Kraus, *Rubber Chem. Technol.* **40**, 1183 (1967).
230. R. E. Cunningham, M. Auerbach, and W. J. Floyd, *J. Appl. Polym. Sci.* **16**, 163 (1972).
231. M. Matsuo, T. Ueno, H. Horino, S. Chujyo, and H. Asai, *Polymer* **9**, 425 (1968).
232. M. Morton, J. E. McGrath, and N. R. Legge, *J. Polym. Sci., Part C* **26**, 99 (1969).
233. J. F. Henderson, K. H. Grundy, and E. Fischer, *J. Polym. Sci., Part C* **16**, 3121 (1968).
234. G. S. Fielding-Russell, *Rubber Chem. Technol.* **45**, 252 (1972).
235. T. Inoue, T. Soen, T. Hashimoto, and H. Kawai, *Macromolecules* **3**, 87 (1970).
236. L. J. Fetters, B. H. Meyer, and D. McIntyre, *J. Appl. Polym. Sci.* **16**, 2079 (1972).
237. G. Riess, J. Kohler, C. Tournut, and A. Banderet, *Rubber Chem. Technol.* **42**, 447 (1969).
238. General Tire and Rubber Co., Br. Patent 1,259,932 (1972).
239. K. Satake, *J. Soc. Rubber Ind. Jpn.* **43**, 992 (1970); **42**, 501 (1969); **42**, 506 (1969).
240. K. Satake, K. Shinki, T. Teraoka, and S. Ibe, *J. Soc. Rubber Ind. Jpn.* **44**, 294 (1971).
241. K. Satake, K. Shinki, T. Teraoka, and S. Ibe, *J. Soc. Rubber Ind. Jpn.* **44**, 299 (1971).
242. K. Van Henten and R. W. Stephens, *Rev. Gen. Caoutch. Plast.* **48**, 131 (1971).
243. Asai Kasei Kogyo Kabu-Shiki Kaisha, Br. Patent 1,298,813 (1969).
244. R. T. L. Flair and J. F. Henderson, U.S. Patent 3,641,205 (1968).
245. A. Noshay, M. Matzner, G. Karoly, and G. B. Stampa, *J. Appl. Polym. Sci.* **17**, 619 (1973).
246. G. Karoly, *Polym. Prepr., Am. Chem. Soc., Div. Polym. Chem.* **10**(2), 837 (1969).
247. L. J. Fetters and M. Morton, *Macromolecules* **2**, 453 (1969).
248. J. C. Saam and F. W. G. Fearon, *Ind. Eng. Chem., Prod. Res. Dev.* **10**, 10 (1971).
249. T. Tanaka and T. Yokoyama, *J. Polym. Sci., Part C* **23**, 865 (1968).
250. T. Tanaka, T. Yokoyama, and Y. Yamaguchi, *J. Polym. Sci., Part A-1* **6**, 2137 (1968).
251. G. M. Estes, S. L. Cooper, and A. V. Tobolsky, *J. Macromol. Sci., Rev. Macromol. Chem.* **4**, 313 (1970).
252. K. C. Frisch, *Rubber Chem. Technol.* **45**, 1442 (1972).
253. D. C. Allport and A. A. Mohajer, *in* "Block Copolymers" (D. C. Allport and W. H. Janes, eds.), Ch. 8C. Appl. Sci. Publ., London, 1973.
254. J. A. Koutsky, N. V. Hien, and S. L. Cooper, *J. Polym. Sci., Part B* **8**, 353 (1970).
255. L. L. Harrell, *Macromolecules* **2**, 607 (1969).
256. G. W. Miller, *J. Appl. Polym. Sci.* **15**, 39 (1971).
257. J. Ferguson, D. J. Hourston, R. Meredith, and D. Patsavoudis, *Eur. Polym. J.* **8**, 369 (1972).
258. R. Rogers, *RAPRA Inf. Rep.* No. 5913 (1968).
259. L. H. Sperling, A. F. George, and V. Huelck, *J. Appl. Polym. Sci.* **14**, 2815 (1970).
260. V. Huelck, D. A. Thomas, and L. H. Sperling, *Macromolecules* **5**, 340 (1972).
261. L. H. Sperling and H. D. Sarge, *J. Appl. Polym. Sci.* **16**, 3041 (1972).
262. V. Huelck, D. A. Thomas, and L. H. Sperling, *Macromolecules* **5**, 348 (1972).
263. P. J. Corish and B. D. W. Powell, *Rubber Chem. Technol.* **47**, 481 (1974).

Chapter **13**

Thermoplastic Elastomers

JAMES C. WEST* and STUART L. COOPER

DEPARTMENT OF CHEMICAL ENGINEERING
UNIVERSITY OF WISCONSIN
MADISON, WISCONSIN

I. Introduction

A. MOLECULAR STRUCTURE

The term "rubberlike elasticity" implies [1] that a material can be extended to several times its original length, yet return rapidly to nearly its initial dimen-

---

* Present address: Allied Chemical, Morristown, New Jersey.

Copyright © 1978 by Academic Press, Inc.
All rights of reproduction in any form reserved.
ISBN 0-12-234360-3

sions upon removal of the deforming force. This phenomenon requires long-chain macromolecules composed of many repeat units which can rotate fairly freely about their backbone bonds. The restoring force is generated by the reduction of conformational entropy during extension of the sample. Rapid recovery requires that the molecular motions experience little restriction, so that the material must be largely amorphous and well above ($\sim 50°C$ above) its glass transition temperature (T_g). In order to prevent liquidlike flow at temperatures above T_g, the macromolecules must be linked together to form a network structure. In conventional rubbers this is accomplished by cross-linking reactions (vulcanization) that insert chemical bonds between chains. The cross-link points must be widely spaced in order to maintain the mobility of the repeat units in the macromolecule. Unfortunately, chemical cross-linking also renders materials insoluble and incapable of being further shaped by heat and pressure.

Thermoplastic elastomers are block copolymers that exhibit rubberlike elasticity [1] without requiring chemical cross-linking. Block copolymers that behave as thermoplastic elastomers are broadly described as either ABA or $(AB)_n$ polymers, according to the number and distribution of similar repeat units per macromolecule. Typical structures are represented in Fig. 1. Intermolecular segregation of similar repeat units in block copolymer molecules gives rise to microphase separation and the formation of domains. For rubberlike solids, the chemical structure of the blocks is chosen such that both a rigid high-modulus phase and a flexible elastomeric phase are formed. Consequently, the hard domains serve as virtual cross-links by providing junction points for the rubbery chain segments. At the same time, the hard domains also serve as reinforcing filler, similar to the effect of carbon black in conventional rubbers.

The reinforcing domains of elastomeric block copolymers are rigid because their chain segments are below their glass transition temperature (T_g) or melting point (T_m) when at the temperature of polymer service. Since the hard domains soften at their transition temperature, a thermally reversible elastomeric system results, and rapid thermoplastic processing methods such as injection moldings and extrusion can be applied. These materials are also easily applied as coatings and are used estensively as additives, processing aids, and compatibilizing agents.

Block Polymer Type	Block Structure	Typical Degree of Polymerization	
		A	B
Vinyl-Diene (ABA)	A ----B---- A	100-200	800-1600
Segmented Polyurethane $(AB)_n$	A --B-- A --B-- A	1-10	15-30

Fig. 1. Thermoplastic elastomer molecular structures. A denotes hard segment, B soft segment.

B. BLOCK COPOLYMER SYNTHESIS

The current understanding and utilization of thermoplastic elastomers derives in part from the development of refined synthesis techniques* which allow careful design of molecular structure. For example, homogeneous anionic polymerization methods [2–4] provide close control of the molecular weight, molecular weight distribution, and composition of each block in ABA copolymers. Certain vinyl, diene, and cyclic monomers can be polymerized by an anionic mechanism with no inherent termination step [3, 4]. Since the growing chain remains active in such "living polymers," different monomers may be added stepwise to build each block in sequence. The number average molecular weight ($\overline{M}_n$) of each block depends only on the concentrations of initiator and polymerizable monomer in the system. In the case of monofunctional initiators, the number average molecular weight is given by Eq. (1).

$$\overline{M}_n = \frac{\text{grams of monomer}}{\text{moles of initiator}} \tag{1}$$

Furthermore, if the initiation step is fast relative to propagation, the molecular weight distribution approaches the Poisson distribution [5], which is virtually monodisperse for chains with more than 100 repeat units.

$$\bar{p}_w / \bar{p}_n \cong 1 + (1/\bar{p}_n) \tag{2}$$

where $\bar{p}_n$ and $\bar{p}_w$ represent number and weight average degrees of polymerization.

Narrow molecular weight distribution is a consequence of anionic polymerizations in which nonterminating chains are initiated at nearly the same time and grow at nearly the same rate until the monomer supply is used up. Ultraclean high-vacuum polymerization systems are frequently used to avoid termination by adventitious impurities.

Block copolymers [6] can also be prepared by combining active chain ends with coupling agents, such as multifunctional organic halides. Alternatively, different low molecular weight polymer segments can be prepared independently, using difunctional initiators, and each chain end can be capped with a functional group [4]. End-capped polymer segments can be combined by means of coupling agents or can be chain extended to form $(AB)_n$ block copolymers. Relatively few monomers [3, 4] are capable of forming living polymers, but the control of molecular structure that is possible by using anionic polymerization techniques has been of great value in correlating the structure and properties in a number of styrene–butadiene and styrene–isoprene block copolymers.

Many polyurethanes undergo microphase separation and domain formation that is characteristic of block copolymers. These polymers are synthesized

* See also Chapter 2.

by condensation polymerization [7]. Soft blocks are composed of linear dihydroxy polyethers or polyesters with molecular weights between 600 and 3000 (Fig. 1). In a typical polymerization the macroglycol is end capped with an aromatic diisocyanate. Subsequently the end-capped prepolymer is further reacted with an excess of monomeric diol and additional diisocyanate. The diol links the prepolymer segments together while excess diol and diisocyanate form short hard-block segments, leading to the $(AB)_n$ structure illustrated in Fig. 1. Block lengths in $(AB)_n$ polymers are frequently much shorter than those in anionically synthesized ABA block copolymers.

Molecular structure can be widely varied by changing the chemical composition of the three reactants (macroglycol, diisocyanate, and monomeric diol) or by changing the method of polymerization. All three reactants can be simultaneously polymerized in a one-step [8] reaction or they can be added sequentially [9] after forming an isocyanate-capped prepolymer. The one-step and two-step procedures can lead to polymers with different segmental molecular weight distributions. Chemical composition determines many molecular properties, such as polarity, hydrogen bonding capability, and crystallizability of the blocks. If monomeric diols are replaced by diamines, highly polar urea linkages are formed in the hard blocks. Polyurethanes have also been synthesized with piperazine replacing the diisocyanate [10], thereby eliminating all possibility of hydrogen bonding.

Vinyl–diene triblock copolymers and segmented polyurethanes are among the most extensively investigated of ABA and $(AB)_n$ thermoplastic elastomers because they were the earliest to be produced and commercialized (Kraton,* Estane†). However, block copolymers can be synthesized from a very broad range of compositions, and elastic noncross-linked materials can be produced by a variety of polymerization mechanisms. In some cases, such as spandex fibers [11], elastomers may owe their properties to phase separation and chemical cross-linking. Virtual cross-linking effects have also been suggested for elastomeric polymers that are not block copolymers, such as carboxylic rubbers [12], ionomers [12], and plasticized poly(vinyl chloride) [13]. This chapter will be limited to block copolymers whose structure is similar to vinyl–diene or polyurethane thermoplastic elastomers.

C. METHODS OF STRUCTURE INVESTIGATION

Thermoplastic elastomers exhibit structural heterogeneity on the molecular, the domain, and in some cases on a larger scale involving periodic or spherulitic texture. An understanding of the properties and morphology in thermoplastic

---

* Trademark, Shell Chemical Company.
† Trademark, B. F. Goodrich Chemical Company.

elastomers requires correlation of structures on each of these levels with their effect on the physical behavior of the material.*

Each level of structural organization is studied by specific methods. Molecular sequence distributions may be studied by chemical methods such as nuclear magnetic resonance or infrared spectroscopy, or they may be predicted by the chemistry of polymerization. Domain structures are studied directly by electron microscopy or small-angle X-ray diffraction, methods that are particularly applicable because of the size range of typical domains. Diffraction methods provide structural information which is averaged over the whole sample and are particularly suited for periodicity studies. Electron microscopy, on the other hand, provides direct information on the structure of specific loci within the sample. These methods are complemented by differential calorimetry and by various techniques for studying dynamic–mechanical behavior, which can be interpreted to give additional, if somewhat less direct, information on domain structure.

Molecular orientation effects have been studied by infrared dichroism, birefringence, and X-ray diffraction. Small-angle light scattering and polarized light microscopy have been employed to clarify spherulitic structure and deformation behavior. This list of techniques that can be applied to structure–property studies of thermoplastic elastomers is not intended to be exhaustive but rather is meant to illustrate the general approaches to structure–property investigations in thermoplastic elastomers.

II. Morphology of Thermoplastic Elastomers

A. Compatibility and Domain Formation

When two components are mixed, the change of free energy is given by the equation

$$\Delta G = \Delta H - T \Delta S \tag{3}$$

where ΔH and ΔS are the enthalpy and entropy changes of mixing, respectively, and T is the absolute temperature. For true molecular solution the net effect of the enthalpy and entropy changes must lead to a negative value for the free energy of mixing. If ΔG is positive, the two components will segregate and form separate phases. This criterion of phase separation is strictly applicable to reversible or equilibrium processes. However, even though the mixing of two polymer molecules is frequently not an equilibrium process, due to restricted large-scale molecular mobility, the analysis is qualitatively useful.

Mixing of two components is accompanied by an increase in the disorder of the system and hence by an increase in entropy, $\Delta S > 0$. The entropy change

* See also Chapter 3.

has the proper sign to favor solution, but its magnitude is much smaller for the mixing of two polymer molecules than for the mixing of lower molecular weight materials with equivalent mass [14]. Although ΔS favors homogeneity, it is a small effect and it will be overwhelmed by anything other than a negative, zero, or only slightly positive enthalpy value. A positive ΔH value is the usual case for polymers where only nonpolar forces are concerned, since it indicates the preference of polymer molecules for their own environment (coil interpenetration) rather than for a mixed system. The usual result is that a positive enthalpy term dominates the free energy expression, so that it is unlikely that a single homogeneous phase will form when two polymer molecules are mixed. Commonly two distinct phases exist in blends, block copolymers, and graft copolymers. It has even been suggested that many random copolymers should be expected to separate into two or more phases as a result of their composition distribution [15–17].

Determination of phase separation on the scale found in block copolymers is not always unambiguous. Phase-separated materials of differing refractive index will generally appear opaque or opalescent due to the scattering of light from the phase boundaries or interface. On the other hand, many phase-separated copolymer systems are very nearly transparent owing to the formation of domains that are smaller (< 500 Å) than the wavelength of visible light. The phenomenon of phase separation and its detection in blends and block or graft copolymers has recently been considered by several authors [18–21]. Evidence for phase separation can be obtained by optical methods, such as electron microscopy, or by analysis of the thermal or dynamic–mechanical properties. A homogeneous copolymer should have a single glass transition (T_g) at a temperature intermediate between that of its component homopolymers. On the other hand, a microphase-separated block copolymer or blend [22, 23] should have two T_g's, at temperatures representative of each of its homopolymer constituents. This is a practical rule, but other factors are known to be significant. For example, experimental work on blends [23] suggests that rate processes [12] may have to be considered in predictions of the degree of polymer inhomogeneity.

B. THERMODYNAMICS OF DOMAIN FORMATION

Several authors [24–43] have developed thermodynamic expressions to describe equilibrium phase separation in block copolymers and to predict the size and shape of the domains that are subsequently formed. Fedors [24] applied the Flory–Huggins liquid lattice theory to estimate minimum molecular weights for phase separation in nonpolar ABA block copolymers. Chemical potentials of mixing were assumed to be affected by chemical structure only through a polymer–polymer interaction parameter, χ_{12}, described by Eq. (4).

$$\chi_{12} = M_1(\Delta\delta)^2/\rho_1 RT \tag{4}$$

where M_1 is the molecular weight of component 1, ρ is the density of component 1, and $\Delta\delta$ is the difference in solubility parameters for the two components. By considering a reasonable arbitrary value for χ_{12}, minimum molecular weights of 6000 for a butadiene segment and 2500 for a styrene segment were predicted in order to achieve immiscibility in a triblock copolymer. Assuming spherical domains, the maximum polystyrene domain diameter was estimated to be about 300 Å.

Krause [25–28] has developed a strictly thermodynamic approach to phase separation, independent of morphology. The treatment predicts microphase separation in systems with crystalline or noncrystalline domains, systems with any number of blocks, and mixtures of block copolymers with one of the corresponding homopolymers. It assumes a block copolymer composed of molecules with the same lengths, average composition, and sequence distribution of blocks. Long block lengths and complete phase separation with sharp boundaries are also assumed.

For a noncrystallizing block copolymer, the entropy change per copolymer molecule due to phase separation is represented by Eq. (5).

$$\Delta S/k = \ln(v_1)^{v_1}(v_2)^{v_2} - 2(m - 1)(\Delta S_{dis}/R) + \ln(m - 1) \qquad (5)$$

where v_1 and v_2 are volume fractions of the respective monomer units, k is Boltzmann's constant, and m is the number of blocks per molecule. The first term on the right-hand side of this equation represents the entropy loss due to the decrease in volume available to each block upon phase separation, assuming no volume change on mixing. The second term represents the additional entropy loss due to repeat unit immobilization at the sharp phase interfaces. The orientation entropy loss per immobilized unit is $(\Delta S_{dis}/R)$. It must be assigned an experimental or theoretical value. The third term expresses the effect of increased repeat unit immobilization at interfaces when more than three blocks are present in the copolymer molecules. A crystallizable block copolymer would require an additional term in Eq. (5) to account for the additional entropy change due to crystallization of one of the phases.

The enthalpy change upon microphase separation is described by a Hildebrand–Van Laar expression, Eq. (6).

$$\Delta H = (-kTV/V_r)v_1v_2\chi_{12}(1 - 2/Z) \qquad (6)$$

where T is the absolute temperature, V the total volume of the system, V_r the volume of a lattice site, Z the lattice coordination number, and χ_{12} an interaction parameter between repeat units of type 1 and type 2. Again, an additional term is required if one of the blocks is crystallizable. The entropy and enthalpy expressions are related to the Gibbs free energy by Eq. (3). Like any new phase formation, microphase separation occurs when, for this process, $\Delta G \leq 0$, so that the critical interaction parameter $(\chi_{12})_{cr}$ can be calculated for various compositions and block lengths by setting $\Delta G = 0$.

Relationships (4)–(6) predict that phase separation in amorphous block copolymers should be favored by high molecular weight of the blocks, or by a minimum number of blocks per molecule. For molecules with the same chain length and m value, a copolymer of 50/50 volume composition is more likely to phase separate than a 25/75 copolymer. When the composition of one block is crystallizable, the enthalpy and entropy of crystallization of the polymer dominates the value of $(\chi_{12})_{cr}$. For a 50/50 block copolymer in which, e.g., one block is crystallizable ethylene oxide, which is only six units long, the theory predicts microphase separation regardless of the chemical composition of the other block. Unfortunately, Krause's theory was derived independently of morphological considerations, so that it cannot be used to predict domain shapes and dimensions.

C. DOMAIN SIZE AND SHAPE

Statistical thermodynamic treatments have been developed which include predictions of domain size and shape. For example, Bianchi et al. [33, 34] related heats of mixing to interfacial energy, and in turn to domain size. The equations developed show how demixing of the system is favored by energetic terms and opposed by entropic terms. A higher interfacial energy leads to a greater tendency for aggregation into domains. Domain size was also predicted to decrease with increasing temperature.

Both the size and shape of block copolymer domains are influenced by conformational entropy and interfacial energy. Since the copolymer chains are made up of chemically dissimilar segments, one must expect different solubility behavior for different parts of the chain. In fact it has been found that block copolymers, in concentrated solutions, behave very much like soaps in that they form ordered micellar structures. The colloidal properties of block copolymers and their morphological ramifications have been reviewed by Molau [44]. When a block copolymer solution is allowed to evaporate, spherical, cylindrical, or lamellar domains may be formed. The structure is presumed to persist from the concentrated solution ($\sim 10\%$) into the solvent-free polymer film. Possible structures for a block copolymer with A and B blocks are shown in Fig. 2 [44]. The specific geometry depends mainly on the volume ratio of the two phases.

The dependence of domain shape on relative molar volumes of the blocks was intuitively described by Skoulios [45] with respect to the behavior of an AB copolymer in a solvent that was selective for one of the blocks. In a lamellar structure, the interface between the domains is a plane and approximately equal space is available to the blocks on either side of the interface. If one block occupies less molar volume than the other, it might be expected to be located on the concave side of a curved plane, eventually forming a cylinder. For a spherical structure, a doubly curved interface, one would expect even more

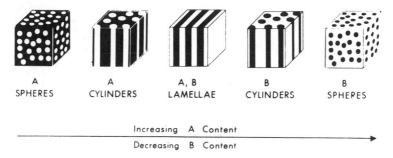

A	A	A, B	B	B
SPHERES	CYLINDERS	LAMELLAE	CYLINDERS	SPHERES

Increasing A Content

Decreasing B Content

Fig. 2. Domain morphology in thermoplastic elastomers [44].

disproportionate molar volumes to be occupied by the blocks. It should be noted that similar morphological structures are formed from the melt as well as by casting from solution.

Thermodynamic analyses of domain shapes and sizes have been developed by Meier [29–32] and by Inoue *et al.* [36, 37]. Meier's derivation considers AB copolymers, but the results are presumed applicable to ABA copolymers that can be represented by $(A - B/2)$ diblock structures. Meier assumed uniform segment density within domains and the applicability of random flight statistics to segmental conformations. It was shown that domain size is limited because of the restriction of A and B units to specific regions of space, with A–B junctions located at the interface. The diffusion equation was used to calculate density distribution functions, which were solved for various ratios of domain size to chain dimensions. Values of this ratio were chosen which gave the most uniform segment densities, resulting in a series of equations of the form shown in Eq. (7), where specific values of k are required for each domain shape.

$$D = k(\sigma l^2)^{1/2} = k\alpha(\sigma l^2)_0^{1/2} = kC\alpha M^{1/2} \tag{7}$$

In Eq. (7), D is the characteristic domain dimension (radius of sphere or cylinder; thickness of lamella), σ the number of statistical elements of length l in the chain, α the chain perturbation parameter (ratio of perturbed to unperturbed root-mean-square end-to-end distance, $\alpha^2 = (\sigma l^2)/(\sigma l^2)_0$), and k is a parameter depending on domain shape and molecular architecture. The chain perturbation parameter α accounts for deviations from random coil dimensions, which are required in order to satisfy the uniform segment density assumption.

The free energy change in forming a domain system from a random solution of a block copolymer includes contributions from interfacial energy, placement and segment constraint, and the free energies of mixing and of chain perturbation. Domain morphologies were considered, in terms of relative free energies, as a function of volume fraction of casting solvent in the system. It was found that, in principle, either component of a block copolymer could be forced to be the dispersed phase by adding a preferential solvent for one of the domains.

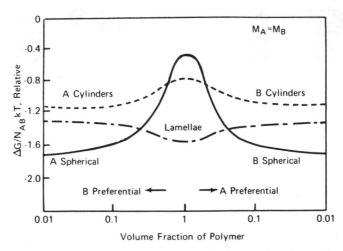

Fig. 3. Free energy and solvent content relationships for different domain morphologies [30].

Figure 3 indicates that in an AB block copolymer with equal block lengths, lamellar morphology should be favored in the absence of solvent, and that spherical domains become stable at concentrations below about 20% polymer. Surprisingly, cylindrical domains are unstable at all concentrations. Although minimum free energy morphologies are thermodynamically most stable, Meier emphasizes the large kinetic barrier against transformation of a metastable domain into a lower free energy state. Once microphase separation has taken place under a given set of conditions, molecular transport through incompatible viscous or glassy media would be required to change morphological form in a solvent-free system. Consequently, the morphology that is first formed upon evaporation of solvent is expected to be retained even when it becomes thermodynamically unstable under severe annealing conditions.

Vinyl–diene block copolymers are particularly suited to electron microscopic investigations because of residual double bonds found in their soft segments. The soft segments can react with osmium tetroxide [46], which increases their electron density, which gives rise to distinct high-contrast electron micrographs. Marker [38] has summarized typical AB and ABA domain dimensions that were experimentally determined. Composition-dependent lamellar thicknesses were observed to range from 150 Å to 500 Å, while spherical domains between 100 Å and 300 Å were found. Other authors [47, 48] have estimated the spherical hard-segment domains in useful thermoplastic elastomers at about 200 Å. Somewhat smaller domains [49], ~ 30–100 Å, have been suggested for segmented polyurethane elastomers.

Electron microscopy [50–53] has shown progressive changes of morphology with block composition, from spheres to rods to lamellae, qualitatively con-

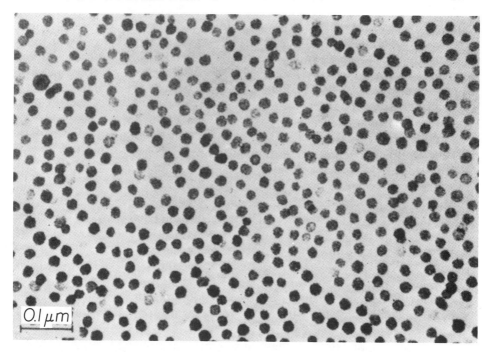

Fig. 4. Spherical butadiene domains in ABA block copolymer (80% polystyrene content) film cast from toluene solution. The same patterns were observed in both normal and parallel sections, confirming the sphericity of butadiene domains [53].

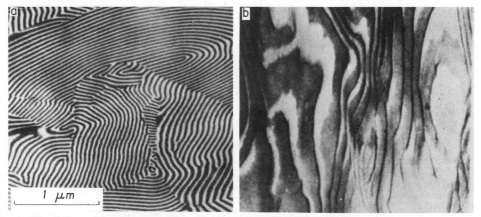

Fig. 5. Lamellar domain in ABA block copolymer (40% polystyrene content) film cast from cyclohexane: (a) normal section; (b) parallel section with lamellar layers orienting their surface parallel to film surface [53].

firming theoretical predictions. Figure 4 illustrates spherical domains formed by casting a film of styrene–butadiene (80% styrene) triblock copolymer from toluene solution. The spherical shape of the minor component, the darkly stained butadiene domains, was confirmed by photographing sections parallel and normal to the film surface. Figure 5 [53] shows striking evidence of lamellar domain formation by normal and parallel views of a 40% styrene–butadiene triblock cast from cyclohexane.

The large effect of sample history suggests that both thermodynamic and kinetic factors are important to domain morphology. The selectivity of the casting solvent [35] as well as the rate of evaporation [48, 51] have been found to affect domain structure significantly. Solvent-cast films that are dried slowly are more likely to approach equilibrium morphology, whereas quickly dried films show less distinct, more irregular structures. These diffuse structures, however, achieved much more uniform shape and size and rearranged into more regular and distinct arrays when annealed above the T_g of their hard segments. Orientation [48] of domain structure along melt flow lines has been found in compression-molded samples. The method of sample preparation, including the type and time scale of solvent and thermal history, must be given careful consideration when evaluating properties and morphology of thermoplastic elastomers.

The presence or extent of phase mixing at domain boundaries is still controversial. Several thermodynamic treatments [25, 31, 33, 36, 41] of phase separation have assumed a discrete interface with no intersegmental mixing. However, Leary and Williams [39, 40], LaFlair [43], and Meier [29] have recently proposed derivations that consider the presence of a mixed interface. Skoulios [45] and Kim [54] have presented X-ray diffraction data that indicate a negligible mixed-phase interface. On the other hand, others, interpreting data from stress–relaxation [55], dynamic–mechanical properties [56], X-ray diffraction [57], and electron microscopy [51] experiments have suggested a significant diffuse interface layer. Meier [29] has extended his thermodynamic treatment of phase separation in block copolymers so that it applies to mixed-interface systems. He found that the fraction of the system that is intermixed at the interface ($\Delta V/V$) should rapidly rise with molecular weight and/or the interaction parameter between the components. This is shown in Fig. 6, where the interfacial fraction is plotted against a molecular weight and interaction parameter product $\Lambda = (\delta_1 - \delta_2)^2$. The theory predicts that very large interaction parameter differences should be required for phase separation in systems with overall molecular weights $\leq 10^4$, and that domain formation in such systems will usually occur only when an additional energy source (e.g., crystallization or hydrogen bonding) is available. Some of the different viewpoints on diffuse phase boundaries may then be a result of different chemical structures and molecular weights in the block copolymers that are studied.

There has also been controversy concerning the type of organization com-

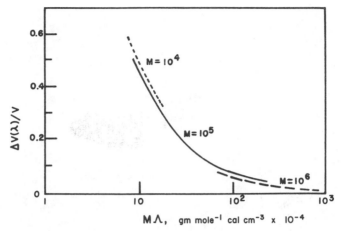

Fig. 6. Interfacial fraction versus product of molecular weight and interaction parameter [29].

monly found in ABA styrene–diene thermoplastic elastomers with polystyrene segment compositions of 10–35 wt%. Spherical domains arranged in cubic arrays have been reported by several authors [47, 54, 58–63]. The regularity of these structures has been attributed [47] to entropic retractive forces in rubbery polymer chains emanating radially from each plastic domain. Other investigators [43, 64–66] have found evidence for hexagonally arranged rodlike domains. Bi *et al.* [62] have suggested that a body-centered cubic lattice can give the appearance of cylindrical shape due to overlap of spherical domains when viewed out of plane. These results emphasize the advantage of experi-

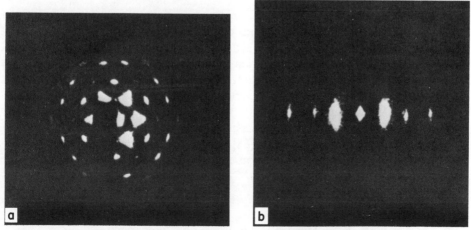

Fig. 7. Low-angle X-ray diffraction patterns from a "single-crystal" sample of extruded and annealed ABA copolymer (Kraton 102): (a) beam parallel to extrusion direction (plug axis); (b) beam perpendicular to the plug axis, which is vertical [67].

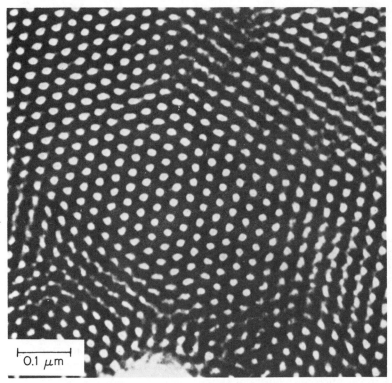

Fig. 8. Ultramicrotome section cut perpendicular to the extrusion direction of a "single-crystal" sample of extruded and annealed ABA copolymer (Kraton 102); [68], with permission of Dietrich Steinkopff Verlag.

mental measurements that combine small-angle X-ray diffraction and electron microscopy, and the value of multidirectional sampling by each technique.

Hexagonal cylindrical domain structures have been found by Keller *et al.* [67] in extruded and annealed styrene–diene ABA block copolymers. Figures 7a and 7b are low-angle X-ray diffraction patterns, taken with the beam parallel and perpendicular to the extrusion direction, respectively, which show isolated reflection patterns similar to those typical of single crystals. Figure 8 is a transmission electron micrograph from a microtomed section cut perpendicular to the extrusion direction [68] of a similar sample. An electron micrograph taken parallel to the extrusion direction had a striated structure.

III. Behavior in the Transition Temperature Range

A. Measurements of Dynamic–Mechanical Properties

Dynamic–mechanical properties [69] provide information about first- and second-order transitions (T_m and T_g, respectively), phase separation, and viscoelastic properties of polymers. A wide variety of experimental techniques are available that measure the response of materials to periodic forces over a broad range of frequencies. Results are conveniently expressed in terms of complex moduli [70], Young's storage modulus (E'), and the loss modulus (E'') as a function of the phase angle between stress and strain (tan δ).

Figure 9 shows typical storage modulus data for several representative polymer systems [71]. Below T_g, the glassy state prevails with modulus values of the order of 10^{10} dyn/cm^2 for all materials. A rapid decrease of modulus is seen as the temperature is increased through the glass transition region (above $-50°C$ for these polymers). A linear amorphous polymer that has not been cross-linked (curve A) shows a rubbery plateau region followed by a continued rapid drop in modulus. Cross-linking (curve B) causes the modulus to stabilize with increasing temperature at about three decades below that of the glassy state. In block copolymers (curve D and E) an enhanced rubbery plateau region appears where the modulus changes little with increasing temperature. Another rapid drop in modulus occurs when the temperature is increased to the hard-segment transition point. In contrast, a semicrystalline polymer (curve C)

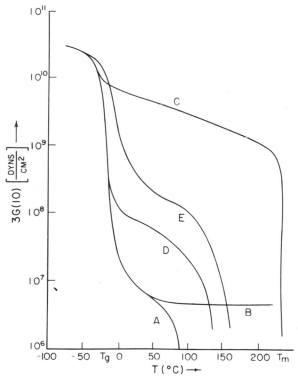

Fig. 9. Storage modulus–temperature curves: (A) linear amorphous polymer; (B) cross-linked polymer; (C) semicrystalline polymer; (D,E) polyester urethanes [71].

maintains high modulus through the glass transition region and up to the crystalline melting point, where the structural identity of the crystallites is destroyed.

In Fig. 9, curves D and E represent thermomechanical spectra for segmented polyurethane (AB)$_n$ block copolymers. Figure 10 [53] shows similar plots of both storage and loss moduli for a series of ABA styrene–diene triblock copolymers. In both cases the storage moduli are higher than those usually found for unfilled cross-linked rubbers and are comparable to those of filled rubber vulcanizates. Also, in both Fig. 9 (curves D and E) and Fig. 10, higher plateau moduli correspond to higher hard-segment concentration. Matsuo [53] showed that the block sequences per se (e.g., AB, ABA, BAB) had little effect on the thermomechanical spectra.

Two distinct transitions are indicated by the precipitous drops in storage moduli and the corresponding presence of two loss peaks. Ideally, for block copolymers these transitions are located at the T_m or T_g of the corresponding homopolymer component. However, the degree of phase separation and the block lengths have been found to influence the shape and temperature location of the dynamic–mechanical transition points. Phase mixing between domains has been indicated by decreased slopes in storage modulus transitions and by broadened loss peaks. Narkis and Tobolsky [72] found that in a dimethyl siloxane–polycarbonate (AB)$_n$ block copolymer, the T_g of the hard segment

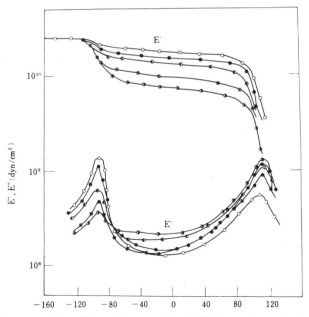

Fig. 10. Storage modulus (E') and loss modulus (E''), dynamic–mechanical properties of styrene–butadiene triblock copolymers. Styrene/butadiene mole ratio decreases from sample SBS-1 (○) to SBS-5 (◑); intervening samples are SBS-2 (●), SBS-3 (◐), and SBS-4 (⊙) [53].

(polycarbonate) is shifted to lower temperatures and becomes more diffuse as the silicone block concentration is increased. Similarly, in nonhydrogen bonded polyurethanes [73], the slope of the storage modulus curve at the hard-segment transition was found to be less for hard segments of broad molecular weight distribution (MWD) than for a narrow MWD one in polymers of similar composition and average block length. This finding was interpreted as indicating that the hard-segment domains of broad MWD were more diffuse, being diluted with interpenetrating soft segments. The overall effect of such diffuse morphology would be to hinder the motion of the soft segments and broaden the distribution of their relaxation times. In the same polyurethanes [10, 74], increasing the average hard-segment length from one to four repeat units causes an increase of 100°C in their T_m. When ABA and AB styrene–diene block copolymers with equivalent compositions (weight fraction) and overall molecular weights were compared [75, 76], the ABA copolymers with the lower hard-segment molecular weight had distinctly lower T_g values.

Narrowing of the segmental polydispersity in styrene–diene block copolymers [75, 76] caused the T_g values of the soft segments to decrease and the loss peaks in the transition region to become narrower. In a series of polyether–polyester $(AB)_n$ block copolymers the T_g of the soft segments [77] has been found to increase with hard-segment content. This was attributed to restricted soft-segment mobility because of more intersegmental contacts and to a greater degree of incorporation of hard-segment units into soft-segment domains. In studies where the morphology was controlled by preferential solvents in the film casting process [56, 78], an additional broad transition peak has been observed at an intermediate temperature between the hard- and soft-segment transitions. This peak has been attributed to incomplete phase separation caused by a solvent which is poor for one of the constituent blocks.

Extensive dynamic–mechanical property studies have been carried out on hydrogen bonded [79] and nonhydrogen bonded [73] polyurethanes. Several secondary relaxations were found in addition to the major hard- and soft-segment transitions. Molecular mechanisms could be assigned to each of these. A low-temperature γ transition ($\sim -100°C$) was attributed to localized motion in methylene sequences, mainly in the soft segments. Similar γ transitions have been found in other block copolymers [77]. In polyurethanes with long soft segments (MW $= 2000–5000$), a soft-segment melting transition was detected. The T_g occurred at lower temperatures when the length of the soft segments was increased. Since the longer segments are expected to produce better-ordered and larger domains, some soft segments can exist in regions well removed from the domain interface and hard-domain interactions, so that their motion may be relatively unrestricted by the hard domains. Soft segment T_g's were lower in nonhydrogen bonded materials than in hydrogen bonded samples with equivalent hard segment content. Presumably this was due to the absence of hydrogen bonding though increased hard segment crystallization undoubtedly contributes to better microphase separation in the piperazine based poly-urethanes [73, 80].

B. THERMAL AND INFRARED TRANSITION STUDIES

The study of transition behavior by various thermoanalytical techniques [differential thermal analysis (DTA), differential scanning calorimetry (DSC), thermal expansion measurement, thermomechanical analysis] has been important to the understanding of morphology and intermolecular bonding in thermoplastic elastomers. The two major block copolymer transitions can be determined by thermal methods [81], although absolute values may differ from those obtained from dynamic–mechanical properties due to differences in measurement frequency. In thermoplastic elastomers, secondary thermal transitions are also found. Early studies [82–84] attributed these to hydrogen bond disruption, e.g., an endotherm about 80°C for dissociation of hard- and soft-segment hydrogen bonds, and an endotherm around 150–170°C for inter-urethane hydrogen bond dissociation. Interurethane hydrogen bonding [85, 86] has been found to occur between secondary NH and carbonyl groups in the hard segments. Hard-to-soft-segment hydrogen bonding occurs between urethane NH and either the ether oxygen of poly(tetramethylene ether) or the carbonyl of the polyester segments.

More recent studies have cast doubt on the interpretation of intermediate DSC transitions in terms of hydrogen bond dissociation. Typical DSC thermal spectra for a segmented polyether urethane are shown in Fig. 11. The transition behavior is strongly dependent on thermal history. The endotherm ascribed to hard- and soft-segment interaction may be moved, by annealing, continu-

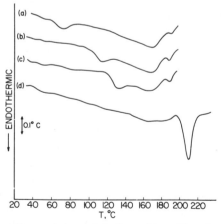

Fig. 11. Effect of annealing on intermediate thermal transitions in a hydrogen bonded poly-urethane (ET-38): (a) control; (b) 80°C for 4 hr; (c) 110°C for 4 hr; (d) 150°C for 4.5 hr.

ously up the scale in temperature until it merges with the presumed interure-thane hydrogen bond dissociation endotherm. Severe annealing can lead to a material's showing only a single microcrystalline peak. Materials of lower di-isocyanate content (and thus short hard segments) are incapable of crystalliza-tion and show only a merging of lower-temperature peaks after annealing.

The thermal lability of polyurethane hydrogen bonds has been investigated by studying the temperature dependence of infrared absorption for the NH vibration [81]. Participation in hydrogen bonding decreases the frequency of the NH vibration and increases its intensity. The hydrogen bonded NH peak is located at about 3320 cm^{-1}, whereas nonassociated NH absorbs around 3450 cm^{-1} with a significantly smaller absorption coefficient [87, 88]. Changes in hydrogen bonding can thus be followed, in principle, by IR frequency or intensity measurements. Because of the intensity decrease with bond disruption, the total area of the NH peak decreases as the number of free NH groups in-creases. For a variety of hydrogen bonding polyurethanes a slope discontinuity in the absorbance–temperature curves at about 80°C was attributed to an acceleration of hydrogen bond dissociation which occurs at the glass transition of the hard segments.

It is important to emphasize that the infrared transition near 80°C is not related to the DSC peak that appears at about the same temperature. The infrared transition is insensitive to annealing, whereas annealing effects dom-inate the DSC transitions. Furthermore, the enthalpy change accompanying thermal hydrogen bond dissociation has been shown [81] to occur at a rate too slow to be measured by the usual DSC experiment. Therefore, it seems that chain mobility controls hydrogen bonding, rather than the reverse, and that the mechanical properties of polyurethanes must be ascribed to phase separa-tion from an incompatibility of segments rather than directly to the presence of hydrogen bonds.

In summary, regions of endothermal activity observed in the DSC for polyurethane elastomers may be ascribed to morphological effects. These effects can be broadly divided into a loss of long-range order and a loss of short-range order which arise in part due to the distribution in hard-segment lengths. Short-range ordering may be continually improved by annealing; hence the merging of the low-temperature endotherms in the DSC trace. This process occurs without affecting any long-range order that may be present. Regions of short-range order can be made microcrystalline by severe annealing, but only if the average hard-segment length is sufficiently great. Recent DSC studies [81] of polyurethane block copolymers without capability for hydrogen bond-ing confirm that DSC endotherms are not related to hydrogen bond disrup-tion. DSC scans of nonhydrogen bonded polyurethanes show virtually the same intermediate transition behavior, including annealing effects, as depicted in Fig. 11.

IV. Orientation

A. X-Ray Diffraction Studies

Investigations of styrene–diene block copolymers [89] and segmented poly-urethanes [90] have shown that small-angle X-ray diffraction (SAXS) can detect domain structure and provide a powerful tool for its characterization. The use of SAXS to demonstrate the hexagonally packed macrolattice structure [67, 68] of Figs. 7 and 8, and SAXS studies of domain periodicity in ABA copolymers [58–63] have been discussed in Section II.C.

Early X-ray diffraction studies by Bonart [91] paved the way for studies of structure–property relationships on a molecular level in segmented polyure-thanes. Using a weak X-ray interference at 12 Å, identified with a particular short-range order brought about by a specific arrangement of hydrogen bonds, Bonart followed hard-segment orientation as a function of strain. The aromatic urethane segments were found to orient perpendicular to the stretch direction at elongations below about 200%. Further stretching moved the X-ray reflection to the meridian, indicating that the hard segments were then orienting into the stretch direction. A model of hard-segment domains in a soft-segment matrix was proposed to account for this behavior. Figure 12a shows the soft segments being stretched initially to different extents depending on their position within the material. In Fig. 12a, local torques acting through "force strands" of soft segment cause the long axes of the hard-segment domains to be oriented into the direction of stretch, therefore leading to orientation of the individual hard segments transverse to the stretch direction. Further elongation caused the hard segments to slip past one another, breaking up the original structure. As the elongation continued, hard segments becamse progressively oriented into the stretch direction (Fig. 12b). Bonart suggested that this restructuring of the hard segments during elongation is related to stress softening and hysteresis phenomena characteristic of these polymers.

Further studies [92] with urethane polymers of longer segment length, chain extended with butane diol, showed the existence of crystallized and of paracrystallized hard segments. Although suitable thermal treatment may induce true crystallization in a paracrystalline sample, this process does not involve a simple reordering within the original domain structure. Rather, small-angle X-ray scattering indicates that a change in form and spatial ar-rangement of the domains accompanies the crystallization process. Bonart also distinguishes between the behavior of crystalline and paracrystalline hard segments during orientation. Paracrystalline segments behave as previously described, showing two distinct mechanisms. Crystalline hard segments, how-ever, apparently behave more like an inert filler. They were found to orient preferentially orthogonal to the stretch direction, even at high elongations.

Based on wide- and small-angle X-ray scattering results, Bonart *et al.* [93]

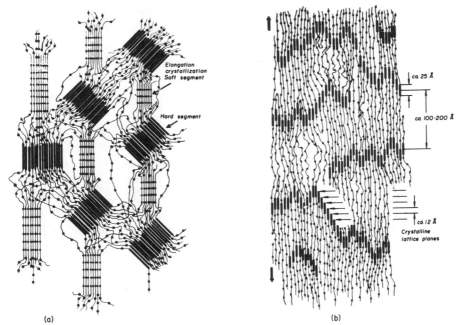

Fig. 12. (a) Schematic drawing of the structure in material stretched approximately 200%. The thick strokes represent hard segments and the thin strokes soft segments. (b) Schematic drawing of the structure in material stretched approximately 500% and subsequently fixed in water at approximately 80°C [91].

have proposed models of molecular arrangement within hard-segment domains in segmented polyurethanes and poly(urea–urethanes). Domain structure was related to mutual hard-segment affinity, enhanced by molecular arrangements which allow the formation of a maximum number of stress-free hydrogen bonds. Possible hard-segment molecular arrangements were correlated with softening temperatures and degrees of hard-segment segregation. Polyurethanes composed of macroglycol, butane diol, and diphenylmethane diisocyanate (MDI) in the molar ratio 1:1:2 had mutually soluble hard and soft segments and showed no evidence of phase separation. However, comparable diamine-extended polymers were phase separated at the same molar composition ratio. Three-dimensional hydrogen bonding, made possible by excess NH groups in the urea linkages, was proposed to explain this difference. Similar three-dimensional structures were suggested for diol-extended polyurethanes which were composed of molar ratios of at least 1:2:3.

Later papers [93, 94] describe the application of small-angle X-ray scattering to measure the state of phase separation with respect to size distribution, phase boundary sharpness, and interphase mixing. This technique was applied to hydrazine-extended polyether and polyester urea–urethanes. The higher softening point of the polyether-based polymer could not be explained by segment affinities but was tentatively attributed to the molecular weight distribution of the soft segments. Phase segregation was found to be diffuse, with a transition zone estimated at 20 Å. Hard-segment domains apparently contained a considerable quantity of soft-segment material.

B. INFRARED DICHROISM

Molecular orientation can also be studied spectroscopically, using the technique of infrared dichroism. It is of particular interest in block polymer systems when the dichroic behavior of bands characteristic of the individual blocks can be followed, i.e., if there are infrared-active groups, peculiar to each segment, that absorb in regions free from other bands. A quantitative description of segmental orientation requires a reasonably accurate knowledge of the transition-moment directions for the vibrations of interest. In polyurethanes, the NH group is characteristic solely of the hard segment and may be conveniently used to study hard-block orientation. The CH_2 group generally can be used for soft-segment orientation, although methylene groups are also present at small concentrations in the hard segment. The transition-moment direction [85] for both of these vibrations can be taken as 90°. The carbonyl group, with a transition moment at an angle approximately 79° to the chain backbone, is also used to characterize hard-segment orientation. Dichroism yields the second moment of the orientation distribution function, which is commonly expressed in terms of the uniaxial orientation function f:

$$f = (3 \overline{\cos^2 \theta} - 1)2 \tag{8}$$

where θ is the angle of orientation referenced to the stretch direction. This function has a value of unity for a sample whose elements are completely oriented in the stretch direction. For ideal orientation transverse to the stretch direction, $f = -\frac{1}{2}$, whereas for $f = 0$ there is no orientation.

Infrared dichroism has been used to study segmental orientation, relative to the degree of uniaxial extension, in polyurethane block copolymers with varied molecular structures and compositions. The viscoelastic nature of these materials is demonstrated by the time dependence of their orientation functions after imposing a 150% step strain increment. Figure 13 shows the time variation of hard- and soft-segment orientation functions for a polyurethane block copolymer [95] composed of 1000 molecular weight poly(tetramethylene ether) glycol (PTMEG), chain extended with butane diol and 31 wt% p,p'-diphenylmethane diisocyanate (MDI). Decreasing values of CH and non-hydrogen bonded (free) C=O orientation functions show that soft segments quickly relax into a more random state. As soft segments relax, they exert tension on the hard segments, represented by NH and hydrogen bonded C=O orientation functions, causing the hard segments to orient further in the stretch

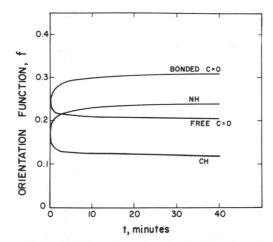

Fig. 13. Hard (bonded C=O, NH) and soft (free C=O, CH) orientation function variation as a function of time in a polyether polyurethane (ET-38-1) after 150% strain. Reprinted with permission from [95] Seymour *et al.*, *Macromolecules* **6**(6), 896 (1973). Copyright by the American Chemical Society.

direction. Free C=O absorptions arise from hard segments which are either dispersed in the soft-segment region or located at the hard-domain interface, so that they are not able to undergo hydrogen bonding. Different behavior for hard- and soft-segment orientation functions supports the existence of separate domains with different viscoelastic responses.

Orientation function values, determined on similar samples as a function of elongation, show that orientation behavior is highly dependent on hard-segment concentration and domain structure. When hard blocks average only slightly more than one MDI unit in length [79] (24 wt% MDI), infrared dichroism indicates a small, and similar, segmental orientation of hard and soft segments. The uniformly very low degree of orientation indicated that most of the molecular alignment induced by the deformation process was able to relax during the time scale of the test. The low level of mechanical properties and orientability was attributed to the inability of this material to form an interlocking domain structure.

A significant change in orientation is seen when the MDI content is increased. Figure 14 summarizes hard-segment (f_{NH}) orientation behavior for samples with 28, 31, and 38 wt% MDI. Sample ET-38-2 has 38 wt% MDI and soft segments of 2000 molecular weight. Thus, hard- and soft-segment lengths are approximately doubled in comparison with ET-38-1. In Fig. 14, the substantial segmental orientation suggests that an important change in morphology occurs when the diisocyanate content is increased from 24 to 28%.

The hard-segment orientation functions (Fig. 14) fall into two groupings. The compositions in the upper group (higher orientation functions) have been

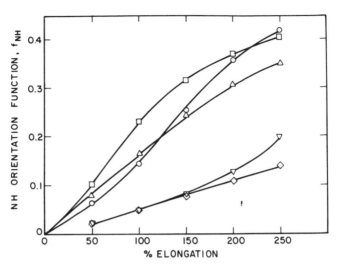

Fig. 14. Hard-segment orientation functions versus strain for polyether polyurethanes with variable hard-segment content: ○, ET-28-1; □, ET-31-1; △, ET-38-1; △, annealed ET-38-1; ◇, ET-38-2. Reprinted with permission from [95] Seymour *et al., Macromolecules* **6**(6), 896 (1973). Copyright by the American Chemical Society.

characterized as having short-range hard-segment order only [81]. Increasing the hard-block length (ET-38-2) leads to significant microcrystallinity and gives rise to initial transverse orientation of some of the hard segments, causing an overall lower degree of hard-segment orientation. This behavior can also be obtained by severe annealing (150°C for several hours) of ET-38-1. No *f* values are shown at very low strains for the polymers containing microcrystallinity because peak splitting in the dichroic spectra indicates that two types of orientation are present at these strain levels. The NH and bonded C=O peaks contain both negatively and positively orienting components. The splitting decreases with increasing strain and at 100% elongation only positive orientation is observed. The peak splitting is found only in samples with partially crystalline hard segments.

Dichroic behavior of the noncrystalline samples suggests that noncrystallized hard segments may be expected to orient positively, implying that it was the crystallized hard segments which were orienting negatively. These results may be explained on the basis of the morphological models of Estes *et al.* [85] and Bonart *et al.* [91, 92] (Fig. 12), which envision lamellar domains whose thickness is determined by hard-segment lengths. The long dimension of the domain is perpendicular to the hard-segment backbone. The observed transverse orientation of hard-segment backbone direction thus implies that at low elongations the long axis of the crystallite was turning into the stretch direction. At higher degrees of orientation the crystalline domain structure is disrupted and therefore only positive orientation is detected. Soft-segment orientation functions varied much less with composition and crystallinity.

Similar results have been found by combined hydrogen bonding and infrared dichroism studies on a series of more crystallizable segmented poly-(urea–urethanes) which exhibited spherulitic texture [89, 96, 97]. Some breaking of interurethane soft-segment hydrogen bonding occurred with increasing elongation, but no change in interurea hard-segment hydrogen bonding could be detected until elongation had reached 100%. Breakage of interurea hard-segment hydrogen bonding began at the elongation where transverse orientation reached a maximum. It was concluded that hydrogen bonding in the soft segments does not act like network junctions in determining the mechanical properties of the poly(urea–urethanes). The unusual mechanical properties were attributed to both the transverse orientation of the hard segments and the disintegration of lamellar hard-domain structure upon elongation.

Nonhydrogen bonded polyurethanes have been prepared [10] with a molecular structure in which MDI is replaced by piperazine. Figure 15 shows typical orientation function versus strain results for these polymers. Since DSC studies [73] have shown these materials to possess substantial hard segment crystallinity, the initial transverse orientation of the hard segments shown in Fig. 15 is not surprising. Comparison of Fig. 15 with Fig. 14 shows that the nonhydrogen bonded polymers develop significantly greater hard-segment orientation when elongated than do the hydrogen bonded polymers. The nonhydrogen bonded hard-segment domains are more easily disrupted than the hydrogen bonded polyurethanes, yet the system may be reinforced by a greater tendency to crystallize at high strains. Thus, piperazine segments, unhindered by interchain hydrogen bonds and by strong interdomain interactions, show a greater orientability than the hydrogen bonded MDI segments. Soft-segment orientation of these polymers is low and comparable to that of the hydrogen bonded polyurethanes.

To investigate the connection between strain history and segmental orientation, orientation functions were measured on films prestrained to various elongations. A virgin sample was stretched to 25% and allowed to come to stress equilibrium, after which dichroic measurements were made. The load was removed and the sample relaxed for 5 min, then stretched to 25% and to 50%, then relaxed, etc., through 250%. Dichroic measurements were made at each strain level generating f values at different prestrains. The results were plotted as cyclic hysteresis curves like those in Fig. 16, representing a typical microcrystalline nonhydrogen bonded polymer. Each succeeding orientation function curve passes through a minimum at about the maximum strain of the preceding run. Also, the first point of each run is at a higher f value than the last point of the preceding run. Finally, a curve drawn through the last points of each run conforms very closely to the control curve. These results can be explained by considering two types of orienting hard-segment species. One is

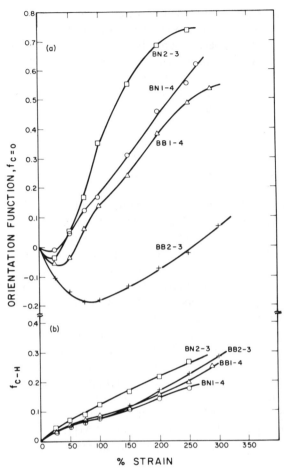

Fig. 15. Orientation function versus strain of control samples of nonhydrogen bonded poly-urethanes: (a) hard segment, $f_{C=O}$; (b) soft segment, f_{C-H}. Reproduced from A. E. Allegrezza, Jr., R. W. Seymour, H. Ng, and S. L. Cooper, *Polymer* (1974) **15**, 1, by permission of the publishers, IPC Business Press Ltd. ©.

the highly ordered, partly crystalline hard segment; the other an ordered, but not crystalline, hard segment. When a uniaxial load is applied, the partly crys-talline lamallae orient with their long dimension turning into the direction of stretching. The hard segments that make up the lamellae are arranged perpen-dicular to the long dimension of the domain and accordingly take up a trans-verse orientation. This results in a negative segmental orientation function. Also, this transverse orientation appears to be reversible when the applied load is removed. For the noncrystalline domains, the applied stress causes a shearing force which disrupts the structure and orients the hard segments in

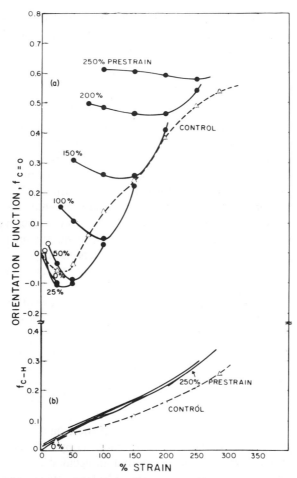

Fig. 16. Orientation hysteresis of a nonhydrogen bonded polyurethane (BB1-4) as a function of strain: (a) hard segment; (b) soft segment. Reproduced from A. E. Allegrezza, Jr., R. W. Seymour, H. Ng, and S. L. Cooper, *Polymer* (1974) **15,** 1, by permission of the publishers, IPC Business Press Ltd. ©.

the stretch direction. At higher elongations, this component outweighs the transverse orientation of the crystalline domains and a positive orientation function is obtained.

In summary, segmental orientation in polyurethane block copolymers passes through three stages as hard-segment content increases. At low content, very little equilibrium orientation is retained in either the hard or the soft segments. At somewhat higher hard-segment content, the segments retain significantly greater orientation. It has been suggested [98] that this composition level represents the point where morphology changes from independent hard domains to an interlocked structure. When the length of the hard segments is sufficiently great to permit microcrystallinity, spherulitic texture is observed and transverse orientation appears at low strains.

C. SMALL-ANGLE LASER LIGHT SCATTERING

Qualitative observations of the shape of small-angle light scattering (SALS) patterns can provide an approximation of the shape and anisotropic nature of scattering elements within a polymer film. Quantitative measurements of intensity as a function of radial scattering angle and azimuthal angle can give more detailed information regarding size and distribution of these elements. Two types of pattern are usually recorded. They are obtained by using polarized light from a laser which passes through an analyzer which may be oriented either parallel to the original polarization direction (Vv scattering) or perpendicular to it (Hv scattering). The polarized light passes through the polymer film sample and analyzer, and the scattering pattern is recorded on photographic film. The sample is usually mounted so that it may be stretched in the polarization direction of the original laser light.

Undeformed cast films of the piperazine polyurethanes with variable hard-segment length have been shown [99] to produce "four-leaf clover" Hv patterns characteristic of spherulitic scattering. The spherulitic structure was confirmed by scanning electron microscopy. Spherulite size was comparable when estimated by either technique. It appeared to increase with increasing hard-segment length. SALS studies on elongated samples showed shifting of the Hv and Vv pattern lobes that was consistent with the ellipsoidal deformation of the spherulites.

Further studies [100] on these polymers and their mixtures were made by scanning (SEM) and transmission electron microscopy (TEM). Polarizability calculations and TEM results suggested that molecular chains were aligned radially in the spherulitic superstructure. Electron microscopy revealed a porous structure whose exact morphology could be controlled by sample preparation conditions.

Spherulitic structures [96, 97] have also been detected in poly(urea–urethanes) by application of small-angle light scattering techniques. Upon deformation the SALS patterns showed that the spherulites distorted into truncated shapes. However, the spherulites survived sample extension and heat treatments, which suggested that they were quite stable, being only slightly affected by reorganization of crystalline lamellae under stress. Segmented poly(ether–ester) thermoplastic elastomers [77], which are also semicrystalline, have been found to possess spherulitic structure which deforms upon sample elongation. The deformation of an amorphous spherical domain structure into ellipsoidal shapes during extension has also been observed [101], by means of electron microscopy, in styrene–diene ABA block copolymers.

The deformation of spherulitic material, which is largely reversible at low strains, may be considered in relation to the infrared dichroism behavior of semicrystalline polyurethanes. Transverse segmental orientation arises from the lamellar radii of the spherulite orienting in the stretching direction. This implies that the chain backbones are roughly perpendicular to the long direction of the lamellae, which is usual for many crystallizable macromolecules.

V. Mechanical Properties

A. STRESS–STRAIN PROPERTIES

In order to function as a thermoplastic elastomer, a block copolymer must possess a molecular structure [102] that includes at least one soft block (B) which is terminated at each end by a hard (A) block [ABA or $(AB)_{n \geq 2}$]. Structures which do not meet this requirement [BAB or AB] have characteristically low ultimate elongations [53, 103] and tensile strengths because their soft segments are not tied into a network structure. Tensile strengths of ABA copolymers are also reduced if small quantities of AB diblock copolymer are present as "impurities." This has the effect of adding network defects to the system. On the other hand, the tensile strength can be increased [104] by the addition of polystyrene to a styrene–butadiene ABA copolymer. If the homopolymer has a molecular weight similar to that of the styrene block in the copolymer, the homopolymer is apparently incorporated into hard-segment domains, leading to a higher weight percentage of hard segments.

Thermoplastic elastomers possess the high values of elastic modulus characteristic of conventional elastomers containing chemical cross-linking and reinforcing filler. The elastic behavior of unfilled, cross-linked elastomers has been described by the semiempirical Mooney–Rivlin equation [105]

$$F = \left(\frac{\rho RT}{M_c} + \frac{2C_2}{\lambda}\right)\left(\lambda - \frac{1}{\lambda^2}\right) \qquad (9a)$$

where F is the force per unit original cross-sectional area, ρ the polymer density, R the gas constant, T the absolute temperature, λ the extension ratio, and M_c the average molecular weight between cross-links. Stress–strain and swelling measurements [102] have showed that M_c values in ABA thermoplastic elastomers are more closely related to soft-segment entanglement lengths than to the actual soft-segment molecular weights. Soft-segment entanglements, trapped in by the hard-segment tie-down points, act as physical cross-links. Hard-segment domains also lead to higher elastic modulus values because they act as filler. For spherical domains the modulus increase has been described by the Guth–Smallwood relation [106]

$$F_F = F(1 + 2.5\phi + 14.1\phi^2) \qquad (9b)$$

where F_F is the stress required to extend a filled polymer and ϕ is the volume fraction occupied by hard-segment domains. The stress–strain behavior of styrene–isoprene triblock elastomers has been represented [102] by combining Eqs. (9a) and (9b). In practice, the shape and morphology of block copolymer domains are highly influenced by segment composition, thermal and solvent history, and the method of sample preparation. The trends of the moduli of thermoplastic elastomers have recently been analyzed by theories originally derived for composite materials [107, 108]. These treatments predict the expected range of values of elastic moduli, taking into consideration the various possible morphological structures.

B. Viscous and Viscoelastic Properties

When tested above the melting temperatures of the A component, ABA thermoplastic elastomers show higher melt viscosities [102, 109–111] than either their constituent homopolymers or random copolymers with equivalent compositions and molecular weight. This effect has been attributed to the retention of phase-separated structures in the molten state. Flow processes then require, it has been suggested, additional energy to transfer hard segments from one domain to another through a thermodynamically incompatible soft-segment matrix, but it is more likely that extra dissipation mechanisms due to residual domains actually raise the melt viscosity.

The Williams–Landel–Ferry [112] relationship is commonly employed to interrelate the time and temperature dependence of viscoelastic properties by means of a shift factor, a_T:

$$\log a_T = \frac{-C_1(T - T_0)}{C_2 + (T - T_0)} \tag{10}$$

where C_1 and C_2 are numerical constants, and T and T_0 are the temperature of measurement and a reference temperature, respectively. For many single-phase amorphous polymers, dynamic–mechanical modulus or stress–relaxation data can be represented by a single master curve when Eq. (10) is applied. However, the viscoelastic behavior of block copolymers is thermorheologically complex [55, 56, 113–117] due to the simultaneous occurrence of multiple relaxations in the two phases. For a two-phase material, it is unlikely [116, 117] that a single master curve could be obtained by simple translation of viscoelastic responses along the logarithmic time or frequency axis, because such a result would mean that all retardation times were equally affected by a change in temperature. This would require identical responses in each phase over the entire time scale. A model that assumes additivity of strain (i.e., compliances) has been applied to describe the contribution of individual phases in thermorheologically complex polymers. Both entanglement slippage and interfacial-domain softening effects have been found to contribute to the complexity of viscoelastic behavior in ABA copolymers. In an ABA thermoplastic elastomer, stress–relaxation behavior at temperatures between the two major transitions has also been interpreted as suggesting the presence of a diffuse mixed-segment layer at domain interfaces [55].

C. STRESS SOFTENING, HYSTERESIS, AND SET

Thermoplastic elastomers exhibit characteristic high levels of stress softening, hysteresis, and permanent set. These properties are also found to a lesser extent in many conventional filled and cross-linked elastomers. Several mechanisms have been described in recent reviews [12, 118, 119]. Unfilled, cross-linked elastomers also show stress softening, which has been attributed to configurational changes resulting in "non-affine displacement of network junctions and entanglements during deformation and their incomplete subsequent recovery to their original positions" [119]. The presence of filler or domains introduces possible additional softening mechanisms [120], including breakage of rubber–filler attachments, disruption of internal or aggregate filler structure, or chain slippage at the filler surface. In situations where intermolecular structures are irreversibly disrupted or reform in new positions while the polymer is extended, permanent deformation (set) occurs.

In thermoplastic elastomers, the shape of the stress–strain curve and the extent of stress softening are dependent on sample composition and the method of sample preparation. Figure 17 shows the stress–strain behavior [102] as a

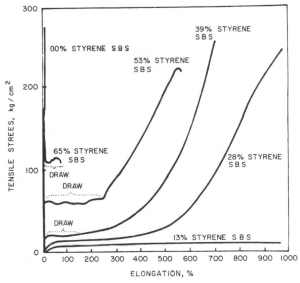

Fig. 17. Dependence of stress–strain properties on weight fraction of styrene units for ABA poly(styrene–b-butadiene) [102].

function of composition in a styrene–diene triblock polymer. At higher styrene contents (> 30%), a yielding phenomenon occurs which has been interpreted to indicate the presence of a semicontinuous polystyrene phase [102, 121]. An interconnected polystyrene phase has also been achieved [56] by casting films of an ABA copolymer from a good solvent for the hard segment. Similar samples cast from a nonselective solvent had fewer connections between hard-segment domains and showed no yield point in their stress–strain curve.

Stress softening has also been reported in polyurethane (AB)$_n$ block copolymers [122, 123]. Stress softening has been attributed to restructuring [91] of domains or to a reduction in the number of network chains [124] when the samples are strained. In segmented polyurethanes, crystallization has been found to increase stress hysteresis, permanent set, and tensile strength. Heat buildup in polyurethanes due to their high hysteresis losses has limited their suitability in applications such as high-speed tires.

D. TENSILE PROPERTIES

Several mechanisms [125, 126] contribute to the high tensile strengths of thermoplastic elastomers. The fracture process can be represented by three steps: initiation of microcracks or cavitation, slow crack propagation, and catastrophic failure. Dispersed phases tend to interfere with the crack propagation step, redistributing energy that would otherwise cause the cracks to reach catastrophic size. Thus, a two-phase morphology is essential to the achievement of high strength in elastomers. The presence of hard-segment domains increases energy dissipation by hysteresis and other viscoelastic mechanisms. Growing cracks may be deflected and bifurcated at phase boundaries. Upon deformation, triaxial stress fields are formed about hard-phase particles, tending to inhibit the growth of planar cracks. Cavities which do form may be limited to small sizes and become stabilized by neighboring hard surfaces. The hard phase can also relieve stress concentration by undergoing deformation or internal structural reorganization. At lower temperatures, strength may be raised due to the greater domain yield stresses, increasing matrix viscosity, or strain-induced crystallization effects. The relative importance of these and other reinforcement mechanisms in two-phase polymer systems depends on the type, size, and concentration of the domains or phases.

In segmented polyurethanes strength is enhanced by long, rigid hard segments with high cohesive energy. Although hydrogen bonding may contribute to domain cohesiveness, hydrogen bonding itself is not directly responsible for high strength. Tensile strength has been found [126] to be remarkably independent of domain concentration when the weight fraction of hard segments is higher than about 20%. This finding indicates that specific domain characteristics are more important than domain concentrations in both ABA and (AB)$_n$ block copolymers. High T_g values in hard- and soft-segment domains tend to favor higher tensile strengths. Polystyrene domains are less effective in strength reinforcement than poly(α-methyl styrene) domains, which have a higher T_g. When the T_g of the soft segments is close to service temperature, viscoelastic strengthening mechanisms are more important. Soft segments can also contribute to the high strength of ABA block copolymers because of possible entanglement slippage that leads to reduced stress concentrations. In general, the presence of a dispersed phase of appropriate size provides additional mechanisms for the dissipation of energy that would otherwise lead to sample fracture.

VI. Application of Thermoplastic Elastomers

Most of the unique applications [11, 111] for thermoplastic elastomers derive from their high-strength elastomeric properties combined with thermoplastic processing capabilities. They can be used in both solution cast or hot melt applications, finding utility as adhesives, caulks, and sealants, as well as extruded or molded articles.

Thermoplastic elastomer solutions may show a maximum in solution viscosity when the solubility parameter of the solvent is matched to that for either of the constituent blocks. This is thought to result from clustering and agglomeration of the less soluble segments. When the solvent is compatible with both blocks, the solution viscosity of ABA copolymers is lower than that of conventional elastomers, which are usually used at higher molecular weight levels.

Thermoplastic elastomers can be mixed or compounded with a variety of materials for cost advantages or property modification. Solution mixing or hot mastication techniques may be employed. ABA copolymers can be blended with many nonpolar resins in order to develop tack or enhance other specific properties. Large amounts of petroleum-based processing oil can be absorbed without significant loss of properties. Fillers may be added to improve many properties, but added filler will not bring about substantial strength reinforcement, as in the case of conventional rubbers. Blends with asphalt can produce useful tacky cements for adhesive applications. Further property modifications can be accomplished by graft polymerization of additional monomers that can be codissolved with ABA copolymers.

Like most polymers containing ethylenic backbone unsaturation, styrene–diene triblock copolymers are susceptible to oxidative and ultraviolet-light-induced degradation. Consequently they require stabilizing additives. ABA block copolymers with saturated chain backbones, and therefore with improved oxidative and chemical stability, have recently become commercially available. In the absence of chemical stability limitation, the upper temperature limit of thermoplastic elastomer utility is generally decided by the phase transition temperature of the hard segment.

A great variety of objects can be produced by extrusion or molding techniques, allowing manufacturers of thermoplastics to expand their operations into the production of elastomers. Short molding cycles, recyclable scrap, and the lack of required mixing with vulcanization ingredients provide significant advantages over conventional elastomer technology. The absence of vulcanization residues permits the preparation of clean, uncontaminated objects for use in contact with food or in pharmaceutical applications. Extrusion, blow molding, injection molding, and vacuum forming can be employed.

Thermoplastic processing techniques are also applicable to linear polyurethane elastomers [127]. A wide range of moduli and other mechanical properties can be obtained by varying the length and chemical composition of the segments. Segmented polyurethanes are resistant to petroleum-based fuels, oils, and greases, but they are attacked by polar or aromatic solvents, hot water, and concentrated acids or bases. Their moisture sensitivity requires that molding compounds and recyclable scrap be kept dry during storage and molding operations. Polyurethane thermoplastic elastomers are characterized by exceptional toughness, tear strength, and abrasion resistance. These properties lead to their utilization as diaphragms and impellers in pumps, gears, conveyor belting, industrial rollers, and abrasion-resistant fabric coatings. They also find use as hot melt coatings and heat-activatable adhesives.

References

1. L. R. G. Treloar, *Rubber Chem. Technol.* **47**(3), 625 (1974).
2. L. J. Fetters, *J. Res. Natl. Bur. Stand., Sect. A* **70**, 421 (1966).
3. L. J. Fetters, *J. Polym. Sci., Part C* **26**, 1 (1969).
4. M. Morton and L. J. Fetters, *Macromol. Rev.* **2**, 71 (1967).
5. P. J. Flory, *J. Am. Chem. Soc.* **62**, 1561 (1940).
6. M. Morton, *in* "Encyclopedia of Polymer Science and Technology," Vol. 15, (N. Bikales, ed.), p. 508. Wiley (Interscience), New York, 1970.
7. J. H. Saunders and K. C. Frisch, "Polyurethanes, Chemistry and Technology, Part I, Chemistry." Wiley (Interscience), New York, 1962.
8. C. S. Schollenberger, U.S. Patent 2,871,218 (1955).
9. R. M. Carvey and D. E. Witenhafer, Br. Patent 1,087,743 (1965).
10. L. L. Harrell, Jr., *Macromolecules* **2**, 607 (1969).
11. D. C. Allport and W. H. Janes, "Block Copolymers" Wiley, New York, 1973.
12. G. M. Estes, S. L. Cooper, and A. V. Tobolsky, *J. Macromol. Sci., Rev. Macromol. Chem.* **4**(2), 313 (1970).
13. L. Nielsen, "Mechanical Properties of Polymers," p. 62. Reinhold, New York, 1962.
14. S. L. Rosen, *Polym. Eng. Sci.* **7**(2), 115 (1967).
15. R. L. Scott, *J. Polym. Sci.* **9**, 423 (1952).
16. W. H. Stockmayer, L. D. Moore, M. Fixman, and B. N. Epstein, *J. Polym. Sci.* **16**, 517 (1955).
17. G. E. Molau, *J. Polym. Sci., Part B* **3**, 1007 (1965).
18. L. Bohn, *Rubber Chem. Technol.* **41**(2), 495 (1968).
19. S. L. Rosen, *Trans. N.Y. Acad. Sci.* **35**(6), 480 (1973).

20. A. J. Yu, *Adv. Chem. Ser.* **99**, 2 (1971).
21. S. Krause, *J. Macromol. Sci., Rev. Macromol. Chem.* **7**(2), 251 (1972).
22. P. J. Corish, *Rubber Chem. Technol.* **40**(2), 324 (1967).
23. P. J. Corish and B. D. W. Powell, *Rubber Chem. Technol.* **47**(3), 481 (1974).
24. R. F. Fedors, *J. Polym. Sci., Part C* **26**, 189 (1969).
25. S. Krause, *in* "Block and Graft Copolymers" (J. J. Burke and V. Weiss, eds.), p. 143. Syracuse Univ. Press, Syracuse, New York, 1973.
26. S. Krause, *in* "Colloidal and Morphological Behavior of Block and Graft Copolymers" (G. E. Molau, ed.), p. 223. Plenum, New York, 1971.
27. S. Krause, *Macromolecules* **3**, 84 (1970).
28. S. Krause, *J. Polym. Sci., Part A-2* **7**, 249 (1969).
29. D. J. Meier, *Polym. Prepr., Am. Chem. Soc., Div. Polym. Chem.* **15**(1), 171 (1974).
30. D. J. Meier, *in* "Block and Graft Copolymers" (J. J. Burke and V. Weiss, eds.), p. 105. Syracuse Univ. Press, Syracuse, New York, 1973.
31. D. J. Meier, *Polym. Prepr., Am. Chem. Soc., Div. Polym. Chem.* **11**, 400 (1970).
32. D. J. Meier, *J. Polym. Sci., Part C* **26**, 81 (1969).
33. U. Bianchi, E. Pedemonte, and A. Turturro, *Polymer* **11**, 268 (1970).
34. U. Bianchi, E. Pedemonte, and A. Turturro, *J. Polym. Sci., Part B* **7**, 785 (1969).
35. H. Kawai, T. Soen, T. Inoue, T. Ono, and T. Uchida, *Mem. Fac. Eng., Kyoto Univ.* **33**, Part 4, 383 (1971).
36. T. Inoue, T. Soen, T. Hashimoto, and H. Kawai, *in* "Block Polymers" (S. L. Aggarwal, ed.), p. 53. Plenum, New York, 1970.
37. T. Inoue, T. Soen, T. Hashimoto, and H. Kawai, *J. Polym. Sci., Part A-2* **7**, 1283 (1969).
38. L. Marker, *Polym. Prepr., Am. Chem. Soc., Div. Polym. Chem.* **10**, 524 (1969).
39. D. F. Leary and M. C. Williams, *J. Polym. Sci., Part B* **8**, 335 (1970).
40. D. F. Leary and M. C. Williams, *J. Polym. Sci., Part A-2* **11**, 345 (1973).
41. J. Pouchly, A. Zivny, and A. Sikora, *J. Polym. Sci., Part A-2* **10**, 151 (1972).
42. H. Krömer, M. Hoffmann, and G. Kämpf, *Ber. Bunsenges. Phys. Chem.* **74**, 859 (1970).
43. R. T. LaFlair, *Pure Appl. Chem.* **8**, 195 (1971).
44. G. E. Molau, *in* "Block Polymers" (S. L. Aggarwal, ed.), p. 102. Plenum, New York, 1970.
45. A. E. Skoulios, *in* "Block and Graft Copolymers" (J. J. Burke and V. Weiss, eds.), p. 121. Syracuse Univ. Press, Syracuse, New York, 1973.
46. K. Kato, *J. Polym. Sci., Part B* **4**, 35 (1966).
47. E. Campos-Lopez, D. McIntyre, and L. J. Fetters, *Macromolecules* **6**(3), 415 (1973).
48. P. R. Lewis and C. Price, *Polymer* **13**, 20 (1972).
49. J. A. Koutsky, N. V. Hien, and S. L. Cooper, *J. Polym. Sci., Part B* **8**, 353 (1970).
50. P. R. Lewis and C. Price, *Polymer* **12**, 258 (1971).
51. M. Hoffmann, G. Kämpf, H. Krömer, and G. Pampus, *Adv. Chem. Ser.* **99**, 351 (1971).
52. M. Matsuo, S. Sagae, and H. Asai, *Polymer* **10**, 79 (1969).
53. M. Matsuo, *Jpn. Plast.* **2**, 6 (1968).
54. H. Kim, *Macromolecules* **5**(5), 594 (1972).
55. M. Shen and D. H. Kaeble, *J. Polym. Sci., Part B* **8**, 149 (1970).
56. J. F. Beecher, L. Marker, R. D. Bradford, and S. L. Aggarwal, *J. Polym. Sci., Part C* **26**, 117 (1969).
57. D. G. LeGrand, *J. Polym. Sci., Polym. Lett. Ed.* **8**, 195 (1970).
58. E. Pedemonte, A. Turturro, U. Bianchi, and P. Devetta, *Polymer* **14**, 145 (1973).
59. D. McIntyre and E. Campos-Lopez, *Macromolecules* **3**, 325 (1970).
60. D. S. Brown, K. V. Fulcher, and R. E. Wetton, *J. Polym. Sci., Polym. Lett Ed.* **8**, 322 (1970).
61. E. Fischer, *J. Macromol. Sci., Chem.* **2**(6), 1285 (1968).
62. L.-K. Bi, L. J. Fetters, and M. Morton, *Polym. Prepr., Am. Chem. Soc., Div. Polym. Chem.* **15**(2), 157 (1974).

63. R. Montiel, C. Kuo, and D. McIntyre, *Polym. Prepr., Am. Chem. Soc., Div. Polym. Chem.* **15**(2), 169 (1974).
64. P. R. Lewis and C. Price, *Polymer* **13**, 20 (1972).
65. C. Price, A. G. Watson, and M. T. Chow, *Polymer* **13**, 333 (1972).
66. W. R. Krigbaum. S. Yazgan, and W. R. Tolbert, *J. Polym. Sci., Part A-2* **11**, 511 (1973).
67. M. J. Folkes and A. Keller, *in* "The Physics of Glassy Polymers" (R. W. Howard, ed.), p. 575. Applied Science Publishers, Barking, England, 1973.
68. J. Dlugosz, A. Keller, and E. Pedemonte, *Kolloid Z. Z. Polym.* **242**, 1125 (1970).
69. L. Nielsen, "Mechanical Properties of Polymers," p. 138. Reinhold, New York, 1962.
70. J. D. Ferry, "Viscoelastic Properties of Polymers." Wiley, New York, 1961.
71. S. L. Cooper and A. V. Tobolsky, *Text. Res. J.* **36**(9), 800 (1966).
72. M. Narkis and A. V. Tobolsky, *J. Macromol. Sci., Phys.* **4**(4), 877 (1970).
73. H. N. Ng, A. E. Allegrezza, R. W. Seymour, and S. L. Cooper, *Polymer* **14**, 255 (1973); A. E. Allegrezza, Jr., R. W. Seymour, H. Ng, and S. L. Cooper, *Polymer* **15**, 1 (1974).
74. L. L. Harrell, *in* "Block Polymers" (S. L. Aggarwal, ed.), p. 213. Plenum, New York, 1970.
75. G. Kraus, C. W. Childers, and J. T. Gruver, *J. Appl. Polym. Sci.* **11**, 1581 (1967).
76. C. W. Childers and G. Kraus, *Rubber Chem. Technol.* **40**, 1183 (1967).
77. M. Shen, U. Mehra, M. Niinomi, J. T. Koberstein, and S. L. Cooper, *J. Appl. Phys.* **45**(10), 4182 (1974).
78. T. Miyamoto, T. Kodama, and K. Shimbayama, *J. Polym. Sci., Part A-2* **8**, 2095 (1970).
79. D. S. Huh and S. L. Cooper, *Polym. Eng. Sci.* **11**(5), 369 (1971).
80. R. W. Seymour, G. M. Estes, and S. L. Cooper, *Macromolecules* **3**, 579 (1970).
81. R. W. Seymour and S. L. Cooper, *Macromolecules* **6**, 48 (1973).
82. S. B. Clough and N. S. Schneider, *J. Macromol. Sci., Phys.* **2**, 553 (1968).
83. G. W. Miller and J. H. Saunders, *J. Polym. Sci., Part A-1* **8**, 1923 (1970).
84. C. M. F. Vrouenraets, *Polym. Prepr., Am. Chem. Soc., Div. Polym. Chem.* **13**(1), 529 (1972).
85. G. M. Estes, R. W. Seymour, and S. L. Cooper, *Macromolecules* **4**, 452 (1971).
86. H. Ishihara, I. Kimura, K. Saito, and H. Ono, *J. Macromol. Sci., Phys.* **B10**, 591 (1974).
86a. S. L. Cooper and R. W. Seymour, *in* "Block and Graft Copolymers" (J. J. Burke and V. Weiss, eds.), p. 208. Syracuse Univ. Press, Syracuse, New York, 1973.
87. G. C. Pimentel and A. L. McClellan, "The Hydrogen Bond." Freeman, San Francisco, California, 1960.
88. M. J. Hannon and J. L. Koenig, *J. Polym. Sci., Part A-2* **7**, 1085 (1969).
89. V. H. Hendus, K. Illers, and E. Ropte, *Kolloid Z. Z. Polym.* **216**, 110 (1967).
90. S. B. Clough, N. S. Schneider, and A. O. King, *J. Macromol. Sci., Phys.* **2**(4), 641 (1968).
91. R. Bonart, *J. Macromol. Sci., Phys.* **2**(1), 115 (1968).
92. R. Bonart, L. Morbitzer, and G. Hentze, *J. Macromol. Sci., Phys.* **3**(2), 337 (1969).
93. R. Bonart, L. Morbitzer, and E. H. Muller, *J. Macromol. Sci., Phys.* **9**, 447 (1974).
94. R. Bonart and E. H. Muller, *J. Macromol. Sci., Phys.* **10**, 177 (1974).
95. R. W. Seymour, A. E. Allegrezza, Jr., and S. L. Cooper, *Macromolecules* **6**(6), 896 (1973).
96. I. Kimura, H. Ishihara, and H. Ono, *IUPAC Macromol. Prepr.* **23**(1), 525 (1971).
97. I. Kimura, H. Ishihara, H. Ono, N. Yoshihara, S. Nomura, and H. Kawai, *Macromolecules* **7**, 355 (1973).
98. R. W. Seymour and S. L. Cooper, *Rubber Chem. Technol.* **47**(1), 19 (1974).
99. S. L. Samuels and G. L. Wilkes, *J. Polym. Sci., Polym. Symp.* **43**, 149 (1973).
100. G. L. Wilkes, S. L. Samuels, and R. Crystal, *J. Macromol. Sci., Phys.* **10**, 203 (1974).
101. J. F. Beecher, L. Marker, R. D. Bradford, and S. L. Aggarwal, *Polym. Prepr., Am. Chem. Soc., Div. Polym. Chem.* **8**, 1532 (1967).
102. G. Holden, E. T. Bishop, and N. R. Legge, *J. Polym. Sci., Part C* **26**, 37 (1969).
103. M. Matsuo, T. Ueno, H. Horino, S. Chujyo, and H. Asai, *Polymer* **9**, 425 (1968).

104. M. Morton, L. J. Fetters, F. C. Schwab, C. R. Strauss, and R. F. Kammereck, *Synth. Rubber Symp.*, *4th* p. 70 (1969).
105. S. M. Gumbrel, L. Mullins, and R. S. Rivlin, *Trans. Faraday Soc.* **49**, 1495 (1953).
106. E. Guth, *J. Appl. Phys.* **16**, 20 (1945).
107. S. L. Aggarwal, R. A. Livigni, Leon F. Marker, and T. J. Dudek, *in* "Block and Graft Co-polymers" (J. J. Burke and V. Weiss, eds.), p. 157. Syracuse Univ. Press, Syracuse, New York, 1973.
108. L. E. Nielsen, *Rheol. Acta* **13**, 86 (1974).
109. G. Kraus and J. T. Gruver, *J. Appl. Polym. Sci.* **11**, 2121 (1967).
110. P. F. Erhardt, J. J. O'Malley, and R. G. Crystal, *in* "Block Polymers" (S. L. Aggarwal, ed.), p. 225. Plenum, New York, 1970.
111. G. Holden, *in* "Block and Graft Copolymerization" (R. J. Ceresa, ed.), Vol. 1, p. 133. Wiley, New York, 1973.
112. M. L. Williams, R. F. Landel, and J. D. Ferry, *J. Am. Chem. Soc.* **77**, 3701 (1955).
113. S. L. Cooper and A. V. Tobolsky, *J. Appl. Polym. Sci.* **10**, 1837 (1966).
114. T. L. Smith and R. A. Dickie, *J. Polym. Sci.*, *Part C* **26**, 163 (1969).
115. T. L. Smith, *in* "Block Polymers" (S. L. Aggarwal, ed.), p. 137. Plenum, New York, 1970.
116. D. G. Fesko and N. W. Tschoegl, *J. Polym. Sci.*, *Part C* **35**, 41 (1971).
117. D. G. Fesko and N. W. Tschoegl, *Int. J. Polym. Mater.* **3**, 51 (1974).
118. J. A. C. Harwood, L. Mullins, and A. R. Payne, *J. Inst. Rubber Ind.* **1**, 17 (1967).
119. L. Mullins, *Rubber Chem. Technol.* **42**(1), 339 (1969).
120. J. A. C. Harwood, L. Mullins, and A. R. Payne, *J. Appl. Polym. Sci.* **9**, 3011 (1965).
121. M. Morton, J. E. McGrath, and P. C. Juliano, *J. Polym. Sci.*, *Part C* **26**, 99 (1969).
122. G. S. Trick, *J. Appl. Polym. Sci.* **3**, 252 (1960).
123. S. L. Cooper, D. S. Huh, and W. J. Morris, *Ind. Eng. Chem.*, *Prod. Res. Dev.* **7**, 248 (1968).
124. D. Puett, *J. Polym. Sci.*, *Part A-2* **5**, 839 (1967).
125. T. L. Smith, *Polym. Prepr.*, *Am. Chem. Soc.*, *Div. Polym. Chem.* **15**, 58 (1974).
126. T. L. Smith, *J. Polym. Sci.*, *Polym. Phys. Ed.* **12**, 1825 (1974).
127. J. H. Saunders and K. C. Frisch, "Polyurethanes, II, Technology," p. 376. Wiley (Interscience), New York, 1964.

Chapter 14

Tire Manufacture and Engineering

F. J. KOVAC

TIRE REINFORCING SYSTEMS
THE GOODYEAR TIRE AND RUBBER COMPANY
DAYTON, OHIO

I. Introduction

The wheel was invented by the Sumerians over 5000 years ago. It was refined with passing years and in 1846 R. W. Thomson was granted a patent in England for an "air tube device"—the forerunner of the modern pneumatic tire. This invention was unused until 1888, when it began to achieve wide application. After nearly a century of use, tires have made possible the evolution of a sophisticated personal transportation system.

To many people a tire is a large rubber donut. Actually a tire is many things. Geometrically, it is a torus; mechanically, a flexible-membrane pressure container; structurally, a high-performance composite; and chemically, a tire consists of materials from long chain macromolecules. These facets and their interrelationship are to be the subject of this chapter.

© Copyright 1978
by The Goodyear Tire & Rubber Company.
Reproduced by permission.

A. TIRE FUNCTIONS

Fundamental to the study of tires is a discussion of the basic functions of a pneumatic tire. Briefly, a tire must:

1. Provide load carrying capacity.
2. Provide cushioning and enveloping.
3. Transmit driving and braking torque.
4. Produce cornering force.
5. Provide dimensional stability.
6. Resist abrasion.
7. Provide steering response.
8. Have low rolling resistance.
9. Provide minimum noise and permit minimum road vibrations.
10. Be durable and safe.

Because of the unique deformability and dampening characteristics of the pneumatic tire, it is the only product which has satisfied all these functions.

B. TIRE REQUIREMENTS

Basically, a tire's functions can be considered in relation to three areas: (a) vehicle mobility, (b) performance and integrity, and (c) esthetics and comfort. Performance, including driving and braking torque and rolling resistance, exerts or transfers forces or moments (tangentially) forward. Vehicle mobility, including cornering, steering response, and abrasion, acts in the lateral direction, and the forces involved with esthetics and comfort act vertically (Fig. 1).

To determine the effect of each of these forces on the specific tire application, many factors must be considered. One primary consideration is the type

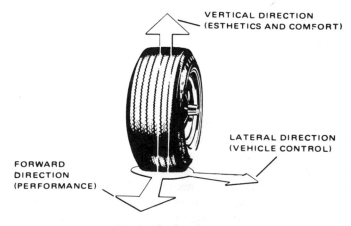

Fig. 1. Tire functions.

TABLE I

TIRE CONSTRUCTIONS

| | Tire type | | | Service requirement | |
Vehicle	Size (in.)	Diameter (in.)	Load (lb)	Speed (mph)	Time (min)
Passenger	LR78-15	29.1	1680	55	30,000
Tractor	7.5L-15	29.5	1590	20	Years
Truck	7.00-15	29.6	1720	55	40,000
Racing	8.20-15	29.0	1200	210	500
Off-the-road	7.50-15	29.8	4200	5	Years
Aircraft	30 × 8.8	29.9	21,000	250	—

of motor vehicle for which the tire is intended: passenger car, tractor, truck, off-the-road vehicle, or aircraft. Another equally important consideration is the performance level required. Six widely different tires are analyzed in Table I. Notice the difference in requirements just for these six tires—load requirements range from 1200 to 21,000 lb; speed requirements vary from 5 to 250 mph; and mileage requirements range from 500 to 40,000 miles.

C. BASIC TIRE DESIGN

A tire is a cord–rubber composite. The tire composite is a network of cord structures arranged in parallel configuration and embedded in a rubber matrix. Rubber as used here is an elastomer compounded by the addition of carbon black, pigments, and other chemicals. The cords reinforce the rubber matrix of a tire in much the same manner as steel reinforces concrete.

There are three basic tire types (Fig. 2). One is the bias or crossply tire. In this construction the reinforcing cords extend diagonally across the tire from bead to bead (inextensible wire-reinforced hoops that anchor the plies). These cords, generally with a bias angle between 30 and 40°, run in opposite

BIAS ANGLE BIAS/BELTED RADIAL PLY

Fig. 2. Basic tire constructions.

directions in each successive layer (called a *ply*) of reinforcing material.

A second design is the bias/belted tire, which consists of a bias-angle carcass with a circumferential restricting belt under the tread. The carcass angle is generally 25–40° and that of the belt between 25 and 30°. In addition, the angle of the belt is at least 5° lower than that of the carcass.

The third type is the radial tire: it has plies of reinforcing cords that extend traversely from bead to bead; on the top of these plies (under the tread) is an inextensible belt composed of several more layers of cord. The belt cords are low angle (10–30°) and act as a restriction on the 90° carcass plies.

Tire Components

A tire comprises a number of parts or subassemblies. Each serves a specific and unique function. These parts and their functions are shown in Fig. 3 and include:

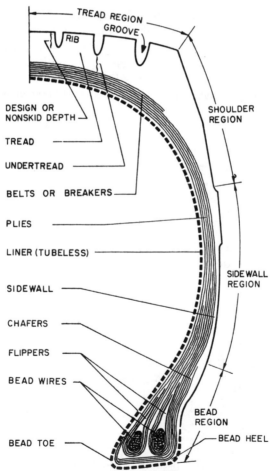

Fig. 3. Structural components.

1. Tread—the wear-resistant component that provides traction, silent running, and low heat buildup. The tread is normally a blend of oil-extended SBR and polybutadiene elastomers and natural rubber that have been compounded with carbon black, oils, curative ingredients, and other chemicals and pigments. The tread consists of (a) ribs—circumferential rows of tread rubber designed for noise suppression and traction, and (b) grooves—circumferential channels essential for traction, directional control, and cool running.

2. Sidewalls—the portion of the tire between the beads and the tread; it controls the ride and offers support. It is compounded of rubber that gives high flex and weather resistance.

3. Shoulder—the upper portion of the sidewall, just below the tread edge; it affects tire heat behavior and cornering characteristics.

4. Bead—a structure composed of high-tensile-strength steel wire formed into inextensible hoops functioning as anchors for the plies and holding the assembly on the rim of the wheel. The shape of the bead conforms to the flange of the wheel to prevent the tire from rocking or slipping on the rim.

5. Plies—layers of fabric cord extending from bead to bead reinforcing the tire.

6. Belts—Narrow layers of tire cord, directly under the tread in the crown of the tire, that resist deformation in the footprint (contact patch on the road) and restrict the carcass plies.

7. Liner—the thin layer of rubber inside the tire that contains compressed air.

8. Chafer—narrow strips of material around the outside of the bead that protect the cord against wear and cutting by the rim, distribute flex above the rim, and prevent moisture and dirt from getting into the tire.

II. Tire Engineering

Tire engineering involves studying the stresses of a moving tire, determining the kind and amount of material that will withstand such forces, distributing these forces while satisfying all conflicting requirements, and comprehending the laws governing changes in the tire during the manufacturing process. To understand the tire structure and its stresses, the purpose of all tire components must be known. To know them, a thorough understanding of the tire reinforcing system, rubber compounding, and tire design is necessary.*

A. TIRE REINFORCING SYSTEM

A tire is a composite structure in the form of a network of cord structures arranged in a parallel configuration and embedded in a rubber matrix. The

* Compounding of elastomers is discussed in Chapter 12 and will not be included in this discussion.

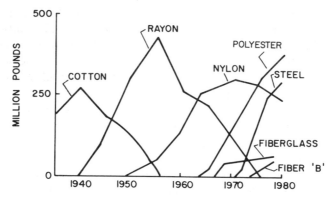

Fig. 4. Tire cord used.

cord reinforces the rubber much as steel strengthens concrete. However, the tire cord is unique in that it must provide fatigue resistance. Tire cords also give the tire shape, size stability, bruise resistance, and load-carrying capacity.

Stringent demands on the reinforcement materials suitable for tires have held the number to only seven. These seven are noted in Fig. 4 along with their history and usages. Of the seven, nylon, rayon, cotton, and polyester are primarily carcass material used in the plies. Steel, aramid, and fiberglass are primarily belt materials.

1. Definitions

Filament. Smallest continuous element of a tire cord material; character-

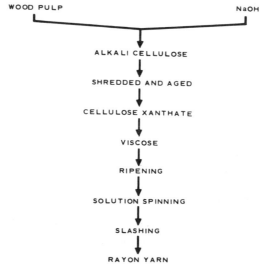

Fig. 5. Rayon manufacture.

ized by high ratio of length to thickness.

Yarn. An assembly of continuous filaments.

Tire cord. A twisted or cabled structure composed of two or more yarns.

Warp. Cord in tire fabric that runs lengthwise.

Fill. Light thread that is placed at right angles to the warp.

Denier. Unit of measurement for textile cords; weight in grams of 9000 m of the material.

Twist. Number of turns per unit length of cord; an S twist is a turn to the right or clockwise from top to bottom, whereas a Z twist denotes counterclockwise spirals.

2. Filament Forming

Rayon was the first man-made tire fiber, cotton being a natural fiber. Being man-made, rayon could be engineered for tires. Basically, rayon is a regenerated cellulose manufactured as a continuous filament by a wet spinning process as outlined in Fig. 5. Its structure is

$$
\left[
\begin{array}{c}
\overset{\displaystyle OH}{\underset{\displaystyle |}{}} \quad H \\
C \text{---} C \\
H{\diagup}H \quad | \quad \diagdown H \\
\text{---}C \quad HO \quad C\text{---}O\text{---} \\
\diagdown H \quad \diagup \\
C\text{---}O \\
| \\
CH_2OH
\end{array}
\right]_n
$$

Nylon is a synthetic thermoplastic fiber derived from petroleum. The first truly synthetic fiber, it is basically a continuous-filament aliphatic polyamide manufactured by the steps shown in Fig. 6. Cyclohexane and ammonia are the basic raw materials. The spinning operation listed consists of extruding the molten polymer through stainless steel spinnerets to form filaments. These filaments are cooled, finish is applied to them, and they are assembled to form a yarn, which is then drawn about 500%. Drawing orients the molecules and regulates the crystallinity.

There are hundreds of nylon polymers. The difference is in the acid used. Only two types are used in tires—nylon 6 and nylon 66. Both have the same empirical formula: $(C_6H_{11}OH)_n$. The process shown in Fig. 6 is for nylon 66. It is composed of two recurring units, each of 6 carbon atoms, whereas nylon 6 consists of only one recurring unit with 6 carbon atoms (derived from caprolactam).

Nylon exhibits thermoplastic shrinkage, resulting in the phenomenon known as flatspotting. Figure 7 shows how this effect is caused. During tire operation the nylon tire cords in the carcass are extended under tension and high temperature. When the car is parked, the nylon cools. The cords out of the area of the tire that contacts the road remain in tension (upper curve). As the cords cool, the shrinkage forces are reduced, allowing these cords to

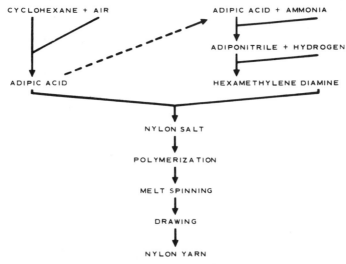

Fig. 6. Nylon 66 manufacture.

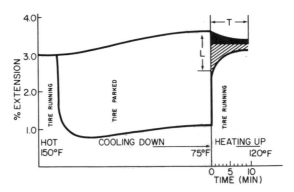

Fig. 7. Flatspotting.

extend further. The cords in the footprint area are under less tension while they cool and shrink substantially. When driving is resumed these cords do not extend to their original length again until the tire is heated up, whereas those cords that were out of the footprint area are longer in length. The distance L represents the difference in length between the cords that were in the footprint area and those out of the footprint area. For rayon and polyester, L is less than the threshold of detection.

Polyester is also a synthetic thermoplastic continuous-filament fiber derived from petroleum; it is manufactured by a melt process similar to nylon. Chemically, however, polyester differs from nylon in that it contains aromatic groups. The starting raw materials are ethylene glycol and dimethylterephthalate (DMT). The DMT is melted with glycol and an ester interchange takes place. [In some processes terephthalic acid (TPA) is used in place of DMT.] The glycol ester is then polymerized through polycondensation to polyethylene terephthalate. The molten polymer is then extruded through stainless steel spinnerets to form filaments about 0.024 mm in diameter. The filaments pass through a long cooling tower, finish is applied, and the filaments are assembled into a yarn. The yarn is then drawn to give the desired orientation and crystallinity. The final product has the chemical structure

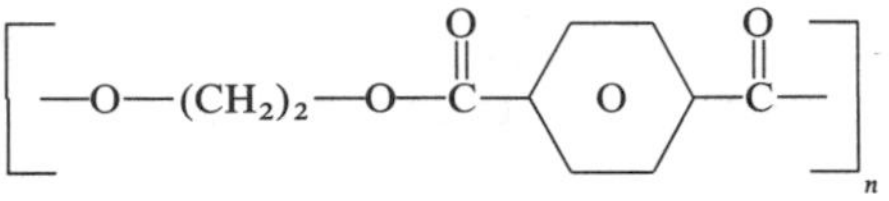

$$\left[-O-(CH_2)_2-O-\overset{\overset{O}{\parallel}}{C}-\underset{}{\bigcirc}-O-\underset{}{\bigcirc}-\overset{\overset{O}{\parallel}}{C}- \right]_n$$

Fiberglass is an inorganic fiber. The typical composition of the fiberglass used in tire cord is silicon dioxide, 53%; calcium oxide, 21%; aluminum oxide, 15%; boron oxide, 9%; magnesium oxide, 0.3%; other oxides, 1.7%. Thus, fiberglass is a lime–alumina–borosilicate glass consisting of a three-dimensional silica network containing oxides. Manufacture of the continuous filament consists of blending sand, clay, limestone, and borax, and melting the blend in a direct gas-fired furnace. The molten glass at a surface temperature of 3000°F is gravity fed through a platinum–rhodium bushing where strands of continuous glass filaments 0.009 mm in diameter are drawn or attenuated. These filaments are then water quenched, after which a coupling agent (adhesive precoat) and forming size (finish) are applied as the filaments pass to a high-speed takeup. Fiberglass for tires is impregnated with an RFL adhesive (consisting of resorcinol–formaldehyde resins and latex). It is used in the form of a yarn with low twist and is not normally cabled into a cord structure.

Like fiberglass, *wire* is an inorganic material, basically iron. Typical steel for a tire cord is 0.7% carbon, 0.5% magnesium, 0.3% silicon, 0.05% (maximum) chromium, 0.02% (maximum) copper, and 0.03% (maximum) sulfur and phosphorus. Manufacture of steel for use as a typical tire cord is outlined in Table II.

TABLE II

- 5.5-mm steel rod. Cleaned in acid bath, rinsed, and coated with lime, which neutralizes the acids used to remove oxide scale.
- Drawn finally to 2.0 mm. Heat treated to change microstructure for next drawing.
- Drawn to 0.8 mm. Brass plated (7 g of brass per kilogram of steel). Brass composition is 70% copper and 30% zinc.
- Drawn to 0.2 mm.
- Stranding combining of drawn wire filaments.

Strands are combined to obtain final wire tire cord. In some cases, an additional filament is spiral wrapped around the entire cord structure. This spiral wrap resists compressive forces when the tire enters the footprint area, improves cord tension uniformity, enhances mechanical adhesion, prevents flaring of the strands when cut, and reduces liveliness for better factory handling.

Aramid is the generic name for a new class of fibers. Two new fibers produced by du Pont are in this category. The first, with the trade name Kevlar (Fiber B), is the first fiber specially developed for tire reinforcement. The second is Nomex, also noted for its heat resistance and valued for use in clothing and electrical goods because of its fire resistance. Kevlar is denoted chemically as poly(*p*-phenylamine terephthalamide), which has the following structure.

It is the polymerization product of terephthalic acid and *p*-phenylamine diamine and is solution spun to form a continuous-filament fiber.

How do these materials differ from one another (other than chemically) and what are their primary areas of use in the tire? The differences in physical properties and uses are shown in Table III, from which several generalizations can be made. First, higher durability fabrics are used in the plies because fatigue is generally a ply phenomenon. Second, higher modulus fabrics (which show low elongation) are generally used in belts or in the plies of heavy-duty tires (i.e., those for off-the-road vehicles and trucks).

3. Tire Cord Construction

To use the fibers described in the tire applications shown in Table III, the tire yarn as manufactured must be twisted and processed to form cord suitable for use in the tire.

The first operation in the fabric mill is called *ply twisting* and consists of the yarn being twisted on itself the desired number of turns per inch. After ply twisting, two or more spools of twisted yarn are twisted into cord. For example, if two 840-denier nylon yarns are twisted together, an 840/2 nylon cord is formed. Similarly, three 1300-denier polyester yarns will form a 1300/3 polyester cord. Generally, the direction of twist of the cord is opposite to that in the yarn. This is called a *balanced* twist.* Twist imparts durability and fatigue resistance to the cord while reducing strength. Thus, an optimum twist

---

* A complete theoretical and mathematical treatment of yarn and cord twisting is given in J. W. S. Hearle, P. Grosberg, and S. Backer, *Structural Mechanics of Fibers, Yarns and Fabrics*, Wiley (Interscience), New York, 1969.

TABLE III

TIRE CORD USES

Fiber type	Tenacity (strength) (g/den)	Ultimate elongation (%)	Modulus (g/den)	Relative durability (cycles)	Use (vehicle)
Rayon	5.0	16	50	300	Plies (passenger car, farm vehicle)
					Belt (passenger car)
Nylon	9.0	19	32	1200	Plies (passenger car, farm and off-the-road vehicles, truck, aircraft)
					Chafer
Polyester	6.5	18	65	400	Plies (passenger car, truck, farm vehicles)
Fiberglass	9.0	4.8	260	5	Belt (passenger car, truck)
					Plies (truck, off-the-road vehicles)
					Belt (passenger car, truck, off-the-road vehicles)
Wire	3.8	2.5	200	3	Bead (all vehicles)
					Belt (passenger car)
Aramid	20.1	4	350	400	Chipper

must be achieved. The improved durability can be explained by considering a bundle of filaments under bending stress. The filaments above the neutral axis are in tension whereas those below the neutral axis are in compression. Without twist, the compressive forces would cause the outer filaments to buckle. As the tire operates, tire cords undergo continuous flex cycles and the repeated buckling caused by these bending stresses results in filament rupture. Increasing twist in the cord reduces filament buckling by increasing the extensibility of the filament bundle. Thus, as the cord is bent, the portion above the neutral axis extends, lowering the neutral axis and reducing the compressive forces in the lower portion.

Durability reaches a maximum and then begins to decrease with increasing twist. This can be explained by considering the effect of stresses on the cord as the twist increases. As twist increases, the helix angle (angle between the filament axis and cord axis) increases (Fig. 8). Thus, tension stresses normal to the cord axis result in greater force parting filaments. The reason for the reduction in strength is also evident in this figure. As cord twist increases, the force in the direction of the yarn axis increases, causing a lower overall breaking strength.

In addition to twists, the cord size may be varied to allow for different applications. Generally, three-ply cords have the best durability.

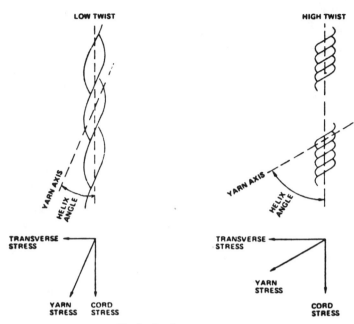

Fig. 8. Cord geometry.

After cable twisting the next production step is called *weaving*. Here multiple cords are woven together by small fill threads to form the fabric. These are required to facilitate factory handling. A typical roll of tire cord fabric is approximately 60 in. wide, 4 ft in diameter, and weighs 2500 lb. The unprocessed fabric is referred to as greige fabric.

The weaving process provides an additional construction variable to the engineer, this being the number of cords per inch that are woven into the fabric and will be used in the tire. High end count fabric results in greater plunger strength or penetration resistance. Low end count fabrics have more rivet (distance between cords) and give better separation resistance because of greater rubber penetration around the cord.

4. Fabric Processes

The most critical stage in preparing the cord or fabric for use in the tire is fabric *treating*, which consists of applying an adhesive under controlled conditions of time, temperature, and tension. This process gives the fabric adhesion for bonding to the rubber; optimizes the physical properties (strength, durability, growth, and elongation) of the cord to meet tire requirements; stabilizes the fabric; and equalizes differences due to the source of the fiber (different yarn producers).

In the days of cotton, fabric was coated with rubber with little or no processing. With the advent of rayon, a method was established to stabilize the cord and reduce growth (extension). With nylon, the problem is greater because of this cord's viscoelastic memory, which seeks to recover from treatment. Thus the current fabrication units, as schematized in Fig. 9, were developed to give constant modulus, maximum strength, minimum growth, and optimum dimensional stability. The unit outlined in Fig. 9 is about seven stories high and can process up to 100,000 miles of tire cord per day. It is the largest piece of equipment in the tire manufacturing process.

These fabric treating units lend themselves to computer control. There are three main computer subsystems for a 3T unit-man–computer, external–computer, and process–computer. The man–computer subsystem consists of links from the operator console, input/output teletype, line printer, card punch, and card reader to the computer. Similarly, the external–computer system links the disk memory and the real-time clock to the computer. The process–computer subsystem provides communication between the computer and the various process controllers and indicators such as thermocouples, speed and stretch, tension, electrical drive, and oven burner valves.

The times, temperatures, and tensions of the process are optimized by the use of experience and regression analysis to tailor the cord physical properties for maximum end use performance. The treating process optimizes strength, durability, dimensional stability, modulus, heat generation, growth, and creep of the fabric. To illustrate this optimization, polyester will be used; the same principles apply to nylon and relationships are similar.

Processing consists of passing fabric through a series of zones, which can be thought of generally as the first dip zone (adhesive application zone), the first drying zone, the first heat treatment zone, the second dip zone, the second drying zone, the second heat treatment (or normalizing) zone, and a cooling zone.

The adhesive is applied in the dip zones by immersion. Excess adhesive is removed by vacuum dewebbers or air blowoff. The correct degree of penetration into the cord is determined by soak or immersion time and dip tensions. The adhesive is generally water based and drying zone dwell times and temperatures are specified to allow complete drying of the cord.

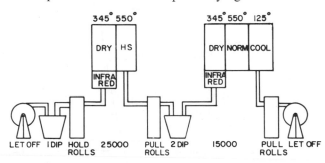

Fig. 9. Fabric processing unit.

To attain optimum cord properties (strength, growth, shrinkage, modulus, heat generation, and durability), specific temperatures and tensions (or stretches) are set. The temperatures and tensions in part determine the ratio of crystalline and amorphous areas in the fiber and the orientation of the crystals, which in turn determine the cord's physical properties.

Polyester (like nylon) is a thermoplastic material in that it has a memory when heated and tends to revert back to its unorientated form; hence, the cords tend to shrink. This phenomenon can be controlled by stretching the cord in the first heat treating zone and then allowing the cord to relax in a controlled manner in the second heat treating zone. This is called stretch relaxation. The greater the stretch relaxation, the lower the shrinkage.

Net stretch refers to the difference between the applied stretch in the first zone and the relaxation in the second (or, generally, the net length change through the unit). Generally, the higher the net stretch and the greater the stretch relaxation, the higher the cord strength and the lower the cord modulus. However, with increased stretch relaxation and increased net stretch, durability decreases.

Another variable is the temperature used. Higher temperatures decrease cord strength and modulus. With higher temperatures, however, fatigue life is increased.

Hysteresis (heat generation) is the work lost through heat during dynamic operation of the cord. The greater the hysteresis, the higher the tire running temperature. Heat decreases cord durability and adhesion. Hysteresis can be determined by measuring the loss modulus (the imaginary component of the complex modulus), since this work or heat loss is associated with the intermolecular drag in the viscous portion of the polymer. The lower the loss modulus, the lower the heat generation. Increasing processing temperature decreases loss modulus, particularly in the 120–140°C range.

A complete relation of cord properties to processing tensions and temperatures is shown in Table IV. Note that not all cord properties behave similarly with varying processing conditions. Thus, it is necessary to find the processing conditions that optimize cord properties for the desired tire end use. When two or more diametrically opposite properties have to be optimized, more difficult mathematical methods must be used. Suppose a minimum breaking strength of 63 lb, maximum total shrinkage of 4.0%, and a minimum fatigue level rating of 90 is desired. First, assume that only the first zone stretch S_1 is varied. Regression equations of 1300/3 polyester for these variables show

$$\text{Break strength} = 0.61S_1 + 60.4$$
$$\text{Total shrink} = 0.45S_1 + 3.3$$
$$\text{Fatigue} = 3.0S_1 + 100$$

Any positive stretch above 4.3% will give a greater than 63-lb breaking strength, but the same stretch will give a 5.2 total shrinkage value and an 87 fatigue

TABLE IV

RELATION OF CORD PROPERTIES TO PROCESSING TENSIONS AND TEMPERATURE

Cord property	Change effected by first increase in		Change effected by second increase in	
	Tension time	Temperature	Tension time	Temperature
Tensile strength	Slightly decrease	Decrease	Slightly decrease	Decrease
Load at specified elongation (5%)	Decrease	Decrease	Decrease	Decrease
Ultimate elongation	Increase	Increase	Increase	Increase
Shrinkage	Decrease	Decrease	Decrease	Decrease
Rubber coverage	Increase	Increase	Increase	Increase
Fatigue	Decrease	Decrease	Decrease	Decrease
SCEF	Decrease	Decrease	Decrease	Decrease
Voids	Decrease	Decrease	Decrease	Decrease

value. Thus, changes in the first zone stretch only will not satisfy all of these properties. To optimize the process, changes in more than the first stretch will be needed. If the second stretch S_2 is variable, the new regression equations will give the equation for the dependent variable as a function of two independent variables as follows:

$$\text{Break strength} = 1.1S_1 - 0.2S_2 + 60.2$$
$$\text{Total shrink} = 0.6S_1 + 0.7S_2 + 2.0$$
$$\text{Fatigue} = 2.2S_1 - 1.5S_2 + 122$$

Solution of these equations for the desired properties is difficult. A computer is utilized to achieve the final results and optimized properties.

Wire and fiberglass, being high modulus inorganic belt cords, receive little processing at the fabric mill. Glass yarn is received from the producer with twist and with adhesive, and generally goes directly to the weaving operation. Wire is received with a plating of brass which serves an adhesive function. Wire normally bypasses the normal fabric operations and is taken directly to the calendering operation.

5. Adhesive

After processing, textile fabrics are coated with rubber. This task of bonding the tire cord and the rubber into a dynamic composite structure is difficult due to the wide differences in modulus between textile and rubber materials (Fig. 10).

There are three aspects to the phenomenon of adhesion—molecular, chemical, and mechanical. Molecular bonding is achieved through adsorption of adhesive ingredient chemicals on the fiber surface, diffusion of other

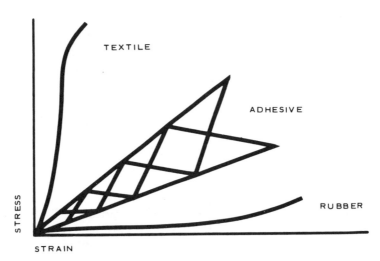

Fig. 10. Role of adhesive.

adhesive components into both the fiber polymer and the rubber compound, and physicochemical activity. Chemical bonding is achieved through chemical reactions between the adhesive and the fabric and rubber. The type of functionality and number of functional groups present are important, as is the mobility of the various molecules in the fiber adhesive and the rubber compound. Finally, steric hindrance must be considered.

The fiber properties of primary importance to adhesion are reactivity, surface characteristics, and finish. Each fiber has a different reactivity. Rayon has many reactive hydroxyl groups. Nylon is less reactive but contains highly polar amide linkages. Polyester is quite inert. An adhesive system must be designed for each fiber type. Properties include intermediate cohesive energy density, a high fiber surface area, and a low contact angle. Finally, since the adhesive must bond to the rubber compound, the type and nature of the rubber plays an important role and must be considered.

Adhesive systems must conform to a rigid set of auxiliary requirements; in addition to being water based, the systems should provide

1. Rapid rate of adhesion.
2. Compatibility with many types of compounds.
3. No adverse effects on cord properties.
4. Heat resistance.
5. Aging resistance.
6. Good tack.
7. Mechanical stability.

The adhesive bond between tire and cord is achieved during the tire cure

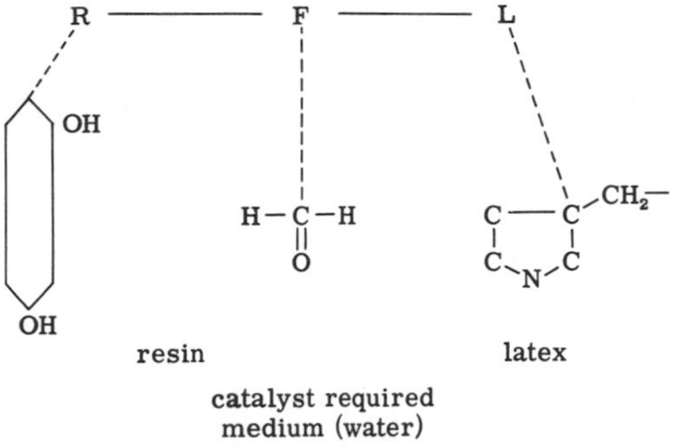

Fig. 11. RFL adhesive system.

cycle. The rate of adhesive formation should give maximum adhesion at the point of pressure release.

Today, most tire cord adhesive systems are based on resorcinol–formaldehyde resins and latex (RFL) in an aqueous medium (Fig. 11). Generally, only RFL is needed for bonding rubber to nylon and rayon. Because polyester is generally inert, more reactive chemicals are added to the RFL to promote adhesion. Polyisocyanates and epoxies are most commonly used.

B. TIRE CONSTRUCTION

The tire engineer has the difficult job of balancing the tire structure and materials with an anticipated environment. There are many construction alternatives, but due to many contradictory effects of each alternative, the design process consists of determining the operable range of a structure and its components and selecting that design which offers the optimum balance of characteristics for the tire under consideration.

1. Definitions

Tire and Rim Dimensions. The terminology used for tire and rim dimensions are shown in Fig. 12. These terms are used to describe size, growth, and clearance factors. They are also used to compute empirical quantities such as load capacity and revolutions per mile.

Aspect Ratio. The ratio of the section height (SH) to the section width (SW) for a specified rim width

$$AR = \frac{SH}{SW} \times 100$$

Structural Dimensions. The dimensions outlined in Fig. 12 describe only the tire envelope and are regarded as boundary limits. Structural geometry

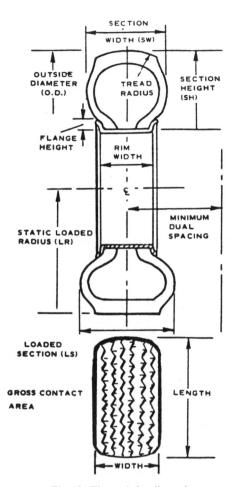

Fig. 12. Tire and rim dimensions.

is used to locate and define actual components within the tire. These dimensions (shown in Fig. 13) provide the basis for the mathematical definition of tire shape, cord curvature, cord path, stress, and other interrelated analytical expressions.

Cord Angle. Cord angle (angle of the cord path to the centerline of the tire) is the predominant parameter affecting the tire shape or contour.

Restricting Components. Belts are the primary component that restricts the carcass. Components such as breakers and cap plies differ from belts in that their cord angles are nearly the same as the carcass plies. Breakers do not restrict circumferential growth but are used primarily to add strength in the tread and shoulder regions.

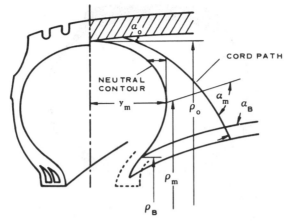

Fig. 13. Tire geometry: y, axial dimension; α, acute angle formed by cord and circumferential line at contour; ρ, radius from axis of rotation to any point on the tire contour; ρ_0, radius from the axis of rotation to the neutral contour at the centerline; ρ_m, radius from the axis of rotation to the point on the tire contour where y is maximum.

2. Design Factors

Basically, a tire is a torus, which can be visualized as the volume of space that a sphere occupies while orbiting some center. For simplicity, the tire toroidal shell is often treated two dimensionally as a circular and as a meridian shape. The structure of the tire defines the type, number, location, and dimensions of the various components used in its composition.

The primary components are the carcass plies, beads, belts, and tread, because they are responsible for the fundamental tire characteristics. Chafers, flippers, and breakers (strips of fabric located in the bead and crown areas) are called secondary components because they protect the primary components by distributing stress concentration.

a. Tire Structure and Size. The initial step in design is sizing the tire. The problem is, given a load and minimum wheel diameter, what configuration will fulfill the envelope requirements best? For a given inflation pressure, dimension determines load (see Fig. 14). These are related by the equation

$$L = K(0.425)(P)^{0.585}(S_{62.5})^{1.39}(D_R + S_{62.5})$$

where L is the load of the tire in pounds, K the load service factor, P the inflation pressure, $S_{62.5}$ the section width of 62.5% rim, and D_R the nominal rim diameter. This is a basic formula, and modifications are specified for certain tires.

Once the diameter and section width have been determined, the actual shape of the elliptical cross section must be defined. The carcass of the tire represents the surface of equilibrium between internal inflation pressure and

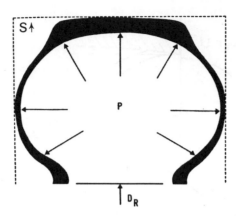

Fig. 14. Tire dimension envelope: P, inflation pressure; S, section width; D_R, rim diameter. Load is proportional to P, S, and D_R.

tension in the cords. Cord tension depends on the curvature of the cord path, which is a function of the cord path at the centerline of the tire and the cord path relationship from the centerline to the bead area. For nonbelted tires this can be expressed as

$$Y = \int_\rho^{\rho_0} \frac{(\rho^2 - \rho_m^2) \exp\left(\int_\rho^{\rho_0} \frac{\cot^2 \alpha}{\rho} \, dp\right)}{\left[(\rho_0^2 - \rho_m^2)^2 - (\rho - \rho_m^2)^2 \exp\left(2\int_\rho^{\rho_0} \frac{\cot^2 \alpha}{\rho} \, dp\right)\right]^{1/2}} \tag{1}$$

where

$$(\cot^2 \alpha/\rho)\alpha\rho = \text{cord path definition} \tag{2}$$

Cord path definition is the most significant factor in the description of tire shape.

In conventional tire building operations, the plies are placed in successive layers in opposite directions to form a matrix (Fig. 15). Once the ply matrix is formed into the cylinder, any change in circumference results in a corresponding angle change. During manufacture this displacement occurs in increasing amounts along the cord, causing the cord angle α to decrease constantly from the bead to the maximum radius at tread center. This relationship is essentially a pantographic action (discussed in more detail on page 603). Thus, the angle change is a direct function of the change in circumferential length (i.e., the radius). The mathematical expression is

$$\alpha_0 = \text{arc cos}\left(\frac{\rho_0 \cos \alpha_1}{\rho_1}\right) \tag{3}$$

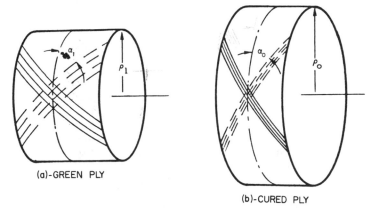

(a)-GREEN PLY

(b)-CURED PLY

Fig. 15. Ply matrix.

This changes Eq. (1) to the following equation for conventional bias ply tires.

$$Y = \int_{\rho}^{\rho_0} - \frac{(\rho^2 - \rho_m^2) \sin \alpha}{[(\rho_0^2 - \rho_m^2) \sin^2 \alpha_0 - (\rho^2 - \rho_m^2)]^{1/2} \sin^2 \alpha} \, d\rho$$

If the conventional tire is considered with a defined cosine cord path and the limits of integration are changed, the following equation represents the contour shape in terms of the top half and bottom half of the tire:

$$Y = \int_{\rho=0}^{\rho=\pi/2} \frac{\rho(\rho^2 - \rho_m^2)}{[(\rho^2 - \rho_m^2) \cot^2 \alpha - 2\rho^2]} \, d\phi$$

Computer capabilities are such that solutions satisfying both halves can be found by iteration methods. The resulting shape is the natural or equilibrium shape for a given centerline angle, as shown in Fig. 16. This information enables us to determine the shape for a given cord angle. Trends with high and low cord angles are given in Table V.

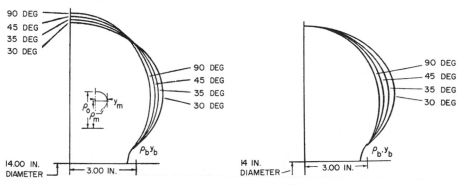

Fig. 16. (a) Equal contour length. (b) Equal ρ_0.

TABLE V

EFFECTS OF CORD ANGLES

Parameter	Cord angle	
	High	Low
Section height	Higher	Lower
Section width	Narrow	Wider
Tread radius	Rounder	Flatter
Rolling resistance	Lower	Higher
Shear stresses	Lower	Higher
Bead stresses	Higher	Lower
Cord tension	Lower	Higher
Deflection	Higher	Lower
Enveloping power	Higher	Lower
Road contact length	Longer	Shorter
Radial cracking in sidewall	More	Less
Lateral stability	Poorer	Better
Treadwear	Poorer	Better
Tread cracking, circumferential	Better	Worse
Cornering power	Poorer	Better
Ride	Softer	Harder

For belted tires, additional input is needed but basically only as constraints on the fundamental natural shape of the carcass as defined by the carcass angle. Figure 17 shows the mold shape of the belted tire. The mold has a fixed contour length from which the theoretical natural shape can be computed for any angle. As the angle increases, the difference between mold and theoretical inflated shape becomes greater, causing greater restriction of the carcass by the belt and greater tension to be carried by the belt. Also, the width of restriction by the belt over the tread width becomes larger, providing more stiffness support to the tread. Thus, the constraint on the shape is belt constriction.

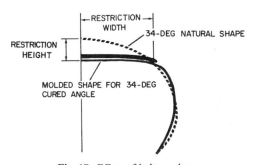

Fig. 17. Effect of belt on shape.

b. Materials. Components used in the manufacture are basically three: rubber, fabrics, and bead wire. These have different mechanical properties and must operate in direct contact for millions of cycles. Figure 18 illustrates the stress–strain relationships among the three. Fabrics, as mentioned, are the primary reinforcing components that afford the tire a high strength-to-weight ratio, flexibility, and size and shape stability.

Rubber compounds are discussed in Chapter 9. Compound requirements vary for different areas of the tire. The tread area must withstand wear and provide traction, requirements that are best met by the selective distribution in the tread of two different compounds. The cap furnishes the desirable wear resistance and traction, while the base compound is highly resilient and cool running.

Bead wire is basically highly tensile, bronze-plated steel wire. The bead bundle is constructed by winding a single wire, or a flat tape containing several wires, a predetermined number of turns. Prior to the winding operation each wire is encapsulated in a special compound that, after vulcanization, aids in uniformly distributing stresses among the wires and combines them into a strong, flexible component.

c. Stress and Strain Relationships. Stresses within a tire are functions of the type and dimensions of the tire structure as well as of such service conditions as inflation, deflection, and speed. Calculations are based on the inflation pressure required to carry the load.

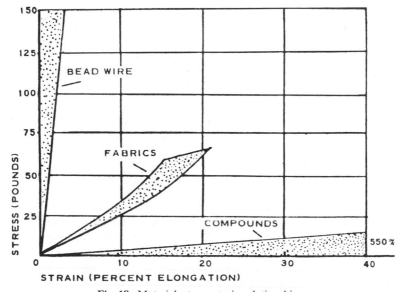

Fig. 18. Materials stress–strain relationships.

Basic calculations include carcass strength, bead strength, and ply strength. All calculations are based on static conditions with stress due to inflation pressure. Safety factors are then applied.

Tension per carcass cord can be expressed in terms of dimensions in the carcass, cord angle, number of cords, number of plies, and contained inflation pressure. From Fig. 13, the following equations are derived.

$$t_0 = \frac{P(\rho_0{}^2 - \rho_m{}^2)}{2\rho_0 n_0 W(\sin^2 \alpha_0)}$$

where t_0 is the cord tension at P, the contained inflation pressure, n_0 is the number of cords per unit length measured normal to the cord direction, and W is the number of plies; and the burst pressure

$$P_{max} = \frac{2T_{max}\rho_0(\sin^2 \alpha_0)n_0 W k}{\rho_0{}^2 - \rho_m{}^2}$$

where T_{max} is the ultimate cord tensile strength and K the empirical constant related to cord strength.

Cord tension varies along the cord; it is greatest at the centerline where ρ is the largest, and is least at the bead. The relationship of cord tension t at any point along the cord to maximum t_0 is

$$t = \frac{t_0 \sin^2 \alpha_0}{\sin^2 \alpha}$$

Figure 19 illustrates the relative tension from centerline to bead of some different-sized bias ply tires.

Bead stress arises from tire inflation, driving on brake torque, side thrust, centrifugal forces, and the various structural effects due to both bead and rim configurations. These are directly related to the beads' primary functions: (1) providing rigid tie-in for carcass plies; and (2) acting as a nonextensible support for aim interlocking.

The first case involves forces caused by the direct pull of the cords on the bead. This tension is directly related to inflation pressure and computed by

$$T = \frac{P(\rho_0{}^2 - \rho_m{}^2)(\sin \alpha_0)\rho_0 k}{2[(\rho_0{}^2 - \rho_\beta{}^2)\cos^2 \alpha_0]^{1/2}}$$

where K is the experimental angle and ρ_β is the perpendicular distance from the axis of rotation to the point on the neutral contour where the flexible structure joins the rigid structure. This represents only one part of the total static tension on the bead. Additional tension is a function of the tire fit on the rim and resulting compression. Dynamic stress effects on tire beads are also important and are illustrated in Fig. 20.

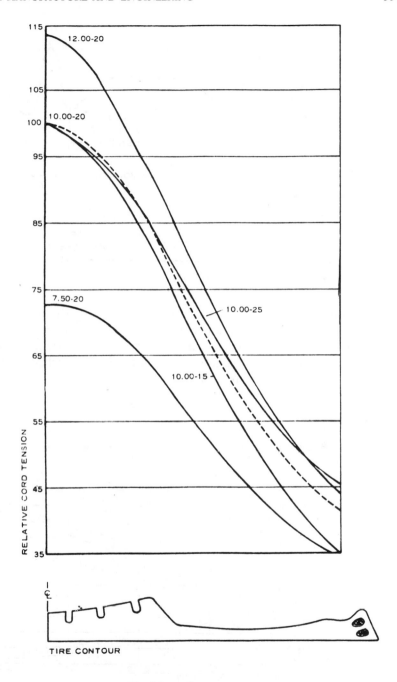

Fig. 19. Cord tension.

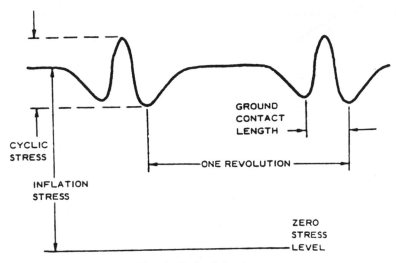

Fig. 20. Dynamic bead stress.

Tension stresses in the belts of a tire are caused by the same factors that affect the carcass plies, but the magnitude of these stresses is modified to the extent that the belts restrict the plies. The degree of restriction r determines the effectiveness of the belt and is a ratio of the inflated belted section height SH_r to natural height SH, related as:

$$r = \frac{SH - SH_r}{SH}$$

Figure 21 shows carcass cord tension, belt cord tension, and total force on the belt for different r values.

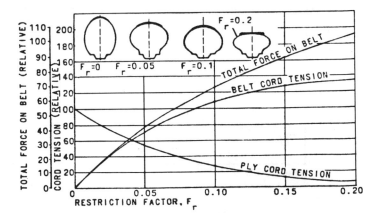

Fig. 21. Belt stress.

Shear stresses also exist throughout the tire; they depend on the tensile stresses as well as on the bending deformation forces caused by tire deflection on loading.

3. Tire Behavior

The primary purpose of the tread design is traction. Tread pattern enhances the ability of the tire to transmit driving and brake torques.

The design of a tread is basically the division or separation of smooth rubber into smaller elements. These elements of the tread pattern are arranged in a symmetric repetitive pattern of voids, ribs, lugs, slots, and grooves.

Once the surface dimensions are determined, the depth is considered. Depth dimensions are considered first with regard to the individual design element and, second, with regard to the relationship of these elements to each other.

There are three basic designs (Fig. 22). The first, with rib design elements oriented circumferentially, is the most common type and provides overall good service for all wheel positions. However, this design is used basically for passenger car tires and for front wheel service on trucks due to uniform wear properties and lateral traction. On drive axles, where forward traction is a prime requirement, the lug or cross rib design offers the best performance. For even greater traction requirements, as for off-the-road service, the third design in Fig. 22 is utilized.

The sidewall of a tire protects the carcass, has esthetic value, and is used for tire identification. Both sidewall thickness and design serve to protect the underlying carcass. A "scuff-rib" region in the mid-sidewall provides extra protection to the carcass from curb or rock damage. The thickness of the sidewall depends on tire type and intended service conditions.

Design of the upper sidewall at the tread is very critical due to heat buildup. Deep radial grooves are designed circumferentially about the tire in this region for cooling and water evacuation.

a. Performance Properties. Desired performance properties of tires can be divided into the following two facets.

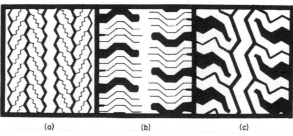

Fig. 22. Basic tread patterns: (a) highway rib; (b) lug or cross rib; (c) special service.

1. Mechanical characteristics, which relate to *load deflection,* the deflection characteristics of a vertically loaded tire; *load capacity,* the maximum load a tire can carry at *any* specified inflation pressure; and *load rating,* the maximum load a tire is rated to carry at a highest permissible inflation pressure. Both load capacity and load rating are dependent on (1) inflation, (2) tire size, (3) deflection, (4) speed, and (5) distance.

Steering properties are mechanical characteristics involving (1) *lateral force,* the amount of steering pull from the direction of travel; (2) *aligning torque,* the restoring force of the tire in slip about the vertical axis at the center of the footprint; and (3) *static steering torque,* the torque required to steer tires on a stationary vehicle.

Road contact pressure is related to (1) *traction,* the ability of a tire to grip the road under various conditions and to generate propulsive forces for driving, lateral forces for cornering, and retarding forces for braking; (2) *power loss,* hysteresis losses within a tire structure and frictional losses in the contact area; (3) *revolution rate,* the number of revolutions per fixed distance; normally less than that calculated for a static loaded tire; (4) *rim effects,* effects on the bead area due to movement of the tire; and (5) *noise and vibration,* sound generated by a tire.

2. Endurance has to do with the response of a tire to *treadwear,* the "wearing-off" of a tire due to abrasion on the road surface; and *fatigue,* weakening or failure from repeated load cycles. *Separation* refers to the inability of the components of a tire to maintain composite integrity in service. Endurance is also determined by the reaction of a tire to *heat,* the buildup of thermal energy in the tire during dynamic operation, and by *bruise resistance,* the ability of a tire to withstand damage upon impact; and *cutting, cracking, and tearing resistance,* the ability to resist the initiation and propagation of fissures in the rubber of a tire.

C. TIRE TESTING

In all fields of technology, a method of measuring performance is essential. The major areas of tire testing include (a) laboratory testing of cords, compound, and composites; (b) laboratory testing of tires; (c) proving grounds testing; (d) commercial evaluation; and (e) quality control testing of production. From these areas there must be constant feedback to tire researchers and developers at all stages.

1. Laboratory Testing

The components of a tire are strained in a cyclic manner in service. In addition, because tire components are not perfectly elastic, some of the energy is absorbed and converted to heat. Thus, the materials must also be able to withstand the effects of heat.

Many tests have been developed to approximate these effects in the laboratory. With the advent of the need for faster and more accurate data it has become necessary to link the laboratory to a computer. It should also be pointed out that the computer allows data calculations of certain properties that could not be achieved otherwise.

To illustrate a computer-controlled laboratory test, the creep behavior of tire cords will be examined. Stability testing (determining the effect of time on tire cord growth and shrinkage) consists of using programmed time, temperature, and load cycles to simulate the shrinkage and growth properties of cords during processing, curing, and tire operation. The graphs in Fig. 23 are plotted by computer and show the effect of times, temperatures, and loads at tire operating conditions. It can be seen that inorganic fibers show little growth, whereas organic fibers under load exhibit growth and creep that level out with time. With increasing temperatures, both shrinkage and growth forces are increased. In the case of nylon the shrinkage force exceeds the growth at low loads and the cord contracts initially at elevated temperatures. With rayon and polyester the cord continues to grow with higher temperatures and heavier load. All fabrics tend to return to their original length when the temperature and load are decreased.

The tire industry is also continuously developing laboratory tire tests to fail tires purposely and thus determine performance limits. Most laboratory tests are (a) developed to produce a specific type of failure; (b) accelerated tests, generally lasting only several days; (c) semiquantitative with respect to service conditions. Thermocouples are often used to record tire temperatures during tire testing.

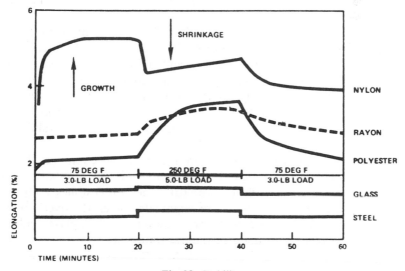

Fig. 23. Stability.

Most laboratory tire-testing machines are based on the principle of a steel flywheel. Many of these flywheels are electronically programmed to act as simulators. In testing aircraft tires, for example, a complete cycle of taxi, takeoff, and landing can be continuously repeated.

2. Nondestructive Testing

Most classical testing techniques seek performance limits and, consequently, result in destruction of the specimen. Thus, considerable effort has been expended to develop nondestructive tests. The following nondestructive testing techniques are being evaluated by the tire industry.

Infrared. A scanning system used to obtain patterns from a tire during operation, it continuously records temperature gradients on the surface and identifies "hot spots."

Ultrasonic. Ultrasonic waves travel a short distance through water and a submerged tire that is not dimensionally uniform will cause changes in the reflectance of the signals.

Microwave. Radar-type frequencies are used to penetrate the tire and detect changes in physical properties.

Holography. An optical technique used to measure tire surface displacement.

3. Proving Grounds

The best method of determining the behavior and integrity of a tire is to examine its performance when subjected to millions of road-tested miles. All types of tires (passenger car, truck, earthmover, farm vehicle, etc.) are torture tested at proving grounds.

An industry proving ground generally will have the following test tracks and road courses: (a) high-speed tracks (circular or oval); (b) interstate simulations (straight and curved); (c) gravel and unimproved roads; (d) cobblestone roads; (e) cutting, chipping, and tearing courses; (f) Baja road hazard courses (to induce bruise and rupture); (g) skid pads (to test wet and dry traction); (h) serpentine and slalom courses (for evaluating esthetics and handling); (i) tethered tracks (for determining farm tire durability); and (j) glass roads.

The 7300-acre Goodyear Proving Grounds in Texas comprises the following tracks: (a) a 5-mile high-speed circle; (b) 8 miles of simulated interstate highway; (c) extensive gravel courses; (d) huge cobblestones embedded in concrete; (e) a 2-mile road of coarse aggregate for earthmover tires; (f) a Baja road course; (g) skid pads with spray equipment; (h) a glass road facility, which permits direct observation of the tire footprint.

4. *Commercial Evaluations*

Commercial value analysis is performed on fleet vehicles. In addition, test data from all stages of tire evaluation are fed into computers for in-depth and regression equations.

Consumer satisfaction is the true test of a tire.

III. Tire Manufacture

The tire manufacture process is shown in Fig. 24. Basically, tire manufacturing consists of mixing elastomers, carbon blacks, and chemicals to form a rubber compound; processing the various fabrics and coating them with rubber in a calendering operation; tubing the rubber treads and side-

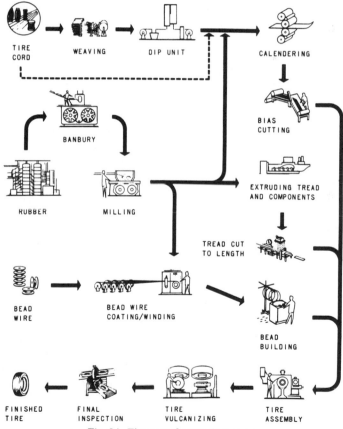

Fig. 24. Tire manufacture flow sheet.

walls; assembling these components on the tire building machine; curing the tire under heat and pressure; and finally finishing and screening the product.

A. RUBBER

The elastomer comprises about 60–70% of the average passenger tire. There are up to eight different compounds per passenger car tire, varying in physical properties and chemical compositions. The number increases rapidly with the complexity of the tire. However, all rubber compounds are prepared in basically the same manner.

1. Breakdown

The first step in preparing compounds for tire building involves the breakdown of the rubber. The objective of the breakdown step is to soften and masticate the compound to a workable stage where other ingredients can be added and to make a more uniform product. In the initial breakdown operation the incoming elastomers, carbon black, and chemicals are automatically weighed at established intervals and transferred into an internal mixer (generally a Banbury type) for 2–3 min of mixing at elevated temperatures and pressures. Breakdown is also accomplished by use of a mill.

Mills operate at lower temperatures than the Banbury mixer (300°F to about 240°F) and thus breakdown in a mill takes considerably longer. Degree of breakdown with both types of equipment is dependent on the "friction ratio" or the difference between the operating speeds of the front and back rolls (or rotors, for internal mixers). In addition, clearances, condition of rotor surfaces, pressure, and speed influence breakdown.

2. Mixing

The purpose of mixing is to obtain thorough, uniform dispersion of all compounding ingredients within the elastomer. For every batch there is a definite time, temperature, and order of addition. Again, both the open mill and internal Banbury mixers are used. Mixing on the mill is a slow operation, with the addition sequence and degree of mixing dependent largely on the mill operator. The Banbury mixing equipment is much faster and with the inception of computer control much more accurate.

Generally, the order of addition is (1) rubber; (2) plasticizers and softeners; (3) fillers; (4) sulfur (curing agents); (5) accelerators and antioxidants. Of course, this order can be and is changed for many rubber formulas. Simple rules are always followed in mixing:

1. Keep sticky resins from contact with dry powders.
2. Use temperatures above the softening point of hard resins.
3. Keep liquids contained so they do not leak out.

4. Make use of rubber's stiffness to grind materials.
5. Avoid scorch and the subsequent formation of crumbs.

This can be done by (a) reducing internal friction by using plasticizers; (b) using well broken-down rubber; (c) withholding sulfur until the end of the mix; (d) selecting the proper accelerator and accelerator-to-sulfur ratio; (e) using masterbatches. After mixing, the compounds are dipped in a cooling bath and stored in a cool-air room to prevent overcure.

3. Caldendering

Calendering is the forming operation in which the rubber compound is spread on the fabric. The calender is a heavy-duty machine equipped with three or more rolls revolving in opposite directions (Fig. 25). Rolls are heated by circulating water or by steam, and gears permit the rolls to operate at varying speed ratios.

The amount of compound deposited onto the fabric is determined by the distance between rolls and is monitored by beta gauges (profiling). Each cord is insulated on all sides by rubber. From here, the fabric is cut on the bias to the desired width and angle.

Compounds used for calendering should have little or no shrinkage; should be tacky enough for building operations and adhesion to fabrics; and should be scorch free. Also, warmup mills must be provided to keep the feed stock uniform and plastic, so that a uniform material is applied to the fabric.

4. Extrusion

Most of the rubber in the tire manufacturing operation goes into forming the tread and sidewall by extrusion from a tuber. Most extruders are of the

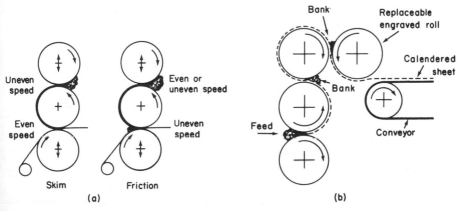

Fig. 25. (a) Applying rubber to fabrics. (b) Profiling by means of a four-roll engraving calendar.

screw type; strips of compounds are fed from a warmup mill into the barrel mouth and forced by the screw through a die.

Proper extrusion depends on the plasticity of the stock, the scorch rate, and the compound cleanliness. Dirt and agglomerates are removed by a straining operation.

In addition, smoothness of the compound is essential. This is controlled by the operating conditions (temperature, speed, feed, and amount of breakdown) and by the compound itself, including the plasticity to which the rubber is broken down, the type of rubber, and the amount and type of loading.

Following extrusion, the continuous stream of formed compound is bevel cut to a predetermined length and is weighed, cooled, and cemented. These strips are then ready for tire building.

Rubber is also used to coat bead wires. Bead wire is received on large spools. The bundles of wire are passed through an extrusion die, where a coat of rubber is added. These rubber-coated wires are wound into a hoop of specific diameter and sent to the tire builder.

B. Tire Building

All the tire components—calendered and cut carcasses and belts; extruded tread, sidewall, and beads—are assembled at the tire building machine. On this machine the "green" or uncured tire is built (Fig. 26).

Tires with carcass angles up to 65° are built on flat-topped drums in a single-stage process. Above this angle (e.g., radial tires), a two-stage process is necessary.

Basically, tire building involves the application of plies and other components onto a cylindrical form. The process begins with the application of a thin layer of special rubber compound called the inner liner. Following this, the plies are placed on the drum, one at a time. Then the beads are set in place. The plies are turned up around the beads. Belts, if any, are applied next. Finally, the tread and sidewalls are added, the drum is collapsed, and the uncured tire removed.

An uncured bias (or bias/belted) tire resembles a barrel with both ends open. Before the belts are applied to a radial tire, the green tire is generally expanded from a cylindrical to a toroidal shape. The belts and tread are then added to this toroidal shape.

The uncured tires are loaded into automatic tire presses and cured (vulcanized) at high temperatures and pressures. Curing may be done by a number of methods, including press cures, open steam cures, dry heat cures, hot water cures, and room temperature cures. Tires are normally cured by press cures.

In press curing, the compound flows into the mold shape to give a design to the tread and the desired thickness to the sidewall. Good flow depends

(a)

(b)

Fig. 26. (a) Bias and bias/belted green tire. (b) Radial green tire.

on the plasticity of the uncured stock. To flow, the compound must resist scorching. Flow must be complete before cure begins or distortion may result. Proper flow is accomplished by (a) compounding fillers according to their effect on plasticity, (b) adjusting acceleration, and (c) utilizing plasticizers. Curing also results in mechanical changes in the tire. When changing from a cylindrical to a toroidal shape, the cord angle in the ply is changed by the pantographing effect (Fig. 27). This law must be altered due to the extensible

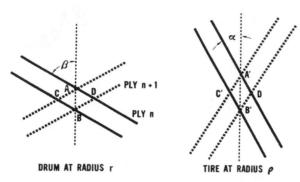

Fig. 27. Pantographic action; $\cos \beta / r = \cos \alpha / \rho$.

nature of the cord and the elongation imparted to the cord by the shaping tension. When tension is applied to a cord during the change from a flat configuration to the double-curved surface of the ply in a molded tire, there is a component of tension normal to the cord path and in a tangent plane to the tire surface. This force acts to slip the cord sideways toward a path of minimum cord tension. Thus, both elongation of the cord and a slip factor must be incorporated in the pantographing law. The relative elongation is ε and ζ is the factor representing sidewise slip. Knowing the cured tire design angle, we can calculate the ply angle in the green tire if the width of the building drum is also known (Fig. 28).

To find the cord length in a cured tire, basic integration as shown in Fig. 29 can be used. Numerical analysis of this equation is necessary in many instances. Once the cord length in the cured tire is calculated, the green tire's cord length is determined by giving proper consideration to angle change and the cord tension desired in the cured tire. The thermal properties of the cord must also be considered in finding cord length.

Another important factor is the uniformity of the cord tension through the carcass. Uniformity is especially important in multi-ply (i.e., truck, aircraft, and earthmover) tires. Control of uniformity in tensions and thus shear stress has great bearing on tire performance. Because of the inextensible

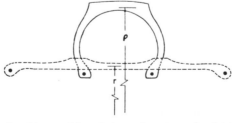

Fig. 28. Green (----) to cured (——) shape change; $\cos \beta = (r/\rho)(1\,\varepsilon)(1 + \zeta) \cos \alpha$.

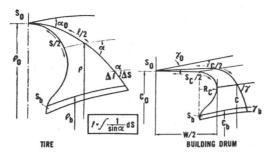

Fig. 29. Determination of cord length in a cured tire.

nature of the tire composite, a pressure gradient exists through the tire thickness that is largest at the first ply and smallest at the last ply. Thus, to maintain uniform tensions in all plies, adjustments in ply angles must be made to balance the gradient.

All the other changes that occur during curing involve a direct relation to drum width and bias angle. One major change is in the end count of the carcass or belt cords. This end count change is defined as

$$n_\alpha = \frac{n_\beta r \sin \beta}{\rho \sin \alpha}$$

Upon formation of the cured tire, this change in end count influences cord tensions and determines what green gauge of rubber is required to achieve designed cured gauges. The blowup ratio and building techniques influence the end count. Blowup ratio influences not only the cord density but also the rubber distribution around the cord. These changes in terms of rubber flow are also important with regard to the physical properties of uncured rubber. The rubber properties must lend themselves to this transition. The equations giving the rubber matrix distribution are, for σ_P, rubber between plies,

$$\sigma_P = \frac{\sigma n_\alpha}{n_\beta(1 + \varepsilon)} - \frac{2K}{(1 + \varepsilon)^{1/2}}$$

(σ is the total gauge of green ply) and for σ_c, rubber between cords,

$$\sigma_c = \frac{1}{n_\alpha} - \frac{2K}{(1 + \varepsilon)^{1/2}}$$

The last important feature that must be considered for curing is the tread. Computation of the green tread follows once the green carcass has been calculated. The carcass specification gives the contour change that occurs during curing. Figure 30 shows the flow of rubber, which is a function of contour change due to pantographing and the stretch due to the circumferential change.

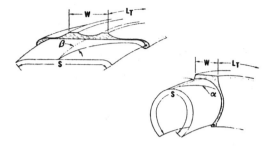

Fig. 30. Change in dimensions that occurs when green carcass is cured.

A third factor, the displacement of rubber caused by the patterned bridgework of the mold, must also be considered.

After curing, the tire can be mounted on a rim and permitted to cool while inflated to reduce internal stresses. This step is called postinflation.

C. Finishing the Tire

Finishing the tire after cure and postinflation involves trimming, buffing, balancing, and inspections by quality control procedures.

In the tire mold, there are small "vents" that permit the removal of any air pockets that may remain from tire building. These vents must be removed from the tire by an operation called trimming.

Because of possible difficulties in the step during which the sidewall is extruded and the requirement for sharp and distinct lines, the white sidewall is extruded with a black overlay. Buffing or rib cutting consists of removing this overlay of rubber from the white sidewall of the tire.

After buffing and trimming, the tire is inspected for imperfections.

Following inspection, the tires are balanced. Balancing ensures dynamic equilibrium by testing the uniformity of mass distribution of a tire relative to its spin and steer axes.

Having been inspected and balanced, the tire is ready to ship to a warehouse for distribution.

BIBLIOGRAPHY

Clark, S. K., ed.. Mechanics of pneumatic tires. *Nat. Bur. Stand.* (*U.S.*), *Monogr.* No. 122 (1971).
Curtiss, W. W., Principles of tire design. *Tire Sci. and Technol.* 1(1), 77–98 (1973).
Dague, M. F., *Tire Reinforcing Syst., 11th Annu. Rubber Lect. Ser., Sci. Technol. Reinforcing Mater., Akron Rubber Group, 1974.*
Davisson, J. A., "Design and Application of Commercial Type Tires," SAE SP-344. *Soc. Automot. Eng.*, New York, 1969.
Kaswell, E. R., "Wellington Sears Handbook of Industrial Textiles." Wellington Sears Co., West Point, New York, 1963.

Kovac, F. J., "Tire Technology," 4th Ed. Goodyear Tire/Rubber Co., Akron, Ohio, 1973.

Morton, W. E., and Hearle, J. W. S., "Physical Properties of Textile Fibers." Textile Inst., Manchester, England, 1962.

Purdy, J. E., "Mathematics Underlying the Design of Pneumatic Tires," 2nd Ed. Hiney Printing Co., Akron, Ohio, 1970.

Winspear, G. G., ed., "Vanderbilt Rubber Handbook." R. T. Vanderbilt Co., New York, 1968.

Author Index

Numbers in parentheses are reference numbers and indicate that an author's work is referred to although his name is not cited in the text. Those numbers prefixed by a Roman numeral refer to references in the tables of Chapter 3.

A

Abagi, V., 510, *528*
Abbott, S., 134, *137*
Abbott, T. N. G., 252, *281*
Abe, K., 509(125), *527*
Abe, M., 140, *142*
Abe, T., 101(IIIA-78b), *135*, *138*
Abel-Alim, A., 143, *144*
Abrams, W. J., 516(191), *529*
Acrivos, A., 267, *282*
Adams, E., 92(IIIA-45b), 134, 135, *137*
Adams, H. E., 135, *137*
Aggarwal, S. L., 519(219), 520(219), *529*, 542
(56), 547(56), 558(101), 560(56, 107), 562
(56), *565*, *566*, *567*
Ahagon, A., 426(15, 19), 427(19), 429(15), *453*
Aklonis, J. J., 157(8), 164, *178*
Akutin, M. S., 476(14), *487*
Alexander, L., 121(V-2), 148, *150*
Alfrey, T., 225, 229(5), 231, 242, 266(256, 257),
281, 287
Alfrey, T., Jr., 19, *21*
Allegra, G., 132, *133*
Allegrezza, A. E., 152, *153*, 547(73), 548(73),
552, 555(73), *566*
Allegrezza, A. E., Jr., 552(95), 553(95), 554(95),
566
Allen, G., 11(21), *20*, 135, *138*, 145, *146*, *147*
Allen, P. W., 66(100), *74*, *487*
Allen, V. R., 259(92), *283*
Allport, D. C., 101(IIIA-76), *127*, 130, 131(13),
132, 133(13), 134, 135, 136(13), 138(13),
140, 142(13), 143, 144(13), 145, 146(13),

148, 150(13), 152, 153(13), 333(102), 335
(102), *337*, *487*, 494, 516(198), 517, 521,
525, *529*, 534(11), 563(11), *564*
Altier, M. W., 36(17), 41(24), *72*
Amberg, L. O., 328(87), *337*
Ambler, M., 135, *137*, *139*, 140, *142*, 143, *144*
Amidon, R. W., 505, *527*
Anderson, A., 145, *146*
Anderson, D. E., 468(8), *487*
Anderson, J., 130, *131*, 145, *146*
Anderson, J. E., 145, *146*
Ando, I., 504(99), *527*
Andrews, E. H., 422(4), 440(36), 450(4), 452
(60), *452*, *454*, 502(94), *527*
Andrews, R., 145, *147*
Andrews, R. D., 174(32), 175, 176(32), 177(32),
178, 225, 231(7, 8), 256(6), 257, 261, *281*
Andries, J. C., 495, *526*
Angier, D. J., 67(107), *74*, 265, *281*, 494, *525*
Angove, S. N., 492, *525*
Arest-Yakubovich, A. A., 60(70), *73*
Arlman, E. J., 62(78), *73*
Armstrong, A., 145, *146*
Armstrong, R. T., 301(25), *336*
Arnett, R., 90(IIIA-31b), 134, *136*
Aron, J., 345(12), *364*
Arridge, R. G. C., 520(225), *530*
Asada, T., 152, *154*
Asai, H., 519(220), 520(231), *530*, 559(103), *566*
Ashare, E., 243(165), *285*
Ashida, M., 509(125), *527*
Ashworth, J. N., 208(32), *221*
Auer, E. E., 31(11), *72*
Auerbach, M., 520(230), *530*
Ayrey, G., 264, *282*

609

G

Wringbaum, W. R., 519(223), *530*
Wun, K., 151, *153*
Wunderlich, B., 145, *146*, 148, 149, *150*
Wylie, C. R., 247(318), 254(318), *289*

Y

Yakubchik, A. I., 460(5), *487*
Yamada, I., 109(IIIB-26b), 141, *142*
Yamagi, I., 498, *526*
Yamaguchi, K., 135, *138*
Yamaguchi, N., 134, *137*
Yamaguchi, Y., 521(250), *530*
Yamaja, I., 498(71), *526*
Yamakawa, H., 75(5), 92(IIIA-44), 93(5), 94
   (5), *127*, 132, 133(5), 134, 135, 136(5), 137
   (5), 138(5)
Yamamoto, A., 135, *139*
Yamamoto, M., 234, 235, 256(323), 261, *289*
Yamamoto, Y., 140, *142*
Yamashita, S., 516(185), *529*
Yamazaki, Y., 194(16), *220*
Yardley, J. F., 493(14, 20), *525*
Yasuda, G., 225(213), *286*
Yazgan, S., 519(223), *530*, 543(66), *566*
Yeh, G., 151, *153*
Yeh, H. C., 243(324), *289*
Yerzley, E. H., 217(53), *221*
Yim, A., 145, *146*
Yoda, O., 148, *150*
Yokoyama, T., 521(249, 250), *530*
Yoon, Y. N., 130, *131*
Yoshihara, N., 555(97), 558(97), *566*

Yoshimiya, N., 498, *526*
Yoshimura, N., 500(87), 504, *527*
Yost, F. L., 449(56), *453*
You, W., 135, *138*
Young, E. J., 306(34), *336*
Young, R. N., 60(69c), *73*
Yozka, S., 135, *138*
Yu, A. J., 536(20), *565*
Yu, T. C., 266, *284*
Yudin, V. P., 254(289b), *288*
Yukshnovich, S. G., 510(138), *528*
Yurchuk, T. E., 516(196), *529*
Yusek, C., 152, *153*

Z

Zachman, H. G., 148, *150*
Zakharenko, N. V., 245, 269(325), *289*
Zakharkin, O. A., 510(138), *528*
Zakharov, N. D., 513(170), *528*
Zamodits, H., 244(326), 277, 279, *289*
Zapas, L. J., 225, 234(24, 327, 328), 236(24, 25,
   327, 328), 258(24), 267(25), *282*, *289*
Zapp, R. L., 504(100), *527*
Zaremba, S., 233(329), *289*
Zbinden, R., 80(I-17), 128, *129*
Zelinski, R. P., 63(79), *73*, 385, *417*
Ziabicki, A., 187(9), *220*
Ziegler, K., 62, *73*
Zimm, B. H., 134, 135, *137*, 139(IIIC-13), 143,
   144
Zivny, A., 536(41), 542(41), *565*
Zugenmaier, P., 132, *133*
Zware, M., 517, *529*

Subject Index

A

ABA copolymers, 69, 259, 334, 532–564
  structure of, 69
(AB)$_n$ copolymers, 532–564
Abbe refractometer, 122
Abradability, *see* Abrasion
Abrasion, 16, 36, 362, 451, 452, 570
Abrasion resistance, 209, 345, 369, 379, 384,
    395, 402, 407, 510, 512, 513, 564
  effect of particle size on, 395
Abrasive wear, *see* Abrasion
Absorbance, 120
Absorption, 80, 113
  of amorphous regions, 120
  of crystallite, 120
  intensity of, 128, 136, 149
    ratios of, 78
Absorption spectroscopy, 128
Accelerator, 513, *see also* specific substances
  effect on vulcanizate properties, 310–315
  Mooney scorch time, 392
  secondary, *see* Kickers
  in sulfur vulcanization, 297–300
  for vulcanization, 41, 295, 395
  triangular system, 395
Accelerator-retarder, 308
Accelerator selection, 306
Accelerator-sulfur ratio, 310–315
Accuracy
  limiting factor, 218
  of measurements, 218
Acetylation, 458
Acid initiator, 50
Acrylonitrile
  block copolymerization, 475
  reactivity ratios in chloroprene copolymer-
    ization, 47
Activation energy, 51
Activation step, 27

Acyl urea linkages, 332
Addition, 331, 464–466
  of ethylene derivatives, 464–466
  intermolecular, 465
  intramolecular, 464
  mechanism, 464, 465
Addition polymer, 25
Addition polymerization
  anionic mechanism, *see* Anionic polymeriza-
    tion
  cationic mechanism, *see* Cationic polymeriza-
    tion
  free radical mechanism, *see* Chain addition
    polymerization, free radical mechanism
Additive, 79–82, 563
  in chromatography, 97
  halogenated, as flame retardants, 385
  phosphorus-containing, 386
Adhesion, 404, 459
  aged, 308, 309
  to brass-plated steel, 308–310
  degree of, 308
  effect of vulcanization parameters on, 309
  mechanical, 578
  resin, 316
  of rubber to copper, 308
Adhesion force, 308
Adhesion index, 357
Adhesive, 458, 464, 563, 564, 583–585
  application to fabric, 580
  application by immersion, 581
  application zone, 581
  precoat, 577
Adiabatic calorimetry, 120, 148
Adsorption
  physical, 347, 348
  of polymer segments on filler, 349, 350, 357
Agglomeration, 112, 352, 354
  in osmometry, 86
  secondary, 348, 350, 356, 359
Agglomeration-deagglomeration, 353

Additions to Subject Index

ERRATA

Page	Location	
113	Line 8	. . . but is of <u>ill-defined order</u>.
	Line 11	<u>ill-defined order</u> . . .
	Line 12	. . . in the <u>ill-defined order</u> . . .
119	Line 5	Add: , θ is the scattering angle, and λ the wavelength
162	Footnote	Add: At large strains, the volume contracts again.
170	Line 1	Replace α with λ
	Line -6	. . . we get <u>in place of Eq. (28)</u>
184	Footnote	Add: For Boltzmann's superposition principle, see [1].
185	Line 6	. . . their <u>conformations (see footnote, p. 2)</u> . . .
226, 227	Eqs. (1)– (8)	m = mass element, ρ = density, dV = volume element, $\mathbf{v}$ = velocity
229	Eq. (15a)	. . . $= G(d\gamma/dt) - (\sigma/\tau) = $ . . .
	Line -9	. . . $d\gamma/dt$ <u>from time s</u> to give
230	Line 7	. . . to Eq. <u>(15a)</u> . . .
231	Line 10	. . . $H(\tau)$, <u>the relaxation function,</u> i.e.,
	Line 17	Add: or the rheological equation of state, e.g., (15a).
232	Eq. (27)	V = macroscopic velocity
	Line -8	Add: (See Eq. (11).)
	Line -1	. . . measures <u>shear</u> strain . . .
234	Line -10	. . . of the <u>memory functions</u> . . .
235	Eq. (40)	This is <u>Newton's Law</u>
	Line -8	viscosity, <u>measured in poises, dimensions m/lt.</u> . . .
	Eq. (41)	β_i = normal stress constants
239	Line -7	. . . rheological <u>experience</u> . . .
241	Eq. (50)	v_i = velocity <u>components</u>
	Line 6	Add: (inflation)
	Line -6	. . . leads to <u>Trouton's viscosity</u>
242	Line 6	. . . manner, <u>leading to flow instability.</u> . . .
244	Eq. (62c)	T = temperature, c = heat capacity, last term is $\boldsymbol{\sigma} : \nabla\mathbf{v}$
246	Line -4	Add: and R is the cup radius
253	Line 7	. . . within the die <u>and</u> . . .
	Line -9	Add: the volume flow rate.
	Line -3	. . . telescoping flow <u>(capillary flow)</u> . . .
254	Line 1	Add: a parabolic velocity field
		Add: $Q = pR^4/8\eta L$ $\qquad\qquad$ (86a) where $R = \frac{1}{2}D$. This is (86) fully integrated, usually called Poiseulle's equation.
255	Line 2	that, <u>with L as filament length,</u>
	Line -11	. . . is the <u>degree of</u> . . .
256	Eq. (89)	$T_f = $ . . .
	Line 17	. . . temperature T_f. . . .
257	Line 18	. . . temperature $\overline{T_f}$. . . .
278	Line 15	Add: (see Eq. (38c))
352	Line 22	Add: , stress magnification,
475	Line 13	The required <u>viscous</u> . . .
582	Line -15	Add: See Chapter 5.
585	Line -3	Add: (avoiding aquaplaning)

A
B
C
D 8
E 9
F 0
G 1
H 2
I 3
J 4